Springer Series in Optical Sciences Volume 4

Editor David L. MacAdam

Springer Series in Optical Sciences

Volume 1 **Solid-State Laser Engineering**
By W. Koechner

Volume 2 **Table of Laser Lines in Gases and Vapors**
By R. Beck, W. Englisch, and K. Gürs

Volume 3 **Tunable Lasers and Applications**
Editors: A. Mooradian, T. Jaeger, and P. Stokseth

Volume 4 **Nonlinear Laser Spectroscopy**
By V. S. Letokhov and V. P. Chebotayev

Volume 5 **Optics and Lasers** An Engineering Physics Approach
By M. Young

V. S. Letokhov V. P. Chebotayev

Nonlinear Laser Spectroscopy

With 193 Figures

Springer-Verlag Berlin Heidelberg New York 1977

Prof. VLADILEN S. LETOKHOV
Institute of Spectroscopy, Academy of Sciences USSR, Akademogorodok,
SU - 142092 Moskow / USSR

Prof. VENIAMIN P. CHEBOTAYEV
Institute of Semiconductor Physics,
Siberian Branch of the Academy of Sciences of USSR
SU - 630090 Novosibirsk / USSR

Dr. DAVID L. MACADAM
68 Hammond Street, Rochester, NY 14615/USA

ISBN 3-540-08044-9 Springer-Verlag Berlin Heidelberg New York
ISBN 0-387-08044-9 Springer-Verlag New York Heidelberg Berlin

Library of Congress Cataloging in Publication Data. Letokhov, V.S. Nonlinear laser spectroscopy. Includes bibliographies. 1. Laser spectroscopy. 2. Nonlinear optics. I. Chebotayev, Veniamin Pavlovich, joint author. II. Title. QC454.L3L46. 535.5'8 76-54499

Offset printing and bookbinding: Zechnersche Buchdruckerei, Speyer. 2153/3130-543210

Preface

The laser as a source of coherent optical radiation has made it possible to investigate nonlinear interaction of optical radiation with atoms and molecules. Its availability has given rise to new research fields, such as nonlinear optics, laser spectroscopy, laser photochemistry, that lie at the boundary between quantum electronics and physical optics, optical spectroscopy and photochemistry, respectively. The use of coherent optical radiation in each of these fields has led to the discovery of qualitatively new effects and possibilities; in particular, some rather subtle effects of interaction between highly monochromatic light and atoms and molecules, in optical spectroscopy, have formed the bases for certain methods of so-called *nonlinear, laser Doppler-free spectroscopy*. These methods have made it possible to increase the resolution of spectroscopic studies from between 10^5 and 10^6, limited by Doppler line broadening up, to about 10^{11}; at present some laboratories are developing new techniques that have even higher resolution. The discovery and elaboration of the methods of nonlinear laser spectroscopy have resulted largely from contributions by scientists from many countries, in particular from the USA (Massachusetts Institute of Technology, Stanford University, National Bureau of Standards in Boulder, Harvard University, etc.), the USSR (P.N. Levedev Institute of Physics, Institute of Semiconductor Physics in Novosibirsk, Institute of Spectroscopy, etc.), France (University of Paris, Ecole Normal Superiere), Germany (University of Heidelberg), Japan (Tokyo University). Their contributions are used as the basis of our book, in which we have tried to describe as fully as possible, and systematically, the present statue of this area of research. In so doing, we used the publications of many scientists and of our own. In some places, our personal interests seem to be rather strong, and we have disturbed the proportions to some extent by centering emphasis on the research work carried out at the laboratories in the Institute of Spectroscopy and the Institute of Semiconductor Physics of the USSR Academy of Sciences. Unfortunately, we could not deal with some aspects of the subject. We live about 4000 km apart; let that be some excuse for us in this joint project. When preparing the English version of this book, we used the Russian publication of our book *Principles of Nonlinear Laser Spectroscopy* published in the Soviet Union by the Nauka (Science) in the summer of 1975. However, in doing so we considered it necessary to change both its form and content. Instead of the six chapters in the Russian edition this book consists of ten chapters, and it is much larger. So, the book is essentially new; we have changed even its title, a little. We have worked out the plan of this book together and discussed the contents of each chapter, so that the whole book could be coordinated. But we had to finish some chapters separately. Chapters from 1 to 4 and 10 were written by V.S. Letokhov and those from 5 to 9 by V.P. Chebotayev.

This monograph is the first attempt to generalize and analyze from a common standpoint this quickly developing area in laser spectroscopy. Recently Springer-Verlag has published the monograph *High-Resolution Laser Spectroscopy* (Volume 13 of its *Topics in Applied Physics* series) to which we have also made a contribution. That has some points in common with our present monograph, in which all of the topics are given more detailed and wider consideration, which should be of use, not only to scientists specialized in this field but also to students and post-graduates as well as to specialists in associated fields of research.

When preparing the English version, we took into consideration, of course, the results of the investigations carried out from 1974 to 1975 and also comments we received after publication of the Russian book. In this connection, we would like to express our deep appreciation to Prof. E.B. Alexandrov, Prof. S.G. Rautian, Dr. B.D. Pavlik for their remarks. Thanks are also due to Dr. E.V. Baklanov and Dr. E.A. Titov who greatly assisted in writing some new chapters for the English publication. We want to express our thanks to Prof. A.L. Schawlow and Dr. H. Lotsch who initiated the publication of our book in English.

August 1976

Professor V.S. Letokhov
Moscow, Podolskii rayon,
Akademgorodok, Institute of
Spectroscopy Akademy of
Sciences USSR

Professor V.P. Chebotayev
Novosibirsk, Akademgorodok
Institute of Semiconductor
Physics Siberian Branch of
Academy of Sciences USSR

Contents

Main Notations

"a"	index denoting an amplifying medium
a	diameter of a light beam
a_k	probability amplitude for Kth level (Schrödinger picture)
$a(w)$	homogeneous line shape normalized to the maximum unit
A_{21}	Einstein A coefficient of $2 \to 1$ transition
A	weight of particle in atomic units
"b"	index denoting an absorbing medium
$C(t)$	cosine (in phase) component of polarization
c	speed of light
c.c.	abbreviation for "complex conjugate"
d	distance between two light beams
e	electron charge
$\underline{e}$	vector of light-field polarization
E	instantaneous strength of the field electric component
E_i	energy of ith quantum level
$\mathcal{E}$	slowly varying amplitude (envelope) of the running light wave
$\mathcal{E}_{st}$	slowly varying amplitude of the standing light wave
$\mathcal{E}_0$	amplitude of the weak probe wave
$f = a,b$	index denoting a medium type
f_p	factor of resonance broadening due to saturation
g_f	coefficient in parameter of saturation G_f ($G_f = g_f \cdot P$)
$G = gP = \left(\frac{P_{12}\mathcal{E}}{\hbar}\right)^2 \frac{1}{\gamma\Gamma}$	degree of saturation or parameter of saturation
$G_f = g_f P$	degree of saturation of amplifying ($f = a$) or absorbing ($f = b$) medium
$\hbar$	Planck's constant/2π
h	relative amplitude of dip
$K, \underline{K}$	wave vector of a light wave
K	Boltzmann's constant
$l_f (f = a,b)$	length of the resonance medium
L	cavity length
$L(x) = (1 + x^2)^{-1}$	lorentzian contour

M atom or molecule mass

N_o density of particles in a gas

N_i density of particles on i level

N_i^o initial density of particles on i level (without saturation)

$n_i(\underline{v}) = N_i W(\underline{v})$ velocity distribution of particles on i level

$n(v) = n_1(v) - n_2(v)$ velocity distribution of difference of populations of levels

$\underline{n}$ unit vector in direction of light wave or observation

$n(\omega), n_f$ index of refraction of medium at the frequency ω

$P = \frac{c}{8\pi\hbar\omega}E^2$ density of photon flow of a light wave (in photon/cm^2 s)

$\underline{P}(t,\underline{r})$ polarization of a resonance medium

$P_{12} = P = \underline{P}_{12}\underline{e}$ projection of matrix element of transition dipole moment

P gas pressure

Q_i rate of particle excitation for i level

Q quality factor of the mode cavity

$q_i(J) = q_i$ relative population of the rotational sublevel J of vibrational i level

r curvature radius of a light wave

S factor of frequency pulling (autostabilization)

$S(t)$ in quadrature component of polarization

$S(\omega)$ shape of the Doppler contour

T temperature, Kelvin

$u = (\frac{2kT}{M})^{1/2}$ velocity of particles in a gas

$\underline{v}$ particle velocity

$v = v_z = \underline{v}\underline{n}$ projection of particle velocity onto the direction of a light wave

v_o average velocity of particle

v_r particle-velocity component perpendicular to the direction of a light wave (radial component of velocity)

$v_{res} = c\frac{\nu - \nu_o}{\nu_o} = c\frac{\omega - \omega_o}{\omega_o}$ resonant projection of particle velocity

V absolute velocity of atom or molecule

$V = \frac{pE}{2\hbar}, V_{12} = \frac{P_{12}E}{2\hbar}$ energy interaction between atom and field in frequency units

$W(v)$ distribution of particle velocities

W_{ik} integral probability of transition $i \rightarrow k$

w rate of induced transitions between levels

$\alpha(\omega) = \kappa_a(\omega)$ coefficient of resonant amplification per unit length

$\alpha_{eff} = \kappa_a - \kappa_b = \alpha(\omega) - \kappa(\omega)$ coefficient of effective resonant amplification per unit length

$\beta = \frac{\kappa_{bo} g_b}{\kappa_{ao} g_a} = \frac{\kappa_{bo} G_b}{\kappa_{ao} G_a}$ parameter in theory of a gas laser with nonlinear absorption

Γ, Γ_f homogeneous half-width of line at half-height (rad/s)

$\tilde{\Gamma}$ parameter of collisional broadening of spectral line ($\Delta\omega_{coll} = 2\tilde{\Gamma}p$, p - gas pressure)

$\Gamma_B = \Gamma\sqrt{1+G}$ half-width of the Bennett hole

γ_i rate of radiation decay of population of i level

$\gamma_{coll}^{(i)}$ rate of collisional relaxation of population of i level

$\tilde{\gamma}_i = \gamma_i + \gamma_{coll}^{(i)}$ total rate of relaxation of population of i level

$\gamma = 2\left(\frac{1}{\gamma_1} + \frac{1}{\gamma_2}\right)^{-1}$ parameter, the reciprocal of the total lifetime of a particle at two transition levels

$\tilde{\gamma} = 2\left(\frac{1}{\tilde{\gamma}_1} + \frac{1}{\tilde{\gamma}_2}\right)^{-1}$ parameter γ, with collisions taken into account

$\gamma' = \frac{1}{2}(\gamma_1 - \gamma_2)$

$\gamma_{12} = \frac{1}{2}(\gamma_1 + \gamma_2)$

γ_o coefficient of linear nonresonance losses per unit length

$\gamma_{rad} = \gamma_1 + \gamma_2$ radiation width of the transition at half-height (rad/s)

$\Delta = \omega_1 - \omega_2$ frequency difference of two fields

$\Delta = \frac{c}{2L}$ frequency interval between axial modes

$\Delta_o = \omega_a - \omega_b$ frequency detuning of line centers of amplifying and absorbing transitions

$\Delta\omega = 2\pi\Delta\nu$ width of resonance dip at half-height

$\Delta\omega_c$ cavity bandwidth

$\Delta\omega_{coll} = 2\pi\Delta\nu_{coll} = 2\tilde{\Gamma}p$ collisional broadening

$\Delta\omega_D = 2\pi\Delta\nu_D = 2\omega_o(2\ln 2\frac{kT}{M})^{1/2}$ Doppler width

$\Delta\omega_{tr} = 2\pi\Delta\nu_{tr}$ transit time width of line

$\Delta\omega_{res}$ width of resonance power peak

$\Delta\omega_{sp}$ resonance broadening due to sphericity of a light wave

$\Delta\omega_g = 2\pi\Delta\nu_g$ geometrical broadening of resonance due to nonparallelism of light waves

$\delta = \frac{\Omega}{\Gamma} = \frac{\omega - \omega_o}{\Gamma}$ dimensionless detuning of the field frequency ω about the transition frequency ω_o

$\delta\omega_f (f = a, b)$ width of mode instability region in a ring laser (width of competing resonance)

θ angle between two plane light waves

$\kappa(\omega) = \kappa_b(\omega)$ coefficient of resonance absorption, per unit length

$\kappa_o(\omega) = \kappa_{bo}(\omega)$ coefficient of absorption of a weak field, per unit length

$\kappa_a(\omega) = \alpha(\omega)$ coefficient of resonance amplification, per unit length

$\kappa_{ao}(\omega) = \alpha_o(\omega)$	coefficient of amplification of a weak field, per unit length
Λ	length of particle free path
λ	length of a light wave
ν_o	frequency of quantum transition between two levels (in Hz)
$\rho = \frac{\gamma}{\Gamma}$	parameter of coherent interaction of light field and gas ($\rho = 1$, coherent interaction; $\rho = 0$, incoherent interaction)
ρ_{ii}	diagonal matrix element which is proportional to the ith level population
$\rho_{ij}(i \neq j)$	off-diagonal matrix elements that are proportional to polarization
$\sigma(\underline{v},\omega)$	cross section of radiation transition between levels of a particle with its velocity $\underline{v}$ in the wave field with the frequency ω
σ_o	cross section of particle transition between levels in the maximum
τ_{coll}	average time between collisions of particles
$\tau_i = \frac{1}{\gamma_i}$	lifetime of i level
τ_{tr}	particle transit time with an average velocity through a light beam
τ_v	relaxation time of population of a vibrational level
τ_r	relaxation time of population of a rotational sublevel
ϕ_o	angle of atomic beam
ϕ	light-wave phase
ϕ_{dif}	diffraction angle
$\chi = \chi' - i\chi''$, χ_f	polarizability of a medium
$\psi(\underline{r},t)$	Schrödinger wave function
$\Omega = \omega - \omega_o$	detuning of field frequency ω about the transition frequency ω_o
$\tilde{\Omega} = [\Omega^2 + (2V)^2]^{1/2}$	
$\omega = 2\pi\nu$	field frequency
$\omega_o = 2\pi\nu_o$	frequency of quantum transition between levels
ω_c	resonant frequency of a mode cavity
$\omega' = \omega - \underline{k}\underline{v}$	
$\omega_f(f = a,b)$	central frequency of amplifying ($f = a$) or absorbing ($f = b$) medium

1. Introduction

1.1 Doppler Broadening of Optical Spectral Lines

The basic part of our knowledge of the structure of matter at the atomic-molecular level has been obtained from optical spectroscopy. But as our knowledge of the atomic and molecular structure becomes deeper, the limit of the potentialities of optical spectroscopy, conditioned by the line broadening of emission and absorption spectra of a substance, becomes essential. The biggest spectral-line broadening caused by particle interaction in a condensed medium or a dense gas can be eliminated by observing spectral lines of a low-pressure gas. Yet we cannot accomplish in this way the ultimate goal, which is to produce spectral lines whose widths are dictated by the properties of quantum transitions of individual particles. The thermal broadening of spectral lines due to the Doppler effect is responsible for this.

A moving particle (an atom or a molecule) emits or absorbs radiation that is not exactly at the quantum-transition frequency $\omega_0 = \omega_{21}$ between two energy levels E_1 and E_2, which is determined by the Bohr quantization condition,

$$\hbar\omega_0 = E_2 - E_1 , \tag{1.1}$$

where $\hbar$ is Planck's constant, but at a frequency somewhat shifted by the Doppler effect (Fig.1.1a). The spectral line of a single particle $a(\omega)$ is shifted by a value that depends on the projection of the particle velocity $\underline{v}$ on the direction of observation $\underline{n}$,

$$a_{\underline{v}}(\omega) = a\left(\omega - \underline{n}\frac{\underline{v}}{c}\omega\right) . \tag{1.2}$$

In a gas, the particles move in all possible directions; that is why the Doppler shift differs for each particle. At thermal equilibrium, all directions are equiprobable; that is, the velocity distribution of the particles is isotropic. Therefore, the projection of the particle velocity on any direction ($v = \underline{n}\underline{v}$) is given by the Maxwell distribution,

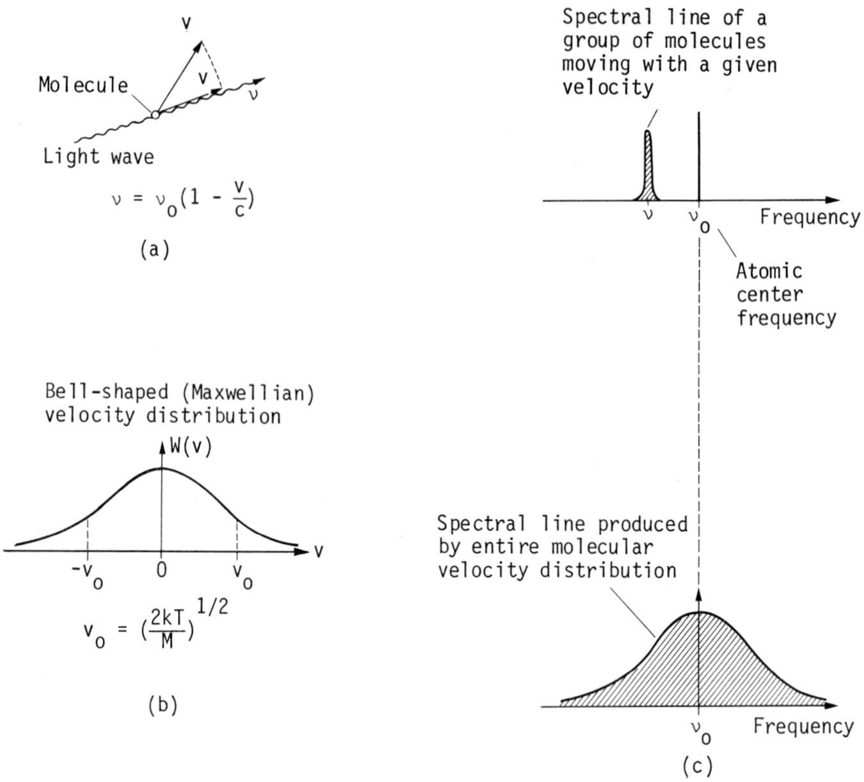

Fig.1.1a-c Influence of the Doppler effect on the shape of a spectral line: (a) Doppler shift due to a molecule moving with velocity $\underline{v}$. The Doppler effect shifts the emission frequency from ν_0 to ν; (b) Thermal distribution of particle velocities; (c) Corresponding spectral lines. The upper curve is the response of particles moving with velocity $\underline{v}$ as in (a). The lower curve is the response due to atoms over the entire thermal distributions

$$W(v) = \frac{1}{\sqrt{\pi}u}\exp[-(v/u)^2] \ , \quad u = \left(\frac{2kT}{M}\right)^{1/2} \ , \tag{1.3}$$

which has the symmetrical form of the Gauss curve (Fig.1.1b). As a result, the spectral line of the assembly of particles has a symmetrical profile with its center at the quantum-transition frequency ω_0 (Fig.1.1c).

In a simple case, when the broadening caused by the Doppler effect and that caused by particle collisions are statistically independent, the spectral-line shape of the whole set of particles $S(\omega)$ is defined by the line-shape convolution of an individual particle with the Doppler-shift distribution, that is, by the distribution of the projections of the atomic velocity on the observation direction,

$$S(\omega) = \int a_{\underline{v}}(\omega) W(\underline{v}) d(\underline{vn}) \ . \tag{1.4}$$

In the case of thermal equilibrium, when the velocity distribution of the particles is maxwellian, given by (1.3), the total half-height broadening of spectral line due to the Doppler effect $\Delta\omega_D$ is given by

$$\Delta\omega_D = \frac{2\omega_o}{c}\left(2 \ln 2 \frac{kT}{M}\right)^{1/2} = 7.163 \cdot 10^{-7} \sqrt{\frac{T}{A}}\,\omega_o \ , \tag{1.5}$$

where M and A denote the mass and atomic weight of a particle, and T is the temperature K. Figure 1.2 shows the dependence of the relative Doppler broadening on the atomic weight of a particle at various temperatures. For atoms and molecules with atomic weight $A \simeq 100$ at normal temperature, the Doppler width $\Delta\omega_D \simeq 10^{-6}\ \omega_o$.

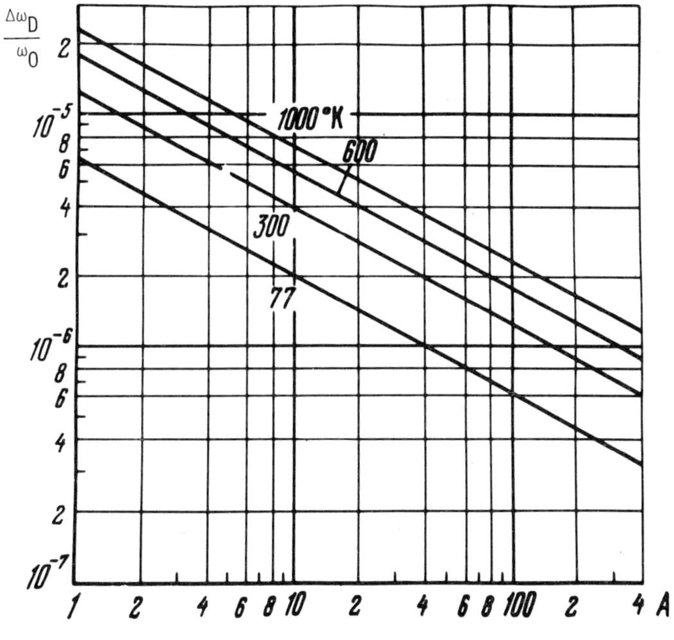

Fig.1.2 Relative value of the Doppler line width for particles with atomic weight A (in at. units) at different temperatures

If an atom or a molecule have several transitions so close that, because of Doppler broadening, their spectral lines overlap, conventional methods of linear optical spectroscopy cannot resolve their structure. High-resolution experiments sometimes indicate, however, that the spectral line under observation is too wide to be explained by only thermal motion of the particles.

Then, the unresolvable splitting can be estimated by thorough analysis; but some details always remain unsolved. Those details sometimes contain information on new physical effects; therefore, they are of particular interest. Here we mention two examples from recent history.

Doppler broadening was a serious handicap to the discovery and measurement of the hyperfine structure of optical transitions. For example, the resonant lines of Na 5890 Å (the transition $3^2S_{1/2} - 3^2P_{3/2}$) and 5896 Å (the transition $3^2S_{1/2} - 3^2P_{1/2}$) have a doublet structure due to hyperfine splitting of the ground state [1.1] by about 0.02 Å. The Doppler broadening of the Na resonant lines exceeds the splitting even at temperature $T \gtrsim 500$ K. Only the method of atomic beams permitted the successful observation of the doublet structure [1.2] and then the full hyperfine structure [1.3] of Na. The Lamb shift in hydrogen atoms is another example of the serious limitation caused by the Doppler effect. As far back as 1934, some scientists began to suspect that the 2 $S_{1/2}$ level of hydrogen was about 0.03 cm^{-1} higher than the 2 $P_{1/2}$ level. But the Doppler broadening of the Balmer line H_α 6563 Å was 0.2 cm^{-1} and made it impossible to observe this effect by the standard methods of optical spectroscopy. That situation existed till 1947 when Lamb and Retherford showed, using the microwave spectroscopy methods [1.4] that the 2 $S_{1/2}$ level was really shifted above the 2 $P_{1/2}$ level by 1062 ± 5 MHz, or 0.034 cm^{-1}. As known, the discovery of the Lamb shift and its further theoretical explanation have promoted the development of quantum electrodynamics.

1.2 Methods of Linear Optical Spectroscopy Without Doppler Broadening

To remove the limitations in optical spectroscopy caused by the Doppler effect, quite a number of methods have been developed to measure the structure of quantum transitions screened by Doppler broadening of spectral lines. The method of atomic or molecular beam developed in the 1930s and successfully applied subsequently, is among these. In the 1950s, some methods of spectroscopy without Doppler broadening (double microwave optical resonance, level crossing, and quantum beats) were developed which were applied mainly to atomic transitions in the visible region. The advent of coherent tunable lasers has culminated in the discovery and wide application of nonlinear laser spectroscopy which, like the method of atomic or molecular beam, is very effective for any atomic and molecular transitions in a very wide frequency range (uv, visible and ir). Though the subject of this monograph is to set forth the principles of nonlinear laser spectroscopy, it is quite opportune to give brief information on all other methods of optical spectroscopy free of Doppler broadening. Doing this, we not only do justice but hope that our readers will be able to develop new methods based on combinations of other methods with nonlinear laser spectroscopy.

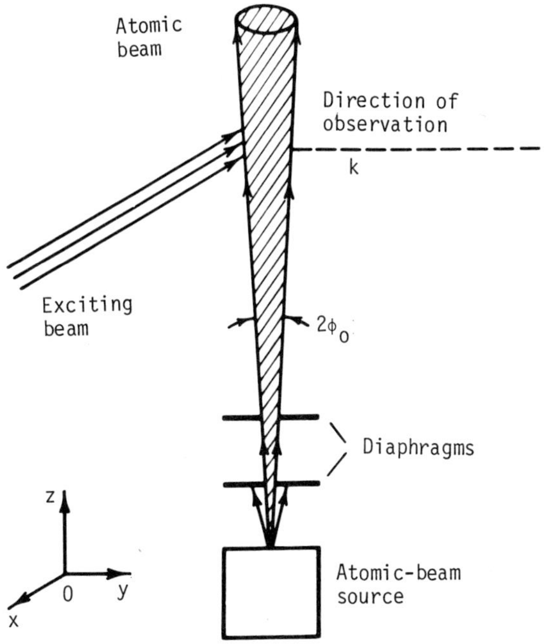

Fig.1.3 Observation scheme of the spectral line of an atomic beam

Method of Atomic or Molecular Beam

In 1928 DOBRETZOV and TERENIN [1.2] used a beam of Na atoms to remove Doppler broadening and to study the hyperfine structure of resonant lines. Their experiment is shown schematically in Fig.1.3. An atomic beam is formed by use of two diaphragms and has, consequently, the definite angular divergence $2\phi_0$. The beam of atoms flying in the Oz direction was excited at a right angle by a well-collimated light beam from a sodium-vapor lamp. The resultant resonant fluorescence of the atoms in the beam was also observed in the Oy direction, strictly perpendicular to the atomic beam. The reduction of Doppler broadening depends on the degree of collimation of the atomic and light beams. The main contribution to the residual Doppler broadening is made by the angular divergence $2\phi_0$ of atomic beam,

$$\delta\omega = 2\phi_0(u/c)\omega_0 \simeq \phi_0\Delta\omega_D \ . \tag{1.6}$$

The difficulties of collimating the exciting light beam are obviated when the atoms are excited by an electron beam [1.3]. Some experiments on observation of light absorption by a beam of potassium atoms and other elements were reported in [1.5]. In this case, a collimated beam of a light source with a broad resonance line, a potassium lamp, was passed through an atomic beam at a right angle; against the background of the broad line, very narrow absorption lines were observed. Of course, spectral equipment with sufficiently high resolution is needed to observe narrow absorption and emission lines in atomic beams. In such cases, the Fabry-Perot êtalon is usually used. More detailed data on the use of the method of beams in optical spectroscopy one can find in [1.6].

Doppler broadening in microwave spectroscopy is much less than in the optical range, but for high-resolution spectroscopes it is substantial. BASOV and PROKHOROV [1.7] and GORDON et al. [1.8] have successfully applied the method of molecular beam to remove Doppler broadening in microwave spectroscopy. In their experiment, the molecules moved perpendicularly to the direction of the microwave field; the line width decreased directly with the degree of molecular-beam collimation.

The potentialities of the method of atomic or molecular beams in optical spectroscopy widen considerably when lasers with a narrow, tunable radiation line are used as sources for exciting the particles in the beam. Because the width of the laser-radiation line can be rather easily reduced to a value much smaller than the atomic-beam line width $\delta\omega$, and can be easily frequency scanned, there is no need in this case to use classical high-resolution spectral instruments like the Fabry-Perot êtalon. The first successful experiments combining a molecular beam (I_2 molecules) with laser radiation (argon laser at 5145 Å) were reported in [1.9]; undoubtedly, this technique should be advanced further.

Method of Double Radiooptical Resonance

In 1952 BROSSEL and BITTER performed an experiment [1.10] in which the Zeeman splitting of an excited state of Hg atom screened by Doppler broadening was recorded by the method of so-called double radiooptical resonance. The atoms of even-number mercury isotopes, with zero nuclear spin, were excited to the $6\ ^3P_1$ state by the linearly polarized radiation of a mercury lamp at $\lambda = 2537$ Å. The mercury vapor was situated in an external constant magnetic field $\underline{H}_0$. The polarization vector of the exciting radiation was directed along $\underline{H}_0$. Such radiation excites the sublevel with $m = 0$ (Fig.1.4a), which results in polarization of the resonant radiation reemitted by the excited atoms that remain undisturbed during the short lifetime (of the order of 10^{-7} s) of the excited state. Now, while the constant magnetic field $\underline{H}_0$ that

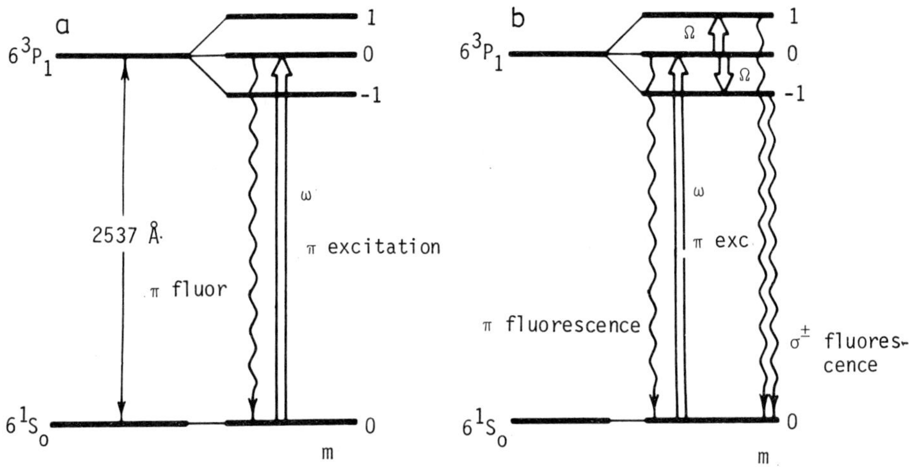

Fig.1.4a,b Explanation of the double radiooptical resonance: (a) Polarized fluorescence without radio frequency field; (b) Appearance of circularly polarized radiation in case of magnetic resonance

results in the magnetic-sublevel splitting is maintained, we introduce perpendicularly an oscillating magnetic field $H_1 \cos \Omega t$ whose frequency is coincident with that of the Zeeman splitting in the $\underline{H}_0$ field, quantum transitions arise between different magnetic sublevels or, as usually stated, a magnetic resonance occurs (Fig.1.4b). The population of magnetic sublevels with $m = \pm 1$ gives rise to $\sigma^{\pm}$ polarization components in fluorescence. So, by variation of the Zeeman splitting or the oscillating-magnetic-field frequency Ω the moment of resonant appearance of circular-polarization components in the radiation of excited atoms can be observed. Such an experiment employs the KASTLER idea of optical atomic pumping [1.11] and the idea of microwave frequency resonance in excited atomic states, applied previously by LAMB and RETHERFORD to the $n = 2$ state of hydrogen atoms [1.4].

An important feature of the double radiooptical resonance is its nonsensitivity to Doppler broadening. The resonant-pumping radiation excited atoms that had all possible velocities. The magnetic-resonance frequency of moving atoms may be shifted, in principle, by a small value

$$\delta\Omega_D = \frac{v}{c}\Omega \tag{1.7}$$

and may cause Doppler broadening of the magnetic resonance line; but this value is very small compared to the value of Doppler broadening for an optical transition. What is more, this Doppler broadening of the magnetic resonance line must disappear if the free path of the atoms is shorter than the wavelength of microwave radiation λ_Ω (the Dicke effect [1.12]). Thus, the method of double radiooptical resonance makes it possible to study the structure of magnetic sublevels that is hidden by the Doppler broadening of an optical spectral line. The method is described in more detail in review [1.13].

The method of double resonance has been used so far only for atoms and levels that can be excited only by resonant radiation. By the use of tunable lasers, it is possible to excite practically any atomic or molecular level, which considerably widens the scope of the method of double radiooptical resonance, particularly its application to vibrational molecular levels [1.14].

Method of Level Crossing

The energy of atomic magnetic sublevels in an external magnetic field depends on the field strength. At certain values of the magnetic field strength, a level crossing may occur, followed by atomic-state interference. Because of quantum-state interference, the polarization and angular distribution of atomic radiation vary. The influence of a magnetic field on the polarization of atomic resonant radiation was first discovered by HANLE [1.15] in 1923. The scheme of his experiment is shown in Fig.1.5a. Even isotopes of Hg were irradiated with the resonant radiation at 2537 Å of a mercury lamp that was linearly polarized in a certain direction $\underline{e}$. When there was no external magnetic field, the reemitted light was strongly polarized (~90%) and the vector of its polarization coincided with the polarization vector of the exciting radiation. The angular distribution of the reemitted light depended on the angular distribution of the electric-dipole radiation, the axis of which was in line with the polarization vector $\underline{e}$ of the exciting radiation. If the constant magnetic field $\underline{H}_0$ was applied parallel to the polarization vector $\underline{e}$, the nature of the radiation did not change. But if $\underline{H}_0$ was perpendicular to $\underline{e}$, it changed radically. With increase of the magnetic field strength the degree of polarization of the radiation emitted in the direction of $\underline{H}_0$ decreased and became zero when the Zeeman splitting of the excited state was greater than the natural width of the level 2Γ.

This effect was explained by BREIT [1.16] in terms of interference of degenerated quantum states. The ground state $6\ {}^1S_0$ of even mercury isotopes is singlet, and the excited state $6\ {}^3P_1$ in the magnetic field is a set of three equidistant magnetic sublevels (Fig.1.4). The excitation conditions of these sublevels are dependent on the polarization and direction of the exciting radiation about the magnetic field $\underline{H}_0$. The wave linearly polarized

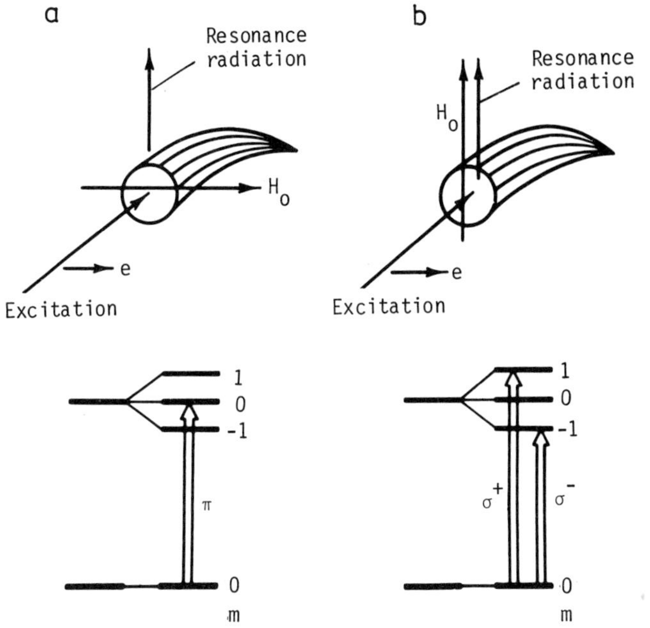

Fig.1.5a,b Observation scheme of the Hanle effect: excitation of pure (a) and superposed (b) quantum state with different orientations of the magnetic field

in the direction of $\underline{H}_0$ excites only one sublevel $m = o$ (π is the transition component), that is, it excites the atom to a pure quantum state (Fig.1.5). If $\underline{e} \perp \underline{H}_0$, the linearly polarized wave, which is a superposition of waves with left- and right-hand polarizations, excites both σ^+- and σ^--components; that is, it excites simultaneously two sublevels: $m = \pm 1$. In this case, the atom is excited to superposition quantum states (Fig.1.5). In this situation, the electric dipole of the atom precesses around the magnetic-field direction. When the Larmor precession period is much longer than the excited-state decay time (the Zeeman splitting is small compared to the natural width of the level), the angle of rotation after the excitation is small and only partial depolarization of the radiation takes place. If the Zeeman splitting exceeds the natural width, the dipoles make many rotations in the process of excitation and the radiation becomes fully depolarized.

The Hanle effect was widely used to evaluate the lifetime of excited states; all of these experiments are well reviewed in the book by MITCHELL and ZEMANSKY [1.17].

In 1959, COLEGROVE et al. [1.18] applied the level-crossing effect in a non-zero constant magnetic field to study the fine structure of the excited state $2\,^3P$ He. In this work, which became, in essence, the second birth of the Hanle effect, they applied practically a new spectroscopic method for studying the structure of excited atomic states screened by Doppler broadening. The limiting resolution of this technique is determined by the natural width of crossing levels and is not dependent upon the thermal motion of atoms. For certain gas systems, the width of level-crossing resonances is some fraction of a Hertz. Unlike the method of double radiooptical resonance, the method of level crossing does not require an external radio-frequency field; hence, it is free from unwanted field shifts and level broadening. Review [1.19] gives more-detailed information on the level-crossing method and on further references to this method.

Method of Quantum Beats

The interference of atomic states forms the basis for the Hanle effect and the level-crossing method. This effect occurs when an atom appears to be in a so-called superposition state, which cannot be characterized by the wave function of a stationary state [1.20]. Such a state may be represented only by the superposition of the ψ_k states with defined energies E_k,

$$\psi = \sum_k a_k \psi_k \exp(-iE_k t/\hbar) \; . \tag{1.8}$$

The averaged squares of complex coefficient modules $\langle|a_k|^2\rangle$ mean the level populations. Assume, for example, that an atom at the instant $t = t_0$ is excited to a superposed state with two close sublevels (Fig.1.6a). Let the radiative transition to the nondegenerated lower state from two excited sublevels be allowed with the same probability. In such a case, the wave function of the excited atom can be written as

$$\psi(t) = \left[a_1\, e^{-i\omega_{01}(t-t_0)} \psi_1 + a_2\, e^{-i\omega_{02}(t-t_0)} \psi_2 \right] e^{-\Gamma(t-t_0)} \; . \tag{1.9}$$

The probability of the transition to the lower state P(t) defined by the matrix element $|\langle\psi_0|\hat{\underline{P}}|\psi\rangle|^2$, where $\hat{\underline{P}}$ is the operator of dipole moment, has apparently an oscillating dependence, with the frequency $\omega_{12} = \omega_{01} - \omega_{02}$. If, for example, $a_1(t_0) = a_2(t_0)$, P(t) has the form given in Fig.1.6b. This effect is called "quantum beats" in spontaneous radiation.

The effect of quantum beats was experimentally detected in 1964 by ALEXANDROV [1.21] and DODD et al. [1.22]. In his experiment, Alexandrov excited Cd atoms at the transition $5\,^3P_1 - 5\,^1S_0$ (3261 Å) by a short pulse (10^{-7} s) of a cadmium lamp. The triplet $5\,^3P_1$ was split by the magnetic field into

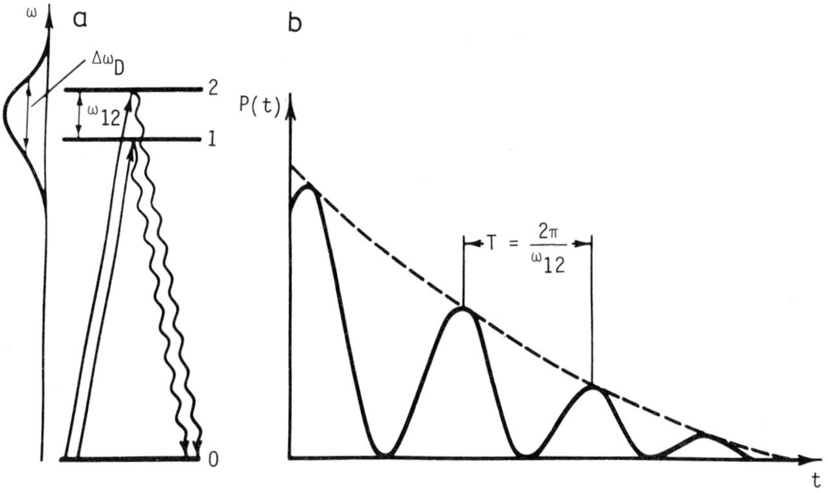

Fig.1.6a,b Method of the quantum beats: (a) Level diagram; (b) Decay kinetics of superposed state

three sublevels with $m = 0, \pm 1$. By selecting the polarization of the exciting light the level with $m = o$ can be left unexcited (the polarization vector $\underline{e} \perp \underline{H}_0$). The width of the $5\,^3P_1$ level is approximately $2\Gamma = 5 \cdot 10^5\ s^{-1}$. The value of level splitting by the magnetic field $\omega_{12}/2\pi = 10^6$ Hz, which is much larger than the natural level width but less than 1/1000 of the Doppler line width. The Doppler effect evidently does not limit the resolution of spectral lines by the method of quantum beats, provided that we ignore a negligible difference of Doppler shifts for the ω_{01} and ω_{02} frequencies. The method has a certain advantage over that of double radiooptical resonance, which is the absence of any disturbance for the system under study. Numerous methods of preparation of superposed atomic states have been developed and successfully used. Among others, the method of state excitation by a pulsed electron beam is very effective. The method of quantum beats is described in detail in a review [1.23] by ALEXANDROV.

Use of pulsed tunable lasers eliminates experimental difficulties in exciting an atom to the superposed state [1.24] with a short light pulse. The method of quantum beats with laser-pulse excitation of atoms must be made very efficient to study the hyperfine structure of highly excited atomic states.

1.3 Saturation Laser Spectroscopy

A Doppler-broadened spectral line is, in essence, a set of a great number of much narrower spectral lines of absorption and emission of particles with different velocities. Therefore the Doppler broadening is often described as *inhomogeneous* broadening. Homogeneous broadening means the width of a spectral line not broadened by the Doppler effect. The homogeneous width of a spectral line of an individual particle in a gas is produced by several effects. Table 1.1 gives values of inhomogeneous broadening of optical spectral lines in gas and the contributions made by various mechanisms to homogeneous broadening. The first and the most fundamental effect is *radiative* broadening due to spontaneous decay of an excited state. A spectral line conditioned by spontaneous decay has a lorentzian shape [1.25],

$$a(\omega) = \frac{(\gamma_{rad}/2)^2}{(\omega - \omega_0)^2 + (\gamma_{rad}/2)^2} , \tag{1.10}$$

where $\gamma_{rad} = \gamma_1 + \gamma_2$ is the radiative half-height width of the quantum transition $2 \to 1$, γ_1 and γ_2 are the decay constants of the higher (2) and lower (1) levels of the transition, and the $a(\omega)$ function is normalized to unity at its maximum. This broadening is usually termed natural. The lifetime of an excited atomic or molecular state depends on the oscillator strength of transition and the wavelength of radiation. For more intensive electronic transitions of atoms and molecules in the visible range of spectrum, $1/\gamma_{rad} \simeq 10^{-8}$ s. For metastable atomic and vibrational molecular levels the radiative lifetime may be as great as $1/\gamma_{rad} = 10^{-1} - 10^{-5}$ s.

Interparticle collisions make a large contribution to homogeneous broadening. Each collision shifts the phase of periodic motion of an electron in an atom or of atomic vibrations in a molecule. Such random shifts result in a quasi-periodic process instead of the regular periodic one that defines the atomic or molecular state. The quasi-periodic process occurs as a sequence of coherent trains with the mean duration τ_{coll} which is the mean time between successive collisions of one particle with others. Besides phase change, the collisions cause state decay. The collision-induced line shape in the simple case when either phase shift or level decay takes place during collisions has also lorentzian shape [1.26]. Collisional half-height width $\Delta\omega_{coll}$ being equal to the collisional frequency of a particle,

$$\Delta\omega_{coll} = 2/\tau_{coll} . \tag{1.11}$$

Table 1.1 Sources of line broadening in gases

Inhomogeneous broadening mechanism

Type	Origin	Line width	Range of values
Doppler broadening	Doppler effect due to thermal molecular motion	$\nu_0 \frac{v_0}{C}$ ν_0 = center frequency of transition v_0 = average speed C = speed of light	10^8-10^{10} Hz

Homogeneous broadening mechanisms

Type	Origin	Line width	Range of values
Natural broadening	Spontaneous decay of an excited state	$\frac{1}{2\pi\tau}$ τ = natural lifetime	Atoms: 10^5-10^7 Hz Molecules: 10-10^3 Hz
Lorentz (collision) broadening	Interparticle collisions	$\frac{1}{\pi\tau_{coll}}$ τ_{coll} = mean time between collisions	$3 \cdot 10^3$-$3 \cdot 10^4$ Hz (at 1 millitorr pressure)
Wall-collision broadening	Particle collisions with the walls of sample cell	$v_0/2\pi L$ v_0 = average speed L = cell diameter	10^3-10^4 Hz
Transit-time broadening	Transit of particles through light beam	$v_0/2\pi a$ v_0 = average speed a = beam diameter	10^3-10^4 Hz
Power broadening	High intensity of laser beam induces high rate of transition	$p_{12}E/\hbar$ p_{12} = transition dipole moment E = strength of laser field $\hbar$ = Planck's constant	10^4-10^5 Hz (for 1 milliwatt/cm^2 intensity)

A comparatively weak particle interaction, when the velocity and direction of particle motion slightly varies, is sufficient to cause phase shifts. Therefore the cross-section of the collisions that result in line broadening is usually much larger than the gas-kinetic cross-section that determines strong interparticle collisions. The collisional-broadening values typical of molecules lie in the range $\Delta\nu_{coll} = \Delta\omega_{coll}/2\pi = (3 \div 30)$ MHz at a gas pressure of 1 torr. This corresponds to the mean time between the collisions that cause broadening $\tau_{coll} = 10^{-7} + 10^{-8}$ s. At such a pressure, the collisional broadening for, say, vibrational transitions of molecules is thousands of times the natural width. Because the collisional line broadening is in proportion to pressure, a pressure less than 10^{-3} torr should be used to reduce the collisional broadening to the natural line width. But at lower pressure, one more mechanism appears, the contribution of which to broadening becomes significant.

At low pressures, the mean length of the free molecular path Λ, which was related to the mean velocity v_0 and the mean free time τ_{coll} by $\Lambda = v_0 \cdot \tau_{coll}$, increases and can, in principle, become comparable with or even larger than the diameter of gas cell. In this case, the time between collisions τ_{coll} is determined not by particle-particle collisions but by collisions of the particles with the cell walls. This effect results in collisional broadening of the spectral line with the value $\sim v_0/L$, where L is the mean distance between the cell walls. When the cross dimension of the cell is several centimeters, the particle-wall collisions induce broadening of the order of 10^4 Hz. This value is rather small, but it exceeds the natural width of molecular transitions in the infrared range.

Consider the interaction of a plane, coherent laser light wave at the frequency ν, with a Doppler-broadened absorption line. Assume that the light-wave frequency coincides accurately with the Doppler-line center, that is with the frequency ν_0 of the quantum transition between molecular levels. Such a wave can interact only with molecules moving almost transverse to the light beam, that is molecules that have very small Doppler frequency shift (Fig.1.7). Otherwise, the molecules cannot interact with the light wave because of the Doppler shift. If the field frequency is not coincident with the line center, a motionless molecule, or one that moves transverse to the beam, does not resonate with the field. Molecules that have significant projections of their velocities onto light beam direction $v_{res} = (\nu - \nu_0)c/\nu_0$ (Fig.1.7b) resonate with the field. Such a velocity is essential for the Doppler shift, to compensate for the detuning of the field frequency from the quantum-transition frequency. When $\nu > \nu_0$, the wave interacts with

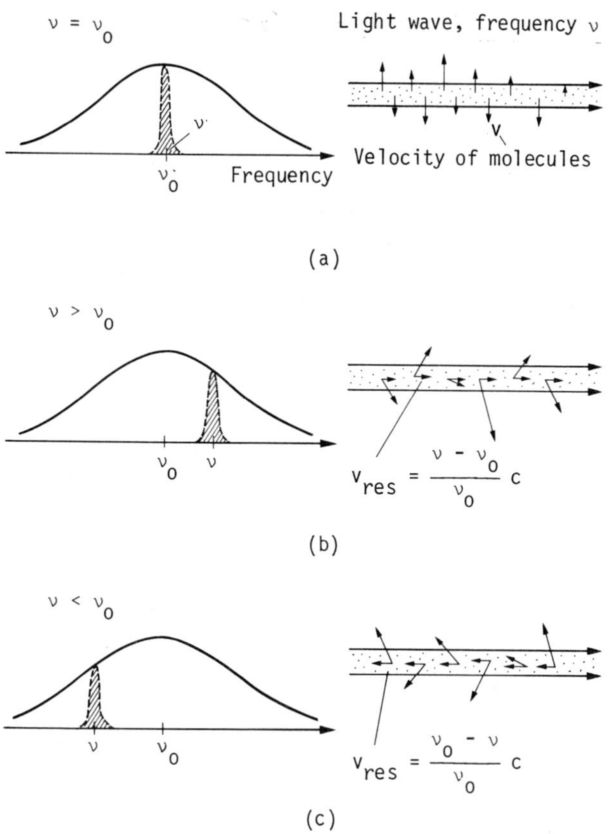

Fig.1.7a-c Atomic-velocity groups resonating with an applied monochromatic field: (a) The monochromatic field is tuned to the atomic center frequency ($\nu = \nu_0$). The resonant atoms are those that have no velocity component in the direction of the light wave. (b) In this case, the light frequency is tuned above the atomic center frequency ($\nu > \nu_0$), so that the atoms that are Doppler shifted into resonance move in the same direction as the light wave. (c) In this case, the light frequency is tuned below the atomic center frequency ($\nu < \nu_0$), so that the atoms that are Doppler shifted into resonance move in the direction opposite to the propagation direction of the light wave

molecules that have positive projections of their velocities onto the wave-propagation direction and, conversely, with those that have negative projections of their velocities when $\nu < \nu_0$ (Fig.1.7c). Therefore, the light wave can interact only with molecules that have the corresponding Doppler frequency shift. Because the spectral line width of each molecule is the homogeneous width, such a group of molecules occupies a narrow spectral range in the Doppler profile; its center is at the field frequency ν and its width is equal to the homogeneous width.

When moving particles interact with a beam of limited diameter, the resonance is broadened by one more effect, apart from those already mentioned.

A molecule that has velocity v_0 crosses a light beam with diameter a in time $\tau_{tr} = a/v_0$. The light beam may be regarded as a measuring tool with which the molecules interact in the finite time $\Delta t = \tau_{tr}$. According to the indeterminacy principle, the energy of the transition between levels cannot be evaluated to better than $\Delta E = \hbar/\Delta t$. This corresponds to an indeterminacy of the transition frequency $\Delta E/\hbar = 1/\tau_{tr}$, i.e., to the spectral line broadening caused by the finite time of particle flight through the light beam (transit-time broadening),

$$\Delta\omega_{tr} \simeq \frac{1}{\tau_{tr}} . \tag{1.12}$$

At relatively high gas pressures, when the free-path length Λ is much shorter than the light-beam diameter, this effect is not significant in comparison with collisional broadening, but at low pressures it is the main mechanism of broadening for long-lived transitions.

Thus, when spectral-line broadening is inhomogeneous, the light wave interacts only with particles with which it is in resonance. The portion of the particles that interact with the field depends on the homogeneous-Doppler-width ratio. Strictly speaking, it depends also on the spatial wave configuration. If a monochromatic field is isotropic (i.e., comprises a series of waves that propagate in different directions inside a cavity with rough walls), all of the particles can interact with the field, whatever their velocity (Fig.1.8a). On the other hand, a plane travelling wave $E\cos(\omega t - \underline{k}\underline{r})$ interacts only with particles located within the spectral range of the homogeneous width 2Γ at the resonance frequency $\omega = \omega_0 + \underline{k}\underline{v}$ (Fig.1.8b). In other words, the field interacts only with particles that have a definite velocity projection on the travelling-light-wave direction,

$$|\omega - \omega_0 + \underline{k}\underline{v}| \lesssim \Gamma . \tag{1.13}$$

The resonance width depends also on the wave-front curvature of the beam. A moving particle responds to the field in particular region of space. For example, in the case of a spherical wave with radius r an additional broadening occurs because of the variable Doppler shift for each particle (Fig.1.8c),

$$E = \Sigma E_i \cos(\nu t + \underline{k}_i \cdot \underline{r} + \phi_i)$$

$$E = E \cos(\nu t - \underline{k} \cdot \underline{r})$$

$$E = E \cos(\nu t - \underline{k} \cdot \underline{r})$$

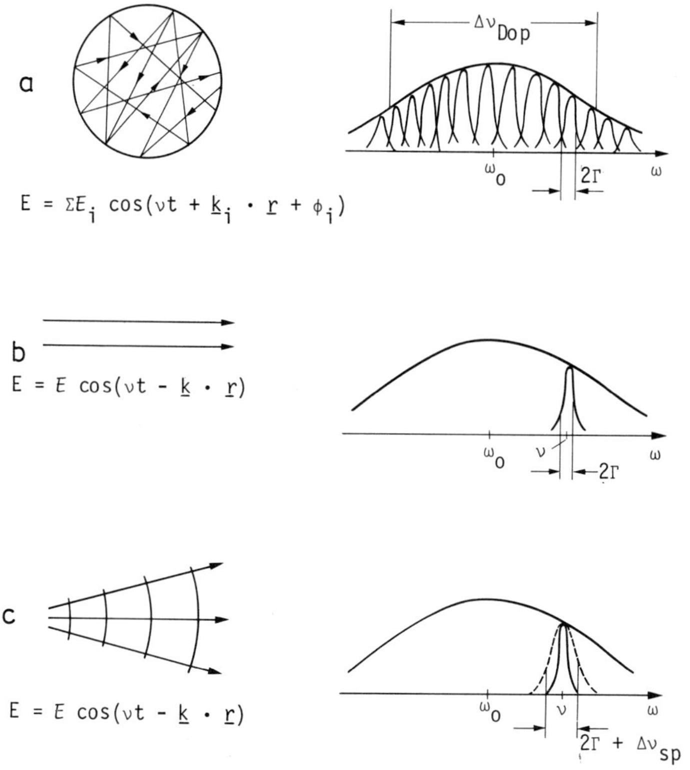

Fig.1.8a-c Influence of the spatial configuration of the light field on resonant interaction width, for a Doppler spectral line: (a) Isotropic monochromatic field; (b) Plane coherent wave; (c) Spherical coherent wave

$$\Delta\omega_{geom} \simeq k v_0 / (kr)^{1/2} , \tag{1.14}$$

where $k v_0 \simeq \Delta\omega_D$. Hence, the motion of a particle causes not only the inhomogeneous Doppler broadening of the whole spectral line but also *spatial* or *geometrical* broadening of the resonance width. The physical explanation for this type of broadening is that, because of the motion of the particle, the field interaction depends on both the particle velocity $\underline{v}$ and its position $\underline{r} = \underline{r}_0 + \underline{v}(t - t_0)$. In the limiting case of a collimated beam with diameter a, the minimum of the wave-front curvature is determined only by diffraction $(r \sim a^2 k)$, and the geometrical broadening in accordance with (1.14) is reduced to the broadening caused by the finite time of flight of the particle through the beam $\Delta\omega_{geom} \simeq v_0/a$. In the limiting case of an isotropic field with mean radius of curvature $r \sim 1/k$ the geometrical broadening (1.14) coincides with the Doppler width.

Thus a monochromatic light wave with a small divergence, i.e., a light field with high spatial and temporal coherence, can interact with a small portion of the atoms or molecules within a Doppler-broadened transition. So this field can change the state of this small part of particles and discriminate them clearly from the rest of the particles, the velocities of which do not comply with the resonance condition (1.13). Let the intensity of the light field be sufficient to transfer a considerable part of the particles to the excited state of the transition. In the simplest case of equal decay constants for two transition levels, the total probability W_{12} of stimulated transition of a particle with velocity $\underline{v}$, under the influence of the travelling wave $E\cos(\omega t - \underline{k}\underline{r})$, to the excited level will be

$$W_{12}(\underline{v}) = \frac{G}{2}\,\frac{\Gamma^2}{(\Omega - \underline{k}\underline{v})^2 + \Gamma^2(1+G)} \quad , \tag{1.15}$$

where $\Omega = \omega - \omega_0$ is the detuning of the travelling-wave frequency ω from the frequency of the fixed-particle transition ω_0, $G = (P_{12}E/\hbar\Gamma)^2$ denotes the saturation parameter of transition, P_{12} is the matrix element of dipole moment of the transition. The probability of transition of the particle to the upper level is determined by the saturation parameter G and by the detuning of the particle velocity from the resonance velocity.

The value $P_{12}E/\hbar$ is the rate of transition of the particle between the levels per unit time and hence causes the so-called *power broadening* of the spectral line (see Table 1.1). The power broadening is given in accordance with (1.15) by the simple expression

$$\Delta\omega = 2\Gamma(1+G)^{1/2} \quad . \tag{1.16}$$

This mechanism of broadening is well-known in microwave spectroscopy [1.27, 28].

Preferential excitation of particles that have a definite velocity of motion alters the equilibrium distribution of particle velocity at each of the transition levels (Fig.1.9). The velocity distribution of particles in the lower level $n_1(\underline{v})$ develops a shortage of particles with velocities that comply with resonance condition (1.13),

$$n_1(\underline{v}) = n_1^0(\underline{v}) - W_{12}(\underline{v})[n_1^0(\underline{v}) - n_2^0(\underline{v})] \quad , \tag{1.17}$$

where $n_1^0(\underline{v})$ and $n_2^0(\underline{v})$ denote the initial equilibrium distributions of velocities of particles that are in the lower and upper levels, respectively. For

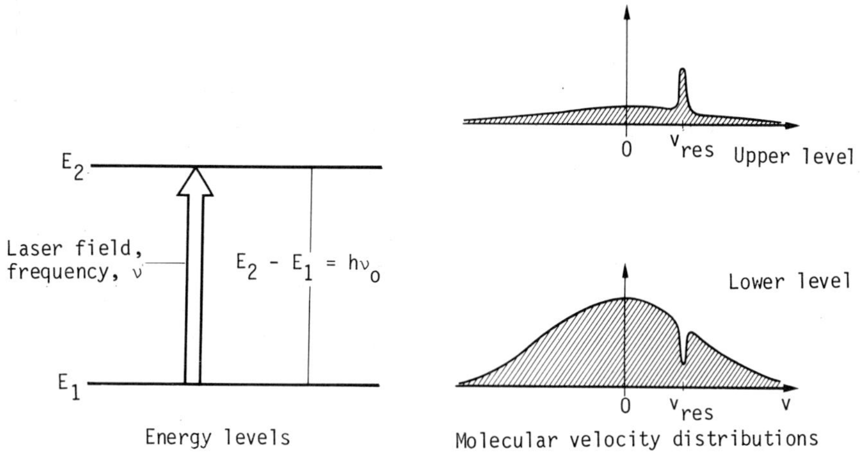

Fig.1.9a,b Changes in the particle velocity distribution for two transition levels under the action of a laser travelling wave of frequency ν: (a) Level diagram; (b) Distribution of Z component of velocity of the particles on lower and upper levels of transition ($v_{res} = (\nu - \nu_0)c/\nu_0$ = resonant projected velocity)

the upper level, the velocity distribution, on the other hand, has excess particles with resónant velocities,

$$n_2(\underline{v}) = n_2^0(\underline{v}) + W_{12}(\underline{v})[n_1^0(\underline{v}) - n_2^0(\underline{v})] \ . \tag{1.18}$$

Figure 1.9b shows the distribution of the projections of particle velocities onto the light-beam direction for the lower and upper levels. In the absence of the light wave, these distributions would be symmetrical, whereas in the presence of a strong light wave a "hole" appears in the velocity distribution for the lower level and a "peak" for the upper level. This hole or peak appears at the particle velocity that depends on the field frequency,

$$v = v_{res} = \frac{\omega - \omega_0}{\omega} c \ . \tag{1.19}$$

The depth of the hole and the height of the peak are determined by the saturation parameter G. Their width equals the homogeneous width 2Γ of the power broadening, which is also determined by G, according to (1.16).

Thus, a light wave changes the velocity distribution of particles on the the various levels, i.e., the distribution becomes essentially nonisotropic.

Indeed, this causes the Doppler-broadened line to be distorted. A "hole" appears in the Doppler profile because of the molecules that have passed into the excited state, and the width of this hole directly determines the homogeneous transition width, which may be many orders of magnitude narrower than the Doppler width. To obtain such a narrow structure inside the Doppler profile, the light wave should satisfy three conditions:

1) monochromaticity, or high temporal coherence;
2) directionality, or high spatial coherence;
3) intensity sufficient to saturate a transition.

These conditions can be met only with laser radiation. Therefore it is quite clear why the development of quantum electronics has inevitably given birth to nonlinear laser spectroscopy inside the Doppler width.

The possibility of "hole burning" in a spectral-amplification line was first treated in the earliest papers on lasers. SCHAWLOW [1.29] discussed this possibility in luminescent crystals, the spectral lines of which are inhomogeneously broadened at low temperatures. BENNETT investigated this effect [1.30] for the Doppler-broadened amplification line of a gas laser. The effect has been found to be most substantial only for gas lasers with a low-pressure (no more than a few torr) amplification medium.

The first gas laser [1.31] operated on a neon-helium mixture with a pressure of about 1 torr at $\lambda = 1.15$ μm. In that case, the homogeneous line width was much narrower than the Doppler width (Table 1.1), and the theoretical study of the operation of such a laser had to allow for the inhomogeneous nature of the broadening. On inhomogeneous broadening, light waves interact only with particles that are in resonance with them. Therefore, a strong light wave that causes amplification saturation "burns" a "hole" (BENNETT [1.30]) at the field frequency in the Doppler contour of the amplification line. In the laser cavity, there is a standing light wave that may be represented as a superposition of two counter-running waves that have the same frequency. In this case, as LAMB shows in his gas-laser theory [1.32], each wave burns its own "hole". Because these two waves run in opposite directions, there arise two holes symmetrical about the center of the Doppler contour (Fig.1.10). In essence, the laser field absorbs the energy from two groups of amplifying particles that have opposite velocities. As the laser frequency is tuned to the center of the Doppler profile, the two holes coincide and the standing light wave interacts with only one group of particles. This results in a resonant decrease of power at the center of the Doppler amplification line (Fig.1.10c). This effect now called the "Lamb dip" was

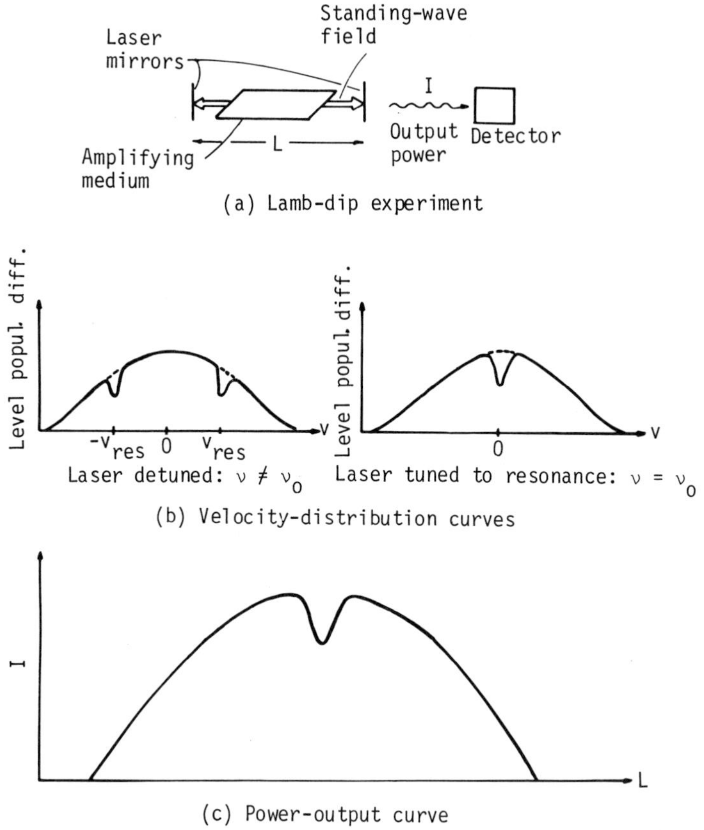

Fig.1.10a-c Lamb-dip experiment: (a) Experimental arrangement; (b) Velocity-distribution curves. Note that the saturated velocity groups can overlap only when the laser frequency is tuned to the center of the Doppler profile; (c) Power-output curve. Laser intensity is plotted as a function of separation between the laser mirrors. The narrow dip in the center is the Lamb dip

first considered by LAMB in his gas-laser theory [1.32]. Experimental observations of the effect were first reported in [1.33,34].

The width of the dip at the center of the Doppler amplification line equals to homogeneous line width 2Γ, which may be considerably smaller than $\Delta\omega_D$. This has opened strong possibilities for spectroscopy inside the Doppler contour; such a method has been used in some experiments on measuring collisional broadening [1.35], isotope shift [1.36] and on stabilizing the laser oscillation frequency to the center of the amplification line [1.37].

Another idea for producing very narrow resonances inside the Doppler contour using a coherent light wave was suggested in 1965 by BASOV and LETOKHOV

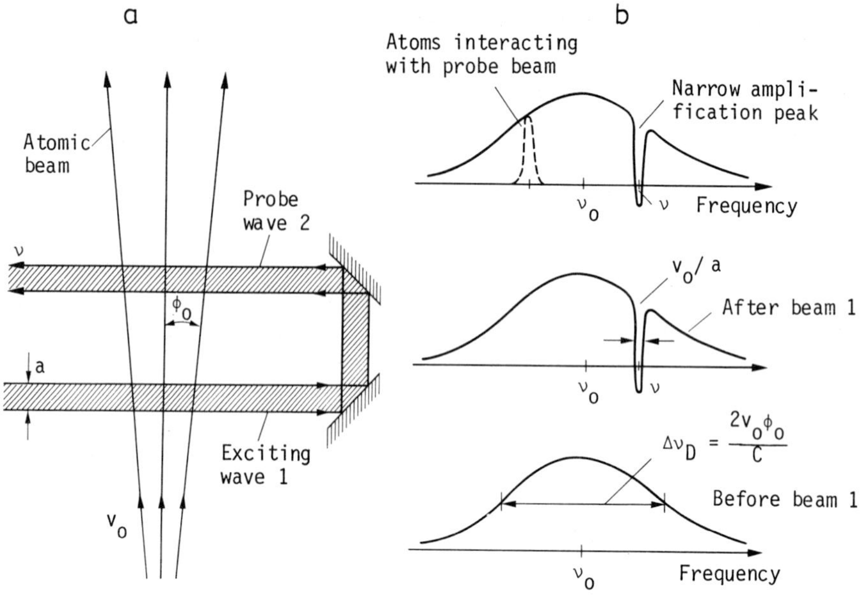

Fig.1.11a,b Production of narrow resonance inside the Doppler-broadened absorption line of an atomic beam, by use of a coherent light wave: (a) Experimental scheme; (b) Line shape of atomic-beam absorption before and after excitation in light wave 1

[1.38]. In this work, they proposed to excite the atoms in a beam with a Doppler-broadened absorption line by use of a coherent light wave. When the saturation parameter $(p_{12}E/\hbar)(a/v_0) = \pi$, a considerable part of the particles that have resonant projections of their velocities on the beam direction become excited and a narrow peak of amplification is formed with width v_0/a determined only by the time of flight of the particle through the beam (Fig. 1.11). To detect such a peak, they suggested using a probe counter-running wave of the same frequency. Such a wave can interact with excited particles, provided that the frequency of the waves coincides with the center of the Doppler absorption line ν_0. In this case, the two waves will interact at the same time with particles flying at right angles to the beams within the limits of a small angle $\delta\phi \lesssim \lambda/a$ (λ is the wavelength of the transition), which is much smaller than the divergence ϕ_0 of particle beams. Therefore a narrow resonance is *induced by the light wave*, and the beam of particles is required just to transfer particles from one light beam to the other without collisions, rather than to narrow the spectral line.

An experiment on observing narrow resonances in spaced light beams has recently been performed at the Institute of Spectroscopy, Academy of Sciences USSR [1.39]. The difference of technique between this experiment and that just described was that instead of a beam of particles they used a molecular gas of such a low pressure (SF_6, a few mtorr) that the free path was comparable with the distance between the spaced light beams. Resonant decrease of probe-wave absorption was observed when the frequency of the exciting and probe waves coincided with that of the center of the Doppler line ν_0. Lower pressures (about 0.1 mtorr) are needed to transfer all of the molecules to the excited level and to form an amplification peak, but such an experiment has not yet been performed.

The method of narrow resonance formation inside the Doppler-broadened line of a particle beam by use of spaced light beams, suggested in [1.38], has not found wide use owing to its considerable complication. The application of the promising Lamb-dip method in lasers is limited to amplifying transitions of low-pressure gas media only; the dip width for real transitions of gas lasers is comparatively large (tens of MHz).

The situation has changed greatly since attention was turned to observation of the Lamb dip in resonant-absorption media. The first suggestions and experiments on using nonlinear absorption resonances were performed separately at the Lebedev Physical Institute Academy of Sciences USSR [1.40], at the Institute of Semiconductor Physics, Siberian Branch of Acad. Sci. USSR [1.41] and in the Perkin-Elmer Lab in the USA [1.42]. These investigations proposed to insert a resonant-absorption gas cell of low pressure into the laser cavity (Fig.1.12a). Absorption saturation in a standing light wave gives rise to a narrow Lamb dip at the center of a Doppler-broadened absorption line. As a result, efficient saturated amplification of the two-component medium in the laser acquires a narrow peak at the center of the absorption line (Fig.1.12b), and the laser output power exhibits a narrow peak (Fig.1.12c), often termed the "inverted Lamb dip". The virtues of this method are that the absorbing gas, at a low pressure and with proper selection of the particle and transition, may have a very narrow homogeneous width, of the order 10^4 to 10^6 Hz (Table 1.1). Particular emphasis is placed upon this essential feature of infrared molecular transitions in papers by LETOKHOV [1.40] and LISITSYN and CHEBOTAYEV [1.41]. There are two circumstances of importance for the application of this method in spectroscopy and for laser-frequency stabilization. First, the dip in the absorption line can be narrowed by factors of 10^{-2} or 10^{-3} times that in the amplification line. Actually, absorption, in contrast to amplification, may take place in transitions

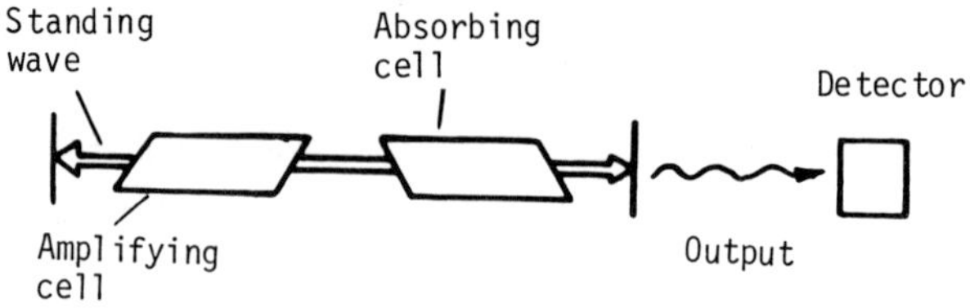

(a) Experiment

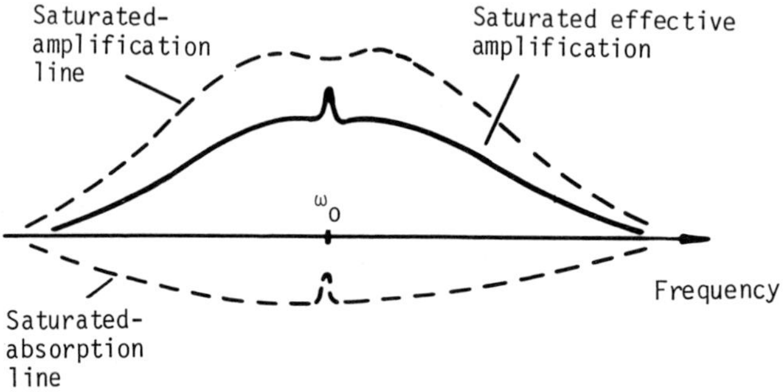

(b) Amplification of two-component media

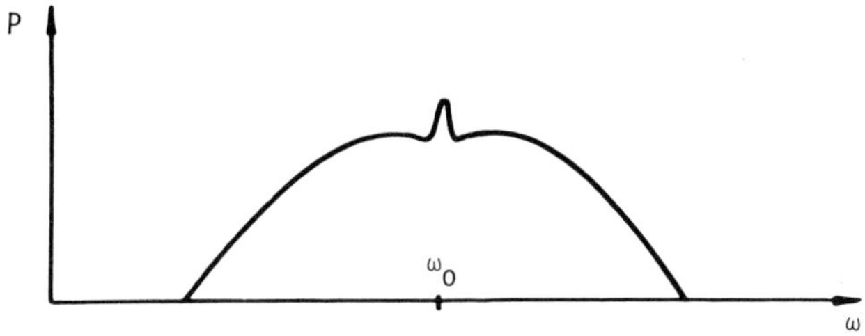

(c) Output laser power

Fig.1.12a-c Production of inverted Lamb dip in gas laser with intracavity low-pressure absorption cell: (a) Experimental arrangement; (b) Frequency dependence of saturated total amplification of two-component medium; (c) Power output as a function of laser-field frequency

from the ground or close-to-ground state into an excited long-lived state. As a result, the radiative width may be negligible. Because the ground-level population is rather large without any excitation, absorption may be observed in a gas at a very low pressure, when the collisional width also becomes small. Second, because of the low pressure and the absence of gas excitation the center of the spectral absorption line may be rather stable. Specifically, in [1.40,41] such an experiment on CH_4 molecules was proposed, in which one component of the rotational-vibrational transition P(7) of the ν_3 band is coincident with the radiation line of a He-Ne laser at $\lambda = 3.39$ μm. An experiment of the kind was performed at the NBS in the USA [1.43], and the resonance obtained was only 0.3 MHz in width, i.e., 10^{-3} times as wide as the Doppler line width of CH_4 absorption.

The method using a nonlinear-absorption molecular cell has assured great progress in generating light oscillations with high frequency stability. Within the short period 1967-1972 the relative long-term stability has improved from 10^{-8} to 10^{-14}. Nonlinear-absorption lasers are an important class of lasers; their output-radiation properties are unusual and essential for applications; the method on which they are based has become a primary method in fast-developing nonlinear high-resolution laser spectroscopy (often called the spectroscopy of absorption saturation).

To obtain a narrow saturation resonance at the center of an absorption line, it is not necessary to use a standing light wave but only to have a strong running wave and a weak counter-running wave (Fig.1.13a). This possibility was clearly suggested in [1.44]. The strong running wave excites molecules whose velocity projections onto the wave direction are $v = (\nu - \nu_0)c/\nu_0$ to the high level. Because the counter-running wave has the same frequency but is opposite in direction, it interacts with molecules that have the same projected velocities but opposite directions relative to the strong wave (Fig.1.13b). If the frequency of the waves does not coincide with that of the center of the Doppler line ν_0, the weak probe wave is not responsive to the strong one. However, when the frequency is coincident with the center of the Doppler profile ν_0 the weak probe wave interacts with molecules the absorption of which is already decreased by the counter-running strong wave. Consequently, the probe-wave absorption has a resonant minimum equal to the homogeneous width and centered exactly in the Doppler-broadened absorption line (Fig.1.13c). Narrow resonances were first observed experimentally by this method as reported in [1.45,46]; this technique is now universally accepted.

The narrow "hole" in the velocity distribution of particles that results when some particles of a gas are stimulated into an excited state by a coherent light wave appears in coupled transitions as well. For example, because of the peak in the velocity distribution in the upper level (Fig.1.9b) the line that corresponds to the transition from this level to a still-higher energy level also has a resonance peak with a width much smaller than the

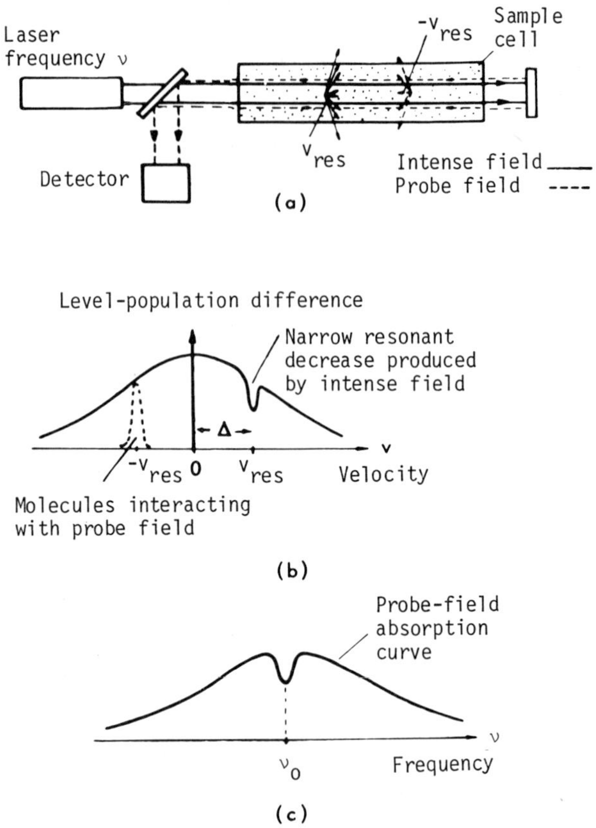

Fig.1.13a-c Observation of the saturation narrow resonance by the single strong coherent travelling wave and counter-travelling weak probe wave: (a) Experimental arrangement. A small part of the intense wave is reflected back through the cell. The attenuation of this weak wave is studied as a function of laser-field frequency. (b) Molecular-velocity distribution, showing velocity groups that resonantly interact with the strong wave and the probe wave. (c) Probe-wave absorption as a function of frequency

Doppler width (Fig.1.14). The line shape for the resulting transition can conveniently be observed by use of an additional probe wave travelling in the same direction as the strong wave. The method of spectroscopy without Doppler breadth on coupled transitions was proposed and experimentally accomplished by JAVAN and SCHLOSSBERG [1.47,48]. For infrared molecular spectroscopy, this method is of interest mostly for making precision measurements of small level splittings, such as those caused by external fields, but it is applicable only to transitions with a common level. For coupled atomic transitions, where the homogeneous line width depends mainly on radiative decay, saturation spectroscopy with a two-frequency field makes it possible to observe narrow resonances with line widths less than the homogeneous width.

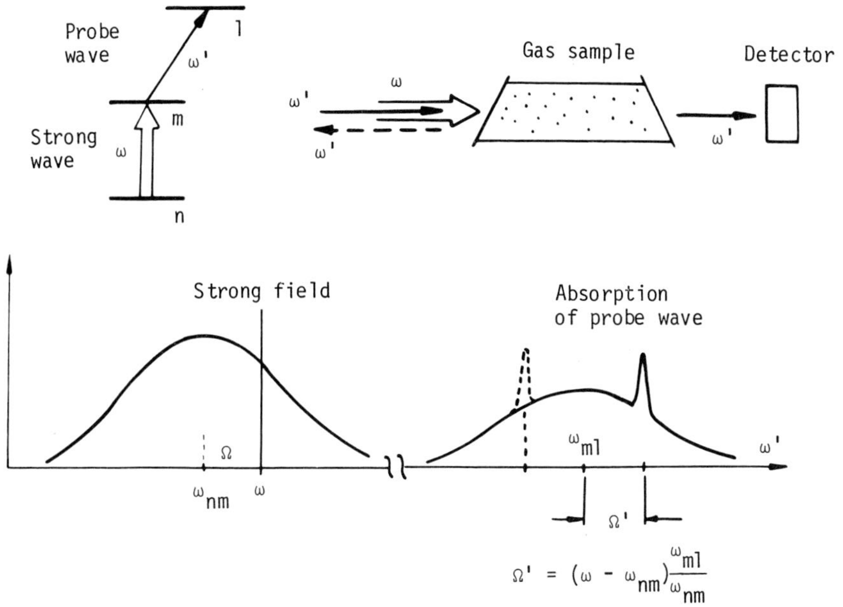

Fig.1.14a-c Observation of the saturation narrow resonances at coupled transitions: (a) Energy-level diagram; (b) Detection scheme; (c) Probe-wave absorption of coupled transition m-l as a function of its frequency ω'

An atom absorbs photons $\hbar\omega$ and $\hbar\omega'$ from two unidirectional travelling waves of frequencies ω and ω' and makes a two-quantum transition from level n to level l (see Fig.1.14) [1.49,50]. In this case, the resonance width, say of the m-l transition is determined by levels n and l rather than by levels m and l. The contribution of level m is small if the difference between the wave vectors of two waves is small. When $\underline{k}_{\omega} = \underline{k}_{\omega'}$, the resonance width is equal to the radiative width of the n-l "forbidden" transition.

In tuning the standing-wave frequency to the center of a Doppler-broadened line, a resonant reduction of the saturated absorption is accompanied also by a resonant change of the total number of particles on every level of a transition, regardless of their velocities (Fig.1.15). There is a resonance minimum in the population of the upper level and a resonance peak in that of the lowest level. This effect widens the potentialities of saturation spectroscopy, because there are many efficient methods for detection of the total population of levels, for example, by measuring the intensity of particle fluorescence from an excited level. Such a method of saturation spectroscopy was proposed by BASOV and LETOKHOV [1.51] and independently in the experiment of FREED and JAVAN [1.52]. An extremely high sensitivity was obtained, which

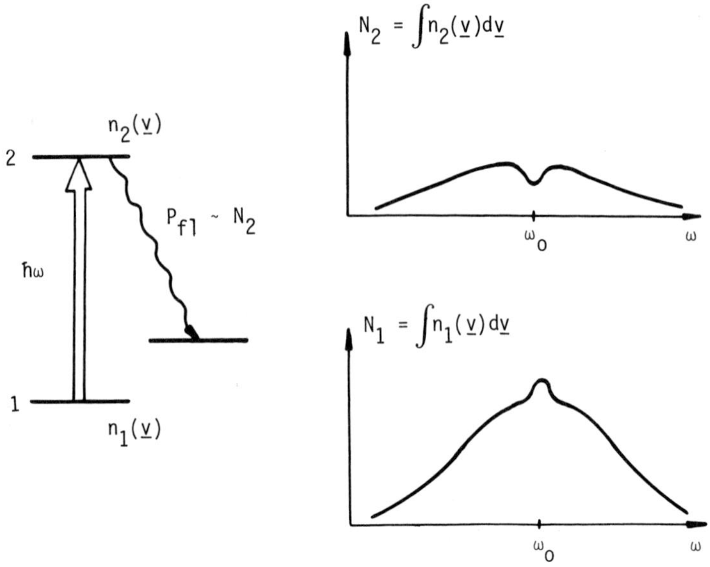

Fig.1.15 Dip formation in the fluorescence intensity P_{f1} under a strong standing wave, due to the dependence of the total number of excited particles N_2 on the wave frequency ω

Table 1.2 Saturation-spectroscopy methods

Method	Proposal	Experiment
1 - Lamb dip	Lamb, 1962 [1.32]	Szöke, Javan, 1963 [1.34] McFarlane, Bennett, Lamb, 1963 [1.33]
2 - Inverted Lamb dip	Letokhov, 1967 [1.40] Lee, Skolnick, 1967 [1.42] Lisitsyn, Chebotayev, 1968 [1.41]	Lee, Skolnick, 1967 [1.42] Lisitsyn, Chebotayev, 1968 [1.41]
3 - Dip with counter-running probe wave	Letokhov, Chebotayev, 1969 [1.44]	Basov, Kompanetz, Letokhov, Nikitin, 1969 [1.45]
4 - Dips at coupled transitions	Javan, Schlossberg, 1966 [1.47]	Schlossberg, Javan, 1966 [1.48]
5 - Dip in total level population (fluorescence dip)	Basov, Letokhov, 1968 [1.51]	Freed, Javan, 1970 [1.52]

makes experiments possible at very low molecular gas pressures (below 10^{-3} torr) or even with molecular beams.

Table 1.2 summarizes presently developed methods of saturation spectroscopy without Doppler broadening, their authors and the performers of the first experiments. Saturation spectroscopy is the best-developed variety of nonlinear laser spectroscopy; its methods have been applied in dozens of experiments with atoms and molecules by many laboratories.

All of the mentioned methods and schemes of absorption-saturation spectroscopy and basic experiments are considered in this book. Chapter 2 is devoted to the theory of the resonant interaction between the light field and atoms and molecules in gases. On the basis of this analysis, Chapter 3 describes theoretically the principal methods of absorption-saturation spectroscopy and contains the results of experimental studies of these methods. The methods based on absorption saturation of two levels of a quantum system are also considered in Chapter 3. Rather complex and fine effects of interaction between a two-frequency field and three-level systems, which are methods of saturation spectroscopy of two coupled transitions, are considered separately in Chapter 5.

1.4 Two-Photon Laser Spectroscopy

This method of nonlinear laser spectroscopy without Doppler broadening was proposed by CHEBOTAYEV et al. [1.53]. Consider a two-quantum atomic or molecular transition in the field of a standing wave of frequency ω (Fig.1.16). For a particle moving with velocity $\underline{v}$, the frequency of the relatively travelling wave is $\omega \pm \underline{k}\underline{v}$. The only particles that can absorb two photons from one travelling wave are those for which $\underline{k}\underline{v}$ complies with the condition of two-photon resonance. However, simultaneous absorption of two photons from equal and opposite travelling waves is possible. In this case, the condition of two-photon resonance is only that the doubled field frequency coincides with the frequency of the two-quantum transition, that is, with the center of the Doppler-broadened line. In this type of resonance, all particles, regardless of velocity, participate in two-photon absorption, resulting in a sharp increase of the absorption signal. The line shape of such two-photon absorption is shown in Fig.1.16c. It is the sum of a wide Doppler contour, representing two-photon absorption from a unidirectional wave, and a narrow resonance corresponding to two-photon absorption by all particles for which $2\omega = \omega_{12}$. The amplitude of the resonance peak at the center of the line has

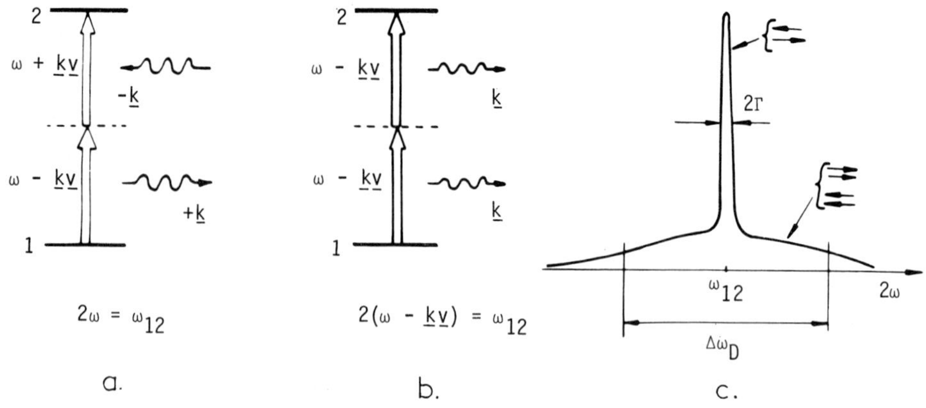

Fig.1.16a-c Two-photon narrow resonance in standing wave: (a) Compensation for the Doppler shift by simultaneous absorption of the photons from two travelling waves that are opposite in direction; (b) Its absence when unidirectional photons are absorbed; (c) The shape of a narrow resonance in two-photon absorption

a high contrast, equal to the ratio of the Doppler width to the homogeneous width.

Narrow two-photon resonances were first observed in 1974 in studies of sodium-atom transitions in the visible [1.54-56] and recently in studies of CH_3F molecular vibrational transitions in the infrared [1.57].

Rapid progress of two-photon spectroscopy can be anticipated because it has several advantages compared to saturation spectroscopy. First, all of the particles on initial level 1 take part in the absorption, regardless of their velocities, whereas in saturation spectroscopy only a small proportion of the particles participate in the production of narrow resonance lines. For extremely narrow resonances, the parameter $\Gamma/ku \simeq 10^{-3}$-10^{-5} and the peak contrast in two-photon absorption is $ku/\Gamma \simeq 10^3$-10^5 times as much as the Lamb-dip contrast in saturation spectroscopy. Second, the two-photon absorption peak is accompanied by a corresponding peak in the density of excited particles. Since sensitive methods are available to detect particles in the excited state, experiments can be performed with a small number of particles, such as in atomic and molecular beams. Third, the width of the resonance peak does not depend on the curvature of the wave front, because the two photons are absorbed simultaneously at the same point in space. In absorption saturation, for narrow resonances to be attained, the wave vector should have strictly the same direction along the whole cross-section of the standing wave. Thus, it is possible to use light beams with large cross-sections (tens of centimeters) for two-photon spectroscopy, and broadening of the resonance peaks due to the transit time can be made very small. Indeed, the probability of a two-photon transition does not depend on the orientation of particle velocities about the standing wave. Therefore, an atom beam moving along a standing wave can be used, and a path length of 1 m or more can be obtained, for the interaction to obtain a line broadening of only 100 Hz [1.58].

The theory and practice of two-photon laser spectroscopy is studied in Chapter 4. Effects of two-quantum transitions in three-level systems, where a considerable contribution is made simultaneously by saturation effects, are studied in detail in Chapter 5. Chapters from 2 to 5 offer the theoretical basis for two trends of nonlinear laser spectroscopy: absorption-saturation spectroscopy and two-photon spectroscopy.

1.5 Problem of Very Narrow Optical Resonances

The discovery of narrow nonlinear resonances in the optical range offers new possibilities for some fields of physics. Indeed, the production of narrow and frequency-stable resonances in absorption or emission spectra of substances over various ranges of electromagnetic radiation has always been an important problem in physics. Every discovery in this direction increases the accuracy of physical experiments and finds wide use in various fields of science.

In the 1940s and 1950s a technique was elaborated to produce narrow resonances in the microwave frequency range using atomic and molecular beams. For example, using two separated electromagnetic fields (RAMSEY [1.59]) interacting with a beam of Cs atoms at the transition between the hyperfine-structure levels of the ground state, resonances can be obtained with the width $\Delta\nu = 50$ Hz at the frequency $\nu_0 = 9.3$ GHz. This corresponds to the relative width $\Delta\nu/\nu_0 = 5 \cdot 10^{-9}$ or to the "quality" (resolution) of resonance $R = \nu_0/\Delta\nu = 2 \cdot 10^8$. Narrow microwave resonances have formed the basis for quantum frequency standards and the universally adopted atomic time scale (atomic clocks).

Extremely narrow resonances in a higher-frequency region of the spectrum were detected at nuclear transitions by MÖSSBAUER [1.60]. For instance, at a γ transition with the energy of 93 keV in Zn^{67} it is possible to obtain a resonance with the resolution of $2 \cdot 10^{15}$. Narrow resonances of nuclear transitions without recoil in a crystal lattice now ensure the highest resolution in physical experiments, of the order of 10^{15}. In the intermediate spectral region, the relative resonance width till recently has been no better than 10^{-6}. The discovery of the subtle effects of resonant nonlinear interaction between the light field and atomic or molecular gases described in this book has allowed the relative widths of optical resonances to be narrowed by factors of 10^{-4} to 10^{-5} or smaller. As an illustration, Fig.1.17 shows the relative widths for the narrowest resonances of quantum transitions in the microwave, optical and γ region of the spectrum.

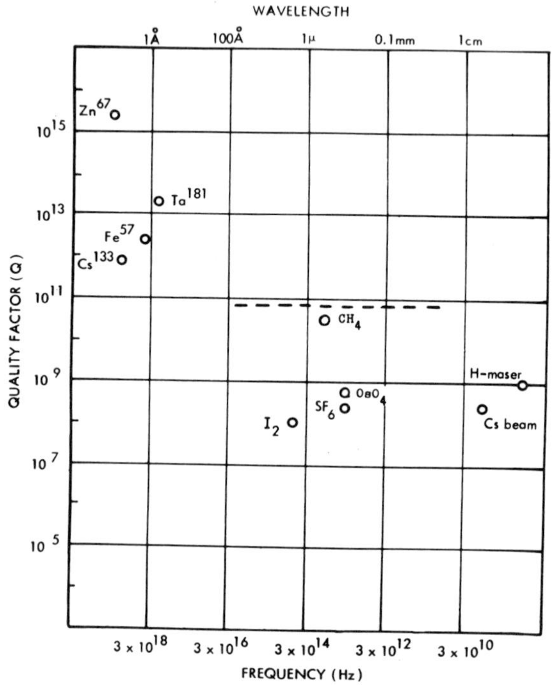

Fig.1.17 Quality factors of narrow resonances in microwave, optical and gamma-ray regions

The dashed line on Fig.1.17 indicates the fundamental limitations of resolving power of all of the previously mentioned methods of nonlinear spectroscopy, due to the second-order Doppler effect for freely moving atoms and molecules, that is, the dependence of the central frequency of a line on the absolute velocity of the particle. The thermal distribution of absolute velocities of free particles inevitably results in a second-order Doppler resonance broadening of

$$\Delta\nu_{\mathrm{sec.Dopp}} \simeq \nu_0 \frac{kT}{Mc^2} \simeq \nu_0 \left(\frac{v_0}{c}\right)^2 , \qquad (1.20)$$

where T is the gas temperature, M is the mass of a particle, and c is the speed of light. This effect limits the resolution obtained by both nonlinear-laser-spectroscopy methods (saturation spectroscopy and two-photon spectroscopy) at the level of $(c/v_0)^2 \simeq 10^{10}$-10^{11} that corresponds to room temperature.

So, search for new methods of nonlinear laser spectroscopy free of this limitation seems to be of great importance. From this standpoint we should focus our attention on the method of spectroscopy without Doppler broadening, based on low-velocity particle trapping in a standing light wave, which was suggested by LETOKHOV [1.61] as far back as in 1968. This method will be discussed briefly in Chapter 6. Unlike the previous two approaches, trapped-particle spectroscopy is still in the theoretical state. However, its potential merits are high. The idea of the method is that Doppler shift of an absorption or emission line can be eliminated if the translational motion of a particle can be transformed into an oscillatory motion with amplitude a less than the radiation wavelength λ. This method is widely used in microwave spectroscopy and is known as the Dicke method [1.12]. In the optical region it requires that the motion of a particle be restricted to a negligible volume $\sim\lambda^3$. LETOKHOV [1.61] suggested that a nonresonant strong standing wave should be employed for this purpose. The wave might act as a potential space-periodical field, trapping particles with very low velocities. A neutral particle (atom or molecule) in a high-frequency electromagnetic field is acted on by a force proportional to the mean square gradient of the electric field intensity. Under this force, the particles are pulled into the standing-wave loops. Recent proposal of a method of laser cooling of freely moving atoms by HÄNSCH and SCHAWLOW [1.62] is very important for practical realization of trapped-particle laser spectroscopy and for elimination of the second-order Doppler effect.

Thus, at least three fundamentally different approaches are used in nonlinear laser spectroscopy without Doppler broadening (Table 1.3). These are outlined below and discussed in great detail in the following chapters:

1) *Saturation spectroscopy* is based on the changes produced by a coherent light wave, in the velocity distribution of atoms and molecules excited from a lower state to a higher energy state.

2) *Two-photon spectroscopy* is based on the simultaneous absorption of photons from two laser beams similar in frequency but travelling in opposite directions.

3) *Trapped-particle spectroscopy* is based on changes in the velocity distribution of atoms and molecules confined to oscillatory motions in a strong, nonresonant light beam.

Discovery of these approaches gave impetus to development of nonlinear ultrahigh-resolution laser spectroscopy and precision optical spectroscopy [1.63].

Table 1.3 Methods of nonlinear laser spectroscopy without Doppler broadening

Methods	Physical phenomena	Author
1 - Saturation spectroscopy	Change of the velocity distribution of atoms or molecules at the quantum levels of the saturated transition	Lamb, 1962 [1.32]
2 - Two-photon spectroscopy	Compensation of Doppler shift due to simultaneous absorption of photons from counter-running light waves	Chebotayev, Vasilenko, Shishaev, 1970 [1.53]
3 - Spectroscopy of trapped particles	Oscillatory motion of slow particles in a strong standing light wave (trapping of atoms or molecules)	Letokhov, 1968 [1.61]

References

1.1 G. Herzberg: *Atomic Spectra and Atomic Structure* (Dover Publications, New York 1944)
1.2 L.N. Dobretzov, A.N. Terenin: Naturwiss 16, 656 (1928)
1.3 K.W. Meissner, K.F. Luft: Ann.Phys. 28, 657 (1937)
1.4 W.E. Lamb, Jr., R.C. Retherford: Phys.Rev. 72, 241 (1947); 79, 549 (1950); 81, 822 (1951); 85, 259 (1952)
1.5 D. Jackson, H. Kuhn: Proc.Roy.Soc.London A154, 679 (1936)
1.6 S. Tolansky: *Fine Structure in Line Spectra and Nuclear Spin* (Methuen, London 1948)
1.7 N.G. Basov, A.M. Prokhorov: Zh.Eksp.i Teor.Fiz. 27, 431 (1954)
1.8 J.P. Gordon, H.J. Zeiger, C.H. Townes: Phys.Rev. 95, 282 (1954)
1.9 S. Ezekiel, R. Weiss: Phys.Rev.Lett. 20, 91 (1968)
1.10 J. Brossel, F. Bitter Phys.Rev. 86, 308 (1952)
1.11 A. Kastler: J.Phys. 11, 255 (1950)
1.12 R.H. Dicke: Phys.Rev. 89, 472 (1953)
1.13 L.N. Novikov, V.G. Pokazan'ev, G.V. Skrotzkii: Uspekhi Fiz.Nauk. 101, 273 (1970)
1.14 K. Shimoda: in *Laser Spectroscopy*, Topics in Applied Physics, Vol.2, ed. by H. Walther (Springer-Verlag, Berlin, Heidelberg, New York 1976)
1.15 W. Hanle: Naturwiss 11, 691 (1923); Z.Phys. 30, 93 (1924)
1.16 G. Breit: Rev.Mod.Phys. 5, 91 (1933); Phys.Rev. 46, 590 (1934)
1.17 A.C.G. Mitchell, M.W. Zemansky: *Resonance Radiation and Excited Atoms* (Cambridge University Press, London, New York 1934)

1.18 F.D. Colegrove, P.A. Franken, R.R. Lewis, R.H. Sands: Phys.Rev.Lett. 3, 420 (1959)
1.19 V.G. Pokazan'ev, G.V. Skrotzkii: Uspekhi Fiz.Nauk 107, 623 (1972)
1.20 L.D. Landay, E.M. Lifshitz: *Quantum Mechanics* (1963)
1.21 E.B. Aleksandrov: Optics and Spectroscopy 17, 957 (1964)
1.22 J.N. Dodd, R.D. Kaul, D.M. Warrington: Proc.Phys.Soc.London 84, 176 (1964)
1.23 E.B. Aleksandrov: Uspekhi Fiz.Nauk 107, 595 (1972)
1.24 S. Haroche, J.A. Paisner, A.L. Schawlow: Phys.Rev.Lett. 30, 948 (1973)
1.25 V. Weisskopf: Ann.Phys. 9, 23 (1931)
1.26 I.I. Sobel'man: *Introduction to the Theory of Atomic Spectra* (Pergamon Press 1970)
1.27 R. Karplus, J. Schwinger: Phys.Rev. 73, 1020 (1948)
1.28 C.H. Townes, A.L. Schawlow: *Microwave Spectroscopy* (McGraw-Hill Publishing Company, New York 1955)
1.29 A.L. Schawlow: in *Advances in Quantum Electronics*, ed. by J.R. Singer (Columbia University Press, New York 1961), p. 50
1.30 W.R. Bennett, Jr.: Phys.Rev. 126, 580 (1962)
1.31 A. Javan, W.R. Bennett, Jr., D.R. Herriott: Phys.Rev.Lett. 6, 106 (1961)
1.32 W.E. Lamb, Jr.: Phys.Rev. 134A, 1429 (1964)
1.33 R.A. McFarlane, W.R. Bennett, Jr., W.E. Lamb, Jr.: Appl.Phys.Lett. 2, 189 (1963)
1.34 A. Szoke, A. Javan: Phys.Rev.Lett. 10, 521 (1963)
1.35 R.H. Cordover, P.A. Bonczyk, A. Javan: Phys.Rev.Lett. 18, 730 (1967)
1.36 A. Szoke, A. Javan: Phys.Rev. 149, 38 (1966)
1.37 K. Shimoda, A. Javan: J.Appl.Phys. 36, 718 (1965)
1.38 N.G. Basov, V.S. Letokhov: Pis'ma Zh.Eksp.i Teor.Fiz. 2, 6 (1965)
1.39 O.N. Kompanetz, V.S. Letokhov: Pis'ma Zh.Eksp.i Teor.Fiz. 14, 20 (1971)
1.40 V.S. Letokhov: Pis'ma Zh.Eksp.i Teor.Fiz. 6, 597 (1967)
1.41 V.N. Lisitsyn, V.P. Chebotayev: Zh.Eksp.i Teor.Fiz. 54, 419 (1968)
1.42 P.H. Lee, M.L. Skolnick: Appl.Phys.Lett. 10, 303 (1967)
1.43 R.L. Barger, J.L. Hall: Phys.Rev.Lett. 22, 4 (1969)
1.44 V.S. Letokhov, V.P. Chebotayev: Pis'ma Zh.Eksp.i Teor.Fiz. 9, 364 (1969)
1.45 N.G. Basov, I.N. Kompanetz, O.N. Kompanetz, V.S. Letokhov, V.V. Nikitin: Pis'ma Zh.Eksp.i Teor.Fiz. 9, 568 (1969)
1.46 Yu.A. Matiugin, B.I. Troshin, V.P. Chebotayev: *Digest of National Symposium on Gas Laser Physics*, Novosibirsk (1969), p. 56
1.47 H.R. Schlossberg, A. Javan: Phys.Rev. 150, 267 (1966)
1.48 H.R. Schlossberg, A. Javan: Phys.Rev.Lett. 17, 1242 (1966)
1.49 G.E. Notkin, S.G. Rautian, A.A. Feoktistov: Zh.Eksp.i Teor.Fiz. 52, 1673 (1967)
1.50 H.K. Holt: Phys.Rev.Lett. 20, 410 (1968)
1.51 N.G. Basov, V.S. Letokhov: Report on URSI Conference on Laser Measurements (Warsaw, September 1968); Electron Technology 2, N2/3, 15 (1969)
1.52 C. Freed, A. Javan: Appl.Phys.Lett. 17, 53 (1970)
1.53 L.S. Vasilenko, V.P. Chebotayev, A.V. Shishaev: Pis'ma Zh.Eksp.i Teor. Fiz. 12, 161 (1970) [JETP Lett. 12, 113 (1970)]
1.54 F. Biraben, B. Cagnac, G. Grynberg: Phys.Rev.Lett. 32, 643 (1974)
1.55 M.D. Levenson, N. Bloembergen: Phys.Rev.Lett. 32, 645 (1974)
1.56 T.W. Hänsch, K. Harvey, G. Meisel, A.L. Schawlow: Optics Comm. 11, 50 (1974)
1.57 W.K. Bischel, P.J. Kelley, C.K. Rhodes: Phys.Rev.Lett. 34, 300 (1975)
1.58 E.V. Baklanov, V.P. Chebotayev: *Report on III Vavilov Nonlinear Optics Conf.* (Novosibirsk, USSR, June 1973); Optics Comm. 12, 312 (1974)
1.59 N.F. Ramsey: *Molecular Beams* (Clarendon Press, Oxford 1956)
1.60 R. Mössbauer: Z.Phys. 151, 124 (1958)
1.61 V.S. Letokhov: Pis'ma Zh.Eksp.i Teor.Fiz. 7, 348 (1968)
1.62 T.W. Hänsch, A.L. Schawlow: Optics Comm. 13, 68 (1975)
1.63 V.S. Letokhov: Science 190, 344 (1975)

2. Elements of the Theory of Resonant Interaction of a Laser Field and Gas

The nonhomogeneous character of broadening due to particle motion in gas complicates, in general, the problem of nonlinear interaction of a resonant transition with a light field even of simple form, like a plane standing wave. However it is possible to gain some insight into the basic qualitative and quantitative features by treating in succession the interaction of a Doppler-broadened transition with travelling and standing waves. Let us consider the basic relations for these cases, which will be essential for our further discussion.

Nonlinear interaction of an electromagnetic field with a two-level quantum system in the microwave case, when the inhomogeneous broadening due to the Doppler effect is small and the relaxation constants are the same for the both levels, was studied in detail by KARPLUS and SCHWINGER [2.1] in connection with microwave spectroscopic problems and by BASOV and PROKHOROV [2.2] in connection with the problem of maser oscillations. The problem for the microwave range is covered comprehensively in monographs [2.3-5]. In the optical range, Doppler broadening gives rise to quite new effects that have no analog and were never studied in the microwave range. Particular emphasis in this chapter is placed on these effects.

Resonance interaction of a two-level particle with the field can be analyzed both by use of the Schrödinger equation with added phenomenological relaxation terms and by use of the density-matrix equation, which is better suited for describing relaxation effects. Let us consider both methods.

2.1 Schrödinger Equation - Transition Probabilities

Let us consider at first the behavior of an isolated system with two energy levels E_1 and E_2 without relaxation and damping under the action of an electromagnetic field. The behavior of the system, that is of its wave function Ψ, is described by the nonstationary Schrödinger equation,

$$i\hbar \frac{\partial \Psi}{\partial t} = \hat{H}\Psi \ , \tag{2.1}$$

where $\hat{H}$ is the total hamiltonian of the system composed of the unperturbated hamiltonian $\hat{H}_0$ and the energy of the quantum system-field interaction $\hbar\hat{V}$

$$\hat{H} = \hat{H}_0 + \hbar\hat{V} . \tag{2.2}$$

The hamiltonian of electric-dipole interaction between a particle and field has the form

$$\hbar\hat{V} = -\hat{\underline{p}}\underline{E} , \tag{2.3}$$

where $\hat{\underline{p}}$ is the operator of electric-dipole moment of a particle, and $\underline{E}$ is the strength of the electric field of wave.

The wave function Ψ can be determined by use of the Dirac nonstationary perturbation theory by expanding in eigenfunctions of the $\hat{H}_0$ operator, i.e., by expressing it as a superposition of the wave functions Ψ_k of the quantum system without a light field,

$$i\hbar \frac{\partial \Psi_k}{\partial t} = H_0 \Psi_k = E_k \Psi_k , \tag{2.4}$$

where E_k is the energy of the quantum system in the kth stationary state. For further calculations it is convenient to isolate the time dependence of Ψ_k, i.e., according to (2.4), to express the wave function in the form

$$\Psi_k = \psi_k \exp(-iE_k t/\hbar) . \tag{2.5}$$

Thus, the wave function Ψ of the field-perturbed quantum system can be found from

$$\Psi(t) = a_1(t)\psi_1 \exp(-iE_1 t/\hbar) + a_2(t)\psi_2 \exp(-iE_2 t/\hbar) , \tag{2.6}$$

where $a_k(t)$ denotes time-dependent coefficients that are to be calculated. These coefficients are called amplitudes of the k-state probability, because the value $|a_k(t)|^2$ is the probability that the system is in the kth state. Substituting (2.6) in (2.1) gives

$$i \frac{da_1(t)}{dt} \psi_1 \exp(-iE_1 t/\hbar) + i \frac{da_2(t)}{dt} \psi_2 \exp(-iE_2 t/\hbar) =$$

$$= a_1(t)\hat{V}\psi_1 \exp(-iE_1 t/\hbar) + a_2(t)\hat{V}\psi_2 \exp(-iE_2 t/\hbar) . \tag{2.7}$$

Taking into account that ψ_k denotes the eigenfunctions of the H_o operator, we cancel $a_k\psi_k E_k \exp(-iE_k t/\hbar)$ from the left and right sides. In order to find the probability amplitude $a_1(t)$, both sides of (2.7) should be multiplied by ψ_1^* and integrated over the whole space of the quantum-mechanical variables, with the orthogonality of the eigenfunctions ψ_1 and ψ_2 and the odd operator of interaction $\hat{V}$ taken into account. As a result, we have

$$i\,\frac{da_1(t)}{dt}\exp(-iE_1t/\hbar) = a_2(t)\exp(-iE_2t/\hbar)\int\psi_1^*\hat{V}\psi_2 dq\ . \qquad (2.8)$$

In a like manner, multiplying by ψ_2^* we obtain an equation for $a_2(t)$,

$$i\,\frac{da_2(t)}{dt}\exp(-iE_2t/\hbar) = a_1(t)\exp(-iE_1t/\hbar)\int\psi_2^*\hat{V}\psi_1 dq\ . \qquad (2.9)$$

If we denote the transition frequency $\omega_{21} = E_2 - E_1/\hbar = \omega_o$ and the interaction-operator matrix element

$$V_{21} = \int\psi_2^*\hat{V}\psi_1 dq\ ,\quad V_{12} = V_{21}^*\ . \qquad (2.10)$$

Eqs. (2.8) and (2.9), in this case, take the form

$$i\,\frac{da_1}{dt} = V_{12}a_2\exp(-i\omega_o t)\ ;\quad i\,\frac{da_2}{dt} = V_{21}a_1\exp(i\omega_o t)\ . \qquad (2.11)$$

The set of equations for probability amplitudes is similar to the initial Schrödinger equation for a two-level system.

To solve the set of (2.11), we should define the perturbation V. Let the electromagnetic field be a linearly polarized coherent travelling wave with the frequency ω and amplitude E, which is given by

$$\underline{E}(t,\underline{r}) = \underline{e}E\cos(\omega t - \underline{k}\underline{r})\ . \qquad (2.12)$$

When turning to the center-of-mass system, we may take into account the particle motion at the $\underline{v}$ velocity. This causes the field frequency ω to be replaced by $\omega - \underline{k}\underline{v}$. In other respects, the interaction of a moving particle with the running-wave field remains the same as in the case of stationary particles. In this case, the matrix element of interaction (2.10) takes the form

$$\hbar V_{21} = -pE\cos\omega' t\ ,\quad V_{12} = V_{21}\ , \qquad (2.13)$$

where $\omega' = \omega - \underline{k}\underline{v}$, and p is the projection of the matrix element of the transition dipole moment on the wave-polarization vector,

$$p = \underline{e}\underline{p}_{21} = \underline{e}\underline{p}_{12} \; . \tag{2.14}$$

Then, in place of (2.11), we have

$$\begin{aligned} i\,\frac{da_1}{dt} &= -V\{\exp[i(\omega' - \omega_0)t] + \exp[-i(\omega' + \omega_0)t]\}a_2 \; , \\ i\,\frac{da_2}{dt} &= -V\{\exp[i(\omega' + \omega_0)t] + \exp[-i(\omega' - \omega_0)t]\}a_1 \; , \end{aligned} \tag{2.15}$$

where $V = pE/2\hbar$, $\omega' - \omega_0 = \Omega$ denotes the field-frequency detuning in the center-of-mass system, with respect to the resonance $(\omega' = \omega_0)$.

Let us solve (2.15) by use of the following approximation. The terms with the frequency $(\omega' + \omega_0)$ on the right side of (2.15) influence the two-level system with a much higher frequency than those with the difference frequency Ω. Therefore, on the average, the action of the high-frequency terms is considerably smoothed over in the time $1/\Omega$ characteristic of the rate of change of the probability amplitude. In other words, the left and right sides of (2.15) are averaged over the time interval $\tau_{av} << 1/\Omega$, $\tau_{av} >> (\omega' + \omega_0)^{-1}$. As a result, the term appropriate to the nonresonant periodic perturbation in (2.15) disappears, and the equations reduce to a much simpler form,

$$i\,\frac{da_1}{dt} = -V \exp(i\Omega t)a_2 \; ; \qquad i\,\frac{da_2}{dt} = -V \exp(-i\Omega t)a_1 \; . \tag{2.16}$$

The set of two equations (2.16) reduces to one second-order equation, for example, for a_1,

$$\frac{d^2a_1}{dt^2} - i\Omega\,\frac{da_1}{dt} + V^2a_1 = 0 \; . \tag{2.17}$$

The general solution of (2.17) has the form

$$a_1(t) = A \exp(i\alpha_1 t) + B \exp(i\alpha_2 t) \tag{2.18}$$

and, similarly,

$$a_2(t) = \frac{1}{V}\,[A\alpha_1 \exp(i\alpha_1 t) + B\alpha_2 \exp(i\alpha_2 t)] \exp(-i\Omega t) \ , \tag{2.19}$$

where α_1 and α_2 are the roots of the characteristic equation

$$\alpha^2 - \Omega\alpha - V^2 = 0 \ , \tag{2.20}$$

determined by the expressions

$$\alpha_{1,2} = \frac{\Omega}{2} \pm \frac{1}{2}\,[\Omega^2 + (2V)^2]^{1/2} = \frac{\Omega}{2} \pm \frac{\tilde{\Omega}}{2} \ , \tag{2.21}$$

where the value with the dimension of frequency $\tilde{\Omega}$ is

$$\tilde{\Omega}^2 = \Omega^2 + (2V)^2 \ . \tag{2.22}$$

The A and B coefficients can be determined from the initial conditions and normalization.

If, at the initial instant ($t = o$), the system is in state 1,

$$|a_1(o)|^2 = 1 \ , \quad |a_2(o)|^2 = 0 \ , \tag{2.23}$$

the A and B coefficients will be

$$A = \frac{1}{2} - \frac{\Omega}{2\tilde{\Omega}} \ , \qquad B = \frac{1}{2} + \frac{\Omega}{2\tilde{\Omega}} \ , \tag{2.24}$$

and the probability amplitudes vary with time,

$$\begin{aligned} a_1(t) &= \left(\cos\frac{\tilde{\Omega}}{2}t - i\,\frac{\Omega}{\tilde{\Omega}}\sin\frac{\tilde{\Omega}}{2}t\right)\exp(i\Omega t/2) \ , \\ a_2(t) &= i\,\frac{2V}{\tilde{\Omega}}\sin\frac{\tilde{\Omega}}{2}t\,\exp(-i\Omega t/2) \ . \end{aligned} \tag{2.25}$$

The square of the probability amplitude $|a_1(t)|^2$, that is the probability of detection of the system in the initial state, at the instant t, is

$$|a_1(t)|^2 = \cos^2\frac{\tilde{\Omega}}{2}t + \left(\frac{\Omega}{\tilde{\Omega}}\right)^2 \sin^2\frac{\tilde{\Omega}}{2}t \tag{2.26}$$

and the square of probability amplitude $|a_2(t)|^2$, that is the probability of finding the system in the upper level, at the instant t, is

$$|a_2(t)|^2 = \left(\frac{2V}{\tilde{\Omega}}\right)^2 \sin^2 \frac{\tilde{\Omega}}{2} t = \left[1 - \left(\frac{\Omega}{\tilde{\Omega}}\right)^2\right] \sin^2 \frac{\tilde{\Omega}}{2} t . \quad (2.27)$$

Expressions (2.26) and (2.27) comply with the condition of conservation $|a_1|^2 + |a_2|^2 = 1$. Therefore, the probability of particle transition to the upper level, that is $|a_1(t)|^2$ when $|a_2(o)|^2 = 0$, is

$$W_{12} = \sin^2 \left[Vt \sqrt{1 + \left(\frac{\Omega}{2V}\right)^2}\right] / \left[1 + \left(\frac{\Omega}{2V}\right)^2\right] . \quad (2.28)$$

In the case of exact resonance ($\Omega << V$),

$$W_{12} = \sin^2 Vt = 1 - \cos 2Vt , \quad (2.29)$$

and far from resonance ($\Omega >> V$),

$$W_{12} = \left(\frac{2V}{\Omega}\right)^2 \sin^2 \left(\frac{\Omega}{2} t\right) = 2 \left(\frac{V}{\Omega}\right)^2 (1 - \cos\Omega t) . \quad (2.30)$$

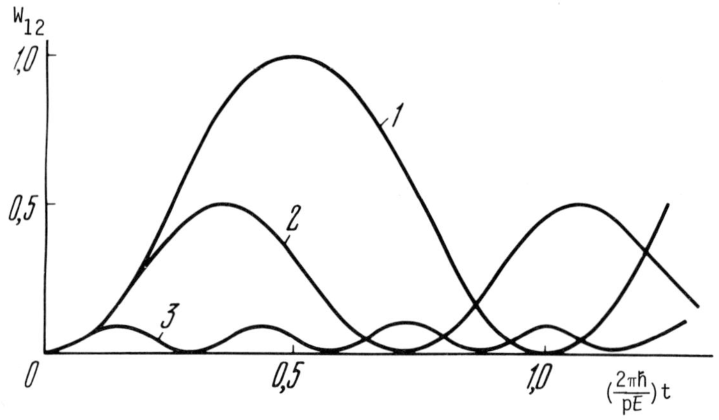

Fig.2.1 Probability of induced transition $1 \to 2$ as a function of time for two-level nondecaying quantum system after switching of light field ($t = o$): 1 - case of exact resonance field and two-level system ($\Omega = o$); 2 - intermediate case of detuning ($\Omega = pE/\hbar$); 3 - case of large detuning ($\Omega = 3pE/\hbar$)

Figure 2.1 shows the temporal dependence of the transition probability for the case of exact resonance, far from resonance and for an intermediate case.

In the case when a particle is initially on the upper level,

$$|a_1(o)|^2 = 0 , \quad |a_2(o)|^2 = 1 , \quad (2.31)$$

the problem is solved in the same way. The change of the probability amplitudes in this case is determined by

$$a_1(t) = i\,\frac{2V}{\tilde{\Omega}}\left(\sin\frac{\tilde{\Omega}}{2}\,t\right)\exp(i\Omega t/2)\ ,$$

$$a_2(t) = \left(i\,\frac{\Omega}{\tilde{\Omega}}\sin\frac{\tilde{\Omega}}{2}\,t + \cos\frac{\tilde{\Omega}}{2}\,t\right)\exp(-i\Omega t/2)\ . \tag{2.32}$$

In this case, $a_1(t)$ means the probability of the $2\to 1$ transition W_{21}. It can be shown that W_{21} is determined by (2.28), as before.

Knowing the amplitudes $a_1(t)$ and $a_2(t)$, we can write, in the obvious form, the time variation of the wave function $\psi(t)$ of the quantum system, which initially was either in the lower or in the upper state. The time variation of system polarization is determined by the wave function,

$$\underline{P} = \int\psi^*\hat{\underline{p}}\psi dq = a_2^* a_1 p\underline{e}\,\exp(i\omega_0 t) + \text{c.c.} \tag{2.33}$$

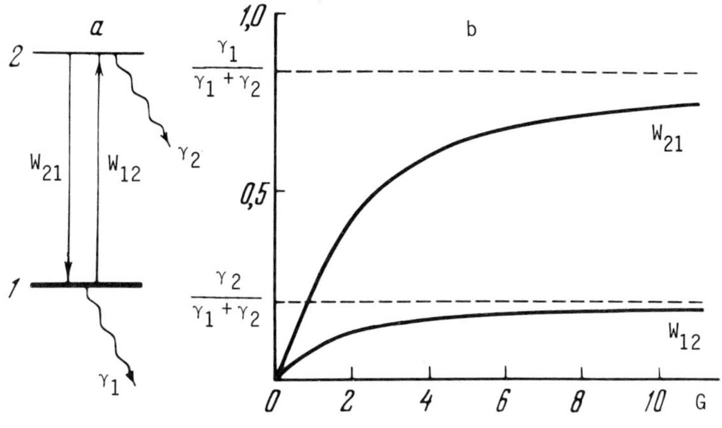

Fig.2.2 Two-level quantum system having decay constants γ_i of each level (a) and the dependence of probability of induced emission W_{21} and induced absorption W_{12} on the saturation parameter G (b)

The behavior of a two-level system in a resonant field substantially depends on the stationary-quantum-state decay. If γ_i is the decay rate of the ith state of the two-level system (Fig.2.2) due to spontaneous transitions, (2.16) for probability amplitudes should be changed to

$$i\left(\frac{da_1}{dt}+\frac{\gamma_1}{2}a_1\right) = -V\exp(i\Omega t)a_2\ ,\quad i\left(\frac{da_2}{dt}+\frac{\gamma_2}{2}a_2\right) = -V\exp(-i\Omega t)a_1\ . \tag{2.34}$$

The behavior of the probability amplitudes is given by cumbersome expressions that can be written in the matrix form [2.6]

$$\hat{a}(t) = \hat{S}(t)\hat{a}(t_0) , \tag{2.35}$$

where $\hat{a}(t)$ is the matrix of the states of the particle at an arbitrary instant,

$$\hat{a}(t) = \begin{pmatrix} a_1(t) \\ a_2(t) \end{pmatrix} ,$$

$\hat{a}(t_0)$ is the matrix of the states at the initial instant of time t_0,

$$a(t_0) = \begin{pmatrix} 1 \\ 0 \end{pmatrix} \quad \text{or} \quad a(t_0) = \begin{pmatrix} 0 \\ 1 \end{pmatrix} .$$

Depending on the initial level of the particle, the $\hat{S}$ matrix is given by

$$\hat{S} = \begin{pmatrix} -\frac{\alpha_2 + \frac{\gamma_2}{2}}{\alpha_1 - \alpha_2} e^{\alpha_1 t} + \frac{\alpha_1 + \frac{\gamma_2}{2}}{\alpha_1 - \alpha_2} e^{\alpha_2 t} ; & -\frac{iV}{\alpha_1 - \alpha_2}(e^{\alpha_1 t} - e^{\alpha_2 t}) \\ -\frac{iV}{\alpha_1 - \alpha_2}(e^{\alpha_1 t} - e^{\alpha_2 t}) e^{i\Omega t} ; & \left(\frac{\alpha_1 + \frac{\gamma_2}{2}}{\alpha_1 - \alpha_2} e^{\alpha_1 t} - \frac{\alpha_2 + \frac{\gamma_2}{2}}{\alpha_1 - \alpha_2} e^{\alpha_2 t}\right) e^{i\Omega t} \end{pmatrix} \tag{2.36}$$

In (2.36) α_1 and α_2 denote the roots of the characteristic equation,

$$\alpha_{1,2} = -\frac{i}{2}\Omega - \frac{\Gamma}{2} \pm \left[\left(i\frac{\Omega}{2} + \frac{\gamma'}{2}\right)^2 - V^2\right]^{1/2} , \tag{2.37}$$

where $2\Gamma = \gamma_1 + \gamma_2$, $\gamma' = \gamma_1/2 - \gamma_2/2$, and, for simplicity, we take $t_0 = 0$.

The integrated probability of a transition of a particle to the upper level W_{12}, that is the probability of photon absorption by the particle, can be calculated from the amplitude of probability of occupation of level 2,

$$W_{12} = \gamma_2 \int_{t_0}^{\infty} |a_2(t)|^2 dt . \tag{2.38}$$

Substitution of the expressions for $a_2(t)$ from (2.35) for the case that at the initial instant the particle is at the lower level, into (2.38) and integration give [2.6,7]

$$W_{12} = \frac{\Gamma}{2\gamma_1} \frac{(2V)^2}{(\Omega - \underline{\underline{k}}\underline{\underline{v}})^2 + \Gamma^2(1+G)} , \tag{2.39}$$

where $\Omega = \omega - \omega_0$ is the detuning of the field frequency with respect to the transition frequency of a stationary particle, i.e., the frequency of the center of the Doppler line, $\underset{\sim}{G}$ is the parameter of absorption saturation,

$$G = \left(\frac{pE}{\hbar}\right)^2 \frac{1}{\gamma_1\gamma_2} = \frac{(2V)^2}{\gamma_1\gamma_2} . \tag{2.40}$$

The probability of stimulated transition to level 1 for a particle excited initially to level 2, i.e., photon-emission probability, is given by an analogous expression,

$$W_{21} = \frac{\gamma_1}{\gamma_2} W_{12} . \tag{2.41}$$

The probabilities of stimulated transitions for particles excited to levels 1 and 2 are unequal, owing to the difference of spontaneous decay of levels. Figure 2.2 shows probabilities of photon emission and absorption W_{21} and W_{12}, which depend on light-wave intensity, expressed in terms of $\underset{\sim}{G}$, the saturation parameter, for the case of exact resonance ($\Omega = \underline{\underline{k}}\underline{\underline{v}}$). At weak saturation ($G \ll 1$), the transition probabilities are proportional to intensity; at strong saturation ($G \gg 1$), they reach their limiting values

$$W_{12} \simeq \frac{\gamma_2}{\gamma_1 + \gamma_2} ; \quad W_{21} \simeq \frac{\gamma_1}{\gamma_1 + \gamma_2} . \tag{2.42}$$

Considering the time dependence of the probability amplitude $|a_i(t)|^2$ given by the general expression (2.35) we can find the conditions under which $|a_i(t)|^2$, as in the case of nondecaying states, are time-oscillating variables. For example, at weak saturation ($G \ll 1$) in the case of exact resonance ($\Omega = 0$) and equal decay constants $\gamma_1 = \gamma_2 = \Gamma$ we have

$$\alpha_{1,2} = -\Gamma \pm iV \tag{2.43}$$

and, if initially the particle was at the excited level,

$$a_1(t) = -i\, e^{-\Gamma t} \sin Vt \ , \quad a_2(t) = e^{-\Gamma t} \cos Vt \ . \tag{2.44}$$

Thus, the probabilities of particle occupation of levels 1 and 2 are oscillating values. These oscillations are substantial, of course, only with $2V > \Gamma$. So probability amplitudes oscillate several times during the lifetime of a particle at a level, i.e., the mean level populations become equalized.

The case of equal level-decay constants ($\gamma_1 = \gamma_2$) is typical of the microwave range and of the optical region for vibrational-rotational molecular transitions in the ir range. The case with $\gamma_1 \neq \gamma_2$ is more typical for the visible range, in which spontaneous decay contributes substantially to level decay. In this case, the behavior of a two-level system in a resonant field is somewhat different. Assume that the upper-level lifetime is longer than that of the lower one,

$$\gamma_2 \ll \gamma_1 \ . \tag{2.45}$$

For exact resonance ($\Omega = 0$), oscillations of probability amplitudes are impossible if the roots of the characteristic (2.37) are real. A condition for this is the inequality

$$G < \frac{(\gamma_1 - \gamma_2)^2}{\gamma_1 \gamma_2} \simeq \frac{\gamma_1}{\gamma_2} \ . \tag{2.46}$$

So even at very strong saturation there are no oscillations of probability amplitudes if the level decay constants differ greatly. The physics of this is quite clear, because with strong decay of one level, the reverse stimulated transition of a particle that has come to this level from the other one becomes highly improbable. From relations (2.42) it follows that at strong saturation the mean population of the long-lived level is much less than that of the short-lived level. At very strong saturation, when the nonperiodicity condition (2.46) is not valid, oscillations of probability amplitudes may occur.

2.2 Density-Matrix Equation

The description of a two-level-system state by the wave function considered in Section 2.1 cannot be used for describing an assembly of two-level systems, because it would require that the initial conditions for each system be known. Usually, we do not know them and have information only concerning mean values of the whole assembly, say, on the mean probability of particle occupation of some levels. In this case, it is advisable to give up trying to describe the behavior of each quantum system, i.e., attempts to find the wave function of the whole assembly, and to restrict the analysis to a less full description in terms of averaged values. To describe the dynamics of a many-particle quantum system in such cases, a density-matrix technique is used [2.8,9].

The quantum-system states described fully, that is with the aid of a wave function, are usually called "pure". In contrast to "pure" states, those with which the wave function cannot be associated are called "mixed", for instance, the state of an assembly of two-level particles that populate the two levels with certain probabilities is a mixed state, and the wave function cannot be associated with it. The mixed (statistical) state may be regarded as a mixture of independent "pure", i.e., quantum-mechanically determined states $\psi^{(n)}$ of a system with statistical weights W(n). An arbitrary pure state can be displayed as a linear superposition of wave functions of the stationary states ψ_k

$$\psi^{(n)} = \sum_k a_k^{(n)} \psi_k \ . \tag{2.47}$$

By definition, statistical weights are real positive numbers that comply with the relation

$$\sum_n W(n) = 1 \ . \tag{2.48}$$

To calculate the mean magnitude with respect to assembly of any physical value, for example, the dipole moment P, we calculate, first, the magnitude probabilities of this value in the pure states $\psi^{(n)}$, that is calculate

$$\langle P^{(n)} \rangle = \int \psi^{(n)*} \hat{P} \psi^{(n)} dq \ , \tag{2.49}$$

and then average the result obtained, using the statistical weight W(n),

$$\langle P \rangle = \sum_n W(n) \langle P^{(n)} \rangle \ . \tag{2.50}$$

Taking into account the possibility of expansion (2.47) of pure states in terms of stationary states, the mean $\langle P \rangle$ can be written in the form

$$<P> = \sum_n W(n) \sum_{kk'} P_{kk'} a_k^{(n)*} a_{k'}^{(n)} , \tag{2.51}$$

where

$$P_{kk'} = \int \psi_k^* P \psi_{k'} dq \tag{2.52}$$

is the dipole-moment matrix element determined by stationary-state eigenfunctions.

Let us introduce the matrix elements

$$\rho_{k'k} = \sum_n W(n) a_k^{(n)*} a_{k'}^{(n)} . \tag{2.53}$$

Hence, by use of the rule of matrix multiplication, (2.52) can be written

$$<P> = \sum_{kk'} P_{kk'} \rho_{k'k} = \sum_k (\hat{P}\hat{\rho})_{kk} , \tag{2.54}$$

or in a shorter form

$$<P> = Sp(\hat{P}\hat{\rho}) = Sp(\hat{\rho}\hat{P}) , \tag{2.55}$$

where the sign "Sp" (spur) denotes the sum of the diagonal elements of the matrix which is the product of the $\hat{P}$ matrix by the matrix elements (2.52) and the $\hat{\rho}$ matrix by the elements (2.53). The $\hat{\rho}$ matrix is called the *density matrix* of states of a quantum-system assembly. The number of lines and columns in the density matrix corresponds to the number of independent states used to describe a pure state, for example the number of levels in an n-level quantum system.

Let us find the equation for time variation of the states described by the density matrix. We should differentiate the time relation (2.52),

$$\frac{\partial}{\partial t} \rho_{k'k} = \sum_n W(n) \left[\frac{\partial a_k^{(n)*}}{\partial t} a_{k'}^{(n)} + a_k^{(n)*} \frac{\partial a_{k'}^{(n)}}{\partial t} \right] . \tag{2.56}$$

In order to find the derivatives $\partial a_k^{(n)}/\partial t$, we must substitute $\psi^{(n)}$ from (2.47) ($\psi^{(n)} = \sum_m a_m^{(n)}(t)\psi_m$) into the Schrödinger equation (2.1). By multiplying the equation derived from $\psi_k^*(q)$ and integrating in region of change of the q variables we have

$$i\hbar \frac{\partial a_k^{(n)}}{\partial t} = \sum_m H_{km} a_m^{(n)} \ . \qquad (2.57)$$

Substitution of the expressions for $\partial a_k^{(n)*}/\partial t$ and $\partial a_{k'}^{(n)}/\partial t$ from (2.57) into (2.56) gives

$$\frac{\partial}{\partial t} \rho_{k'k} = \sum_m \sum_n W(n) \left[- \frac{1}{i\hbar} H^*_{km} a_m^{(n)*} a_{k'}^{(n)} + \frac{1}{i\hbar} H_{k'm} a_k^{(n)*} a_m^{(n)} \right] \ . \qquad (2.58)$$

Taking into account the hermite property of the matrix ($H_{mk} = H^*_{km}$) and the definition for the density matrix (2.53), we obtain

$$i\hbar \frac{\partial}{\partial t} \rho_{k'k} = \sum_m (H_{k'm}\rho_{mk} - \rho_{k'm}H_{mk}) \ . \qquad (2.59)$$

With the matrix notations used, this equation can be written in the form

$$i\hbar \frac{\partial \hat{\rho}}{\partial t} = \hat{H}\hat{\rho} - \hat{\rho}\hat{H} \qquad (2.60)$$

or

$$i\hbar \frac{\partial \hat{\rho}}{\partial t} = \{\hat{H},\hat{\rho}\} \ , \qquad (2.61)$$

where $\{\cdots\}$ are Poisson brackets. The matrix equation (2.60) allows determination of the density matrix for any instant, provided that it is known at any initial moment.

Let us consider the equation for the density matrix of an assembly of independent two-level particles interacting with an electromagnetic field. For this purpose (2.60) with the full hamiltonian $\hat{H}$ in (2.2) and (2.3) should be used. Because the density matrix operates with averaged values for many quantum systems, phenomenological rates of decay or relaxation of its elements can be introduced into the equation. The diagonal elements ρ_{kk} of the two-level system describe the level populations which relax to equilibrium populations ρ^0_{kk} at the rates γ_k $(k = 1.2)$. Relaxation of level population is termed longitudinal relaxation. This term has been introduced in nuclear magnetic resonance, where $\gamma_1 = \gamma_2 = \gamma$ and the notation $\gamma = 1/T_1$ is used for the longitudinal relaxation rate. The off-diagonal elements $\rho_{k'k}$ $(k' \neq k)$ describe the high-frequency dipole moment and decay at the rate Γ. When spontaneous transitions are the only relaxation mechanism in an assembly, then $2\Gamma = \gamma_{rad} = \gamma_1 + \gamma_2$. In the general case, owing to collisions, the rate of decay of nondiagonal elements is greater than that of population relaxation, and so $2\Gamma > \gamma_1 + \gamma_2$. Relaxation of high-frequency polarization, i.e., relaxation of nondiagonal elements of the density matrix, is called transversal relaxation, as in nuclear magnetic resonance, and its rates are designated by $T_2^{-1} = \Gamma$. As a result, we obtain a kinetic equation for the density matrix $\hat{\rho}$ with phenomenologically introduced longitudinal and transversal relaxations,

$$i\hbar \frac{\partial \hat{\rho}}{\partial t} = [\hat{H}_0 + \hbar\hat{V}, \hat{\rho}] - i\hbar(\Gamma\hat{\rho}) \ , \qquad (2.62)$$

where the $(\Gamma\hat{\rho})$ matrix describes the relaxation,

$$(\Gamma\hat{\rho}) = \begin{pmatrix} \gamma_1(\rho_{11} - \rho_{11}^0) & \Gamma\rho_{12} \\ \Gamma\rho_{21} & \gamma_2(\rho_{22} - \rho_{22}^0) \end{pmatrix} . \qquad (2.63)$$

Instead of the matrix equation (2.63), we may write four equations that describe the time variation of the matrix elements. The left side of (2.62) is apparent, the last term of the right side is determined by (2.63). The Poisson bracket has the form

$$[\hat{H}_0 + \hbar\hat{V}, \hat{\rho}] = [\hat{H}_0, \hat{\rho}] - \hbar\underline{E}[\hat{\underline{P}}, \hat{\rho}] \ , \qquad (2.64)$$

where the two brackets can be written in the form

$$[\hat{H}_0, \hat{\rho}] = (\hat{H}_0\hat{\rho}) - (\hat{\rho}\hat{H}_0) = \begin{pmatrix} E_1 & 0 \\ 0 & E_2 \end{pmatrix} \begin{pmatrix} \rho_{11} & \rho_{12} \\ \rho_{21} & \rho_{22} \end{pmatrix} -$$

$$- \begin{pmatrix} \rho_{11} & \rho_{12} \\ \rho_{21} & \rho_{22} \end{pmatrix} \begin{pmatrix} E_1 & 0 \\ 0 & E_2 \end{pmatrix} = \begin{pmatrix} 0 & \rho_{12}(E_1 - E_2) \\ \rho_{21}(E_2 - E_1) & 0 \end{pmatrix} ; \qquad (2.65)$$

$$[\hat{\underline{P}}, \hat{\rho}] = (\hat{\underline{P}}\hat{\rho}) - (\hat{\rho}\hat{\underline{P}}) = \begin{pmatrix} 0 & \underline{P}_{12} \\ \underline{P}_{21} & 0 \end{pmatrix} \begin{pmatrix} \rho_{11} & \rho_{12} \\ \rho_{21} & \rho_{22} \end{pmatrix} -$$

$$- \begin{pmatrix} \rho_{11} & \rho_{12} \\ \rho_{21} & \rho_{22} \end{pmatrix} \begin{pmatrix} 0 & \underline{P}_{12} \\ \underline{P}_{21} & 0 \end{pmatrix} = \begin{pmatrix} \underline{P}_{12}\rho_{21} - \underline{P}_{21}\rho_{12} & \underline{P}_{12}(\rho_{22} - \rho_{11}) \\ \underline{P}_{21}(\rho_{11} - \rho_{22}) & \underline{P}_{21}\rho_{12} - \underline{P}_{12}\rho_{21} \end{pmatrix} . \qquad (2.66)$$

By the use of (2.63), (2,65) and (2.66) the matrix (2.62) may be presented as equations for the diagonal elements,

$$\frac{\partial \rho_{11}}{\partial t} = \frac{i}{\hbar} \underline{E}(\underline{P}_{12}\rho_{21} - \underline{P}_{21}\rho_{12}) - \gamma_1(\rho_{11} - \rho_{11}^{o}) ,$$
$$\frac{\partial \rho_{22}}{\partial t} = -\frac{i}{\hbar} \underline{E}(\underline{P}_{12}\rho_{21} - \underline{P}_{21}\rho_{12}) - \gamma_2(\rho_{22} - \rho_{22}^{o}) \tag{2.67}$$

and off-diagonal elements,

$$\frac{\partial \rho_{12}}{\partial t} = i\omega_o\rho_{12} + \frac{i}{\hbar} \underline{E}\underline{P}_{12}(\rho_{22} - \rho_{11}) - \Gamma\rho_{12} ,$$
$$\frac{\partial \rho_{21}}{\partial t} = -i\omega_o\rho_{21} - \frac{i}{\hbar} \underline{E}\underline{P}_{21}(\rho_{22} - \rho_{11}) - \Gamma\rho_{21} , \tag{2.68}$$

where $\omega_o = \omega_{21} = 1/\hbar\ (E_2 - E_1)$. The projection of the transition dipole moment $\underline{P}_{12}$ onto the direction of the polarization vector $\underline{e}$ of a light wave ($\underline{E} = \underline{e}E$) will be denoted henceforth by $P_{12} = \underline{P}_{12}\underline{e}$.

The density-matrix technique considerably simplifies the procedure for calculation of the medium polarizability and makes it possible to introduce phenomenological relaxation constants due to collisions and to obtain averaged equations for microscopic media. Let us illustrate briefly the characteristics of a gaseous medium that interacts with a light wave, calculated by use of the density-matrix method.

The equations for the density matrix of the particles $\rho_{k'k}$ (z,v,t) that have a definite projection of their velocities onto the chosen direction OZ (light-beam direction), located at the point Z, averaged over moments of particle excitation to the levels, have the form (FELDMAN and FELD [2.10])

$$(\frac{\partial}{\partial t} + v\frac{\partial}{\partial z} + i\omega_o)\rho_{21} + \Gamma\rho_{21} = -i2V(z,t)(\rho_{22} - \rho_{11}) , \tag{2.69}$$

$$(\frac{\partial}{\partial t} + v\frac{\partial}{\partial z})\rho_{22} + \gamma_2\rho_{22} = -i2V(z,t)(\rho_{21} - \rho_{12}) + \gamma_2\rho_{22}^{o} , \tag{2.70a}$$

$$(\frac{\partial}{\partial t} + v\frac{\partial}{\partial z})\rho_{11} + \gamma_1\rho_{11} = +i2V(z,t)(\rho_{21} - \rho_{12}) + \gamma_1\rho_{11}^{o} , \tag{2.70b}$$

$\rho_{21} = \rho_{12}^{*}$, $\rho_{11}^{o}(v)$ and $\rho_{22}^{o}(v)$ denote the populations of fields 1 and 2 in the absence of field, averaged over the duration of excitation.

2.3 Polarization and Level Populations. Rate Equations

Diagonal elements of density matrix determine the difference of level populations,

$$n(z,v,t) = N_0[\rho_{11}(z,v,t) - \rho_{22}(z,v,t)] \tag{2.71}$$

where N_0 is the number of particles in a volume unit, for which the density matrix is introduced. Nondiagonal elements determine the polarization of unit volume of the medium in a light field,

$$P(z,v,t) = P_{21}\rho_{12}(z,v,t) + P_{12}\rho_{21}(z,v,t) \ . \tag{2.72}$$

Decay constants of nondiagonal elements are determined by the homogeneous half-width Γ.

The differences of level population and medium polarization averaged over particle velocities are

$$N(z,t) = \int n(z,v,t)W(v)dv \ , \qquad P(z,t) = \int P(z,v,t)W(v)dv \ , \tag{2.73a,b}$$

where $W(v)$ is the distribution of projections of particle velocities onto axis OZ ($v = v_z$). The dependence of $N(z,t)$ and $P(z,t)$ on the Z coordinate is determined by the spatial configuration of light field. If a light field is spatially inhomogeneous in structure, i.e., a standing wave, N and P depend on Z. In the case of a travelling wave, there is no spatial dependence, but the polarization has the high-frequency time dependence

$$P(z,t) = C\cos(\omega t - kz) + S\sin(\omega t - kz) \ , \tag{2.74}$$

where C and S are polarization coefficients, cophasal and quadratic with respect to the field. The coefficients C and S are proportional to the field amplitude E. Therefore the polarization of the medium is often presented by another form,

$$P(z,t) = \chi E(z,t) = (\chi' - i\chi'')E(z,t) \ , \tag{2.75}$$

where χ is polarizability of the medium, and the field is taken in the complex form $E(z,t) = E\exp[i(\omega t - kz)]$.

We shall not derive differential equations for polarization $P(z,t)$ and level-population difference $N(z,t)$ because in each particular case, as will be seen, it is more convenient to use nonaveraged equations for the density-matrix elements $\rho_{ij}(z,v,t)$. In one important case, however, we can substantially simplify (2.69) and (2.70) and get instead of them rather simple rate equations for the level population.

Assume that the change of the medium parameters under field action comes about more slowly than the transversal relaxation rate Γ, that is,

$$2V = \frac{P_{12}E}{\hbar} << \Gamma \tag{2.76}$$

and the rate of level-population relaxation is much less than that of polarization relaxation, that is

$$\gamma_1, \gamma_2 << \Gamma . \tag{2.77}$$

In this case, the off-diagonal elements of the density matrix, which are the medium polarizations, follow quasistatically the field amplitude, and so a simple algebraic relation between ρ_{21} and V can be written instead of the differential equation (2.69).

Let us consider a light field as a running wave (2.12); then the function $\rho_{21}(z,v,t)$ must have the form

$$\rho_{21}(z,v,t) = \tilde{\rho}_{21}(v)\, e^{-i(\omega t - kz)} . \tag{2.78}$$

By substituting (2.78) into (2.69), we derive

$$\tilde{\rho}_{21}(v) = \frac{V}{(kv - \Omega) - i\Gamma}(\rho_{22} - \rho_{11}) , \tag{2.79}$$

where $V = P_{12}E/2\hbar$. Neglecting the spatial derivative in (2.70), that is, considering an assembly of particles with the given projected velocity v, we can derive at once the equations for level population density,

$$\begin{aligned} \frac{\partial n_1(v)}{\partial t} + \gamma_1[n_1(v) - n_1^0(v)] &= \frac{2V^2}{\Gamma} L(kv - \Omega)[n_2(v) - n_1(v)] , \\ \frac{\partial n_2(v)}{\partial t} + \gamma_2[n_2(v) - n_2^0(v)] &= -\frac{2V^2}{\Gamma} L(kv - \Omega)[n_2(v) - n_1(v)] , \end{aligned} \tag{2.80}$$

where $n_i(v) = N_0\rho_{ii}(v)$, $n_i^0(v) = N_0\rho_{ii}^0(v)$ and the lorentzian form factor is designated as

$$L(x) = \Gamma^2/(\Gamma^2 + x^2) . \tag{2.81}$$

In a particular case of equal level-decay-rate constants $(\gamma_1 = \gamma_2 = \gamma)$ we can derive directly from (2.80) the equation for level-population-difference density $n(v) = n_2(v) - n_1(v)$,

$$\frac{\partial n(v)}{\partial t} + \gamma[n(v) - n_0(v)] = \gamma G L(kv - \Omega)n(v) \ , \tag{2.82}$$

where the saturation parameter G of level-population difference is expressed by

$$G = \frac{4V^2}{\gamma\Gamma} = \left(\frac{P_{12}E}{\hbar}\right)^2 \frac{1}{\gamma\Gamma} \ . \tag{2.83}$$

Often, other concepts are used: coefficients of absorption per unit length $\kappa(\omega)$ and the refractive index of the resonant medium $n(\omega)$, which are related to the defined quantities by

$$\kappa(\omega) = -4\pi \frac{\omega}{c}\frac{S}{E} = 4\pi \frac{\omega}{c} \chi''(\omega) \ , \tag{2.84}$$

$$n(\omega) - 1 = 2\pi \frac{C}{E} = 2\pi\chi'(\omega) \ . \tag{2.85}$$

The values $\chi'(\omega)$ and $\chi''(\omega)$ can be calculated by velocity averaging of $\chi'(\omega,v)$ and $\chi''(\omega,v)$ values, which are directly found from (2.73),

$$\chi'(\omega,v) - i\chi''(\omega,v) = \frac{P_{21}}{E(z,t)} [\rho_{12}(z,v,t) + \rho_{21}(z,v,t)] \ . \tag{2.86}$$

2.4 Absorption Coefficient of Running Wave

Saturation deforms the Doppler line shape. It follows from (2.39) that particles, the velocities of which satisfy the resonance condition

$$\underline{kv} - \omega - \omega_0 = 0 \ , \tag{2.87}$$

interact with the field most effectively. The resonance half-width is

$$\Gamma_B = \Gamma(1+G)^{1/2} \ , \tag{2.88}$$

i.e., increases with increase of degree of saturation G. The quantum-transition saturation for the particles that interact resonantly with the field gives rise to a shortage of particles that comply with resonance condition (2.87) at the lower level, i.e., "hole burning" occurs; at the upper level, particle excess occurs at the same velocity, i.e., a peak is formed in the velocity distribution (see Fig.1.9b). As a result, the velocity distribution of the population difference that describes the absorption-line shape takes the form

$$n(v) = n_o(v)\left(1 + \frac{G\Gamma^2}{(\Omega - kv)^2 + \Gamma^2}\right)^{-1} . \tag{2.89}$$

Relation (2.89) follows directly from steady-state solution of (2.82). In the population difference distribution (2.89) "a hole" appears, corresponding to particles that comply with the resonance condition. This corresponds to Bennett "hole burning" in the Doppler profile of absorption, provided that it is measured with another probe light wave.

The light-wave absorption coefficient per unit length in an absorption medium is equal to the ratio between the power absorbed by all particles in volume unit and the flux density of the incident radiation,

$$\kappa(\omega) = \hbar\omega[Q_1\langle W_{12}(v)\rangle - Q_2\langle W_{21}(v)\rangle]\left(\frac{c}{8\pi}E^2\right)^{-1} , \tag{2.90}$$

where Q_i is the number of particles excited to the ith level in a unit volume per unit time, which is related to the steady-state density of population N_i^o by $N_i^o = Q_i/\gamma_i$. In (2.90), the probabilities of the induced transitions W_{12} and W_{21}, given by (2.39) and (2.41), must be averaged over the velocity distributions. For the Maxwell velocity distribution

$$W(v) = \frac{1}{(\sqrt{\pi}u)^3}\exp\left(-\frac{|\underline{v}|^2}{u^2}\right) , \quad u = \left(\frac{2kT}{M}\right)^{1/2} , \tag{2.91}$$

the probability of stimulated transition $1\rightarrow 2$ has the form [2.6]

$$\langle W_{12}\rangle = \frac{\sqrt{\pi}}{ku}\frac{\gamma_2}{2}\frac{G}{(1+G)^{1/2}}\,U\left(\frac{\Omega}{ku}, \frac{2\Gamma}{ku}\sqrt{1+G}\right) , \tag{2.92}$$

where

$$U(x,y) = \frac{1}{\pi}\int_{-\infty}^{\infty}\frac{y\,e^{-t^2}dt}{(x-t)^2 + y^2} = \mathrm{Re}\{\exp[-i(x+iy)^2][1 - \Phi(x+iy)]\} , \tag{2.93}$$

where $\Phi(z)$ is the probability integral, and the function $U(x,y)$ is tabulated in [2.11]. In the limiting case of Doppler broadening

$$\Gamma_B = \Gamma(1 + G)^{1/2} \ll ku , \tag{2.94}$$

which is of most interest for problems of nonlinear spectroscopy inside the Doppler width, the function $U(x,y) \approx \exp(-x^2)$ and (2.92) reduce to the simpler form

$$\langle W_{12}\rangle = \frac{\sqrt{\pi}}{ku}\frac{\gamma_2}{2}\frac{G}{(1+G)^{1/2}}\exp[-(\frac{\Omega}{ku})^2] \ . \qquad (2.95)$$

An analogous expression holds for $\langle W_{21}\rangle$. As a result, we get the expression for the coefficient of absorption per unit length for a strong field [2.6,12]

$$\kappa(\omega) = \kappa_0(\omega)(1+G)^{-1/2} \ ; \qquad \kappa_0 = \frac{1}{\hbar u}\,4\pi^{3/2}P_{21}^2(N_1^0 - N_2^0)\exp[-(\frac{\Omega}{ku})^2] \ , \qquad (2.96)$$

where κ_0 is the coefficient of absorption per unit length for a weak field, i.e., a field that causes no transition saturation.

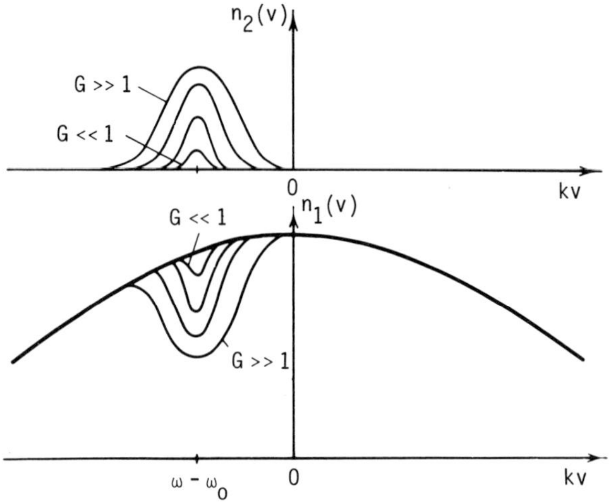

Fig.2.3 Velocity distribution of atoms (molecules) on lower (n_1) and upper (n_2) levels of transition at various degrees of saturation

The dependence of the absorption coefficient of a strong field $\kappa(\omega)$ on its frequency ω is given by the Doppler-contour shape, but the absorption coefficient value decreases, because of saturation. It is significant that, when the line broadening is inhomogeneous, the absorption coefficient at strong saturation is proportional to $\sqrt{G}$, i.e., to the field amplitude. In the case of homogeneous line broadening, the absorption coefficient decreases in proportion to G, i.e., to the field intensity. This distinction is related to the increased number N_{inter} of particles that interact with the field as the intensity increases, in case of inhomogeneous broadening. According to

(2.88), the hole width in the velocity distribution of particles increases in proportion to the field amplitude, i.e., $N_{inter} \sim (1+G)^{1/2}$. Figure 2.3 illustrates particle-velocity distribution at the lower and upper transition levels with different degree of saturation. The absorption coefficient for particles that are in resonance with the field is proportional to $N_{inter}/(1+G)$; hence the resultant absorption coefficient is proportional to $(1+G)^{-1/2}$.

In conclusion, let us return to the influence of other (nonradiative) relaxation mechanisms on saturation of Doppler-broadened absorption lines. If state decay is determined, not only by spontaneous decay, but also by collisions of moving particles, the problem of interaction between field and a Doppler-broadened transition becomes more complicated. Calculation can be done, for instance, with the model of relaxation constants [2.6], in which collisions are accounted for phenomenologically by corresponding variations of longitudinal and transverse relaxation constants. The constants of level-population relaxation (diagonal elements of the density matrix) are added to with the terms $\gamma_{coll}^{(1)}$ and $\gamma_{coll}^{(2)}$, whereas those of medium-polarization relaxation (off-diagonal elements of the density matrix) are added with $\Delta\omega_{coll}$. In this approximation, the inclusion of collisions causes a corresponding increase of the constants of level-population decay,

$$\tilde{\gamma}_i = \gamma_{coll}^{(i)} + \gamma_i \tag{2.97}$$

and of the homogeneous width,

$$2\Gamma = \tilde{\gamma}_1 + \tilde{\gamma}_2 + \Delta\omega_{coll} \quad . \tag{2.98}$$

The saturation parameter becomes

$$G = \left(\frac{P_{12}E}{\hbar}\right)^2 \frac{1}{\tilde{\gamma}\Gamma} \ , \quad \frac{2}{\tilde{\gamma}} = \frac{1}{\tilde{\gamma}_1} + \frac{1}{\tilde{\gamma}_2} \ . \tag{2.99}$$

Formulas (2.89) and (2.96) hold with due regard for the new expressions for the homogeneous transition width and the saturation parameter.

The formulas derived correspond to the interaction of a plane wave with a gas. In real lasers, we usually deal with a light beam limited in diameter and with gaussian intensity distribution over the cross-section. If the free-path length of atoms is shorter than the beam diameter, we may neglect the transverse changes of the field amplitude. The intensity inhomogeneity causes only absorption-saturation inhomogeneities over the beam cross-section.

If the length of the free path of a particle is comparable with the light-beam diameter, the transit-time effects studied separately in Chapter 6 become significant. Finally, cases are possible when the light beam is a spherical wave rather than a plane wave. In these cases, the degree of saturation depends also on the wave-front curvature. Such effects are considered in Chapter 9.

2.5 Absorption Saturation by Standing Wave

2.5.1 Weak Saturation. Lamb Dip

Assume that the light field is a standing plane wave, which can be presented as a superposition of two counter-running waves that have the same frequency,

$$E = E_{st} \cos\omega t \cos\underline{\underline{k}}\underline{\underline{r}} = E \cos(\omega t - \underline{\underline{k}}\underline{\underline{r}}) + E \cos(\omega t + \underline{\underline{k}}\underline{\underline{r}}) \quad , \tag{2.100}$$

where $E_{st} = 2E$ is the standing-wave amplitude, and E is the amplitude of each running wave. Such a field interacts with two groups of particles, the velocities of which satisfy one of the resonance conditions,

$$\omega - \omega_0 \pm \underline{\underline{k}}\underline{\underline{v}} = 0 \ . \tag{2.101}$$

The velocity distribution and the Doppler profile of the absorption line of these two groups are located symmetrically about the center. If the field-frequency detuning, with respect to the center of line, $\Omega = \omega - \omega_0$, is much larger than the "hole" half-width $\Gamma_B = \Gamma(1+G)^{1/2}$, then each of the running waves independently burns its own "hole" during absorption saturation (Fig. 1.10b). The parameters of each "hole" and the saturated absorption of each running wave are given by the expressions presented in Section 2.4, in which the field amplitude in the formula for the parameter of saturation G means the running-wave amplitude E.

As the field frequency is tuned to the center of the Doppler line ($|\omega - \omega_0| \lesssim \Gamma_B$), when the holes begin to overlap each other, the same group of atoms interacts with both light waves. Each of the two running waves is "responsive" to the decrease in level-population difference that is caused by the field of the counter-running wave. The absorption saturation by one wave is added to the absorption by the other wave. In the frame of the atomic center-of-mass, the light waves have the different frequencies $\omega \pm \underline{\underline{k}}\underline{\underline{v}}$. This corresponds to the fact that in the laboratory coordinate frame an atom moves in a space-modulated standing light wave. Nonmonochromaticity (in the center-of-mass system) or field nonuniformity (in the laboratory frame) make it more difficult to investigate nonlinear resonant interaction. However the main effect that arises in a standing wave--the production of a resonant dip at the center of Doppler line (Lamb dip) for the saturated absorption coefficient of a standing wave--can be understood in simple terms of "hole burning". Indeed, when the field frequency is tuned to the center of the line, the efficient field that acts on particles with $\underline{\underline{kv}} = 0$ is doubled. Accordingly, the absorption-saturation parameter is doubled and the absorption coefficient decreases resonantly. This corresponds to the coincidence of two holes at $\Omega = 0$ and the production of a deeper hole at the center of the Doppler profile (Fig.1.10b).

This effect was first described by LAMB in the approximation of weak saturation [2.13], for which case the perturbation theory with a small saturation parameter can be used. The saturated absorption coefficient of a standing wave with the frequency ω is determined by

$$\kappa(\omega) = \kappa_0(\omega)\left[1 - \frac{G}{2}\left(1 + \frac{\Gamma^2}{\Gamma^2 + \Omega^2}\right)\right] , \quad G \ll 1 , \tag{2.102}$$

where G is the parameter of absorption saturation by one running wave. According to (2.102), the degree of absorption saturation is G at the center of the Doppler line and G/2 away from the resonance. The width of the dip at the center of the line is 2Γ, i.e., coincides with the width of the interaction resonance or that of the Bennett hole in velocity distribution when it is expressed in units of kv.

It is easy to deduce (2.102) by solving density-matrix equations (2.69) and (2.70) by perturbation theory with a small parameter V and averaging to obtain (2.73).

We are not going to carry out these calculations by use of (2.69,70) because in Subsection 2.5.2 the calculations are given in a more general form for strong level-population saturation in a standing-wave field. Here, we shall restrict ourselves to the simplest derivation of (2.102).

In the case of weak saturation, the derivation can be simplified greatly by use of the effect of "hole burning" in the velocity distribution. The velocity distribution of difference of populations of levels in the standing-wave field is described by an expression similar to (2.89), but with two running waves taken into account,

$$n(v)/n_0(v) = [1 + GL(\Omega - \underline{k}\underline{v}) + GL(\Omega + \underline{k}\underline{v})]^{-1} \ , \tag{2.103}$$

where a convenient notation (2.81) is introduced for the Lorentz contour that describes homogeneous line broadening.

The absorption coefficient for a wave with frequency ω' and wave vector $\underline{k}'$ is given by

$$\kappa(\omega',\underline{k}') = \int d\underline{v}n(\underline{v})\sigma(\omega_0 - \omega' + \underline{k}'\underline{v}) \ , \tag{2.104}$$

where $\sigma(\omega',\underline{k}')$ denotes the induced-transition cross-section for a particle that has velocity $\underline{v}$ in a light field with frequency ω' and wave vector $\underline{k}'$, which is related to the Lorentz shape of homogeneous transition broadening by

$$\sigma(\omega_0 - \omega' + \underline{k}'\underline{v}) = \sigma_0 L(\omega_0 - \omega' + \underline{k}'\underline{v}) \ , \tag{2.105}$$

in which σ_0 is the induced transition cross-section at exact resonance ($\omega' - \omega_0 = \underline{k}'\underline{v}$). The absorption coefficient of a standing wave that causes absorption saturation is given by

$$\kappa(\omega) = \int d\underline{v}n(\underline{v})[\sigma(\omega_0 - \omega + \underline{k}\underline{v}) + \sigma(\omega_0 - \omega - \underline{k}\underline{v})] \ , \tag{2.106}$$

where the population difference in the standing-wave field is determined by (2.103). In the approximation $G \ll 1$, the denominator of (2.103) can be simplified; then, (2.106) is calculated with the Maxwell distribution (2.91). This results in (2.102).

2.5.2 Strong Saturation. Level-Population Effects

The case of strong saturation of absorption of atoms or molecules is of special importance, because just such a situation is often realized in experiments with a saturated-absorption cell placed inside a laser cavity. Saturation in the strong field of a standing wave was investigated theoretically [2.6,10,14-21]. In the general case of arbitrary values for degree of saturation, detuning, and relaxation constants, the problem can be solved only by use of electronic computers. Analytical solutions can be obtained in the particular cases of exact resonance with $\omega = \omega_0$, for equal relaxation constants $\gamma_1 = \gamma_2 = \Gamma$, and when approximate calculation methods are used which, nevertheless, give an understanding of the standing-wave interaction, and cover practically important situations.

The solution of such problems is complicated because of changes of the atomic line shape and the level-population difference in strong field. The two effects cannot be treated separately. When there are several fields, with frequencies ω_1 and ω_2 interacting, combination frequencies $\omega_1 \pm n(\omega_1 - \omega_2)$ appear in polarization, where $n = 1,2,\cdots$. The polarization at these frequencies gives rise, in turn, to modulation of population difference. Density-matrix equations (2.69) and (2.70) are related by time-dependent off-diagonal and diagonal elements that are directly connected with polarization and population of levels, respectively.

The use of rate equations suggests that the influence of one field on the transitions caused by another field should be explained by level-population variation. The known processes of absorption or amplification line-shape distortion, which occur under the action of another field and are caused by transition kinetics (oscillations of probability of particle occupation of one of the levels), are neglected. It is obvious that the main distortions of the line shape caused by these oscillations of probability take place when $V \gg \Gamma$. This inequality means that throughout coherent atom-field interaction $1/\Gamma$ the atom passes from one state into another many times. With $V \ll \Gamma$, the oscillations may be neglected and the line-shape distortions disregarded. But the condition $V \ll \Gamma$ does not mean that there are no saturation effects. If one of the levels is long-lived, the saturation parameter may be, nevertheless, large, and yet the level-population difference will be below saturation.

Let us calculate the strong-standing wave (2.100) absorption in the rate-equation approximation, neglecting the spatial inhomogeneity of level-population-difference distribution (all of the derivatives with respect to Z in (2.70) for ρ_{11} and ρ_{22} become zero). The standing-wave absorption (amplification) in such an approximation has been studied by several authors (GREENSTEIN [2.15], BAKLANOV and CHEBOTAYEV [2.19], UEHARA and SHIMODA [2.20]). The expressions for standing-wave field absorption found by them are identical and differ only in form.

In this approximation, (2.69) can be solved in the simple form

$$\rho_{21} = iV(r_{(+)}\, e^{ikz} + r_{(-)}\, e^{-ikz}) \; . \tag{2.107}$$

Instead of a set of differential equations (2.69,70), we have the set of algebraic equations,

$$\rho_{11} - \rho_{22} = r_{(\pm)}(\Gamma \pm ikv - i\Omega) , \tag{2.108a}$$

$$\rho_{22} = \frac{1}{\gamma_2} |V|^2 (r^*_{(+)} + r^*_{(-)} + r_{(+)} + r_{(-)}) + \rho^0_{22} , \tag{2.108b}$$

$$\rho_{11} = \frac{1}{\gamma_1} |V|^2 (r^*_{(+)} + r^*_{(-)} + r_{(+)} + r_{(-)}) + \rho^0_{11} . \tag{2.108c}$$

After eliminating $r_{(+)}$ and $r_{(-)}$ from these equations and subtracting (2.108c) from (2.108b), we obtain the expression for population difference per particle $(\rho_{11} - \rho_{22})$,

$$\frac{\rho_{11} - \rho_{22}}{\rho^0_{11} - \rho^0_{22}} = \left\{1 + G\left[\frac{\Gamma^2}{\Gamma^2 + (kv - \Omega)^2} + \frac{\Gamma^2}{\Gamma^2 + (kv + \Omega)^2}\right]\right\}^{-1} . \tag{2.109}$$

This expression can be transformed to

$$\frac{\rho_{11} - \rho_{22}}{\rho^0_{11} - \rho^0_{22}} = \frac{1}{D}[\Gamma^2 + (kv - \Omega)^2][\Gamma^2 + (kv + \Omega)^2] , \tag{2.110}$$

where

$$D = [(kv)^2 + \Gamma^2 a^2][(kv)^2 + \Gamma^2 b^2] ,$$

$$a^2 = \Gamma^2(1 + G) - \Omega^2 + \Gamma[(\Gamma G)^2 - 4\Omega^2(1 + G)]^{1/2} , \tag{2.111}$$

$$b^2 = \Gamma^2(1 + G) - \Omega^2 - \Gamma[(\Gamma G)^2 - 4\Omega^2(1 + G)]^{1/2} .$$

Taking into account the relation between $r_{(\pm)}$ and $(\rho_{11} - \rho_{22})$:

$$r_{(\pm)} = (\rho_{11} - \rho_{22})/(\Gamma \pm ikv - i\Omega) \tag{2.112}$$

and using (2.107) and (2.110), we can directly derive the exact expression for the amplitude of any off-diagonal density-matrix element

$$\rho_{21} = i(\rho^0_{11} - \rho^0_{22}) \frac{V}{D} \{[\Gamma - i(kv - \Omega)][\Gamma^2 + (kv + \Omega)^2] e^{ikz} +$$

$$+ [\Gamma - i(kv + \Omega)][\Gamma^2 + (kv - \Omega)^2] e^{-ikz}\} . \tag{2.113}$$

The medium polarizability $\chi(\omega)$ for each running wave $\exp[i(\omega t \pm kz)]$ in the presence of a counter-running wave with the same amplitude and frequency can be given by use of (2.113) and (2.86) in the form

$$\chi(\omega) = i\,\frac{|P_{21}|^2}{\hbar}\,(\rho_{11}^0 - \rho_{22}^0)N_0 \int_{-\infty}^{\infty} \frac{1}{D}\,[\Gamma - i(kv - \Omega)][\Gamma^2 + (kv + \Omega)^2]dv\,. \quad (2.114)$$

Then, by use of (2.84), we evaluate the coefficient of absorption by one strong running wave with a strong counter-running wave,

$$\kappa(\omega) = \kappa_0(\omega)\,\frac{\Gamma}{\pi}\int_{-\infty}^{\infty} \frac{y^2 + \Gamma^2 + \Omega^2}{(y^2 + a^2)(y^2 + b^2)}\,dy\;, \quad (2.115)$$

where the dimensionless variable $kv = y$ is introduced, and the limiting case of Doppler broadening $ku >> \Gamma_B$ is under consideration.

When evaluating the integral by subtraction theory, we assume that $\mathrm{Re}\,a > 0$, $\mathrm{Re}\,b > 0$. Then the integration contour is closed, for example, in the upper half-plane, we have

$$\frac{\kappa(\omega)}{\kappa_0(\omega)} = \frac{\Gamma}{\pi}\,2\pi i\left[\frac{(ia)^2 + \Gamma^2 + \Omega^2}{2ia(ia + ib)(ia - ib)} + \frac{(ib)^2 + \Gamma^2 + \Omega^2}{(ib + ia)(ib - ia)2ib}\right]. \quad (2.116)$$

From here we can easily derive [2.19]

$$\frac{\kappa(\omega)}{\kappa_0(\omega)} = \frac{\Gamma}{(a + b)}\,(1 + \frac{A}{B})\;, \quad (2.117)$$

where

$$A = (\Omega^2 + \Gamma^2)^{1/2}\;, \qquad B = [\Omega^2 + \Gamma^2(1 + 2G)]^{1/2}\;. \quad (2.118)$$

Even though a and b are complex values, $(a + b)$ is a real value. Because of the complex formulas for a and b, (2.117) is somewhat unwieldy for numerical calculations. Therefore, this formula may be transformed [2.20] by use of (2.117),

$$\frac{\kappa(\omega)}{\kappa_0(\omega)} = \Gamma[(A + B)^2 - 4\Omega^2]^{-1/2}(1 + \frac{A}{B})\;. \quad (2.119)$$

Far from resonance, the absorption coefficient is

$$\kappa(\omega) = \kappa_0(\omega)(1 + G)^{-1/2}\;, \qquad |\Omega| >> \Gamma_B\;, \quad (2.120)$$

i.e., agrees with that of a strong running wave (2.96). This corresponds to independent passing of running waves through a gas. In the case of exact resonance, the absorption coefficient is

$$\kappa(\omega) = \kappa_0(\omega)(1+2G)^{-1/2}, \quad |\Omega| \ll \Gamma_B. \tag{2.121}$$

At the center of the Doppler line, the saturated-absorption coefficient decreases because of increasing saturation parameter. The real part of the susceptibility χ, which determines the index of refraction of the medium in a strong field, can be derived from (2.114),

$$\mathrm{Re}\{\chi(\omega)\} = \frac{|P_{21}|^2}{k\hbar} N_0 \int_{-\infty}^{\infty} \frac{1}{D}\, dy(y-\Omega)[\Gamma^2 + (y+\Omega)^2] \exp\left[-\frac{y^2}{(ku)^2}\right]. \tag{2.122}$$

This integral can be written as

$$\mathrm{Re}\{\chi(\omega)\} = \frac{|P_{21}|^2}{k\hbar} N_0 \left\{\int_{-\infty}^{\infty} dy \frac{(y-\Omega)}{(y-\Omega)^2+\Gamma^2} \exp[-y^2/(ku)^2] - \right.$$

$$\left. - 2G\Gamma^2 \int_{-\infty}^{\infty} \frac{(y-\Omega)(y^2+\Omega^2+\Gamma^2)}{D[(y-\Omega)^2+\Gamma^2]} \right\}. \tag{2.123}$$

After integrating in the limiting case $\Omega \ll ku$, we derive the finite expression for the real part of the polarizability

$$\mathrm{Re}\{\chi(\omega)\} = -\frac{\Omega}{\Gamma}\frac{\kappa_0}{4\pi k}\left\{\frac{2\Gamma}{\sqrt{\pi}ku} - \Gamma(1-\frac{A}{B})[(A+B)^2 - 4\Omega^2]^{-1/2}\right\}. \tag{2.124}$$

According to (2.85), (2.124) determines the gas refraction factor at the Doppler-broadened absorption line in the strong standing-wave field.

Figure 2.4 shows curves that characterize the Lamb-dip shape at varying degrees of saturation G. The dependence of half-height dip half-width $\Delta\omega$ on field found by the use of (2.119) is given in Fig.2.5. At strong saturation, the dip shape is a function of the parameter $\Omega/\Gamma\sqrt{G}$. It is similar to the Lorentz shape with half-width $\Gamma\sqrt{G}$. The relative dip depth, in this approximation, depends simply on G,

$$h = \frac{\Delta\kappa}{\kappa_0} = (1+G)^{-1/2} - (1+2G)^{-1/2}. \tag{2.125}$$

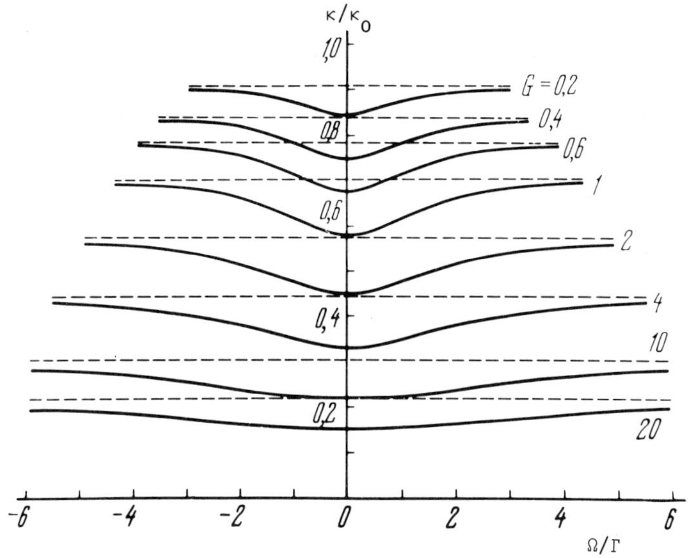

Fig.2.4 Shape of the resonance dip with various degrees of saturation G in the approximation of rate equations

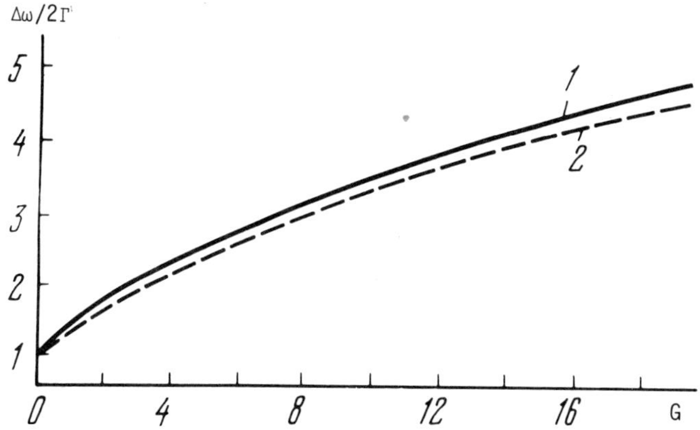

Fig.2.5 Dependence of the dip width Δω on the degree of saturation G: 1 - exact calculation; 2 - in the approximation of rate equations

It is maximum with $G = \sqrt[3]{4} - 1/2 - \sqrt[3]{4} = 1.4$ and equals $h_{max} = 0.133$.

The relaxation constants being equal, when the greatest contributions of coherent processes should be expected, the value of saturated absorption at the center of the line differs by about 20% from the exact result. When they differ greatly, this value turns out to be less. This circumstance makes it rather convenient to use the results obtained with rate equations in calculating the characteristics of lasers with large excess of amplification over threshold and those of saturated-absorption lasers in particular.

2.5.3 Coherence and Nonuniformity of Standing-Wave Field Effects

Approximation Method

Neglect of spatial nonuniformity of standing-wave-field and coherence effects when using rate equations leads to loss of some effects that are important for solving certain problems that are especially significant for nonlinear-absorption lasers. The calculation of the line shape for weak-signal absorption in the presence of the strong field of a standing wave calls, for instance, for studies of single-frequency oscillation stability, laser frequency and power fluctuations. The inclusion of coherent processes is very important here; the analytical solution of the problem is of vital importance. The approximation method for including the contribution of coherence effects has been developed by BAKLANOV and CHEBOTAYEV [2.19]. The basic idea of their approach is to find coherence corrections by use of the parameter $(\tilde{\gamma}/\Gamma)G$ $(2/\tilde{\gamma} = 1/\tilde{\gamma}_1 + 1/\tilde{\gamma}_2)$ for the solution of the rate equations. In the optical region, the relaxation constants differ markedly, as a rule. Therefore, the parameter $\tilde{\gamma}/\Gamma << 1$ and hence the condition $(\tilde{\gamma}/\Gamma)G << 1$ can be fulfilled with large G, which is sufficient to analyze rigorously the effects in many cases. The presence of collisions, which cause phase interruption and have no effect on level lifetime, decreases the ratio $\tilde{\gamma}/\Gamma$, also.

The contribution of coherent processes, when several fields interact, is determined first of all by polarization at combination frequencies. For rather weak fields $((\tilde{\gamma}/\Gamma)G << 1)$ we can restrict the analysis to take account of polarization only at the first harmonics of the combination frequency $\omega_1 \pm (\omega_1 - \omega_2)$. During resonant interaction of two counter-running waves it is better to deal with the spatial harmonics of polarization and the population of the medium. The amount of the first spatial harmonic appears to be equivalent to that of atomic polarization at the first combination frequency, in the coordinate frame related to the atom.

Appearance of spatial harmonics of population, related to the structure of the standing wave, can be simply explained for atoms that have the projected velocity $v_z = 0$. The atoms that are excited in the standing-wave loops do not interact with the field; their density remains constant. The atoms excited in the wave nodes interact with the field. So, along the Z axis, a periodic spatial nonuniformity of the medium appears under the action of the field and, hence, spatially harmonic level populations.

This rough qualitative explanation does not give, of course, any full idea of the nature of spatial inhomogeneity of a gas medium. The atoms excited in the loops of a standing wave can actually turn out to be in its nodes due to motion, and vice versa. In other words, the particle motion compensates for spatial inhomogeneities of gas-medium saturation. Calculations show that spatial inhomogeneity of gases occurs only when the second-order terms of saturation (G^2 and more) are taken into account, whereas in a medium with stationary particles it occurs to an approximation of weak saturation G. Another qualitative difference of a saturation medium with moving particles from that with stationary particles is that maximum saturation-of-level population difference occurs not in the loops of a standing wave, as might be expected at first sight, but in its nodes.

The maximum effect of spatial inhomogeneity in gas saturation is expected to occur when the field frequency ω is tuned to near the center of the Doppler-broadened line, because, in this case, the field interacts with particles whose projected velocities v_z are very small. When the field frequency is detuned away from the center of the spectral line, the field interacts with those atoms whose projected velocities v_z are large. The atoms with large projected velocities $v_z > \Gamma/k$ move through many nodes and loops and are affected mainly by an average field. For these atoms, it is sufficient to take account of the first spatial harmonic only, provided that $(\tilde{\gamma}/\Gamma)G \ll 1$. Atoms with small projected velocities make the main contribution to higher spatial harmonics. Even when $(\tilde{\gamma}/\Gamma)G \ll 1$ for atoms with $v_z \sim \Gamma/k$, all spatial harmonics should be taken into account.[1] In the low-velocity region, the problem can be solved only by use of (2.69,70), with which the contributions of all harmonics can be allowed for. The net result of the problem is obtained by combining the solutions for two velocity ranges.

The contributions of coherence effects are maximum at the line center. With increase of detuning, the relative contribution decreases and tends to zero when $|\Omega| \gg \Gamma_B$. The standing-wave absorption at the center of line is given by

$$\frac{\kappa}{\kappa_0} = (1+2G)^{-1/2} + \frac{\tilde{\gamma}}{2\Gamma}(1+2G)^{-3/2} - \frac{\tilde{\gamma}}{\Gamma} GA(1+2G)^{-2} \ , \qquad (2.126)$$

where

---

[1] For counter-running waves of different intensities the contributions of higher harmonics decrease drastically.

$$A = 1 + \frac{1}{4} P + \frac{11}{96} P^2 + \cdots , \qquad P = \frac{4G^2}{(1+2G)^2} ,$$

which holds with $(\tilde{\gamma}/\Gamma) \cdot G \ll 1$ and $\Gamma_B \ll ku$. The first term in (2.126) determines the main contribution to the absorption coefficient. It differs from that for the case of large detuning by the doubled saturation parameter. The two last terms are conditioned by spatial modulation, i.e., they are bound by spatial harmonics that arise from the difference of populations and polarization of the medium. The second term is related to the contribution of atoms whose velocities $kv \simeq \Gamma_B$. In this velocity range, the amplitudes of the spatial harmonics decrease with the parameter $(\tilde{\gamma}/\Gamma) \cdot G$, and it is necessary to take into account the zeroth and second spatial harmonics of the population difference to solve the problem with the desired accuracy. The third term in (2.126), conditioned by velocities $v \simeq (\tilde{\gamma}/\Gamma)u$, makes a negative contribution to absorption. Its origin is connected with specific features of the interaction between the standing wave field and the atoms. At the nodes, excessive population difference appears, which is equivalent to a decrease of the strong-field absorption.

The quantitative contribution of coherence effects to standing-wave absorption is not large compared to the absorption given by the rate equations. Compare (2.126) with (2.119). However, when the velocity distribution of atoms or the weak-wave absorption in the presence of a standing wave at the same transition is considered, inclusion of coherence effects is of crucial importance and gives qualitatively new corrections.

Exact Solution $(\omega = \omega_0,\ \gamma_1 = \gamma_2 = \Gamma)$

With arbitrary field intensities it is impossible to solve analytically the problem of interaction between a standing-wave field and a gas. The case of exact resonance $(\omega = \omega_0)$ and equal relaxation constants $(\gamma_1 = \gamma_2 = \Gamma)$ is an exception. It has been considered in [2.10,14,17,20]. The absorption in the case of the Doppler limit $(\Gamma_B \ll ku)$ is given by [2.6]

$$\frac{\kappa}{\kappa_0} = \frac{1}{\pi} \int_{-\infty}^{\infty} \frac{dx}{x} (\sin^2 x) \left(1 + \left[1 + 4G \frac{\sin^2 x}{x^2}\right]^{1/2}\right)^{-1} . \qquad (2.127)$$

At low saturation, we have the expansion [2.6]

$$\frac{\kappa}{\kappa_0} = 1 - G + \frac{11}{6} G^2 - \frac{151}{36} G^3 . \qquad (2.128)$$

In strong fields ($G \gg 1$) the absorption coefficient has the form

$$\frac{\kappa}{\kappa_0} = \frac{8}{\pi^2 \sqrt{G}} . \tag{2.129}$$

The exact solution (2.129) and the solution (2.121) derived from rate equations differ by only 15% at high saturation.

Numerical Solution

In papers [2.10] by FELDMAN and FELD and [2.16] by LAMB and STENHOLM, the problem has been solved by the density-matrix method with arbitrary intensities, by use of computers. The interaction energy in (2.69,70) is written in the form

$$\hbar V(z,t) = -P_{12}E_{st} \sin kz \cos\omega t . \tag{2.130}$$

The solution of (2.69,70) can be given in the form of the series

$$\rho_{21}(v,z,t) = e^{-i\omega t}\sum_{n=0}^{\infty}\left[\prod_n^{(+)}(v)\, e^{inkz} + \prod_n^{(-)}(v)\, e^{-inkz}\right] , \tag{2.131}$$

$$\rho_{22}(v,z,t) = \sum_{n=0}^{\infty}[a_n(v)\, e^{inkz} + \text{c.c.}] + \rho_{22}^0 , \tag{2.132}$$

$$\rho_{11}(v,z,t) = \sum_{n=0}^{\infty}[b_n(v)\, e^{inkz} + \text{c.c.}] + \rho_{11}^0 . \tag{2.133}$$

After (2.131-133) are substituted into (2.69,70), it turns out that instead of (2.69) and (2.70) we can obtain recurrence relations, only, for y_n,

$$y_n = \frac{P_{12}E_{st}}{4\hbar R_n}(y_{n+1} - y_{n-1}) , \quad y_o = \frac{P_{12}E_{st}}{2\hbar}\frac{\Gamma}{\gamma_1\gamma_2}(y_1 + y_1^*) + 1 , \tag{2.134}$$

where

$$\frac{1}{R_n} = \begin{cases} \dfrac{1}{\gamma_1 + inkv} + \dfrac{1}{\gamma_2 + inkv} , & n = 0,2,4,\cdots \\[2ex] \dfrac{1}{\Gamma + i(nkv - \Omega)} + \dfrac{1}{\Gamma + i(nkv + \Omega)} , & n = 1,3,5,\cdots \end{cases} \tag{2.135}$$

with which (2.71) and (2.72) can be expressed,

$$P(v,z,t) = -P_{12}N_o \,\mathrm{Re}\left\{e^{i\omega t} \sum_{n=1,3,5\cdots}^{\infty} \left[\left(1 - \frac{i\Omega}{\Gamma + inkv}\right)y_n\, e^{inkz} - \right.\right.$$

$$\left.\left. - \left(1 - \frac{i\Omega}{\Gamma - inkv}\right)y_n^*\, e^{-inkz}\right]\right\} , \tag{2.136}$$

$$n(v,z,t) = N_o\left[y_o + \sum_{n=2,4,6}^{\infty}(y_n\, e^{inkz} + \text{c.c.})\right] . \tag{2.137}$$

By averaging over velocities, we may write the medium polarization in the form

$$P(z,t) = \mathrm{Re} \sum_{n=1,3,5}^{\infty}(\chi_n^{(+)} \sin nkz + i\chi_n^{(-)} \cos nkz)E_{st}\, e^{i\omega t} , \tag{2.138}$$

where $\chi_n^{(\pm)}$ denotes spatial harmonics of medium polarizability. The polarizability averaged over the medium region ($0 \le z \le L$, with $L \gg \lambda$) is determined by

$$P(t) = \frac{1}{L}\int_0^L P(z,t) \sin kz dz = \mathrm{Re}\{\bar{\chi}E_{st}\, e^{i\omega t}\} . \tag{2.139}$$

The imaginary and real parts of $\bar{\chi}$ are

$$\mathrm{Re}\{\bar{\chi}\} = \chi_o\left(\frac{2\hbar ku}{\sqrt{\pi}P_{12}E_{st}}\right)(\omega_o - \omega)\,\mathrm{Re}\int_{-\infty}^{\infty}\frac{W(v)y_1 dv}{\Gamma + ikv} , \tag{2.140}$$

$$\mathrm{Im}\{\bar{\chi}\} = -\chi_o\left(\frac{2\hbar ku}{\sqrt{\pi}P_{12}E_{st}}\right)\mathrm{Re}\int_{-\infty}^{\infty} W(v)y_1 dv , \tag{2.141}$$

where

$$\chi_o = \frac{P_{12}^2 N_o}{\hbar ku}\sqrt{\pi} .$$

A computer solution of the problem has shown that there are no qualitative changes of the Lamb-dip shape, compared with the rate-equations approximation. In the region of strong fields, the dip depth, with equal level-relaxation constants $\gamma_1 = \gamma_2 = \Gamma$, decreases about 20% compared with that calculated from (2.119). Figures 2.6 and 2.7 give the results of numerical calculation of the imaginary and real parts of the medium polarizability with varying

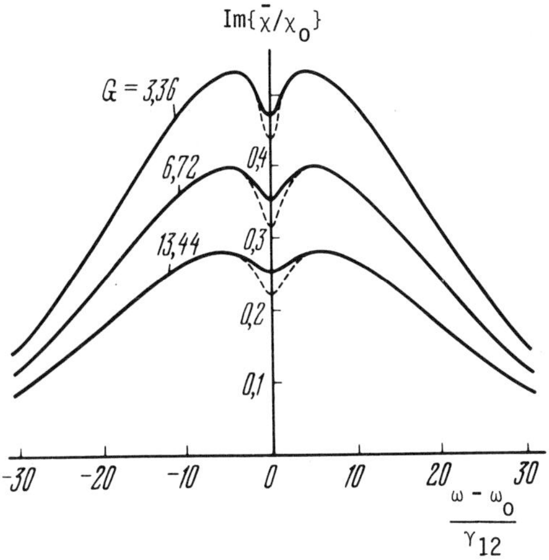

Fig.2.6 Frequency dependence of the imaginary part of the susceptibility $\bar{\chi}/\chi_0$ of the two-level Doppler-broadened transition for various values of standing wave intensity expressed in units of the saturation parameter G. The calculation is performed for the case $ku = 25 \cdot \gamma_{12}$, $\gamma_1 = \gamma_2$, $\gamma_{12} = 1/2(\gamma_1 + \gamma_2)$. The dotted lines show the result of a calculation in the rate-equations approximation (FELDMAN and FELD [2.10])

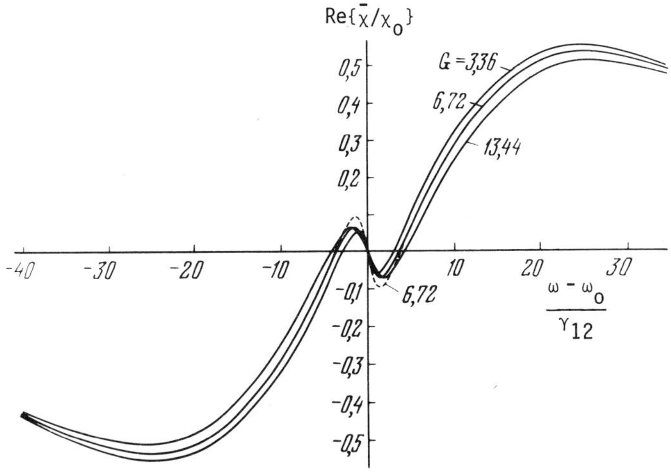

Fig.2.7 Frequency dependence of the real part of the susceptibility $\bar{\chi}/\chi_0$ of the two-level Doppler-broadened transition for the same values of parameters as in Fig.2.6 (FELDMAN and FELD [2.10])

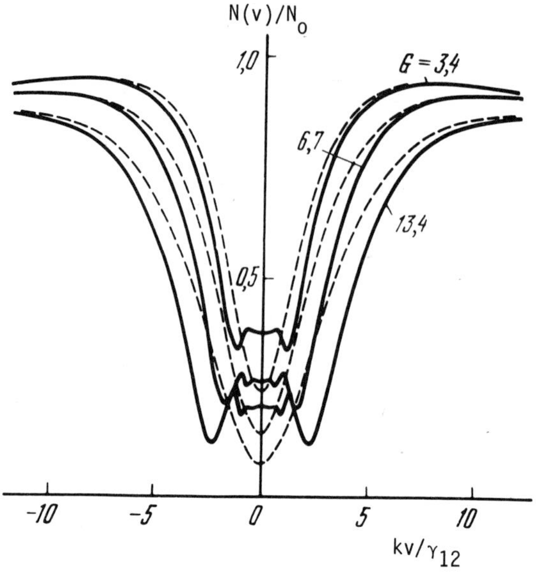

Fig.2.8 Spatially averaged distribution of population difference as a function of the projection of atom velocity on the standing-wave direction with various degrees of saturation. The function was calculated by FELDMAN and FELD [2.10] for the case $ku = 40 \cdot \gamma_{12}$, $\gamma_1 = \gamma_2$, $\omega = \omega_0$. The designations are the same as in Fig.2.6

degree of saturation. The dashed line shows the results of calculations with the rate-equations approximation. The additional structure that appears as a "hole" in the velocity distribution (Fig.2.8) is a qualitatively new phenomenon. The appearance of this structure is explained by the fact that particles with $v_z \ll \Gamma/k$ move almost parallel to the wave surfaces of a standing wave, either in its nodes or in its loops. A considerable fraction of the particles falls within the nodes and is not saturated by the field. At the same time, particles with $v \gtrsim \Gamma/k$ cross several nodes and loops and respond to the field. In this case, a stronger saturation can be attained. When calculating in the approximation of weak saturation, or for arbitrary saturation, in the rate-equations approximation no anomalies occur in the velocity distributions of particles.

2.6 Strong and Weak Counter-Running Waves

Assume that a light field consists of two counter-running waves with the same frequency but with different amplitudes,

$$E(t) = E_0 \cos(\omega t + \underline{k}\underline{r}) + E \cos(\omega t - \underline{k}\underline{r}) \tag{2.142}$$

The running wave with the E amplitude is strong and saturates absorption. The counter-running wave E_0 is weak and causes no saturation. The weak probe wave interacts linearly with molecules within the homogeneous width at the mirror-symmetrical frequency $\omega_0 + (\omega_0 - \omega)$. In this case, the weak-wave transmittance resonantly increases as the wave frequency passes through the center of the Doppler line and interacts with the particles that are saturated by the direct, strong wave.

2.6.1 Level-Population Effects

It is easy to calculate the parameters of the resonance transmission peak in the rate-equations approximation in which only the variation of the velocity distribution of the level population is allowed for and effects of coherent interaction are neglected. The change of velocity distribution of level-population difference $n(\underline{v})$ under the action of the strong wave is given by (2.89). The linear absorption coefficients for the weak counter-running wave will be determined by

$$\kappa(\omega) = \int \sigma(\underline{v},\omega) n(\underline{v}) d\underline{v} \ , \tag{2.143}$$

where $\sigma(\underline{v},\omega)$ is the cross section of induced transition for a particle having the velocity $\underline{v}$ in the field $E_0 \cos(\omega t + \underline{k}\underline{r})$ determined by

$$\sigma(\underline{v},\omega) = \sigma_0 \frac{\Gamma^2}{(\omega - \omega_0 + \underline{k}\underline{v})^2 + \Gamma^2} \ , \tag{2.144}$$

where $\sigma_0 = 4\pi\omega P_{21}^2/c\hbar\Gamma$ is the induced-transition cross section for exact resonance.

Substituting (2.89) for distribution $n(\underline{v})$ in the strong-wave field into (2.143), we have after simple calculations,

$$\kappa(\omega) = \kappa_0(\omega) \left[1 - \left(1 - \frac{1}{\sqrt{1+G}}\right) L\left(2\frac{\omega - \omega_0}{\tilde{\Gamma}}\right)\right] \ , \tag{2.145}$$

where $\Delta\omega$ denotes the half-height dip width determined by

$$\tilde{\Gamma} = \Gamma(1 + \sqrt{1+G}) \ . \tag{2.146}$$

As seen, the dip width equals the half-sum of the hole width $2\Gamma\sqrt{1+G}$ burnt by the strong wave and the homogeneous width 2Γ that corresponds to the range of frequencies with which the weak wave interacts.

2.6.2 Coherence Effects

The absorption-line shape of a weak counter-running wave found in connection with the change of level-population difference is not correct. The strong field that affects the level-population difference changes also the radiated line shape of the atoms and molecules. To allow for both factors, the problem of line shape during interaction of two counter-running waves should be solved rigorously. In this case, the rate equations cannot be used, of course, and it is necessary that density-matrix equations (2.69,70) should be employed. This problem has been solved to a two-level approximation by BAKLANOV and CHEBOTAYEV [2.22] with no limitations on field intensity or relaxation constants. Analogous results for the case of equal relaxation constants for the two levels have been obtained lately by HAROCHE and HARTMANN [2.23].

Let us derive an expression for the weak-wave absorption coefficient in the presence of a strong counter-running wave, according to [2.22]. First, we shall consider the case when a stationary particle is in the field of a strong wave with frequency ω and in the field of a weak wave with frequency ω'. Then the density-matrix (2.69,70) can be written as

$$\frac{\partial\rho_{21}}{\partial t} + (i\omega_0 + \Gamma)\rho_{21} = i(V\, e^{-i\omega t} + V_0\, e^{-i\omega' t})(\rho_{11} - \rho_{22}) \;, \tag{2.147a}$$

$$\frac{\partial\rho_{11}}{\partial t} + \gamma_1(\rho_{11} - \rho_{11}^0) = -i(V\, e^{-i\omega t} + V_0\, e^{-i\omega' t})\rho_{21}^* + \text{c.c.} \;, \tag{2.147b}$$

$$\frac{\partial\rho_{22}}{\partial t} + \gamma_2(\rho_{22} - \rho_{22}^0) = i(V\, e^{-i\omega t} + V_0\, e^{-i\omega' t})\rho_{21}^* + \text{c.c.} \;, \tag{2.147c}$$

where

$$V = P_{21}E/2\hbar \;, \quad V_0 = P_{21}E_0/2\hbar \;.$$

Because the wave with the E_0 amplitude is weak, (2.147) may be solved by use of the perturbation theory, in the form

$$\begin{aligned} \rho_{nn} &= r_n + ic_n V_0\, e^{-i\Delta t} + ic_n^* V_0\, e^{i\Delta t} \;, \\ \rho_{21} &= e^{-i\omega t}(\delta + i\alpha V_0\, e^{-i\Delta t} - i\beta^* V_0\, e^{i\Delta t}) \;, \end{aligned} \tag{2.148}$$

where $\Delta = \omega' - \omega$, $n = 1,2$. For the seven undetermined variables (r_n, c_n, δ, α, β) we can get the following set of algebraic equations by use of the initial set of differential equations (2.147)

$$\gamma_2 r_2 = iV\delta^* - iV\delta + \gamma_2\rho_2^0 ,$$

$$\gamma_1 r_1 = -iV\delta^* + iV\delta + \gamma_1\rho_1^0 ,$$

$$(\Gamma - i\Omega)\delta = iV(r_1 - r_2) ,$$

$$(\gamma_1 - i\Delta)c_1 = -\delta^* + iV\alpha - iV\beta , \tag{2.149}$$

$$(\gamma_2 - i\Delta)c_2 = \delta^* - iV\alpha + iV\beta ,$$

$$(\Gamma - i\Omega - i\Delta)\alpha = iV(c_1 - c_2) + (r_1 - r_2) ,$$

$$(\Gamma + i\Omega - i\Delta)\beta = -iV(c_1 - c_2) ,$$

where $\Omega = \omega - \omega_0$ is the detuning of the strong-wave frequency from the transition frequency ω_0. For further calculations, we need an expression for the value α, which can be derived from (2.149) in the form

$$\alpha = \frac{\rho_{11}^0 - \rho_{22}^0}{\Gamma - i(\Delta + \Omega)}\left(1 - \frac{G\Gamma^2}{\Omega^2 + \Gamma_B^2}\right) +$$

$$+ (\rho_{11}^0 - \rho_{22}^0)|V|^2 \frac{(\Delta + 2i\Gamma)}{\gamma(\Delta)} \frac{(\Omega + i\Gamma)(\Delta - \Omega + i\Gamma)}{(\Delta + \Omega + i\Gamma)(\Omega^2 + \Gamma_B^2)(\Omega^2 + Q^2)} , \tag{2.150}$$

where

$$Q^2 = (\Gamma - i\Delta)\left[\Gamma - i\Delta + \frac{2|V|^2}{\gamma(\Delta)}\right] \tag{2.151}$$

$$\frac{1}{\gamma(\Delta)} = \frac{1}{\gamma_1 - i\Delta} + \frac{1}{\gamma_2 - i\Delta} . \tag{2.152}$$

At low transition saturation ($G \ll 1$) instead of (2.150) we have the simpler expression,

$$\frac{\alpha}{\rho_{11}^{0}-\rho_{22}^{0}} = \frac{1}{\Gamma - i(\Omega+\Delta)}\left(1 - \frac{G\Gamma^2}{\Gamma^2+\Omega^2}\right) - |V|^2 \frac{\Delta + 2i\Gamma}{\gamma(\Delta)(\Delta+\Omega+i\Gamma)^2(\Omega - i\Gamma)} \; . \quad (2.153)$$

Let us consider now a particular case of two counter-running waves of the form of (2.142). For a particle moving in the direction OZ at the velocity v, we should make the substitutions in expression for α (2.150): $\omega \to \omega - kv$ and $\omega' \to \omega + kv$. In this case, according to the general relations, (2.84), (2.86) and (2.73), we have for the weak-wave absorption coefficient $\kappa_{(-)}(\omega)$,

$$\frac{\kappa_{(-)}(\omega)}{\kappa_0} = \frac{k}{\pi} \operatorname{Re} \int_{-\infty}^{\infty} dv \, \frac{\alpha}{(\rho_{11}^{0}-\rho_{22}^{0})} \exp\left(-\frac{v^2}{u^2}\right) \; . \quad (2.154)$$

This expression can be reduced to the sum of the integrals,

$$\frac{\kappa_{(-)}(\omega)}{\kappa_0} = \frac{1}{\pi}\int_{-\infty}^{\infty} dx \, \exp\left[-\left(\frac{x-\Delta/2}{ku}\right)^2\right] \frac{\Gamma}{(x+\Omega_{(+)})^2+\Gamma^2}\left[1 - \frac{G\Gamma^2}{(x-\Omega_{(+)})^2+\Gamma_B^2}\right] -$$

$$- \frac{4|V|^2}{\pi} \operatorname{Im} \int_{-\infty}^{\infty} dx \, \exp\left[-\left(\frac{x-\Delta/2}{ku}\right)^2\right] \frac{(x-\Omega_{(+)} - i\Gamma) f(x)}{[(x-\Omega_{(+)})^2+\Gamma_B^2](x+\Omega_{(+)}+i\Gamma)} \; , \quad (2.155)$$

where

$$f(x) = (3x - \Omega_{(+)} + i\Gamma)(x+i\Gamma)\left(2x + i\,\frac{\gamma_1+\gamma_2}{2}\right)$$

$$\cdot \left[(3x - \Omega_{(+)} + i\Gamma)(x+\Omega_{(+)}+i\Gamma)(2x+i\gamma_1)(2x+i\gamma_2) - \right.$$

$$\left. - 4|V|^2(2x+i\Gamma)\left(2x + i\,\frac{\gamma_1+\gamma_2}{2}\right)\right]^{-1} \; , \quad (2.156)$$

where κ_0 denotes the unsaturated absorption coefficient at the center of the line, $\Omega_{(+)} = \Omega + \Delta/2$.

The first integral in (2.155) gives an expression that corresponds to the rate-equation approximation. The second integral may be regarded as the contribution of coherence effects.

In the Doppler case ($ku \gg \Gamma_B$), we may consider two limiting cases: 1) $\Delta \gg \Gamma_B$, when the two waves do not interact with each other, and 2) $\Delta \lesssim \Gamma_B$, when the weak wave interacts with the same particles with which the strong wave interacts. In the first case, the weak-wave absorption coefficient is evidently equal to the unsaturated absorption coefficient. In the second case, in (2.155) the exponent that describes the Doppler profile of the absorption line can be placed in front of the integral sign; it is assumed therewith that $X = \Omega + \Gamma/2$. After this reduction, the two integrals can be calculated; when the second integral is calculated, it is convenient to close the integration contour in the upper half-plane, because in the upper half-plane the integrand has just one pole. After these calculations, we have

$$\frac{\kappa_{(-)}(\omega)}{\kappa_0} = \exp\left[-\left(\frac{\Omega}{ku}\right)^2\right]\left[1 - \frac{b\tilde{\Gamma}^2}{(\Delta + 2\Omega)^2 + \tilde{\Gamma}^2}\right] + $$

$$+ 4|V|^2 \frac{\Gamma_B - \Gamma}{\Gamma_B} \operatorname{Re}\left[-\frac{f(\Omega + \frac{\Delta}{2} + i\Gamma_B)}{2\Omega + \Delta + i(\Gamma_B + \Gamma)}\right] , \qquad (2.157)$$

where the abbreviations are introduced,

$$b = \frac{G}{1 + G + \sqrt{1+G}} , \quad \tilde{\Gamma} = \Gamma(1 + \sqrt{1+G}) . \qquad (2.158)$$

It is easily seen that the first term is identical with (2.145); hence, the second term describes additively the contribution of coherence effects.

For instance, in a very strong field ($G \gg 1$), at the center of the Doppler contour ($\Omega = 0$) we have

$$\frac{\kappa}{\kappa_0} = \frac{3}{2}\frac{\gamma}{\Gamma}\frac{1}{(3 + \frac{\gamma}{\Gamma})} \exp\left[-\frac{1}{3}\frac{\gamma}{\Gamma}\left(\frac{\Gamma_B}{ku}\right)^2\right] , \qquad (2.159)$$

where

$$\frac{2}{\gamma} = \frac{1}{\gamma_1} + \frac{1}{\gamma_2} .$$

Figure 2.9 shows the absorption-line shape of the weak wave with coherence effects taken into account, which is determined by (2.157). The dashed line indicates the results of calculation to a rate-equations approximation. Coherence effects cause additional broadening of a narrow saturation resonance. Physically, it is caused by coherent oscillations in a two-level system (optical nutations) under a strong field. Figure 2.10 shows the results of calculation for the narrow resonance half-width in probe-wave absorption by use of (2.157) with variable values of the parameter $\rho = \gamma/\Gamma$. The case $\rho = 0$ corresponds to a very great difference between the level-decay rates γ_1 and

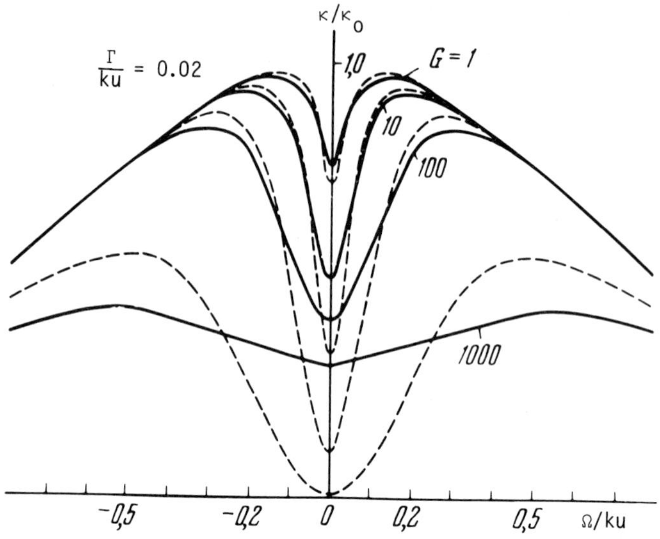

Fig.2.9 Shape of the absorption line for a weak probe wave in the presence of a strong counter-running wave at different degrees of saturation G. The calculation was carried out with the coherence effects (the solid curves, $\gamma = \Gamma$) and rate-equations approximation (the dotted curves, $\gamma/\Gamma \to 0$)

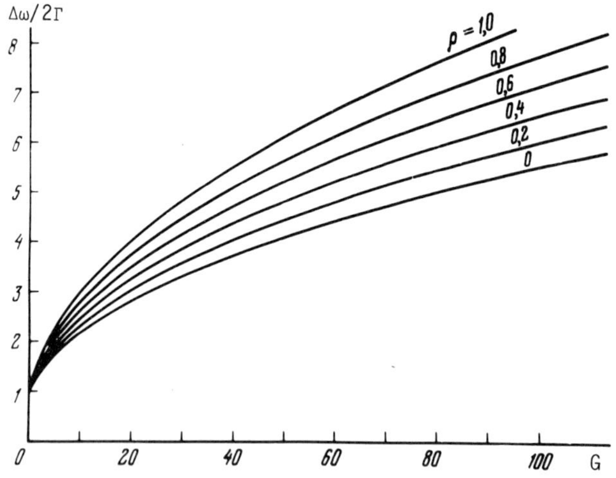

Fig.2.10 Dependence of narrow resonance width $\Delta\omega$ in the weak probe counter-running-wave method on the degree of saturation of absorption by strong running wave at the various values of parameter $\rho = \gamma/\Gamma$ when coherence effects in the strong field are taken into account

γ_2, when the rate-equation approximation holds. The case $\rho = 1$ corresponds to $\gamma_1 = \gamma_2$; coherence effects make the greatest contribution in this case. Thus, an increase of the resonance width, with an increase of the parameter ρ directly explains the contribution of the coherence effects.

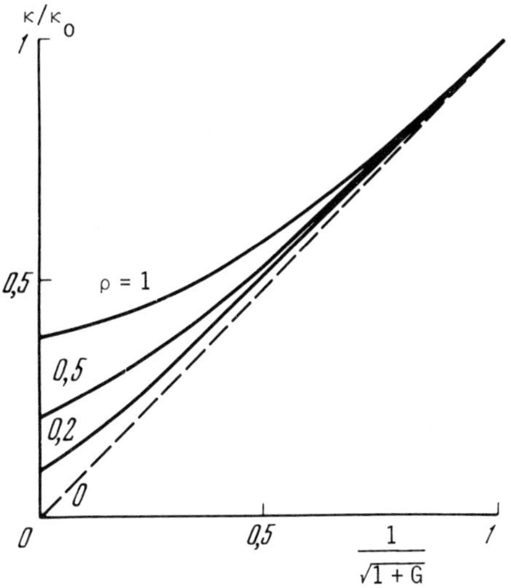

Fig.2.11 Relation between the absorption of the weak wave at the line center $(\omega = \omega_0)$ and the intensity of the strong wave for various values of the parameter $\rho = \gamma/\Gamma$ which characterizes the difference between the constants of level relaxation. The case $\rho = 1$ corresponds to $\gamma_1 = \gamma_2$; $\rho = 0$ corresponds to the greatly different γ_1 and γ_2, which is to the case of incoherent interaction (rate-equations case)

The contribution of coherence effects is proportional to the parameter γ/Γ and appears just in the second order of the saturation parameter. Figure 2.11 shows the dependence of weak-wave absorption at the center of the line on strong-wave intensity which is given by (2.159). It is seen that, first, the weak-wave absorption coefficient is always greater than the saturated absorption of the strong wave. Second, with increase of the strong-field intensity, the weak-wave absorption tends to a constant that is determined by (2.159) and depends only on the relation between the relaxation constants. For example, if the level-relaxation constants γ_1 and γ_2 are equal, which corresponds to $\rho = 1$, the weak-wave absorption tends to the constant $[3/8(\kappa_0)]$. If the relaxation constants differ greatly and there are collisional interruptions of the phase state of an absorbing particle, the contribution of coherence effects is small, and (2.145), derived from the rate equations, may be applied with fair accuracy.

The features of weak-wave absorption considered appear when the Doppler broadening markedly exceeds the homogeneous width Γ. When the Doppler width is comparable to the homogeneous width, one more effect arises, which is well known in microwave spectroscopy: the absorption-frequency dependence may change its sign, that is, in a certain frequency range, instead of absorption, amplification of the counter-running weak wave may occur [2.23].

2.7 Unidirectional Weak and Strong Waves

When unidirectional waves interact in a gas, the absorption-line shape has new features. Apart from a wide "Bennett hole" caused by population-difference decrease, in the line shape additional resonances appear whose widths are equal to the level-decay rate. Resonances that contain information on the level-decay rate γ_1 and γ_2 are characteristic of only unidirectional-wave interactions; they do not arise when counter-running waves interact with gases. Their physical nature can be understood from simple qualitative considerations.

Two unidirectional waves with the close frequencies ω_1 and ω_2 set up at each point a resultant field, the amplitude of which varied with the frequency Δ equal to the frequency difference $|\omega_1 - \omega_2|$. If the field is rather intensive, the level population changes under its action. The time-dependent field amplitude results in level-population modulation. The modulation of population difference leads to common modulation of the absorption coefficient and hence to signal-amplitude modulation. The additional field components that arise due to amplitude modulation at side-band frequencies are summed with the initial fields, which can be regarded as a decrease of their absorption. The depth of population-difference modulation depends on the relation between the modulation frequency Δ and the level-decay rates γ_1 and γ_2. If $\Delta \ll \gamma_1$ and γ_2, the level population "follows" the variation of the resultant field amplitude, and the amplitude-modulation effect is maximum. If $\Delta \gg \gamma_1, \gamma_2$, the medium has no time to respond to the change of the instantaneous value of the field amplitude. In this case, it is only the variation of the average level population under the action of the field. So, the additional resonances are related to the temporal modulation of the level population. In such an interpretation, the method turns out to be very close by its physical nature to the widely used phase method of level-lifetime measurement. In our case, the time variation of level population occurs because of the use of two waves with different frequencies, one of them serving as the level-population probe.

With relaxation constants that differed greatly ($\gamma_1 \gg \gamma_2$) and with a limitation imposed on the strong-wave field ($\gamma G/\Gamma \ll 1$), the absorption (amplification) coefficient of a probe wave was found by RAUTIAN [2.6] by use of equations for probability amplitudes. Probe-signal absorption was found by use of the difference between atomic-transition probabilities when two fields and one strong field were present. The probe-wave absorption coefficient in a two-level gas medium was determined by BAKLANOV and CHEBOTAYEV [2.24] from medium polarizability, without limitations on the field intensity or relaxation constants, in the presence of quenching and phase shifting of states by collisions. The presence of two fields with frequency difference Δ gives rise to a modulation of level population with the frequency Δ. This modulation, in turn, causes the polarization at the frequencies ω and $\omega+\Delta$ to modulate and results in additional polarization at the frequency $\omega-\Delta$. With $\Delta \sim \gamma_1, \gamma_2$, the contribution of this polarization defines the main distinctive features of absorption.

Using (2.148-153) for the state of a two-level particle in the presence of a strong and a weak field, we can evaluate the absorption coefficient $\kappa_{(+)}$ for a weak field at the ω' frequency in the presence of a unidirectional strong field at the frequency ω. For this purpose, we must, in the expression for α (2.150), make the substitutions $\omega \to \omega - kv$ and $\omega' \to \omega' - kv$. Then, for the case of unidirectional waves, we get

$$\kappa_{(+)}(\omega) = \kappa_0 \frac{k}{\pi} \mathrm{Re} \int_{-\infty}^{\infty} dv \frac{\alpha}{(\rho_{11}^0 - \rho_{22}^0)} \exp\left(-\frac{v^2}{u^2}\right) , \tag{2.160}$$

which is quite analogous to that for the previously considered case of counter-running waves. After the variables are substituted $y = kv - \Omega$ the problem reduces to calculating two integrals,

$$\begin{aligned} \frac{\kappa_{(+)}(\omega)}{\kappa_0} &= \frac{1}{\pi} \int_{-\infty}^{\infty} dy \exp\left[-\left(\frac{y+\Omega}{ku}\right)^2\right] \frac{\Gamma}{(y-\Delta)^2+\Gamma^2} \left(1 - \frac{G\Gamma^2}{y^2+\Gamma_B^2}\right) + \\ &+ \frac{\Gamma}{2\pi} \mathrm{Re}\, G\,(\Delta + 2i\Gamma) \frac{\gamma(0)}{\gamma(\Delta)} \int_{-\infty}^{\infty} dy \exp\left[-\left(\frac{y+\Omega}{ku}\right)^2\right] \frac{(y-i\Gamma)(y+\Delta+i\Gamma)}{(y-\Delta-i\Gamma)(y^2+\Gamma_B^2)(y^2+Q^2)} . \end{aligned} \tag{2.161}$$

In the Doppler-limiting case, the exponential factor may be placed in front of the integral sign. After this reduction, the two integrals can be calculated; for the second integral, it is more convenient to close the integration contour in the lower half-plane. As a result, we have

$$\frac{\kappa_{(+)}(\omega)}{\kappa_0} = \exp\left|-\left(\frac{\Omega}{ku}\right)^2\right| \mathrm{Re}\left\{1 - b\frac{(\Gamma_B+\Gamma)^2}{\Delta^2+(\Gamma_B+\Gamma)^2} + G\frac{i\Gamma(\Delta+2i\Gamma)}{2(\Gamma_B^2-\beta^2)}\frac{\gamma(0)}{\gamma(\Delta)}\cdot\right.$$

$$\left.\cdot\left[\frac{\beta+\Gamma}{\beta}\frac{\Delta-i(\beta-\Gamma)}{\Delta+i(\beta+\Gamma)} - \frac{\Gamma_B+\Gamma}{\Gamma_B}\frac{\Delta-i(\Gamma_B-\Gamma)}{\Delta+i(\Gamma_B+\Gamma)}\right]\right\} , \tag{2.162}$$

where the abbreviations from (2.152) and (2.158) are used and some new ones are introduced

$$\beta^2 = (\Gamma - i\Delta)[\Gamma(1 + G\frac{\gamma(0)}{\gamma(\Delta)}) - i\Delta] . \tag{2.163}$$

When the frequencies of the two waves are coincident ($\Delta = 0$), we have

$$\frac{\kappa_{(+)}}{\kappa_0} = [(1+G)^{-1/2} - \frac{G}{2}(1+G)^{-3/2}] \exp\left[-\left(\frac{\Omega}{ku}\right)^2\right] . \tag{2.164}$$

If the absorption saturation by the strong wave is small ($G \ll 1$), in the Doppler-limiting case, from (2.162) we derive a simpler expression, which is, nevertheless, rather cumbersome,

$$\frac{\kappa_{(+)}(\Delta)}{\kappa_0} = \exp\left[-\left(\frac{\Omega}{ku}\right)^2\right]\left\{1 - \frac{G}{2}\frac{(2\Gamma)^2}{(2\Gamma)^2+\Delta^2}\left[1 + \left(\frac{\gamma_1}{\gamma_1^2+\Delta^2} + \frac{\gamma_2}{\gamma_2^2+\Delta^2}\right)\frac{\gamma}{2} + \right.\right.$$

$$\left.\left. + \frac{\gamma}{4\Gamma}\left(\frac{\Delta^2}{\gamma_1^2+\Delta^2} + \frac{\Delta^2}{\gamma_2^2+\Delta^2}\right)\right]\right\} . \tag{2.165}$$

Equation (2.165) describes several dips, the width of which are determined by the relaxation constants γ_1, γ_2, Γ. If there are phase-shifting collisions and $\Gamma \gg \gamma_2 \gg \gamma_1$, the line shape is the sum of three Lorentz dips with half-widths 2Γ, γ_1, γ_2 and depths $G/2$, $(G/2)(\gamma_2/\gamma_1+\gamma_2)$, $(G/2)(\gamma_1/\gamma_1+\gamma_2)$, respectively, on the background of the Doppler profile. This is of particular interest for spectroscopy, because each resonance contains direct information on damping of off-diagonal and diagonal density-matrix elements, that is on line width and level lifetime.

The first two terms in (2.165) describe the absorption line that would be conditioned only by saturation of level-population difference under the strong field. The next terms in (2.165) define the contribution of coherence effects which, in contrast to the case of counter-running waves, occur in the first order of the saturation parameter. With increase of field amplitude,

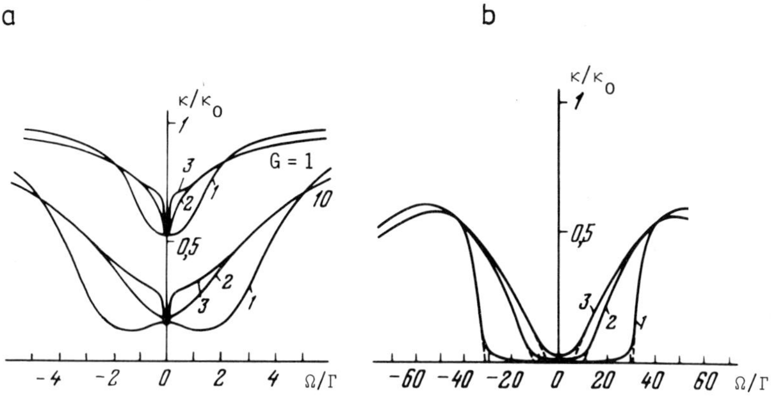

Fig.2.12 Shape of the absorption line of a probe wave in the presence of a strong parallel wave for the saturation parameters $G = 1$ and $G = 10$ (a) and $G = 10^3$ (b). Curves 1 correspond to $\gamma_1/\gamma_2 = 1$; 2 to $\gamma_1/\gamma_2 = 10$; 3 to $\gamma_1/\gamma_2 = 100$; $\gamma/ku = 10^{-2}$

some changes take place in the line shape of weak-wave absorption. The width and depth of the dip in the absorption line, conditioned by changes of level population, increase. The relative width of sharp resonances with the constants γ_1 and γ_2 increases and then begins to decrease. Their width depends in a complicated way on the strong-field amplitude, increasing with decrease of the latter. In very strong fields ($pE \gg \hbar\Gamma$) the fine structure in the line shape of weak-wave absorption practically disappears, and the absorption coefficient in a wide frequency range tends to zero. The absorption-line shape calculated for different relaxation constants is illustrated in Fig.2.12. In strong fields, the expression for the weak-wave absorption coefficient takes a simple (to an accuracy of $1/\sqrt{G}$) form,

$$\frac{\kappa}{\kappa_0} = \begin{cases} 0 & \text{for} \quad \Delta < \frac{pE}{\hbar} \\[2ex] \dfrac{|\Delta| \left[\Delta^2 - \left(\frac{pE}{\hbar}\right)^2\right]^{1/2}}{\Delta^2 + \Gamma_B^2\left(1 - \frac{\gamma}{\Gamma}\right)} & \text{for} \quad \Delta \geq \frac{pE}{\hbar} \; . \end{cases} \tag{2.166}$$

The specific features described can be explained by the Stark effect in a light field at a Doppler-broadened transition.

2.8 Stark Effect in a Light Field at Doppler-Broadened Transitions

For a Doppler-broadened transition, the frequency of the field that acts on a particle depends on its velocity. Therefore the value of Stark splitting in a light field varies for different particles. This, at first sight, might seem to impede observation of the Stark effect in a light field when the Stark splitting is smaller than the Doppler width. The previously described qualitative peculiarities of the line shape of weak-wave absorption when a strong wave is present, however, can be explained by the Stark effect in a high-frequency field [2.22,23].

In a strong light field the "m" and "n" levels are split into two sublevels each,

$$E_m^{(1,2)} = E_m + \frac{\delta}{2} \pm \left[\left(\frac{\delta}{2}\right)^2 + V^2\right]^{1/2} ,$$
$$E_n^{(1,2)} = E_n - \frac{\delta}{2} \pm \left[\left(\frac{\delta}{2}\right)^2 + V^2\right]^{1/2} , \qquad (2.167)$$

where $\delta = \Omega - kv$ is the field-frequency detuning with respect to the resonance frequency for a moving particle.

Under the action of a field an atom acquires three resonant frequencies that correspond to the transitions between the sublevels "m" and "n",

$$\omega_{1,2} = \omega_{mn} + \delta \pm (\delta^2 + 4V^2)^{1/2} \quad ; \quad \omega_3 = \omega_{mn} + \delta \ . \qquad (2.168)$$

The weak counter-running-wave field has the frequency $\omega' = \omega_{mn} + \Omega + kv$ for a particle. The velocities of particles resonantly interacting with the field can be found from the resonance condition $\omega' = \omega_{1,2,3}$,

$$(kv)_{1,2} = -\frac{\Omega}{3} \pm \frac{2}{3}\,(\Omega^2 + 3V^2) \quad ; \quad (kv)_3 = 0 \ . \qquad (2.169)$$

At the center of the line, the main contribution to weak-counter-running-wave absorption is made by atoms the velocities of which satisfy the condition

$$(kv)_{1,2} = \pm \frac{pE}{\hbar\sqrt{3}} \ . \qquad (2.170)$$

The velocity distribution of atoms (in relative units) at the strong field takes the form

$$n(v) = \frac{(kv)^2}{(kv)^2 + \Gamma_B^2} \ , \quad \Gamma_B \ll ku \ . \qquad (2.171)$$

The absorption coefficient is proportional to the number of particles. Taking into account (2.170) and (2.171), it is easy to evaluate the absorption for the weak counter-running wave,

$$\frac{\kappa}{\kappa_0} = 2\,\frac{\gamma}{\Gamma}\left(3+\frac{\gamma}{\Gamma}\right)^{-1} . \qquad (2.172)$$

This result agrees with exact expression (2.159) apart from a numerical coefficient.

The level-splitting models make it possible, now, to understand why the weak-wave absorption tends to a constant as the counter-running-wave intensity increases. With increase of field, the range of velocities of resonant atoms increases, that is the Bennett-hole width in the velocity distribution of atoms builds up, proportionally. Therefore the number of resonant atoms and weak-wave absorption are independent of the field. The relation between the splitting ($pE/\hbar$) and the Bennett-hole width being dependent on the parameter γ/Γ defines the contribution of coherence effects to the absorption of the weak wave in the presence of the strong one. The resonant velocities of the atoms in the case of unidirectional waves can be determined in much the same way. For the detuning Δ, they equal

$$(kv)_{1,2} = \pm[\Delta^2 - (2V)^2]^{1/2} ; \qquad (2.173)$$

with $\Delta < 2V$ there are no resonant atoms. In the frequency range $\Delta < 2V$, the absorption equals zero. It appears only when $\Delta > 2V$.

The peculiarities in the line shape of absorption of a weak signal in the presence of a strong one may be used for determination of level-relaxation constants, direct measurement of the matrix dipole moment P_{12} and studies on the level fine structure.

References

2.1 R. Karplus, J. Schwinger: Phys.Rev. 73, 1020 (1948)
2.2 N.G. Basov, A.M. Prokhorov: Uspekhi Fiz.Nauk 57, 485 (1955)
2.3 A.A. Vuylsteke: *Elements of Maser Theory* (D. Van Nostrand Co., Inc., Princeton, NJ 1960)
2.4 V.M. Fain, Ya.I. Khanin: *Quantum Radiphysics* (Moscow 1965)
2.5 R.H. Pantell, H.E. Puthoff: *Fundamentals of Quantum Electronics* (John Wiley, NY 1969)

2.6 S.G. Rautian: Proc. Levedev Physical Institute 43, 3 (1968)
2.7 W.R. Bennett, Jr.: Phys.Rev. 126, 580 (1962)
2.8 L.D. Landau, E.M. Lifsh'tz: *Quantum Mechanics* (Moscow 1963)
2.9 U. Fano: Rev.Mod.Phys. 29, 74 (1957)
2.10 B.J. Feldman, M.S. Feld: Phys.Rev. A1, 1375 (1970)
2.11 V.N. Fadeev, N.M. Terent'ev: *Tables of Probability Integrals for Complex Variable* (Moscow 1954)
2.12 D.H. Cloze: Phys.Rev. 153, 360 (1967)
2.13 W.E. Lamb, Jr.: Phys.Rev. 134A, 1429 (1964)
2.14 S.G. Rautian, I.I. Sobel'man: Zh.Eksp.i Teor.Fiz. 44, 834 (1963)
2.15 H. Greenstein: Phys.Rev. 175, 438 (1968)
2.16 S. Stenholm, W.E. Lamb, Jr.: Phys.Rev. 181, 618 (1969)
2.17 S. Stenholm: Phys.Rev. B1, 15 (1970)
2.18 H.K. Holt: Phys.Rev. A2, 233 (1970)
2.19 E.V. Baklanov, V.P. Chebotayev: Zh.Eksp.i Teor.Fiz. 62, 541 (1972)
2.20 K. Uehara, K. Shimoda: Japan.J.Appl.Phys. 10, 623 (1971)
2.21 K. Shimoda, K. Uehara: Japan.J.Appl.Phys. 10, 460 (1971)
2.22 E.V. Baklanov, V.P. Chebotayev: Zh.Eksp.i Teor.Fiz. 60, 551 (1971)
2.23 S. Haroche, F. Hartmann: Phys.Rev. 6A, 1280 (1972)
2.24 E.V. Baklanov, V.P. Chebotayev: Zh.Eksp.i Teor.Fiz. 61, 922 (1971)

3. Narrow Saturation Resonances on Doppler-Broadened Transition

Nonlinear narrow resonances at Doppler-broadened transition can arise in a variety of cases. They were observed for the first time when an absorption medium with a Doppler-broadened transition was placed inside a laser cavity [3.1-3]. The scheme of such a version was described in the Introduction (Chap.1, Fig.1.12). This technique is applied when the light field outside the cavity is too weak to saturate absorption and the absorption itself is small. Observation of narrow resonances outside the laser cavity, by use of a probe counter-running wave in particular, is a more flexible method [3.4]. Its scheme is considered also in the Introduction (Fig.1.13). When the absorption medium is outside the cavity, it is convenient to select the desired intensity of each counter-running wave to achieve maximum sensitivity and resolution. The method of probe counter-running wave is the most widely applied method now in high-resolution laser saturation spectroscopy. In a multimode laser, the production of narrow resonances gives rise to resonance effects of wave competition and anticompetition in different modes. This is also of interest for nonlinear spectroscopy. Finally, in a standing light wave a narrow resonance takes place not only for the absorption coefficient but also for the fluorescence intensity of excited particles [3.5,6], that is, for the density of particles excited by the standing light wave. All of these methods of production of narrow saturation resonances at Doppler-broadened transitions are considered in Chapter 3.

3.1 Resonance Effects in a Single-Mode Laser with Saturated Absorption

A saturated-absorption gas laser contains two components in its cavity: an amplifying medium that resonantly interacts with the field (a gas medium usually, but, in principle, it may be any active medium that generates under cw conditions) and a saturated-absorption gas cell. For the resonance effects to arise, the absorption medium should satisfy the following conditions:

1) The absorption line must coincide with the amplification line of the active medium or be inside the amplification line;

2) The absorption line must be broadened inhomogeneously by the Doppler effect;

3) The laser-radiation intensity must be sufficient for absorption saturation.

In this case, at the center of the absorption line a Lamb dip appears; in the effective amplification of two media a resonance peak is formed. The absorption saturation affects appreciably the characteristics of the laser; this results in drastic changes of the properties of the laser radiation. Some new effects can be observed: a resonance peak in the output-power-versus-frequency curve, nonlinear frequency pulling to the center of the absorption line, i.e., frequency self-stabilization, mode self-locking and selection, separate oscillation on clockwise and counterclockwise polarizations in a laser in a magnetic field, output-power pulsations, and various hysteresis-like effects in the laser parameters. All of these effects are of interest for physics and applications and stimulate experimental and theoretical studies.

In this monograph, consideration is given only to problems that are directly related to saturation-resonance effects.

3.1.1 Equations of Single-Mode Laser

The theory of a laser with a saturated-absorption cell is analogous to that of the common low-pressure gas laser. The resulting medium polarization is the sum of polarizations of the amplifying and absorbing media,

$$\underline{P} = \Sigma \underline{P}_f \ , \tag{3.1}$$

where

$$\underline{P}_f(t) = \underline{e}E\chi_f' \cos\omega t - \underline{e}E\chi_f'' \sin\omega t \ ,$$

the index $f = a$ or b, a here and subsequently refers to the amplifying medium and b to the absorbing medium.

The steady-state amplitudes E and the oscillation frequency ω are determined by

$$\begin{aligned} &\mathrm{Im}\{\bar{\chi}(\omega,E)\} = (4\pi Q)^{-1} \ , \\ &\omega - \omega_c = -2\pi\omega \ \mathrm{Re}\{\bar{\chi}(\omega,E)\} \ ; \end{aligned} \tag{3.2}$$

$\bar{\chi}$ means the total susceptibility of the amplifying and absorbing media averaged over the cavity volume.

The reverse influence of polarization of the light field of a cavity mode is described by the standard equation for the electric-field component $\underline{E}(t)$

$$\frac{d^2\underline{E}}{dt} + \frac{\omega}{Q}\frac{d\underline{E}}{dt} + \omega_c^2\underline{E} = -4\pi\left\langle\frac{d^2\underline{P}_a}{dt^2}\right\rangle - 4\pi\left\langle\frac{d^2\underline{P}_b}{dt^2}\right\rangle , \qquad (3.3)$$

where ω_c and Q are the frequency and quality of a cavity mode, and the averaging is over the cavity length. In (3.3), the amplification and losses are supposed to be uniformly distributed in the cavity volume. For most cases, this approximation is rather good, because the amplification and losses for one pass through the cavity are usually small and the transverse intensity distribution of an axial mode is substantial only in the particular case of a low-pressure gas when the free-path length is comparable to the field cross section.

Equation (3.3) for the high-frequency-field component $\underline{E}(t)$ can be reduced to two equations for the slowly varying (envelope) amplitude $E(t)$ and phase $\phi(t)$ of the field,

$$\frac{dE}{dt} + \frac{\omega}{2Q}E = 2\pi\omega E(\chi_a'' + \chi_b'') , \qquad \omega - \omega_c + \frac{d\phi}{dt} = -2\pi\omega(\chi_a' + \chi_b') . \qquad (3.4,5)$$

The field-amplitude equation (3.4) can be reduced to the equation for the radiation intensity of one running wave $P = (c/8\pi\hbar\omega)E^2$

$$\frac{dP}{dt} + \frac{\omega}{Q}P = cP[\kappa_a(P) - \kappa_b(P)] , \qquad (3.6)$$

where κ_a and κ_b denote the coefficients of saturated amplification and absorption per unit length averaged over the cavity length and related to the susceptibility by (2.84). Eq. (3.6) describes the radiation-energy balance in the cavity at the damping rate ω/Q, the rates of saturated amplification of the active medium $c\kappa_a$ and those of saturated absorption $c\kappa_b$. By use of (2.85), the formula for oscillation frequency (3.5) may be expressed in terms of the indexes of refraction of the amplifying and absorbing media averaged over the cavity length,

$$\omega - \omega_c + \frac{d\phi}{dt} = \omega(1 - n_a) + \omega(1 - n_b) . \qquad (3.7)$$

The steady state of the oscillation frequency ($d\phi/dt = 0$) is shifted with respect to the frequency of the empty cavity ω_c, by the change of phase of the field, caused by the refractive indexes of the medium in the region of anomalous dispersion $n_f \neq 1$.

In order to study the properties of the laser radiation, we need to know the dependence of κ_f and n_f $(f = a,b)$ on the field intensity and frequency. These dependences are given by the relations in Section 2.5. In the simplest case, the approximation of weak saturation can be used. However, in many cases the theory must include the condition of strong saturation. This is because some effects appear clearly when the saturation parameters of absorption and amplification differ greatly. Therefore, with small saturation in the amplifying medium, absorption saturation may be large. Besides, some effects, for instance hysteresis-like effects, cannot be described in the weak-saturation approximation; to describe them, we should at least take into account the terms that allow for saturation in the next orders of perturbation theory. It is desirable, of course, to use analytical expressions when analyzing the laser operation. So the use of the approximate methods for evaluating saturated absorption in the strong standing-wave field considered in Section 2.5 proves to be very efficient.

3.1.2 Output Power

Under steady-state operation $(dP/dt = 0)$ the laser output power is determined, according to (3.6), by the condition of effective amplification equality for two media $\alpha_{eff} = \kappa_a - \kappa_b$ and nonsaturable (linear) losses per unit length $\gamma_0 = \omega/Qc$,

$$\alpha_{eff} = \kappa_a(P,\omega) - \kappa_b(P,\omega) = \gamma_0 \ . \tag{3.8}$$

Equation (3.8) determines the dependence of output power $\underset{\sim}{P}$ on laser parameters, provided that the behaviors of the saturated amplification and absorption are known. The functions κ_a and κ_b decrease monotonically with increase of intensity (Sec.2.5). Without the saturated-absorption cell, the solution of (3.8) is very simple and corresponds to the point of intersection of the curve $\kappa_a(P)$ with the line of linear losses γ_0 (Fig.3.1a). The slope of the curve $\kappa_f(P)$ at the point $\underset{\sim}{P} = 0$ is determined by

$$g_f = -\frac{2}{\kappa_f}\frac{d\kappa_f}{dP}\bigg|_{p=0} \ , \tag{3.9}$$

which, according to (2.120), is related to the parameter of amplification (absorption) saturation by one running wave by

$$G_f = g_f P \ . \tag{3.10}$$

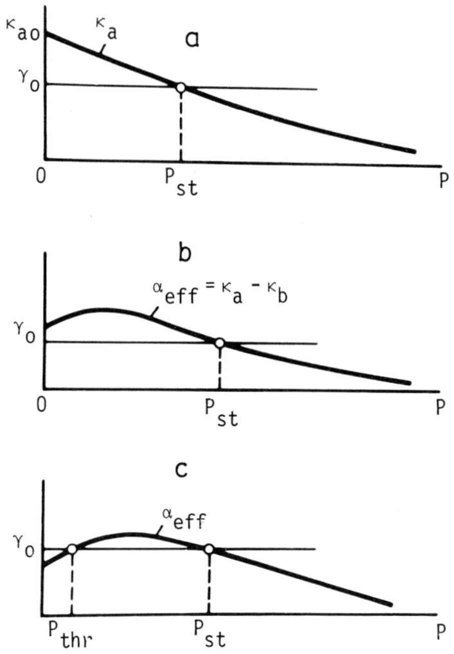

Fig.3.1a-c Possible ways to solve (3.8): a - in the absence of absorption; b,c - in the presence of saturable absorption (b - soft regime of self-excitation, c - hard regime)

With saturable absorption, the dependence of α_{eff} on intensity may vary drastically. If absorption is saturated much earlier than amplification ($g_b \gg g_a$), α_{eff} increases, owing to absorber bleaching, and then decreases monotonically, owing to amplification saturation (Fig.3.1). When the relation between the linear losses γ_0 and initial effective amplification α^0_{eff} is variable, the curve $\alpha_{eff}(P)$ may intersect the straight line γ_0 either at one point ($\alpha^0_{eff} > \gamma_0$) or at two points ($\alpha^{max}_{eff} > \gamma_0 > \alpha^0_{eff}$). In the latter case, the conditions for laser threshold are rigid, that is, the field should have an intensity $P > P_{th}$ for initiation of laser action. If absorption is saturated much later than amplification ($g_b \ll g_a$), saturable absorption has no qualitative effect on the effective amplification-intensity relation. In an intermediate case, the appearance of a maximum of α_{eff} depends on the relation between saturation parameters and the initial values of the amplification and absorption coefficients. These peculiarities of excitation conditions of a saturated-absorption gas laser are typical of all types of lasers with saturable absorbers. For example, the conditions of rigid threshold in lasers were studied as far back as in [3.7]. But the dependence of effective amplification on field frequency is a specific feature for saturated absorption of a gas medium, which determines the most important properties of such a laser.

The power-frequency dependence can be found from (3.8). The main effect is a resonant increase of α_{eff} when the laser frequency ω passes through the center of the absorption line ω_b ($\omega = \omega_b$). This results in a resonant change of the steady-state output power, that is a peak of the laser power output. This is explained, of course, by decreasing saturated absorption at the center of the line.

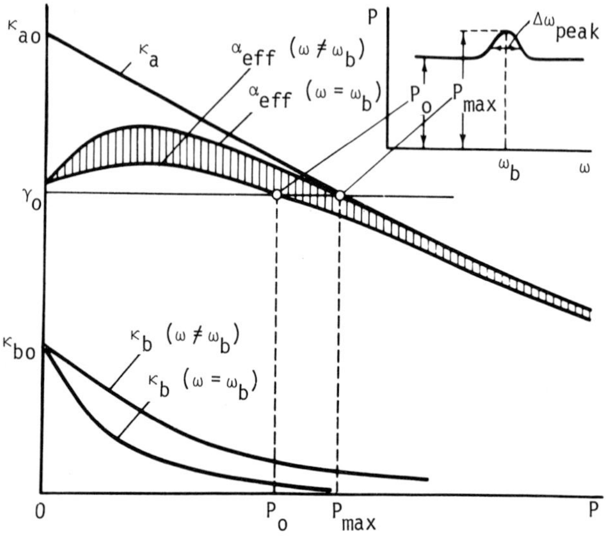

Fig.3.2 Curves of effective amplification at the center of the absorption line and outside the Lamb dip in the absorption line, illustrating the power-peak value for steady-state generation

The power-peak parameters are illustrated in Fig.3.2, which shows the curves of effective amplification at the center of the absorption line ($\omega = \omega_b$) and outside the Lamb dip in the absorption line ($|\omega - \omega_b| \gg \Gamma_b$). If the Lamb dip in the amplifying medium is much wider than in the absorbing medium ($\Gamma_a \gg \Gamma_b$), which is a case of particular interest, the dependence $\kappa_a(P)$ can be regarded as invariable as it passes through the Lamb dip of the absorption line. At the same time, the coefficient of saturated absorption off the resonance in terms of (3.10) in the general case will be

$$\kappa_b = \kappa_{bo}(1 + g_b P)^{-1/2} \ , \qquad |\omega - \omega_b| \gg \Gamma_b \ ; \tag{3.11}$$

when in exact resonance ($\omega = \omega_b$), it can be given to sufficient accuracy by (3.11) with the doubled saturation parameter. This is true in the rate-equation approximation and is accurate to 20% even when coherence effects and spatial inhomogeneity of saturation in a standing wave are allowed for (Subsec.2.5.3). Therefore, the curve $\alpha_{eff}(P)$ in resonance differs from the corresponding curve away from resonance by the doubled compression of $\kappa_b(P)$ through the abscissa axis (Fig.3.2). The output powers in the limiting cases of P_{max} and P_0 are determined by the points of intersection of the $\alpha_{eff}(P)$ curves with the straight line of linear losses γ_0. All of the intermediate curves corresponding to gradual pass through the resonance regions lie within the shaded area.

The power-peak parameters depend in a complicated way on the characteristics of the amplifying and absorbing media, and the peak shape, generally speaking, does not coincide at all with the Lamb-dip shape in the absorption line. This presents certain problems when the power peak is used for spectroscopic measurements. There is a region of parameters where the power peak is of lorentzian shape. When the conditions

$$g_a P, g_b P \ll 1 \ , \tag{3.12}$$

indicating a small degree of saturation for amplification and absorption are met, the output power is determined by [3.2]

$$P(\omega) = \frac{\gamma_0}{g_a \kappa_{ao}} \frac{(\eta - 1)}{(2 - \beta) - \beta L\left(\frac{\omega - \omega_b}{\Gamma_b}\right)} \ , \qquad |\omega - \omega_b| \ll \Gamma_a \ , \tag{3.13}$$

where $L(x) = 1/(1 + x^2)$ is the lorentzian contour, $\eta = \alpha^0_{eff}/\gamma_0$ is the excess of initial effective amplification over linear losses, $\beta = \kappa_{bo} g_b / \kappa_{ao} g_a$ denotes the parameter that characterizes the variation of effective amplification at $P = 0$ (with $\beta > 1$, $d\alpha_{eff}/dP < 0$, and with $\beta < 1$, $d\alpha_{eff}/dP > 0$). The power peak conditioned by the term $L[(\omega - \omega_b)/\Gamma_b]$ in the denominator has a lorentzian shape with the half-height width

$$\Delta\omega_{peak} = 2\Gamma_b \left[\frac{2(1 - \beta)}{2 - \beta}\right]^{1/2} \ . \tag{3.14}$$

Therefore, when experimental data on peak width are analyzed by extrapolation in the field of weak intensities the peak width $\Delta\omega_{peak}$ can be obtained; in the general case, it is not equal to the homogeneous width. They are equal only under the additional condition

$$\beta = \frac{g_b \kappa_{bo}}{g_a \kappa_{ao}} \ll 1 \ . \tag{3.15}$$

This condition can be met only if the absorption in use is very small.

The relative amplitude or the contrast of the peak, that is, the amplitude-to-base ratio

$$h = \frac{1}{P_o}(P_{max} - P_o) \tag{3.16}$$

under condition (3.12) is

$$h = \frac{\beta}{2(1-\beta)} \; . \tag{3.17}$$

So the power-peak shape coincides with the Lamb-dip shape within the limits of peaks with very small contrast. A considerable peak contrast can be obtained by increase of the parameter β, for example, by increase of the absorption coefficient κ_{bo}. In order for the initial effective amplification to exceed the threshold $\eta = \alpha_{eff}/\gamma_o$ (i.e., the conditions of soft self-excitation), κ_{ao} must be increased as κ_{bo} increases. In this case, with $g_b \gg g_a$, the steady-state output power will increase greatly and induce strong saturation of the absorption. Of course, approximation (3.12) then is no longer valid, and it is necessary to use the next-in-intensity terms of the expansion $(gP)^n$ [3.8] of the expression for the absorption coefficient in a strong field, like (3.11).

Figure 3.3 presents dependences of the absolute value of power peak $\Delta P = P_{max} - P_o$ on the initial absorption coefficient κ_{bo} with the amplification coefficient $\kappa_{ao} = 3\gamma_o$ constant, for two values of the parameter g_b/g_a. This figure shows also the dependence of the output power peak P_{max} on the saturation parameter of amplification $G_a = g_a P$. At strong saturation of absorption, the peak contrast falls off, owing to a decrease of the Lamb-dip depth of absorption. The absolute Lamb-dip depth h is

$$h = \kappa_{bo}[(1+G_b)^{-1/2} - (1+2G_b)^{-1/2}] \sim \frac{\kappa_{bo}}{\sqrt{G_b}}\left(1 - \frac{1}{\sqrt{2}}\right) \; . \tag{3.18}$$

It is precisely at large peak contrasts h and strong saturation of absorption that the power-peak shape considerably differs from the lorentzian form. At small saturations and with $\beta \ll 1$, the peak width is narrower than the Lamb-dip width $2\Gamma_b$ in the absorption line, whereas at high saturation the power peak is broadened and its amplitude decreases. In the general case, at strong saturation, the peak broadening due to saturation can be expressed in the form [3.9]

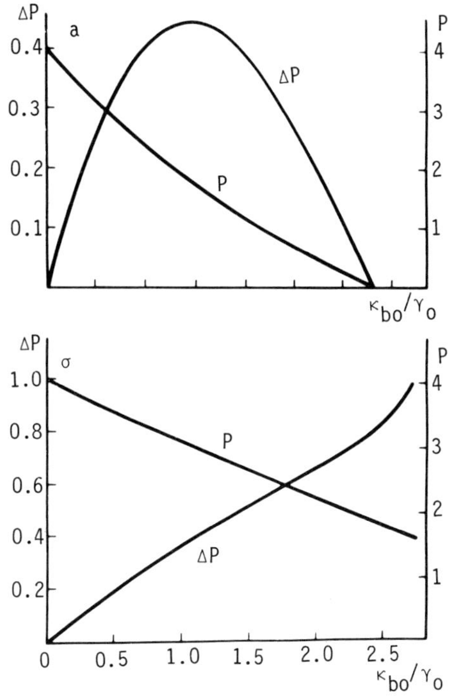

Fig.3.3a,b Relation between the absolute values of the power peak $\Delta P = P_{max} - P_0$ and the output power at maximum P_{max} (in units of $G_a = g_a P$) and the coefficient of initial absorption κ_{bo} (in units γ_0) for cases $g_b/g_a = 1$ (a) and $g_b/g_a = 10$ (b). The calculations were performed for the case $\kappa_{ao} = 3\gamma_0$ (GREENSTEIN [3.9])

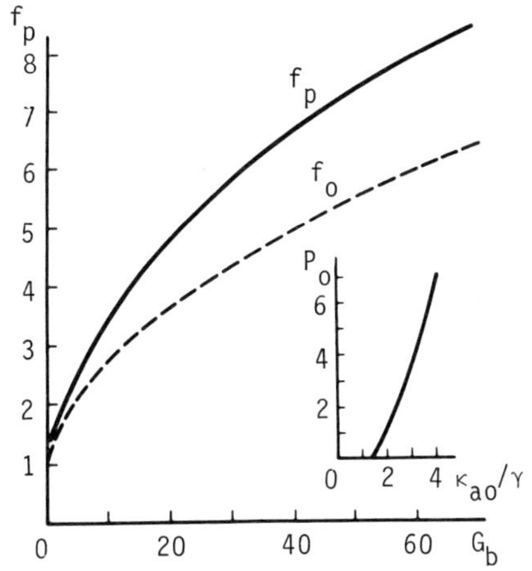

Fig.3.4 Broadening coefficient of the power peak in the strong field. The dotted line shows the coefficient of the field-induced broadening for the dip at the center of the absorption line. The power increase is attained by an increase of the amplification coefficient κ_{ao} with $\kappa_{bo} = 0.5\ \gamma_0$. Parameter $g_b/g_a = 10$ (GREENSTEIN [3.9])

$$\Delta\omega_{peak} = f_p \cdot 2\Gamma_b \, , \tag{3.19}$$

where f_p denotes the broadening factor of power peak. The behavior of this factor is illustrated in Fig.3.4 for the case $g_b/g_a = 10$ and $\kappa_{bo} = 0.5\ \gamma_o$.

Table 3.1 Inverted-Lamb-dip experiments with saturated-absorption cells inside laser cavities

Laser	Wave length	Absorbing particle	References
He-Ne	0.6328 μm	$^{20}Ne^a$	[3.2,3,10-16]
		$^{127}I_2$	[3.17-23]
		$^{129}I_2$	[3.19,23]
		$^{79}Br_2$	[3.24]
He-^{22}Ne	0.6328	$^{22}Ne^a$	[3.25]
		$^{81}Br_2$	[3.24]
He-Ne	1.52	$^{20}Ne^a$	[3.26]
	3.39	$^{12}CH_4$	[3.27-36]
		$^{12}CH_4{}^b$	[3.37]
		$^{12}CH_4{}^c$	[3.38-40]
		$^{12}CH_4{}^d$	[3.41]
He-Ne^b	3.39	$^{13}CH_4$, CH_3OH, C_2H_6	[3.35,42]
		C_2H_4, C_2H_6, C_3H_8	
		C_4H_{10}	[3.43]
		CH_3F^c	[3.44]
He-Xe	3.507	H_2CO	[3.45]
CO_2	10.6	CO_2	[3.46-49]

a - in discharge,
b - in magnetic field,
c - in electric field,
d - in molecular beam.

The first observations of the power peak used a He-Ne laser at $\lambda = 6328$ Å with a Ne nonlinear-absorption cell [3.2,3]. Later, the power peak was observed with different lasers and absorbers, which are summarized in Table 3.1. In these experiments, two types of saturated absorbers were used: 1) atoms and molecules were used in an amplifying medium in which there was absorption instead of amplification; 2) molecules with their absorption frequencies accidentally coincident with the laser amplification line. We will discuss in more detail the results obtained by use of a He-Ne laser at $\lambda = 6328$ Å with a Ne absorption cell and a He-Ne laser at $\lambda = 3.39$ mcm with a CH_4 absorption cell, which are typical of the two classes.

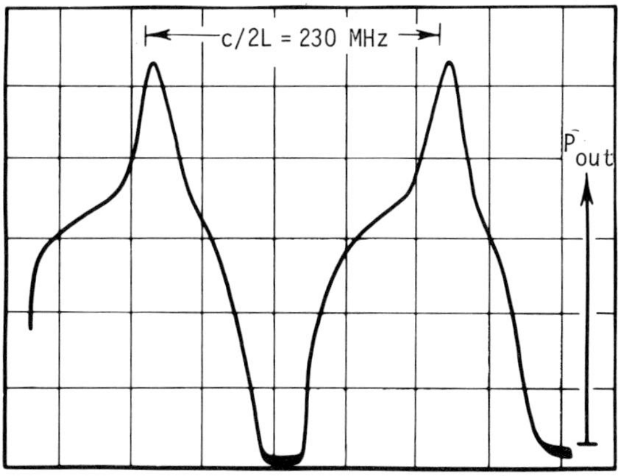

Fig.3.5 Peak of the output power in the He-Ne laser at $\lambda = 6328$ Å with an Ne absorbing cell observed when the frequency of axial modes passes through the center of the absorption line (c/2L is the spacing between the axial modes) (LISITSYN and CHEBOTAYEV [3.2])

He-Ne Laser with Ne Cell

Figure 3.5 shows the output power peak observed in a He-Ne/Ne laser (here and subsequently, this abbreviated designation is used in place of "amplifying medium/absorbing medium"). In a He-Ne/Ne laser at $\lambda = 6328$ Å it is easy to produce an output power peak with the contrast $h = 0.1 - 1.0$. This is because at a small pressure of Ne (0.1 torr) there is marked absorption in Ne, and the homogeneous width of the transition absorption (25 MHz) is much less than in the amplifying medium (Ne, 0.1-0.3 torr; He, 1-3 torr) because there is no contribution of collisional broadening by helium (100-150 MHz/torr He). The difference between the homogeneous widths $2\Gamma_b$ and $2\Gamma_a$ provides a distinct separation of the output power peak and a desired excess of the saturation parameter of absorption g_b over that of amplification g_a ($g_b/g_a = \Gamma_a/\Gamma_b \simeq 10$). The peak width of the laser power output does not coincide with the homogeneous width of the absorption line; it may be either smaller or larger than $2\Gamma_b$. According to the results of independent measurements, the homogeneous width for the 6328 Å line ranges between 40 and 60 MHz. The dependence of radiation-power peak width on different laser parameters was studied experimentally in [3.2,10-15]. The dependence of $\Delta\omega_{peak}$ on the pressure in the absorbing cell, which makes possible determination of the collisional-broadening constant for the absorption transition, is of great interest.

The radiation power peak is shifted with respect to the center of the amplification line (see Fig.3.5) by the amount that is given by

$$\Delta' = (\omega_a - \omega_b) \frac{\kappa_{ao}}{\kappa_{ao} - \kappa_{bo}} , \qquad (3.20)$$

where $(\omega_a - \omega_b)$ is the shift of the center of the amplification line about the absorption line. All shifts are assumed to be less than the Doppler width. At small absorption ($\kappa_{bo} \ll \kappa_{ao}$), the frequency shift of the laser-power peak as a function of pressure makes it possible to measure directly the collisional line shifts ($\Delta'_{Ne-He} = 20 \pm 3$ MHz/torr [3.2,3]).

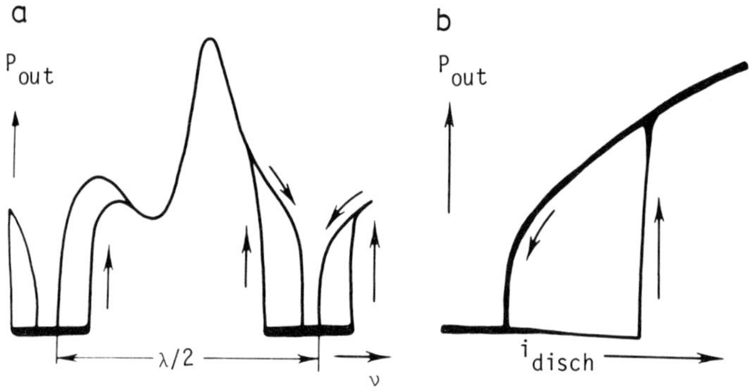

Fig.3.6a,b Peak and hysteresis of generated power as the cavity frequency is varied (a) and hysteresis of the generated power as the discharge current in the amplifying tube is varied (b) (LISITSYN and CHEBOTAYEV [3.55])

In HE-NE/NE lasers, the saturation parameter of neon absorption $G_b = g_b P$ may vary over a very wide range (from 0.1 to 10^2). Because of this, the power peak becomes substantially broadened and practically disappears because of strong saturation. With increase of absorption of the Ne cell in the laser hysteresis effects occur [3.12]: output-power hysteresis as a function of cavity frequency (Fig.3.6a) and output-power hysteresis as a function of discharge current in the amplifying cell, that is amplification coefficient (Fig.3.6b). The hysteresis effects can be explained [3.8] by the use of the effective-amplification curves shown in Fig.3.2.

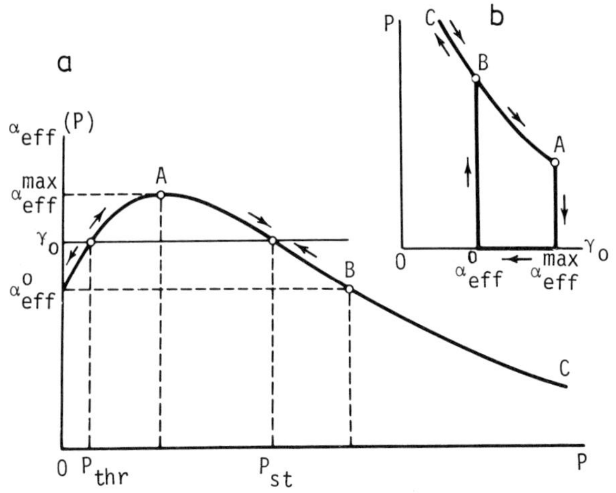

Fig.3.7 Explanations of hysteresis of output power when the linear losses γ_0 are varied over the effective-amplification curve $\alpha_{eff}(P)$

An effective-amplification curve inevitably has a maximum if $d\alpha_{eff}/dP > 0$ at the point $P = 0$. This is possible when the parameter $\beta > 1$. In this case, (3.8) has two solutions (Fig.3.7a). The solution $P_1 = P_{thresh}$ is unstable because an increase of radiation intensity inside the cavity ($P > P_1$) is responsible for further increase of amplification, whereas a decrease of intensity ($P < P_1$) leads to decreasing amplification and further intensity damping. Therefore, the point $P_1 = P_{thresh}$ is the threshold value of the intensity required for laser excitation. The fact that $P_{thresh} > 0$ means a rigid regime of self-excitation. The solution $P_2 = P_{st}$ is stable in the approximation considered and corresponds to steady-state operation. Hysteresis effects occur only when the curve $\alpha_{eff}(P)$ has a maximum, that is $\beta > 1$. In order to gain some insight into these effects, it is sufficient to observe the change of laser power with varying coefficient of linear losses γ_0 (Fig.3.7b). With $\gamma_0 > \alpha_{eff}^{max}$, there is no oscillation, because of a high threshold value; in the region $\alpha_{eff}^0 < \gamma_0 < \alpha_{eff}^{max}$ no laser action occurs, because of the rigid conditions for self-excitation; only at the point $\gamma_0 = \alpha_{eff}^0$ does soft self-excitation take place, which transfers the laser into the B state. With further decrease of losses ($\gamma_0 < \alpha_{eff}^0$), the stationary-oscillation point moves along the section BC of the amplification curve $\alpha_{eff}(P)$. Conversely, if we start with small losses $\gamma_0 < \alpha_{eff}^0$, laser action arises at the section BC and continues, even with increased losses in the region $\alpha_{eff}^0 < \gamma_0 < \alpha_{eff}^{max}$. It is terminated at the A point, with $\gamma_0 = \alpha_{eff}^{max}$. The steady-state laser power in the hysteresis region can be evaluated only with allowance made for the terms of the order of $(g_b P)^2$ and more [3.8].

He-Ne Laser with CH_4 Cell

The laser power peak of a He-Ne/CH_4 laser at $\lambda = 3.39$ mcm was first reported in [3.27]. With a methane pressure of about 10^{-3} torr the half-height width of the power peak was 300 kHz. With specific conditions, that are essential for a minimum width of the power peak to be produced (a low pressure of CH_4 - 0.5 mtorr, and a light beam of diameter $a \simeq 1$ cm), a power peak 50 kHz wide can be obtained [3.30]. Its shape is shown in Fig.3.8. Near threshold a power peak as narrow as about 5 kHz can be observed. According to (3.14), this is because the value of the β parameter approximated unity. The center of the vibrational-rotational absorption line P(7) of the ν_3 band of CH_4 is 50-80 MHz distant from that of the amplification line of the He-Ne laser at $\lambda = 3.39$ μm. Their exact coincidence is best attained by increasing the pressure of He in the amplifying medium [3.27] or by using the isotope Ne^{22} [3.28]. The main feature of the laser power peak is its small width, which is only 0.1-0.3 MHz, whereas the Doppler widths of lines used for Ne amplification and for CH_4 absorption are about 300 MHz. Power-peak narrowing is achieved because of the small homogeneous line width of CH_4 absorption at a pressure between 10^{-2} and 10^{-3} torr. The coefficient of CH_4 absorption at the center of the line is 0.18 cm^{-1} $torr^{-1}$; with the typical cell length of 50 cm, the absorption amounts to 1% per 1 mtorr of CH_4. Under the same conditions, the output-power-peak contrast amounts to several %. But by choice of laser operating conditions (increase of the cell length up to 300 cm, increase of the light-beam diameter inside the cavity, matching of the saturation parameters in the amplifying and absorbing cells in low pressures of methane) increases the power-peak contrast up to 100%; the laser-power output is several mW [3.30]. In He-Ne/CH_4 lasers, the parameter $\beta \ll 1$; therefore, no hysteresis effects can be observed in them.

Because $\beta \ll 1$, the laser power-peak width is equal to the dip width at the center of absorption line $2\Gamma_b$. Power broadening of the peak has been observed, which suggests considerable absorption saturation ($g_b P \simeq 1$). The power peak is broadened with increasing CH_4 pressure, so the collision-broadening constant can be evaluated; it equals 32.6 ± 1.2 MHz $\cdot$ $torr^{-1}$. In some investigations, the power-peak width $\Delta\omega_{peak}$ was assumed, incorrectly, to be equal to the Lamb-dip half-width Γ_b. Therefore the broadening constants were

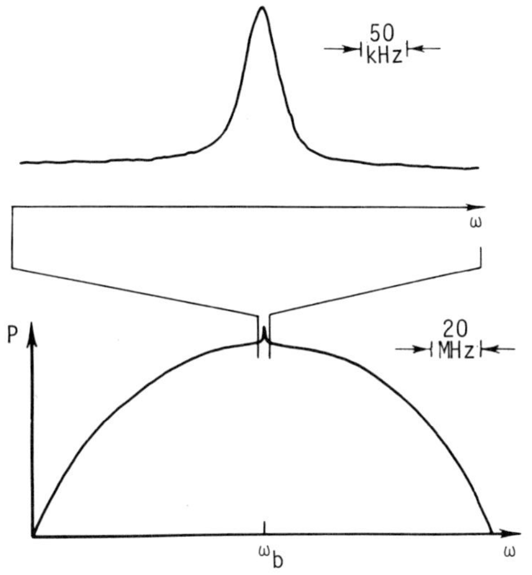

Fig.3.8 Shape of the inverted Lamb dip in He-Ne laser with CH_4 saturable absorption cell. Methane pressure P = 0.5 mtorr, diameter of light beam = 1 cm

underestimated by half.[2] At low pressures (less than 10 mtorr) the collisional-broadening is small, and the contribution of finite time of passage of the particle through the light beam to the Lamb-dip broadening becomes significant. Assume that a light beam has a gaussian profile of intensity over its cross section and the half-height diameter a. The transit half-height line width for molecules that have velocity v and which pass through the center of beam is equal to

$$\Delta\nu_{tr} \simeq \frac{1}{2\pi}\frac{v}{a} . \tag{3.21}$$

For CH_4 molecules, the value of $\Delta\nu_{tr}$ equals 60 kHz with a = 1 cm, if v is taken to be the average thermal velocity of molecules at 300 K $\bar{v} = (8kT/\pi M)^{1/2} = 6.8 \cdot 10^4\ cm \cdot s^{-1}$.

[2] It might seem, at first, that the width of the Lamb dip formed at the center of a spectral line, when two "Bennett holes" of lorentzian form and widths $2\Gamma_b$ overlap each other, would be equal to twice the width of "holes" $2 \cdot (2\Gamma_b)$. Because "hole" overlapping is given mathematically by the convolution of two lorentzian contours, the dip width would be expected to be doubled. But the detuning between two symmetrical "Bennett holes" on the frequency scale is twice the distance of each "hole" from the center of line, that is twice the detuning of the field frequency ω with respect to the center of line ω_b. Therefore, the two "holes" come together twice as fast as the change of the frequency difference $(\omega - \omega_b)$, and the factor 1/2 appears. Quite a different situation arises when two waves are independent and the frequency of only one of the waves varies. In such a case, the dip width in the absorption line would be $2 \cdot (2\Gamma_b)$.

With low gas pressures in the absorbing cell, when the molecules cross the beam almost without collisions, the dip broadening at the center of the absorption line, due to wave-front curvature, becomes significant. This broadening is evaluated by

$$\Delta\omega_{sph} \simeq \frac{ku}{(kr)^{1/2}} , \tag{3.22}$$

where r is the radius of curvature of the wave front. When the laser-output power-peak curve is plotted as a function of the cavity mode frequency ω_c (rather than of the oscillation frequency ω) the power peak in the He-Ne/CH_4 laser is deformed, owing to nonlinear frequency pulling to the center of dip in the absorption line (frequency autostabilization or self-stabilization). In more detail, this effect is considered in Subsection 3.1.3.

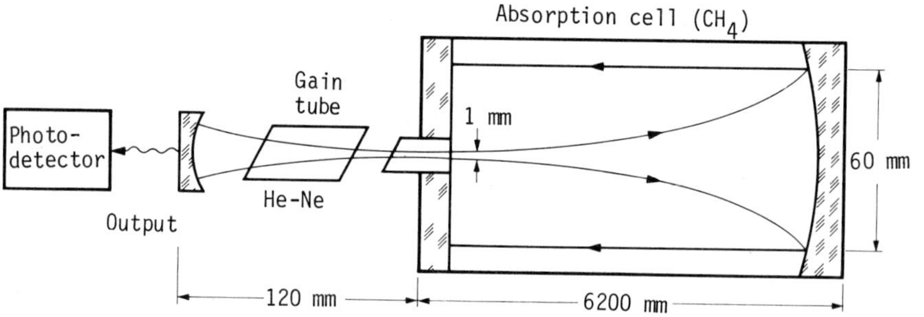

Fig.3.9 Scheme of three-mirror telescopic laser with a methane cell (BAGAYEV et al. [3.50])

Telescopic-cavity lasers (Chap.6) can be used to increase the time of field-molecule interaction and to decrease the diffraction divergence of the light-wave field in the cavity. Figure 3.9 shows one possible version of a telescopic laser described in [3.50]. The laser cavity is made of three coaxial mirrors: short-focus, long-focus and a plane mirror with a hole. The distance between the spherical mirrors is approximately equal to the sum of the focal lengths of these mirrors. The plane mirror is placed near the beam waist in the cavity. The field distribution in such a cavity is illustrated in Fig.3.9. In the narrow section of the light beam an amplifying tube is placed, and a methane absorbing cell is placed in the broad section. The principal merits of the telescopic laser are its simple construction, abscence of astigmatic effect on the intensity distribution over the beam cross section, small losses in the cavity (the diameter of the hole in the plane mirror is much smaller than the diameter of the wide section of the light beam).

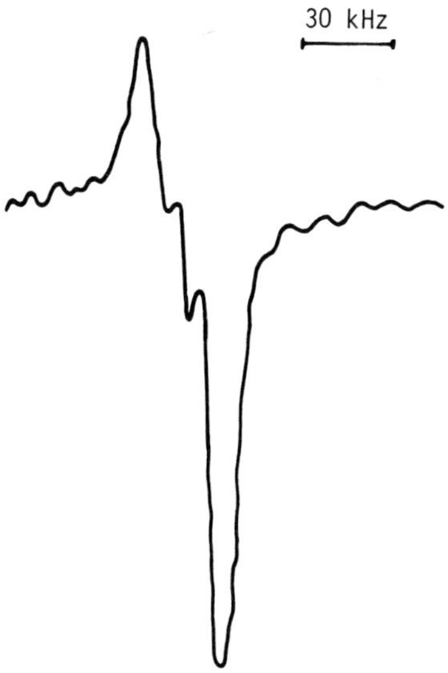

Fig.3.10 Typical record of the first derivative of the inverted Lamb dip in a telescopic He-Ne laser with methane absorber (BAGAEV et al. [3.50])

In experiments with the use of three-mirror telescopic lasers, the radii of curvature of the mirrors were such as to produce a beam in a methane absorbing cell with a cross section of 6 cm. The field cross section in a He-Ne tube was ~1 mm, the hole diameter in the plane mirror was ~3 mm. The methane cell was 1200 cm long. The internal diameter of the absorbing cell was 15 cm. The use of such a telescopic laser made it possible to record the magnetic hyperfine structure in methane at the transition $F_1^{(2)}$ of the line P(7) of the band ν_3. The range of methane pressure in the absorbing cell was from $5 \cdot 10^{-5}$ to $2 \cdot 10^{-4}$ torr. Figure 3.10 shows a characteristic record of the first-derivative curve for the power peak at a pressure $\sim 10^{-4}$ torr. There are well-observed characteristic "resonances" in the first-derivative shape, which correspond to three strong components of the methane magnetic hyperfine structure (see Chap.8). The resonance contrast observed in the telescopic laser was ~1%, and the total peak width, corresponding to the three HFS components, was ~30 kHz. The absence of deeper component resolution was caused by insufficiently high short-term laser frequency stability, which resulted in HFS component broadening.

He-Ne Laser with I_2 Cell

Some investigators tried to find a gas molecular absorber to produce narrow resonances in the visible region of the spectrum, mainly for a He-Ne laser at $\lambda = 6328$ Å. In [3.17,18] they demonstrated narrow resonances in a ^{3}He-^{20}Ne laser with a $^{127}I_2$ saturated absorption cell. Near the line R (127) of the 11-5 band of the $^{127}I_2$ molecule, they observed 14 narrow resonances that corresponded to the hyperfine structure of the electron transition. In [3.19] resonances were observed in a ^{3}He-^{20}Ne laser with a $^{129}I_2$ absorbing cell. In that case a number of resonances were observed; some of them had much larger amplitudes than was the case with the $^{127}I_2$ cell.

Because the absorption of I_2 at $\lambda = 6328$ Å was very small, the power-peak contrast amounted to tenths of per cent. Resonances with such small amplitudes are observed by recording the first derivative of laser output power with respect to oscillation frequency. Several overlapping lines of I_2 absorption, located within the Doppler width of the laser amplification line, decrease the relative resonance intensity at the center of each absorption line. Owing to the small amplitudes of the resonances, the contribution of the first derivative of the curve of the Doppler-contour shape is dominant. In order to eliminate the Doppler-profile effect, the third derivative may be recorded, that is, the modulation of the laser output power at the third harmonic of the modulation frequency. In this case the signal that results from a narrow resonance dip increases about $(\Delta\omega_D/\Gamma_b)^2$ times as much as the signal from the Doppler profile. This can be clearly seen in Fig.3.11, which shows the signals of output-power modulation of a He-Ne laser with a $^{127}I_2$ cell at the first and third harmonics of modulation frequency.

Another method for increasing the resonance amplitude was described in [3.51]; a delay line inside the cavity increased the optical-beam path in the cell and the resonance contrast by approximately one order of magnitude. In a He-Ne/$^{127}I_2$ laser with such an absorber the resonance contrast was from 0.2 to 0.3.

Not every absorbing gas inserted in the laser cavity is suitable for production of saturation resonances. The saturation parameter of absorption $G_b = g_bP$ should be of the same order as or higher than that of amplification $G_a = g_aP$. When they differ greatly ($g_b \gg g_a$), too great saturation of absorption occurs because of which, according to (3.18), the Lamb dip and hence the output-power peak disappear. CO_2 lasers with SF_6 saturated-absorption cell inside the cavity are a typical example of such an unsuitable amplifier-absorber pair. Another unfavorable factor for production of Lamb dips is a long life of molecules in the excited state within the laser light beam. Such a situation may arise with molecules that have long times of vibrational relaxation in some pressure range for which the free path of molecules, due to collisions, is much shorter than the laser-beam diameter. In this case accumulation of excited molecules with different velocities and strong

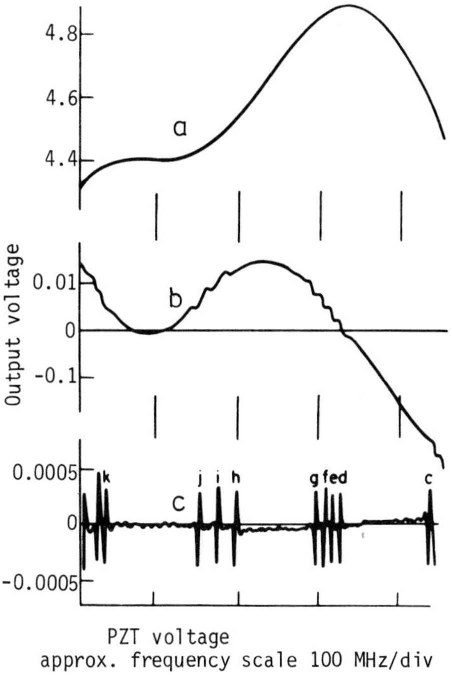

Fig.3.11 Power peaks observed in the He-Ne/I_2 laser by recording the first (b) and third (c) derivatives of the signal. The steady-state signal (a) of the output power is given for comparison (HANES et al. [3.21])

saturation of the whole vibrational band occur. Homogeneous saturation appears to be predominant in this case. As a result, the Lamb-dip amplitude decreases sharply and the power peak practically disappears, despite the fact that the homogeneous width of the absorption line is much smaller than the Doppler width. In such a laser, it is difficult to observe the power peak under cw. It can appear under transient conditions [3.46], with fast switching of the laser, so that excited particles are accumulated and diffused in the velocity space.

3.1.3 Frequency and Spectrum

Under steady-state conditions, the oscillation frequency is given by (3.5) or (3.7) with $d\phi/dt = 0$ and $E = E_0$. In order to understand the effect of saturated absorption on oscillation frequency, we should consider the frequency behaviour in a gas laser with an inhomogeneously broadened amplification line, without absorber. The oscillation frequency, in this case, is determined by a formula derived from (3.5),

$$\frac{\omega - \omega_c}{\omega} = -2\pi\chi'_a(\omega) \ . \tag{3.23}$$

When the degree of amplification saturation is small ($G_a \ll 1$),

$$\chi'_a(\omega) = \frac{\kappa_a(\omega)}{2\pi k} \frac{(\omega - \omega_a)}{ku_a\sqrt{\pi}} \left[1 - G_a \frac{ku_a}{\Gamma_a} L\left(\frac{\omega - \omega_a}{\Gamma_a}\right)\right] \ , \tag{3.24}$$

or

$$\chi'_a(\omega) = \frac{(\omega - \omega_a)}{2\pi\omega} \left[p_a - q_a L\left(\frac{\omega_a - \omega}{\Gamma_a}\right)\right] \ , \tag{3.25}$$

where $\kappa(\omega) = \kappa_{ao} \exp[-(\omega - \omega_a/ku_a)^2]$ is the shape of Doppler amplification line, $2\Gamma_a$ is the homogeneous width of the amplification line ($2\Gamma_a \ll ku_a$), ω_a is the center of the amplification line, p_a and q_a are the linear and nonlinear frequency factors

$$p_a = \frac{\kappa_a c}{\sqrt{\pi} ku_a} \ , \qquad q_a = G_a \frac{\kappa_a c}{\Gamma_a} \ , \tag{3.26}$$

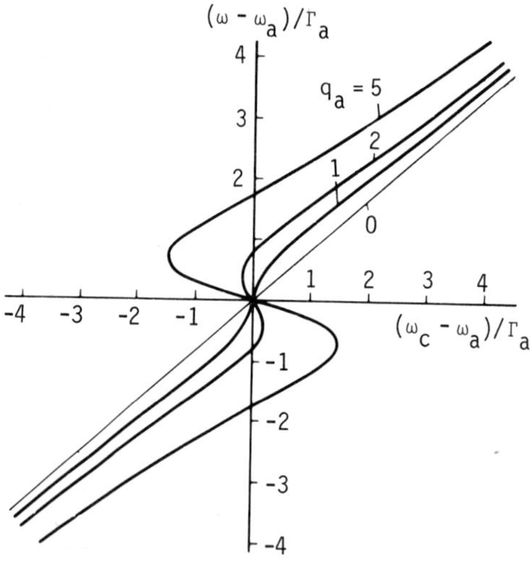

Fig.3.12 Dependence of oscillation frequency of gas laser ω on detuning of the cavity frequency ω_c in the vicinity of the center of the Doppler amplification line ω_a (parameter $p_a = 0.2$)

and G_a is the degree of amplification saturation for each running wave. The typical shape of the curve $\chi'_a(\omega)$ is illustrated in Fig.3.12. The distortion of $\chi_a'(\omega)$ by the field is a result of a narrow dip formed at the center of the amplification line. It should be noted that, even at low saturation, the frequency dependence of $\chi_a'(\omega)$ may vary greatly, for because of the relation $ku_a \gg \Gamma_a$ the q_a factor may be comparable to or higher than P_a, despite the condition $G_a \ll 1$.

The points of intersection of the curve $2\pi\chi_a'(\omega)$ with the line $(\omega - \omega_c)/\omega$ determine the steady-state oscillation frequency. First, the oscillation frequency is pulled towards the center of the amplification line; the value of linear pulling (with $G_a = 0$) is determined by the factor P_a. Second, during saturation, the inclination of the curve $\chi_a'(\omega)$ near ω_a decreases and even changes its sign with $q_a \sim p_a$. In the latter case, the oscillation frequency is "pushed" by the amplification line. The dependence of the laser frequency ω on detuning of the cavity frequency ω_c about the center of amplification line ω_a with a varying degree of nonlinearity, which illustrates transition from linear frequency pulling to nonlinear pushing, is given in Fig.3.12. These dependences are obtained from (3.23).

A saturable absorber drastically changes the frequency dependence in this manner. First, because of the change of the sign of $\chi_b'(\omega)$ linear pulling is replaced by linear pushing from the center of the absorption line ω_b. Second, nonlinear pushing is replaced by nonlinear pulling towards the center of the absorption line ω_b. With low saturation of absorption (3.25) may be used for $\chi_b'(\omega)$, in which the amplification coefficient $\kappa_a(\omega)$ should be replaced by the absorption coefficient $-\kappa_b(\omega)$. The laser frequency, in this approximation, is given by

$$(\omega - \omega_c) = (\omega_a - \omega)\left[p_a - q_a L\left(\frac{\omega_a - \omega}{\Gamma_a}\right)\right] - (\omega_b - \omega)\left[p_b - q_b L\left(\frac{\omega_b - \omega}{\Gamma_b}\right)\right] \tag{3.27}$$

where

$$p_b = (\kappa_b c)/(\sqrt{\pi} k u_b) \ , \quad q_b = G_b \kappa_b c/\Gamma_b \ . \tag{3.28}$$

If $q_b \gg q_a$, which is quite possible with $\Gamma_b \ll \Gamma_a$, the main nonlinear effect is nonlinear frequency pulling towards the center of the absorption line, which is called frequency autostabilization [3.1,52]. Figure 3.13 shows the dependences of oscillation frequency on detuning of the cavity frequency ω_c about the center of the absorption line ω_b with various values of parameter q_b, for the case $\omega_a = \omega_b$ and $q_a \ll 1$. The effect of autostabilization (self-

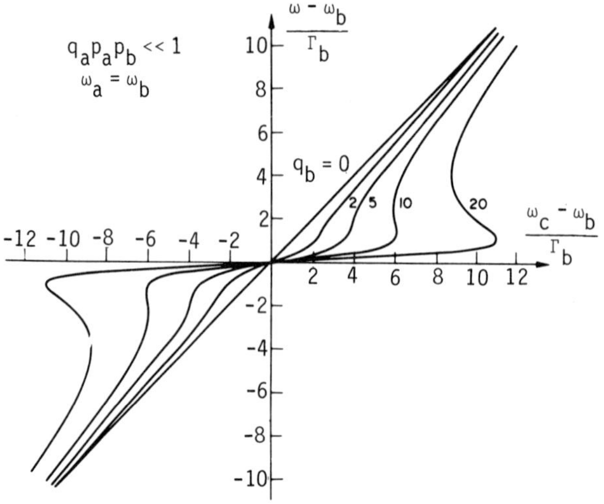

Fig.3.13 Ratio between oscillation frequency ω of a gas laser with saturable absorption and the cavity frequency ω_c in the vicinity of the center of the absorption line (parameters: $\omega_a = \omega_b$; q_a, p_a, $p_b \ll 1$)

stabilization) depends on the slope of the curve at the point $\omega_c = \omega_b$ which is given by

$$\frac{\omega - \omega_b}{\omega_c - \omega_b} = [1 + (q_b - p_b) - (q_a - p_a)]^{-1} = \frac{1}{1+S} , \qquad (3.29)$$

where $S = (q_b - p_b) - (q_a - p_a)$ is the autostabilization factor. Frequency hysteresis is a characteristic feature of frequency dependence when the autostabilization factor is considerable. In this case, the cavity frequency changes by a value that is much greater than Γ_b, and nonlinear pulling will hold the laser frequency within the narrow dip in the absorption line.

The effect of frequency autostabilization was experimentally observed in [3.27] using a He-Ne/CH_4 laser at $\lambda = 3.39$ µm. The autostabilization factor was $S \simeq 2$. The experimental dependence of laser frequency on cavity frequency detuning, which illustrates the effect of autostabilization in a He-Ne/CH_4 laser is shown in Fig.3.14.

To obtain a maximum factor of autostabilization, it is necessary that, first, the Lamb-dip effect in the amplification line should be eliminated (p_a, $q_a \ll 1$), and second, a narrow possible dip in the absorption line ($\Gamma_b \ll ku_b$) and a sufficient degree of absorption saturation G_b should be used, which enable the condition $q_b \gg p_b$ to be fulfilled. Then, the autostabilization factor S is practically equal to the coefficient of nonlinear

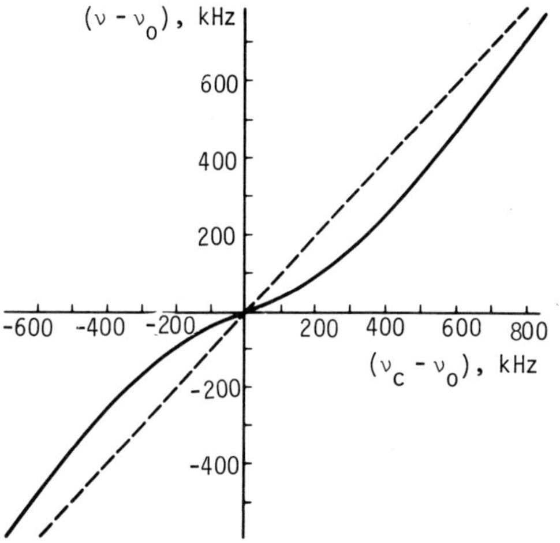

Fig.3.14 Experimental dependence of oscillation frequency of He-Ne laser with CH_4 saturable absorption cell. The dotted line is for the laser without an absorbing cell (BAGAEV et al. [3.56])

frequency pulling q_b. The magnitude of q_b, according to (3.28), depends on many factors. When calculating the coefficient q_b with $G_b \ll 1$ we can use the expression

$$q_b = G_b \frac{\tilde{\kappa}_b pc}{2\pi(\Gamma_b + \tilde{\Gamma}_b p)} \frac{l_b}{L} , \tag{3.30}$$

where $\tilde{\kappa}_b$ is the coefficient of absorption per unit length and pressure; $2\Gamma_b$ is the homogeneous width (Hz) without collision broadening; $2\tilde{\Gamma}_b$ is the coefficient of homogeneous-width increase due to pressure ($Hz \cdot torr^{-1}$); P is the gas pressure in the absorbing cell (torr); l_b is the absorbing-cell length, and L is the cavity length. With increase of pressure, when the dip width depends on collisional broadening, the coefficient q_b tends to a maximum possible value

$$q_b^{max} = G_b \frac{\tilde{\kappa}_b c}{2\pi\tilde{\Gamma}_b} \frac{l_b}{L} . \tag{3.31}$$

For example, for He-Ne/CH_4 lasers the value of $\tilde{\kappa}_b$ equals 0.18 cm^{-1} $torr^{-1}$, $2\tilde{\Gamma}_b = 32$ mHz $\cdot$ $torr^{-1}$ for CH_4 and $l_b \simeq L/2$. For such lasers $q_b^{max} = 13\ G_b$. The maximum for the degree of absorption saturation G_b, which can be substituted into (3.30) and (3.31) varies from 0.3 to 0.5; hence $q_b^{max} \simeq 4$-6. Thus, the evaluation by (3.35) explains experimentally obtained values of $S = q_b^{max} = 4$-6.

It follows from (3.31) that, in order to obtain a maximum factor of autostabilization, absorbing gases that have the maximum values for the ratio $\tilde{\kappa}_b/\tilde{\Gamma}_b$ should be used. This ratio is hundreds of times smaller for molecular than for atomic gases, because of distribution of molecules over many rotational sublevels. Therefore, the effect of laser-frequency autostabilization can be used to stabilize the frequency of tunable dye cw lasers by use of absorption lines of atomic transitions. The autostabilization effect must increase by S times the short-term frequency stability of such a laser, that is the laser-oscillation line width.

With increase of the degree of absorption saturation $G_b \gg 1$ the frequency pulling decreases. This follows from (3.27), which is true only with $G_b < 1$, but can be understood qualitatively from the following considerations. At strong saturation, the Lamb-dip broadening by a strong light field becomes significant. The dip broadening decreases the steepness of the dispersion curve of saturated absorption and the value of nonlinear laser frequency pulling. Therefore, the magnitude of G_b that is optimal for autostabilization lies in the range of $G_b \sim 1$. With the degree of saturation arbitrary, the expressions for the nonlinear coefficients q_a and q_b have the form [3.9]

$$q_f = \frac{\kappa_f c}{\Gamma_f} \frac{(1+2G_f)^{1/2} - 1}{(1+2G_f)^{1/2}[1+(1+2G_f)^{1/2}]} = g_f^0 \frac{\kappa_f c}{\Gamma_f} . \quad (3.32)$$

With $G_f \ll 1$, this is transformed to (3.26) and (3.28). The coefficient g_f^0 reaches its maximum, $(3+2\sqrt{2})^{-1} = 0.17$, when $G_b = 1+\sqrt{2}$.

It is possible, in principle, to construct a laser with what is termed absolute frequency autostabilization [3.4] when nonlinear frequency pulling into the peak of effective amplification remains so that the cavity frequency is detuned by a frequency equal to the spacing between the axial modes of the Fabry-Perot cavity. Such an effect can be more easily calculated for a laser with a quasi-travelling wave in the cavity when one running wave in the laser is strong and the other is weak for the nonlinear absorber. Such conditions can be realized only in a laser with large amplification when one of the mirrors has a small reflection coefficient and an absorbing medium can be placed at this mirror. This regime is of interest because, first, it allows production of laser-output-power peaks with a great contrast even with a considerable difference of the saturation parameters of the amplifying and absorbing media G_a and G_b and, in particular, with strong saturation of absorption of one running wave [3.53]. Second, in this case it is possible to obtain considerable autostabilization factors S and even to reach absolute frequency autostabilization. Note that the latter effect can be successfully used to stabilize the frequency of tunable dye lasers without any servo system for frequency control.

Let us consider now the spectrum of a gas-laser output with an intracavity saturated-absorption cell. When the effect of the active medium is neglected, the frequency of a laser with saturable absorption is given by (3.27),

$$\frac{\omega_c - \omega}{\omega - \omega_b} = G_b \frac{\kappa_{bo} c}{\Gamma_b} \frac{1}{1 + x^2} , \quad G_b \ll 1 , \tag{3.33}$$

where $X = (\omega - \omega_b)/\Gamma_b$ is the relative detuning of the oscillation frequency from the center of the absorption line.

Let the laser-cavity frequency ω_c be subjected to random perturbation, so that the spectral density of fluctuations is given by $P(\omega_c)$. Assume that the characteristic fluctuation frequency of the cavity frequency ω_c is much smaller than the homogeneous line width and, hence, that the steady-state equation (3.33) is true for the instantaneous oscillation frequency. Then the laser-frequency fluctuation distribution $P(\omega)$ is determined by

$$P(\omega) = P(\omega_c) \frac{d\omega_c}{d\omega} , \tag{3.34}$$

where $d\omega_c/d\omega$ can be found from (3.33). Carrying out simple calculations, we get

$$P(\omega) = P(\omega_c) \left(1 + q_b \frac{1 - x^2}{(1 + x^2)^2}\right) . \tag{3.35}$$

When the width of the distribution of $P(\omega_c)$ is much larger than Γ_b, the spectral density of the laser-frequency fluctuations near ω_b has a resonant peak the parameters of which depend on the properties of the saturated absorption medium. The resonance shape is shown in Fig.3.15.

The physical cause of resonance is simple. Nonlinear absorption has an "autostabilization" effect on the laser frequency. That is, it initiates nonlinear pulling of the laser frequency to the center of the absorption line. This effect results because the cavity frequency may change by a large value while nonlinear pulling holds the oscillation frequency near the center of the absorption line. This means that with permanent disturbances of the

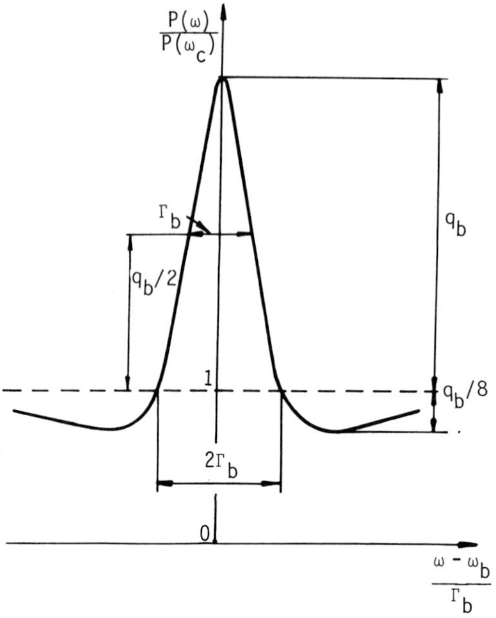

Fig.3.15 Resonance shape of the spectral distribution of oscillation frequency of laser with saturable absorption cell. $P(\omega_c)$ is the spectral density of cavity-frequency fluctuations

laser frequency, the value of which is independent of the frequency position about the line center, the probability of the laser frequency at the center of the absorption line increases. This is equivalent to an increase of the spectral radiation density near ω_b.

A resonance in the spectral-frequency distribution has some properties of particular importance for narrow-resonance detection. Its width is approximately equal to the homogeneous width of the absorption line. Its amplitude and contrast are fully determined by q_b, that is, by the nonlinear-pulling factor. The maximal resonance amplitude depends on the ratio κ_{bo}/Γ_b only, since by choosing a value of the field we can ensure an optimal saturation parameter $G_b \simeq 1$ at any Γ_b ((3.35) holds for $G_b \ll 1$, but when working to an accuracy of 50% we may use it for $G_b \lesssim 1$). Both κ_{bo} and Γ_b are proportional to the atom density. That is why the amplitude of this resonance does not depend on gas pressure and depends only slightly on the absorption-cell length. This important feature of the resonance in the spectrum of the output radiation is of particular importance in detecting supernarrow resonances at very low gas pressures ($\sim 10^{-6}$ torr) when the intensity of laser-power resonances becomes very small.

The laser-radiation spectrum can be studied by use of the usual heterodyne with a reference laser. For such measurements, the line width of the reference laser should be much narrower than the spectral-resonance width Γ_b and the cavity-frequency perturbation should be such that the spectrum of laser-frequency oscillation can be considerably narrower than the homogeneous width $2\Gamma_b$. Under laboratory conditions, when care is taken to prevent mechanical and acoustic perturbations, the laser line width is usually 10^3 to 10^4 Hz. Because of this, when a spectral resonance with width 10^4 to 10^5 Hz, the cavity frequency should be artificially perturbed by applying, for example, alternating voltage to a piezoceramic on which one of the mirrors is mounted. The value and nature of the perturbation are such that the spectral density of the laser radiation in the absence of absorption is a smooth function of frequency, and the spectrum width should be larger than the homogeneous width of the absorption line.

This method of detection can be modified by measuring the alternating component of the spectrum during the artificial periodic modulation of the cavity frequency and by scanning the laser frequency slowly near the absorption-line center. The method was used to observe spectral resonances in a He-Ne laser with a methane absorber at $\lambda = 3.39$ μm [3.54]. The laser setup consisted of two He-Ne lasers with methane absorption cells and a He-Ne laser heterodyne. The frequency of one of the lasers (the reference laser) was stabilized with saturation resonance of the power in methane. The frequency discriminator of the second laser ω_2 was fed with a perturbation signal of 700 Hz, which changed the oscillation frequency of the laser under study. The value of the change was kept constant at 10 kHz. The difference between the frequency of the laser under study and the frequency of the heterodyne laser was measured. Figure 3.16 shows a typical record. The resonance half-width is ~100 kHz at $P_{CH_4} = 3 \cdot 10^{-3}$ torr. The gas-pressure variation in the cell from $3 \cdot 10^{-3}$ to 10^{-2} torr has almost no effect on the resonance amplitude. The spectral resonance shape is given rather accurately by (3.35).

Frequency resonance can be applied to superhigh-resolution spectroscopy in low-pressure gases when the laser-radiation-spectrum width is larger than the homogeneous width of the absorption line. This is quite impossible by detection of the inverted Lamb dip.

3.2 Resonances in Absorption Saturation by Independent Waves

In order to observe resonance effects with a saturable absorber inside a laser cavity, it is necessary that special saturation parameters and

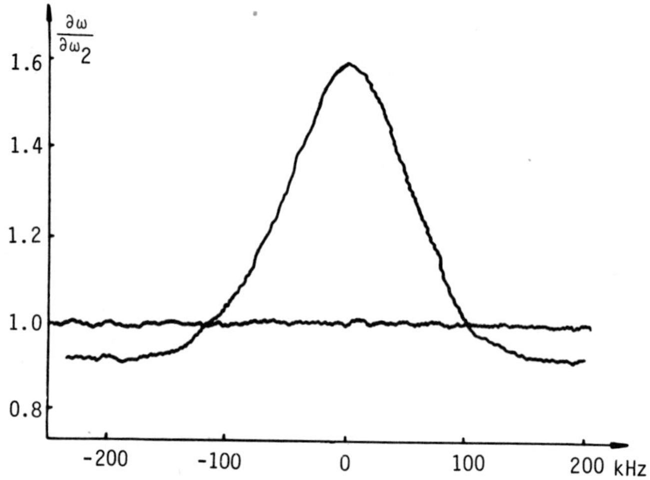

Fig.3.16 Record of shape resonance in the radiated spectrum of a He-Ne laser with a methane absorption cell. Methane pressure in the cell is $\approx 3 \cdot 10^{-4}$ torr (CHEBOTAYEV [3.54])

coefficients of absorption and amplification should be chosen and matched. In many cases, this is not practicable; then an attempt might be made to saturate the absorption by use of laser light outside the cavity [3.12]. By use of an external absorption cell, any influence of the saturated absorber on the oscillation amplitude and frequency of the laser is eliminated. The spatial configuration of the light field can be controlled and changed, to observe how one wave affects the absorption of another, for both counter-running and unidirectional waves. Such an experimental technique is now universally accepted. Table 3.2 gives a list of some atoms and molecules in which narrow resonances have been observed by absorption saturation in an external cell, as well as the lasers by use of which these resonances have been produced. It is evident that, by use of uv, visible and ir tunable lasers, this table can be extended almost *ad infinitum*. The first successful experiments in this direction were reported by SCHAWLOW et al. [3.60] and PATEL [3.68]. In a few years, experimenters will be puzzled by this table which will remind them of our troubles in choosing suitable pairs of lasers and absorbers for our first experiments on laser spectroscopy of absorption saturation.

When absorption is saturated by the field of a laser (or lasers) outside the cavity, the light field must have at least one strong running wave for absorption saturation. To detect the resonant deformation of the Doppler contour, a second (probe) wave should be used. This technique makes it

Table 3.2 Experiments on saturation resonances in external absorbing cells

Laser	Wavelength (μm)	Absorbing particle	References
Ar^+	0.5017	$^{127}I_2$	[3.57]
	0.5145	$^{127}I_2$, $^{129}I_2$	[3.58]
	0.5208	$^{127}I_2$, $^{129}I_2$	[3.57]
Kr^+	0.5682	$^{127}I_2$, $^{129}I_2$	[3.57,59]
Dye laser	0.5890	Na	[3.60]
He-^{20}Ne	0.6328	$^{20}Ne^a$	[3.61-64]
	0.6563	H^a	[3.65,66]
		D^a	[3.66]
	3.39	CH_4	[3.29,33]
		CH_4	[3.67]
He-Ne[b]	3.39	CH_3OH, CH_3Br, CH_3I, $^{13}CH_4$ CH_3F, CH_3Cl	[3.35]
Spin-flip laser	5.3	H_2O	[3.68]
CO	5.714	NH_3, H_2CO^c	[3.69]
CO_2-N_2-He	9.6	PF_5, CF_2Cl_2	[3.70]
		CH_3F^c	[3.69,71-73]
	10.6	SF_6	[3.70,74-77]
		NH_2D^c	[3.78-80]
		SiF_4	[3.81,82]
		OsO_4	[3.83]
		$^{189-192}OsO_4$	[3.84]
		$CH_3{}^{35}Cl$	[3.85]
	9.6-10.6	CO_2	[3.6]
N_2O-N_2-He	10.8	C_2H_4	[3.70]
		NH_3	[3.70,86]
		N_2O	[3.87]
Microwave oscillator	$8.2 \cdot 10^3$	OCS, CH_3CN	[3.88]
	$(1\text{-}3) \cdot 10^3$	OCS, CH_3F, $^{35}ClCN$	[3.89]

a - in discharge,
b - in magnetic field,
c - in electric field.

possible to use some new experimental potentialities. The probe wave may have the same frequency as the strong wave, but then it should propagate in the opposite direction to the strong one (Fig.1.13). It is evident that, when the light-field frequency ω coincides with the center of the absorption line ω_0, the counter-running probe wave interacts with molecules whose absorption is saturated by the direct, strong wave; as a result, the weak-wave absorption at the center of line decreases sharply. Consequently, a resonant dip appears in the absorption curve of the weak (probe) counter-running wave (LETOKHOV and CHEBOTAYEV [3.4]). This method is quite successful; it is widely used in spectroscopy of absorption saturation. Resonances in probe counter-running-wave absorption were first observed in experiments on narrow resonances in SF_6 [3.74] and Ne [3.61].

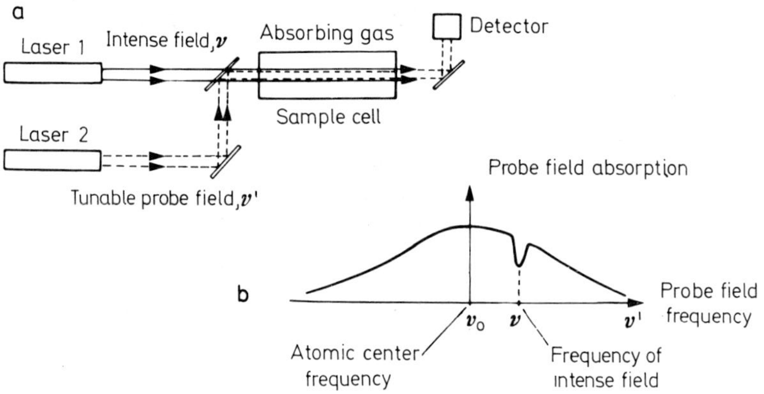

Fig.3.17a,b Observation scheme of narrow saturation resonance by the method of unidirectional probe wave (a) and absorption of probe wave (b)

A probe wave may propagate collinearly with a strong wave, but then its frequency must be scanned, for a resonance to be revealed in the Doppler contour (Fig.3.17). A resonant dip in this case appears not at the center of line but at the strong-wave frequency ν. The method of probe and strong unidirectional waves is theoretically discussed in detail in Section 2.7.

Many other useful modifications of the methods of absorption-saturation spectroscopy have been elaborated recently. In particular, we wish to mention an original method of counter-running waves of different frequency (JAVAN et al. [3.86]). In this case, when absorption is saturated by the strong wave, a narrow dip is developed in the absorption of the probe counter-running wave with another frequency. Finally, the probe wave can be separated spatially from the strong wave. In this case, resonance effects are possible if the free path of the particles exceeds the diameters of the beams and the distance between them (LETOKHOV [3.1]). We will consider each of these methods in more detail.

3.2.1 Probe Counter-Laser-Wave Method

The method of weak probe counter-running wave is theoretically analyzed in Section 2.6. Here we consider only some experimental features of the method and certain modifications of it.

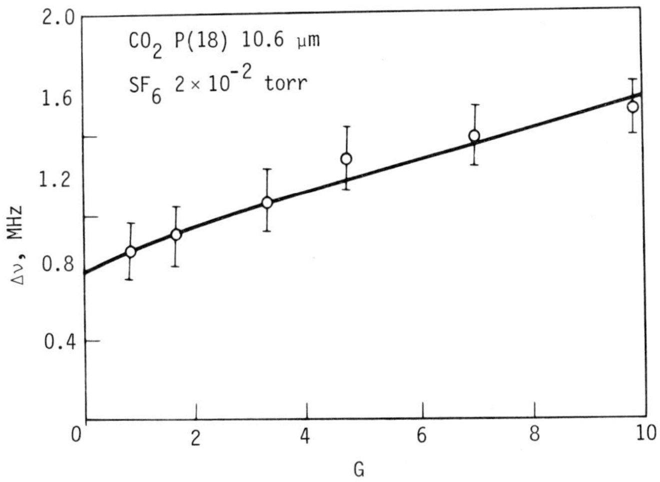

Fig.3.18 Experimental dependence of the narrow dip width inside the Doppler line of SF_6 on the degree of saturation. The curve shows the theoretical dependence (LETOKHOV [3.91])

Incoherent-Interaction Approximation

The parameters of narrow resonances of probe-wave absorption in the field of a strong counter-running wave of the same frequency have been studied in detail by BASOV et al. [3.77] and by KOMPANETZ and LETOKHOV [3.90]. Figure 3.18 shows the experimental dependence of the narrow resonance width inside the Doppler line of the SF_6 molecule on the degree of saturation. In this experiment, they used the line P(18) of a CO_2 laser at 10.6 μm. The frequencies of this line coincided with one of the numerous rotational-vibrational molecular transitions. The solid curve shows the theoretical dependence calculated from (2.146).

The value of contrast, that is the relative value of the dip in the absorption line h, according to (2.145), is

$$h = \frac{\Delta\kappa}{\kappa_0} = 1 - (1+G)^{-1/2} \ . \tag{3.36}$$

It is determined by the difference of the absorption coefficients of the counter wave away from and in resonance. It is significant that the amplitude of the transmission peak increases monotonically with increase of strong-wave intensity, as opposed to the case when absorption is saturated by the standing wave. The resonance amplitude can be increased considerably by use of optically dense absorbing cells with $\kappa_0 l_b \gg 1$. In such a case [3.77], the value of transmission-peak contrast may be of the order of 200%. In the more general case of counter-running waves with arbitrary intensity the peak amplitude, in the simplest rate-equation approximation is

$$h = (1+G)^{-1/2} - (1+G+G_0)^{-1/2} \ , \tag{3.37}$$

where G and G_0 are the saturation parameters of absorption by running waves with amplitudes E and E_0, respectively. Experimental results agree quite well with the calculations in this approximation [3.90]. When the experimental were compared with the calculated results, the radial distribution of the field over the beam cross-section was not taken into account and the contribution of coherence effects was ignored. Nevertheless, we may conclude that the simplest model of dip production for weak-wave absorption describes correctly, for the most part, the important regularities.

Coherence Effects

The contribution of coherent interaction to narrow resonance broadening was measured in an elegant experiment [3.92], in which a narrow resonance was observed by use of two pulse waves at transitions of the D lines of Na. A probe counter-running light pulse was delayed slightly after a strong pulse passed. Because the pressure of the Na vapour was very low, the dip broadening in the velocity distribution of the atoms, due to elastic collisions could not play a significant part. The results of experimental observation of resonance width as a function of delay time of the probe pulse, after the strong-field pulse of several ns duration are shown in Fig.3.19. The resonance narrowing observed is conditioned by disappearance of the coherence-effect contribution that is related to the presence of a strong field.

Experimental Schemes

The method of probe counter-running wave has been very effective in detecting and studying narrow resonances inside Doppler lines of atoms and molecules;

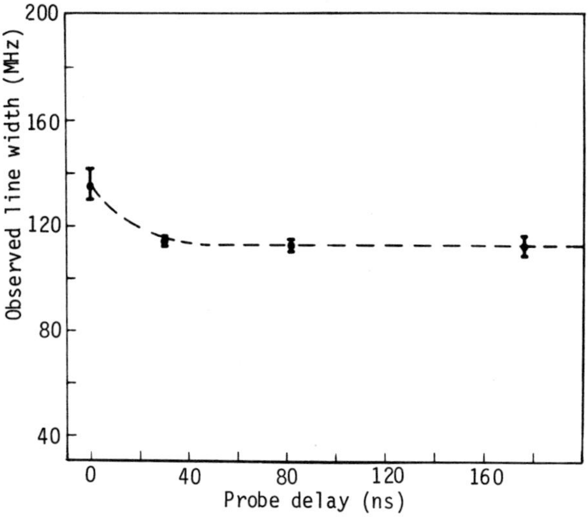

Fig.3.19 Experimental dependence of the dip width $\Delta\omega$ D line of Na observed by the probe-counter-wave method from delay Δt of the probe pulse relative to the strong laser pulse (SHAHIN and HANSCH [3.92])

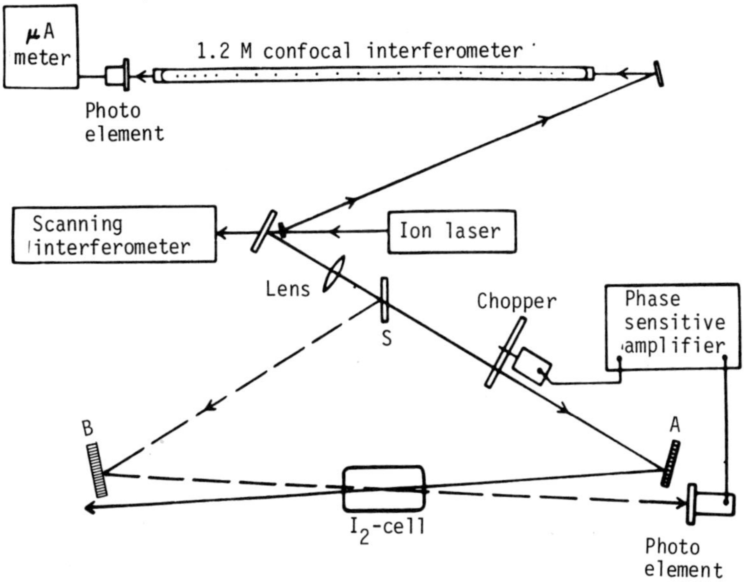

Fig.3.20 Saturation spectrometer using the modulated intensity of the intense wave and synchronous detection of the induced modulation of counter probe wave (measurement of hyperfine structure of I_2 (from [3.59])

it has been applied in many experiments. Figure 3.20 shows a very effective and frequently employed scheme of saturation spectrometer that uses the method of probe counter-running wave. The scheme employs amplitude modulation of the strong running wave, which induces absorption saturation; it was used in [3.62,93]. The probe counter-running-wave intensity is modulated only when the probe wave is responsive to the hole burnt in the Doppler contour by the strong modulated light wave. The modulated part of the probe-wave absorption coefficient is described by the second term in (2.145). In this case, the modulated part of the probe-wave intensity recorded by a phase-locked amplifier depends only on saturated absorption and is not sensitive to the unsaturated part of the Doppler contour. Such a scheme is highly sensitive for detecting the saturated-absorption spectrum.

Usually the weak probe wave is directed towards the strong wave at a small angle θ, in order not to enter the laser mode cavity and act on it. Assume that the counter-running probe wave has the wave vector $\underline{K}_2$, the direction of which differs from that of the wave vector $\underline{K}_1$ of the strong wave by $180+\theta$. Then it is evident that, owing to the first-order Doppler effect, the narrow resonance for the counter-running wave will be broadened by the small value

$$\Delta\omega_g = (\underline{K}_1 - \underline{K}_2)\underline{v} = 2kv \sin\frac{\theta}{2} \; . \tag{3.38}$$

For $\theta \ll 1$, the value of geometric broadening can be written in the form

$$\Delta\omega_g \simeq \frac{\theta}{\pi}\,\Delta\omega_D \quad \text{or} \quad \Delta\nu_g \simeq \theta\,\frac{u}{\lambda} \; , \tag{3.39}$$

where λ is the light wavelength. When the angle of wave-direction detuning is of the same order as the diffraction angle $\phi_{dif} \cong \lambda/a$, where a is the light-beam diameter, the geometrical broadening is $\Delta\nu_g \simeq u/a$; it is of the same order as the resonance broadening caused by the finite time of particle flight through the light beam. Because of this, the diffraction effect becomes significant at the limit of dip width that is determined by the finite time of particle-field interaction.

Here we encounter a rather fundamental limitation of the resolution of saturation spectroscopy. By placing an absorbing cell in the narrowest section of a gaussian beam, where the wave front is quite plane, we can overcome geometric broadening (3.37), but some broadening (3.21) remains, caused by the finite time of flight of the particles through the light beam.

The effect of a weak reflected wave is reduced materially when an optical decoupling filter consisting of a prism polarizer and a quarter-wave plate is used. When the reflected wave passes through this plate in the forward and opposite directions, its polarization changes by 90°, with the result that the wave is deflected by the polarizer prism and does not enter the cavity.

If the dip depth is not large, the power fluctuations of the strong wave reduce the measurement sensitivity. The sensitivity of observation of resonance can be increased by recording the derivative of the absorption-line contrast with respect to frequency. This is accomplished by modulating the laser-radiation frequency or by scanning the absorption line with an oscillating magnetic field. The same sensitivity increase can be achieved also in strong-wave amplitude modulation (see Fig.3.20).

Adiabatic-Passage Resonance

Recently Ref. [3.94] reported another original method for observing narrow resonances without Doppler broadening. The method is based on the well-known effect of level-population inversion during adiabatic fast passage of resonances. In this method, the strong running-wave frequency ω should be scanned along the Doppler contour at the speed,

$$\left(\frac{1}{T}\right)^2 \ll \left|\frac{d\omega}{dt}\right| \ll \left(\frac{pE}{\hbar}\right)^2 \ . \tag{3.40}$$

Under condition (3.40), molecules are left in an inverted state after interaction with the field [3.95]. Consequently, when the strong-field frequency $\omega(t)$ is scanned, very peculiar dynamic deformation of the Doppler contour occurs: molecules with $kv < \omega(t) - \omega$ are in an inverted state ($d\omega/dt > 0$), whereas those with $kv > \omega(t) - \omega_0$ are in the initial state. Let the backward weak wave of the same frequency $\omega(t)$ probe the molecular state. It is obvious that at the instant when the backward probe wave passes the center of Doppler contour ($kv = 0$) it begins to interact with the inverted medium. Such interaction is substantially coherent. That is why, according to the theory in Subsection 2.6.2 there is no marked amplification of the probe wave in an inverted medium and the transparency of the medium, for the probe wave, increases greatly. With further frequency scanning, transparency remains throughout the molecular-relaxation time. So the probe-wave transmission at various sections of the Doppler contour permits determination of the rates of processes of fast molecular relaxation.

Experimental spectral data on saturated absorption, obtained by the method of strong and counter-running weak waves, are presented in Chapters 7 and 8.

3.2.2 Counter Laser Waves of Different Frequencies

An interesting modification of the Lamb-dip method, which enables us to obtain narrow resonances at any point of a Doppler-broadened absorption line, was suggested by JAVAN et al. [3.86]. The method is based on absorption saturation by a running light wave with frequency ω_1 and probing the resultant hole with a counter-running light wave with frequency ω_2. When the frequencies of the two running waves are tuned symmetrically on opposite sides of the center frequency ω_0, a narrow resonance, similar to the Lamb dip, appears. In contrast to the Lamb dip, however, the particles that give rise to the resonance are those that have a nonzero velocity component v_{res} along the laser direction, given by $v_{res} = [(\omega_i - \omega_0)/\omega]c$, where c is the speed of light, ω_i is ω_1 or ω_2. For $\omega_0 - \omega_1 = \omega_2 - \omega_0$, only one velocity group of particles interacts with both travelling waves. The narrow absorption resonance centered at $\omega_0 - \omega_1 = \omega_2 - \omega_0$ appears when the absorption saturation by at least one wave becomes appreciable. The expression for the narrow resonance in the two-frequency field (one strong ω_1 wave and second counter probe wave ω_2) is similar to (2.145).

The kinetic energy of these molecules with "selected" v_z projected velocity

$$v_{res} = \left(\frac{\omega_1 - \omega_0}{ku}\right) u \tag{3.41}$$

equals

$$E_{kin} = \frac{1}{2} M(v_x^2 + v_y^2) + \frac{1}{2} Mv_{res}^2 \tag{3.42}$$

or

$$E_{kin} = kT + \frac{1}{2} Mv_{res}^2 = kT\left[1 + \left(\frac{v_{res}}{u}\right)^2\right] \ . \tag{3.43}$$

The mean kinetic energy of the selected molecules can be changed from (3/2)kT (at $v_{res} = u$) to kT (at $v_{res} = 0$) by frequency tuning of strong-wave frequency ω_1 and the appropriate probing of narrow resonance by weak wave at $\omega_2 = \omega_0 - (\omega_1 - \omega_0)$ frequency. The effective temperature T_{eff} can be defined for the selected molecules by [3.86]

$$T_{eff} = \frac{2}{3} kT\left[1 + \left(\frac{\omega_1 - \omega_0}{ku}\right)^2\right] \quad . \tag{3.44}$$

In the regime where collision broadening exceeds the natural broadening and other contributions to the homogeneous broadening the width of the resonance observed for ω_1 not equal to ω_0 can differ from that of the Lamb dip, owing to the dependence of the collision-broadening cross-section on particle velocity at fixed room temperature.

This method has been applied to the observation of the velocity dependence of collision broadening of an infrared transition of NH_3 [3.86]. Operation of a cw N_2O laser on the P(13) line at 10.8 μm, which is in close coincidence with the ν_2 [as Q (8,7)] transition of $^{14}NH_3$ was utilized in first experiments. Part of the laser output was sent to a standing-wave Ge acoustooptic modulator, which produced light shifted symmetrically above and below the laser frequency. In the experiment [3.86] the frequency shift was about 75 MHz, which at room temperature was $1.5\Delta\omega_D$, corresponding to $v_{res} = 1.5u$. The spatially separated frequency-shifted radiation was split into a strong saturating field and a weaker probe field, which were sent in opposite directions through a NH_3 absorption cell.

3.2.3 Laser Waves with Different Polarization

The results given in the preceding paragraphs involved the simplest case of a two-level system. This is the interaction of unidirectional waves with the same circular polarization at transitions with $J = 0 \rightarrow J = 1$. In other cases, the degeneracy of levels and hence the polarization properties of radiation should be considered. Even in the simplest scheme of levels with $J_2 = 0$ and $J_1 = 1$, depending on the polarization of the strong and probe waves, two different situations may take place, which are illustrated in Fig.3.21. Figure 3.21a corresponds to the case when the strong and probe waves have the same polarization and interact with two levels. This was the case that was considered in the preceding paragraph. If the probe and strong waves have different (left and right) circular polarization they interact with three levels, one of which is common. The line shape of probe-signal absorption for the two cases will be different. The last-mentioned case corresponds to the three-level scheme of field interaction, which is considered in detail in Chapter 5.

The absorption of the ω_2-frequency probe wave in the presence of a field with the ω_1 frequency and inverse circular polarization (Fig.3.21b) at low saturation is given by the expression for counter-running waves,

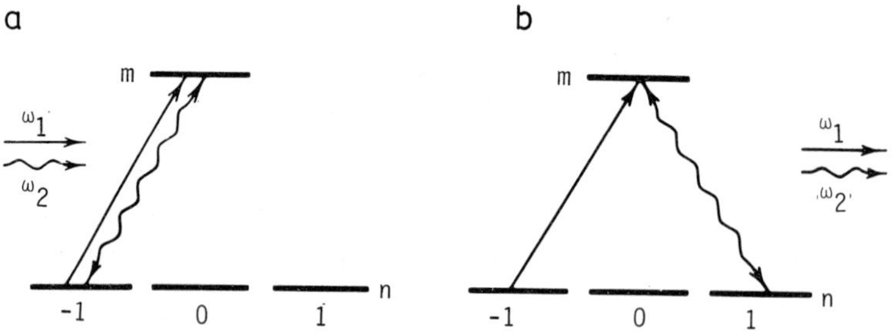

Fig.3.21a,b Interaction of two waves with (a) the same and (b) opposite circular polarization; the lower level of the transition is degenerate

$$\frac{\kappa}{\kappa_0} = \exp[-(\frac{\Omega + \Delta}{ku})^2] - G \frac{\gamma_n \Gamma}{(2\Gamma)^2 + (\Omega + \Delta)^2} \exp[-(\frac{\Omega}{ku})^2] \tag{3.45}$$

and for unidirectional waves,

$$\frac{\kappa}{\kappa_0} = \exp[-(\frac{\Omega + \Delta}{ku})^2] - \frac{G}{2} \frac{\gamma_n^2}{\gamma_n^2 + \Delta^2} \exp[-(\frac{\Omega}{ku})^2] \ , \tag{3.46}$$

where $\Delta = \omega_2 - \omega_1$. The narrow resonance shape described by (3.45) and (3.46) differs from the analogous expressions for the two-level system. For counter-running waves, the dip depth depends on the relation between the level relaxation constants. In the case of unidirectional waves, the narrow resonance width is determined by the width of one level only. Thus, the change of polarization of the probe wave with respect to the incident wave may substantially change the absorption-line shape.

Qualitative features of polarization properties in the absorption (amplification) line shape have been observed in [3.96,97]. In [3.96], particularly, the amplification (absorption) line saturation for the transition $J = 1 \rightarrow J = 2$ was calculated with allowance for depolarizing collisions. The saturated line structure experimentally observed was narrower than the natural width at $\lambda = 3.39$ µm ($3S_2 - 3P_4$) that qualitatively agreed with the calculation for different polarizations of probe and strong waves. In [3.97], the absorption contour was studied at the transition $3S_2 - 2P_4$ ($J = 1 \rightarrow J = 2$) in the presence of a laser field at the same transition with different field polarizations. The polarization dependence of line shape and width was disclosed, and the results obtained were qualitatively interpreted. As would be expected the absorption line of probe signal turned out to be narrower for different circular polarizations.

3.2.4 Spatially Separated Laser Waves

At a very low gas pressure, the mean free molecular path with respect to collisions Λ may be much longer than the light-beam diameter a. In this case, a "hole" burnt in the Doppler line by a running light wave is apparently transferred in space for a distance that is of the order of the free path length. Consequently, two spatially separated running waves may resonantly interact with each other on account of molecules crossing the both beams without collisions, in other words, absorption of one beam depends on the field in the other. To do this, the sum of the distance between the beams d and their diameters a should be less than the free-path length Λ.

Two Spatially Separated Beams

Formulation of the problem is similar to that of the interaction between a beam of atoms or molecules and two oscillating fields studied by RAMSEY [3.98]. There are essential differences in the optical case, however. First, due to Doppler broadening, just a small portion of the molecules interacts with each beam and, second, the dimensions of the interaction area are several orders larger than the wavelength. Therefore, the results obtained by Ramsey cannot be applied to the optical case and calculation should be carried out anew. Such a calculation has been made in [3.99,100] by LETOKHOV and PAVLIK.

The polarization of the molecules in the second beam, previously passed through the first beam, may be presented in three parts,

$$P_2(E_1,E_2) = P_2^{own}(E_2) + P_2^{conv}(E_1) + P_2^{mix}(E_1,E_2) . \tag{3.47}$$

The "own" polarization P_2^{own} depends on the field of the second beam only. It describes the interaction ("hole" burning and polarizability) with molecules the velocity of which complies with the condition,

$$|\omega_0 - \omega + \underline{k}_2 v| \lesssim \Gamma , \tag{3.48}$$

where 2Γ is the line width dependent on the molecular transit time. The convective polarization P_2^{conv} is brought to the second beam from the first by molecules the velocity of which meets the resonance condition,

$$|\omega_0 - \omega + \underline{k}_1 v| \lesssim \Gamma . \tag{3.49}$$

The mixed polarization P_2^{mix} depends on both of the fields. The radiation power absorbed from the second beam by the molecules that pass through the both beams is

$$W = \langle E_2 \frac{dP_2}{dt} \rangle_{t,\underline{r},\underline{v}} \quad , \qquad (3.50)$$

where averaging should be carried out over time and interaction area $\underline{r}$ (diameter and length of beams) and over molecular velocities $\underline{v}$ (transit time and Doppler contour).

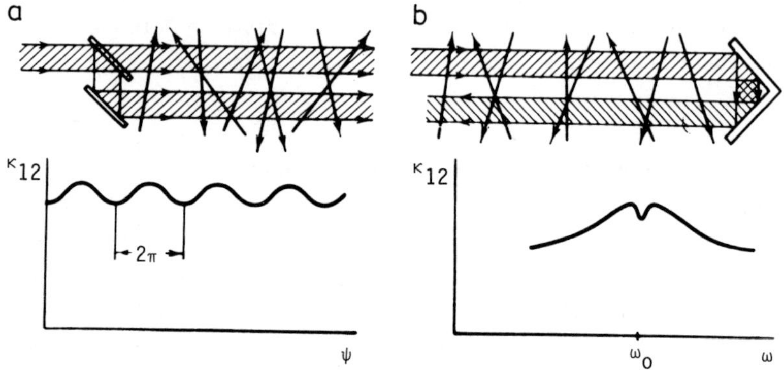

Fig.3.22a,b Diagrams showing cross-absorption (a) in unidirectional and (b) counter-running spatially separated light waves

In cases of unidirectional and counter-running waves the beam interaction differs greatly. In the case of unidirectional running waves ($\underline{K}_1 = \underline{K}_2$) the cross-absorption, that is, the part of the nonlinear absorption that is dependent on the fields of both waves at the same time, is sensitive to the beam phase difference ψ (Fig.3.22a) and given by

$$\frac{\kappa_{12}}{\kappa_0} = 1 - (G_1 + G_2) + f(\frac{d}{a})\frac{E_1}{E_2}(G_1 + G_2)\ \cos\psi \quad , \qquad (3.51)$$

where G_i is the degree of saturation by the wave E_i. The value of the interference term depends on the distance between the beams d and the field-amplitude ratio. The distance dependence is rather strong. The function $f(d/a) \approx 0$, even with $d \approx a$. This is due to the fact that the molecules that belong to the "hole" of the first beam, when flying to the second beam, diverge in a transverse direction, the value of which $\Delta z \approx d(\lambda/a)$, even if they start from the same point of first beam. With $d > a$, the value of expanding $\Delta z > \lambda$, and the interference effect due to spatial averaging is eliminated. Increase of the effect with $E_1 \gg E_2$ is explained by the fact that the molecules in the strong field of the first beam are strongly polarized, and this polarization prevails over that induced by the second weak beam. At the same time, cross-absorption of unidirectional waves is independent of wave-frequency detuning about the center of Doppler line, because both beams interact with the same velocity group of molecules at the Doppler contour ($\underline{K}_1 = \underline{K}_2$).

In case of counter-running waves ($\underline{K}_1 = -\underline{K}_2$) there is no interference effect because of averaging over the interaction area with its dimensions $\gg\lambda$, but there appears a dependence of cross-absorption on detuning $\Omega = \omega - \omega_0$ of the wave frequency about the center of the line (Fig.3.22b),

$$\frac{\kappa_{12}}{\kappa_0} = 1 - G_2 - G_1[1 - \phi(\Omega\frac{2a}{u})] \ , \qquad (3.52)$$

where $\phi(x)$ is the even function with its maximum ($\phi = 1$) at the point $x = 0$ declining with increase in x (the half-width $x \simeq 1$). This function describes the resonant decrease of cross-absorption at the center of the line, which may be considered as a Lamb dip in spatially separated counter-running waves. The resonance width is half the dip width in a standing wave of the same diameter. Distinct from the case of unidirectional waves, coherent interaction is not needed for interaction of spatially separated counter-running waves. A hole burnt in the first wave is enough for such an interaction, which in this case is similar to that of waves discussed in Sections 2.6 and 2.7.

It is rather difficult to observe nonlinear resonance and interference effects in spatially separated beams because very low pressures are required. Under these conditions, the absorption is very small, and the apparatus used should have very high sensitivity and stability. Nevertheless, in [3.90,101] experiments have been reported on counter-running resonance beams from a CO_2 laser and a low-pressure SF_6 cell. The experiment is schematically shown in Fig.3.23. A CO_2 laser operated at the P(16) line of 10.6 μm band. For this line, SF_6 molecules have a large absorption coefficient ($\kappa_0 = 1.3$ cm^{-1} torr^{-1}) that allows operation at low pressures of SF_6 when the free-path length Λ of the molecule is large. The laser radiation, as two separated counter-running waves, passed through an external SF_6 absorbing cell 120 cm long. The intensity-modulated wave I_1 (Ω) saturated absorption. The counter-running weak wave I_2 probed the molecular state close to the strong wave. The frequency was gradually scanned along the Doppler contour of SF_6. A narrow transmission peak appeared at the center of line. But this effect could arise without "spatial transfer" of the hole because, owing to diffraction there is always an area of counter-running-wave overlap, that is an area of ordinary standing wave. To discriminate the effect associated with spatial trans-

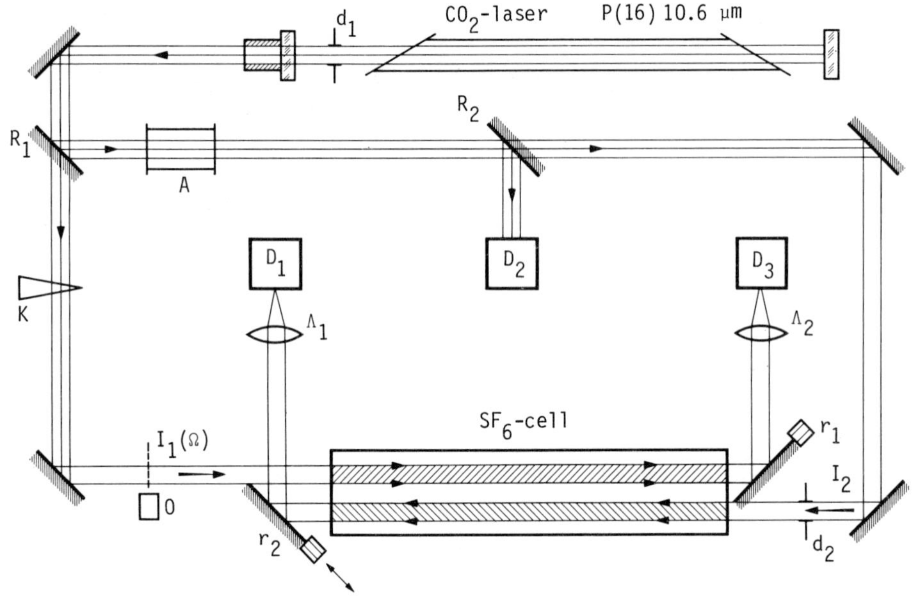

Fig.3.23 Scheme of experimental observation of narrow dip, induced by two spatially separated counter-running waves of CO_2 laser (P(16) line, 10.6 μm band) in SF_6 low-pressure gas cell (d_1 and d_2 - diaphragms, D_i - photodetectors, R_i - semitransparent mirrors, r_i - movable mirrors, L_i - lens, K - optical wedge, A - SF_6 absorbed cell as attenuator)

portation of a hole in the velocity distribution, the dependence of resonance amplitude on the distance between the beams was measured at various SF_6 pressures. The results of the experiment are shown in Fig.3.24. As the pressure was decreased, resonance was observed with a larger distance between the beams. This is explained by spatial transfer of the hole in the velocity distribution of the molecules. This effect has made it possible to evaluate the free-path length of SF_6 molecules $p\Lambda = 0.4$ cm/mtorr. Interference effects have not yet been observed in spaced beams. They can be observed at pressures of the order of 10^{-4} torr.

The resonance effects in separated beams, which arise from the collisionless molecular transit from one beam to another, can be also used for molecular coupling of lasers [3.102].

Nonlinear Ramsey Resonance in Three Beams

We saw previously that the coupling of two fields through polarization transfer could be observed when the distance between the beams was shorter than the diameter of each beam. The influence of polarization effects does not

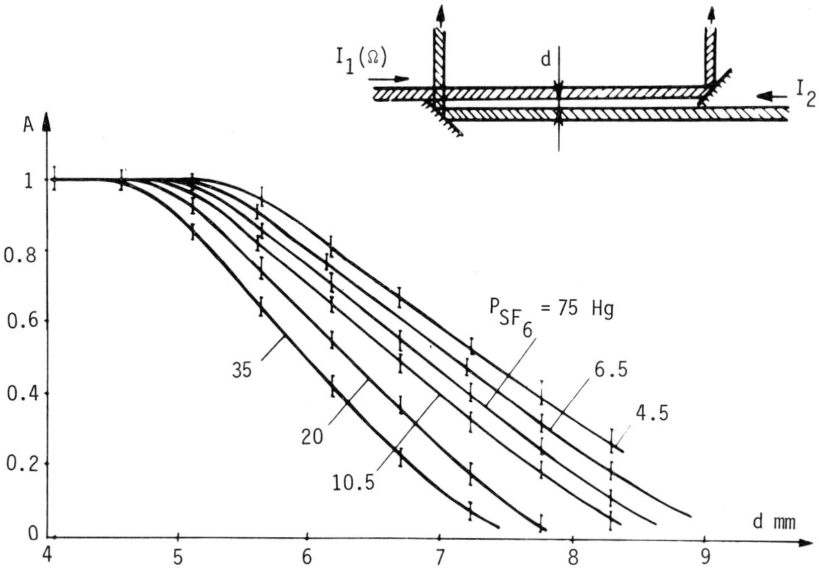

Fig.3.24 Relation between the relative amplitude of the resonant dip when the absorption of SF_6 is saturated by spatially separated counter-running waves and the distance between the beams at various pressures of SF_6 (KOMPANETZ and LETOKHOV [3.101])

result in any qualitative peculiarities of resonance in the two beams, compared to the Lamb dip: its width is comparable to the Lamb-dip width in one beam of the same diameter. In [3.103] particular emphasis is placed upon new physical features of the nonlinear interaction between a group of atoms and widely spaced light beams. After an atomic beam passes through two light beams, 1) polarization is transferred for a distance comparable to that between the beams which may be much longer than the beam width; 2) a periodic beam structure is formed in the velocity distribution of atoms with the period $v_z \simeq \lambda/2T$ (v_z is the projection of the velocity on the wave-propagation direction, λ is the wavelength, T is the transit time between the beams). In the third beam, which may be considered as a probe wave (Fig.3.25), there appears a resonance of absorption with a width inversely proportional to the transit time. It is similar in properties to the Ramsey resonance [3.98].

Let us consider an atomic gas in which the atoms interact resonantly with the field of three standing waves (Fig.3.25a),

$$E(x,z,t) = 2\mathcal{E}(x,z)\cos\omega t \ ,$$
$$\mathcal{E}(x,z) = \mathcal{E}_1 g(x)\cos(kz+\phi_1) + \mathcal{E}_2 g(x-d)\cos(kz+\phi_2) + \mathcal{E}_3 g(x-2d)\cos(kz+\phi_3) \ , \tag{3.53}$$

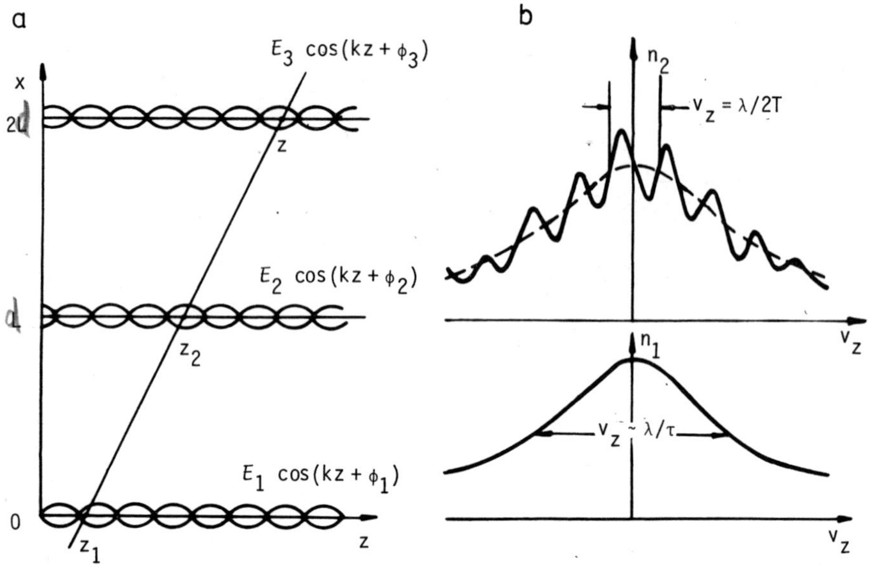

Fig.3.25a,b Three spatially separated beams: (a) geometry of atom flight through the three beams and (b) velocity distribution of polarized atoms after first (n_1) and second (n_2) beams

where

$$g(x) = \begin{cases} 1 , & \text{when } 0 < x < a , \\ 0 , & \text{when } x < 0 , \quad x > a . \end{cases}$$

The equations for atomic probability amplitudes have the form,

$$\begin{aligned} \dot{a}_1 &= \frac{i}{\hbar} P_{12} E(x,z)\, e^{i\Omega t}\, a_2 , \\ \dot{a}_2 &= \frac{i}{\hbar} P_{21} E(x,z)\, e^{-i\Omega t}\, a_1 , \end{aligned} \tag{3.54}$$

where a_1 and a_2 are the probability amplitudes for the lower 1 and upper 2 levels, respectively, and $\Omega = \omega - \omega_0$ is the detuning of the field frequency ω with respect to the atomic frequency ω_0.

We are interested in the energy absorbed by an atom in the third beam by virtue of the polarization induced after passing through the first and second beams. To determine the polarization, we should find the dipole moment brought by one atom into the point Z and then velocity average. For simplicity, the beams are assumed to be thin, that is, $kv_z\tau \ll 1$, $\Omega\tau \ll 1$, $k = 2\pi/\lambda$, where $\tau = a/v_x$ is the time of flight through a beam, v_x is the atomic velocity along the axis X. The initial conditions at the instant of particle entry into the first beam ($t = 0$) are $a_2 = 0$, $a_1 = 1$. In a weak field, the probability amplitudes, after an atom passes through the first beam, have the form

$$a_1^{(1)} \simeq 1\ , \quad a_2^{(1)} = i2V_1\tau \cos(kz_1 + \phi_1) \tag{3.55}$$

and after passing through the second beam,

$$\begin{aligned} a_1^{(2)} &= 1 + 2iV_2\tau \cos(kz_2 + \phi_2)\, e^{i\Omega T} a_2^{(1)} - 2V_2^2\tau^2 \cos^2(kz_2 + \phi_2)\ , \\ a_2^{(2)} &= a_2^{(1)} + 2iV_2\tau \cos(kz_2 + \phi_2)\, e^{-i\Omega T} - 2V_2^2\tau^2 \cos^2(kz_2 + \phi_2) a_2^{(1)}\ , \end{aligned} \tag{3.56}$$

where $V_i = pE_i/2\hbar$; $i = 1, 2, 3$; z_1 is the coordinate of atom entry into the first beam; $z_2 = z_1 + v_z T$ is the coordinate of atom entry into the second beam (see Fig.3.25a); $T = d/v_x$ is the time of flight between the first and second beams. After passing through the second beam, the atom falls on the point Z, at the distance $x = v_x t$ from the first beam, its polarization $P(z,v_z,t) = P_{21} a_2^{(2)} a_1^{(2)*}\, e^{-i\omega_0 t}$, to a desired accuracy will equal

$$\begin{aligned} P(z,v_z,t) = P_{21}\, e^{-i\omega_0 t} \Bigg\{ & 2iV_1\tau \cos\left(kz - \frac{v_z}{v_x} kx + \phi_1\right) + \\ & + 2iV_2\tau\, e^{-i\Omega T} \cos\left[kz - \frac{v_z}{v_x} k(x - d) + \phi_2\right] - \\ & - 2iV_2^2 V_1 (1 + e^{-2i\Omega T})\tau^3 \Bigg[\cos\left(kz - \frac{v_z}{v_x} k(x - 2d) + 2\phi_2 - \phi_1\right) + \\ & + \cos\left(3kz - \frac{v_z}{v_x} k(3x - 2d) + 2\phi_2 + \phi_1\right) + \\ & + 2\cos\left(kz - \frac{v_z}{v_x} kx + \phi_1\right)\Bigg]\Bigg\}\ . \end{aligned} \tag{3.57}$$

The polarization at the point Z can be calculated by averaging $P(z,v_z,t)$ with distribution function $f(v_z)$ over v_z, retaining only the third term of (3.57) for the wide velocity distribution,

$$P(z,t) = -P_{12}\, e^{-i\omega_0 t}\, 2iV_2^2 V_1 (1 + e^{-2i\Omega T})\tau^3 \cdot$$

$$\cdot \int_{-\infty}^{\infty} dv_z f(v_z) \cos\left(kz - k\frac{v_z}{v_x}(x - 2d) + 2\phi_2 - \phi_1\right) . \qquad (3.58)$$

We want to concentrate our attention on a very important feature of the system under discussion. Polarization is finite only near $x = 2d$ in a very narrow range $\Delta x \simeq v_x/k\Delta v_z \ll d$, where Δv_z is the characteristic width of the distribution function $f(v_z)$. Thus, we can focus the first spatial harmonic at the distance $x = 2d$.[3]

Physical interpretation of the result explains the fact that after an atom interacts with the first beam, the polarization of the second beam is zero, owing to averaging over v_z. However, after nonlinear interaction with the second beam, a "grating" structure appears in the velocity distribution of polarized atoms, with the characteristic dimension $v_z = \lambda/2T$ (Fig.3.25b). Because of this, in the third beam, the contribution of these atoms to absorption and amplification with averaging over v_z turns out to be uncompensated, which gives rise to a resonance.

The energy absorbed in the third beam at $x = 2d$ is

$$W = 2\mathrm{Re}\left\{\frac{E_3}{L}\right\} \int_0^L dz \int_{2T}^{2T+\tau} dt \frac{dP(z,t)}{dt} \cos(kz + \phi_3)\, e^{i\omega t} =$$

$$= \hbar\omega 8 V_1 V_2^2 V_3 \tau^4 \cos^2\Omega T \cos(2\phi_2 - \phi_1 - \phi_3) . \qquad (3.59)$$

Formula (3.59) produces a very narrow resonance with the width $1/T$ characteristic of the Ramsey method.

In the methods of nonlinear absorption used now, the resonance width is determined by the light-beam diameter. Beams of 10 cm diameter achieved now are at the practical limit. The possibility being described opens a new way for production of supernarrow resonances and does not need the use of broad light beams.

[3] For real light beams, the distribution-function width of excited atoms after flight through two spaced beams equals the Bennett "hole" width $\Delta v_z \simeq 1/k\tau$, hence $\Delta x \simeq a$.

3.3 Density Resonance of Excited Particles

A narrow resonance in the saturated absorption coefficient of a standing light wave brings about obviously a resonant change in the excitation rate and hence in the particle density at the lower and upper transition levels (Fig.1.15). Resonant changes of excited-particle density can be revealed by using more ways than direct changes of the absorption coefficient. For this purpose, for example, fluorescence of excited particles, absorption by excited particles, electron emission when excited atoms collide with a metal surface, and excitation transfer to other particles can be used. The main advantage of excited-particle-density resonances is that they make it possible to study low-absorption transitions for which, for example, $\kappa_{bo} l_b \ll 10^{-2}$ (κ_{bo} is initial absorption per unit length, l_b is the absorbing-cell length).

3.3.1 Resonances in Two-Level Atom and Rotational-Vibrational Molecular Transitions

The density of excited particles is simply related to the saturated absorption coefficient $\kappa(\omega)$ by

$$N_2(\omega) = \kappa(\omega) P \tau_2 \ , \tag{3.60}$$

where ω is the standing-wave frequency, P is the wave intensity, and τ_2 is the lifetime of particles at the excited level. For example, to an approximation of weak saturation when the dependence of $\kappa(\omega, P)$ on field frequency is given by (2.102), $N_2(\omega)$ is

$$N_2(\omega) = \kappa_0(\omega) P \tau_2 \left[1 - \frac{G}{2}\left(1 + \frac{\Gamma^2}{\Gamma^2 + (\omega - \omega_0)^2}\right)\right] \ . \tag{3.61}$$

Thus, against the background of a gaussian absorption curve $\kappa_0(\omega) P \tau_2$ there is a narrow resonance-density minimum of excited particles with width 2Γ. When the saturation G is about P, at small saturation ($G \ll 1$) the absolute value of the dip in $N_2(\omega)$ is proportional to the square of the intensity; its relative value is given by the same formulas as the Lamb dip.

The effect of resonant changes in $N_2(\omega)$ can be also obtained from simple considerations of "hole" burning. In the standing-wave field away from resonance ($|\Omega| = |\omega - \omega_0| \gg \Gamma$) two "holes" are formed in the velocity distribution of level population difference due to upward transition of particles (Fig. 3.26). The total number of particles at excited level N_2 is proportional to the total area of holes $S_1 + S_2$. In the case of exact resonance ($|\Omega| \ll \Gamma$)

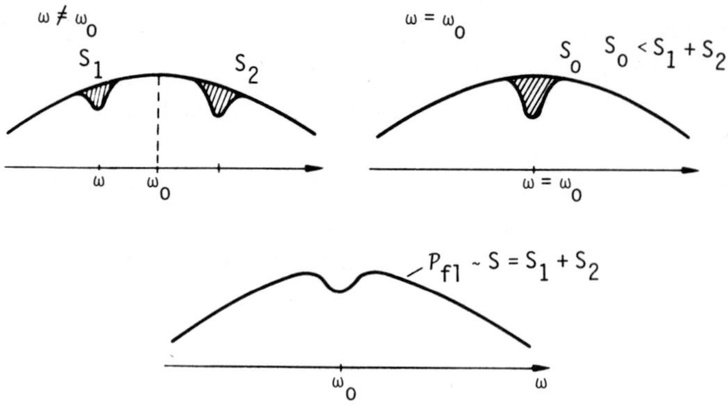

Fig.3.26 Explanation in terms of "hole burning" of a narrow resonance of the fluorescence intensity P_{fl} when the absorption is saturated by the standing wave (S_1 and S_2 are the areas of each hole with $\omega \neq \omega_0$; S_0 is the area of the joined hole, when $\omega = \omega_0$)

the two holes coincide, and the number of excited particles is proportional to the resultant hole area S_0. If $S_0 \neq S_1 + S_2$, the total number of excited particles changes resonantly. It follows from this that no resonance appears in the fluorescence intensity when saturation is caused only by a running wave while the counter-running wave is weak. Actually, in this case, one of the holes is absent ($S_2 = 0$), and the hole area at the center of line S_0 is always equal to the area of holes off the resonance $S_1 + S_2 = S_1$. At the same time, in this case, there is a resonant dip of absorption (Subsec.3.2.1).

Two-Level System

Let us consider at first density or fluorescence resonances when a two-level system is under saturation in a standing light wave. Naturally, we shall restrict ourselves by only level-population effects when the rate equations can be applied (see Subsec.2.5.2). The variation of the population distribution of particles with the projection of v velocity on wave direction $n_1(v)$ and $n_2(v)$ is given by formulas of the type of (2.80),

$$\frac{d}{dt} n_1(v) = \gamma_1[n_1^0(v) - n_2^0(v)] + w[n_2(v) - n_1(v)] ,$$
$$\frac{d}{dt} n_2(v) = \gamma_2[n_2^0(v) - n_1^0(v)] - w[n_2(v) - n_1(v)] , \qquad (3.62)$$

where w is the probability of induced transition of a particle with the v velocity in the standing wave field with the ω frequency determined by

$$w = \sigma_0 P[L(\Omega - kv) + L(\Omega + kv)] , \qquad (3.63)$$

where $\Omega = \omega - \omega_0$, σ_0 is the cross-section of stimulated transition at a maximum for a particle in exact resonance with one running wave; P is the density of running-wave radiation flux (photon/cm^2 · s).

From (3.62) we can easily derive the expression for density of excited particles in a steady-state case,

$$N_2 = \int_{-\infty}^{\infty} n_2(v)dv = N_2^0 + (N_1^0 - N_2^0)\frac{G}{1 + \frac{\gamma_2}{\gamma_1}} f(G,\Omega) , \tag{3.64}$$

which includes the function

$$f(G,\Omega) = \int_{-\infty}^{\infty} dvW(v)\frac{L(\Omega + kv) + L(\Omega - kv)}{1 + G[L(\Omega + kv) + L(\Omega - kv)]} . \tag{3.65}$$

The resonant change of intensity of excited particles with $\Omega = 0$ is given by the integral function $f(G,\Omega)$. The general analysis of this function can be carried out by calculating the integral of $f(G,\Omega)$ and expressing it with the plasma dispersive function. Let us consider here only two limiting cases.

At exact resonance ($|\Omega| \ll \Gamma \ll ku$), integral (3.65) is precisely calculated by

$$f(G,0) = f_0(1 + 2G)^{-1/2} , \quad f_0 = 2\sqrt{\pi}\frac{\Gamma}{ku} . \tag{3.66}$$

Far from resonance ($\Gamma \ll |\Omega| \ll ku$) integral (3.65) can be divided into the sum of two integrals, each mainly contributed to by particles with velocities $v = \pm\Omega/k$. As a result, we have

$$f(G, |\Omega| \gg \Gamma) = f_0(1 + G)^{-1/2} . \tag{3.67}$$

Comparing (3.66) with (3.67), we can see that the total number of particles at the upper level resonantly decreases as the standing-wave frequency scans through the center of Doppler line. With $G \ll 1$, the depth of the narrow dip of fluorescence intensity increases directly with G^2 whereas with $G \gg 1$ it increases directly with $\sqrt{G}$. When $G \gg 1$, the relative dip depth (dip contrast) tends to the constant $h = [1 - (1/\sqrt{2})] = 0.29$. The saturation-parameter value $G \simeq 1$ is optimal. At this value of G, the peak contrast is 0.185, that is 64% of the maximum, and the coefficient of resonance broadening by the strong field equals $\sqrt{2}$.

Vibrational-Rotational Transition

In the first experiment by FREED and JAVAN [3.6] resonances of fluorescence intensity of CO_2 molecules were observed in a low-pressure cell at the 4.3 μm transition with one rotational-vibrational line of the 10.6 μm band saturated by CO_2 laser radiation. An important feature of the case of a molecular cell lies in the fact that the molecules at all the rotational levels of an excited vibrational state coupled by collisions contribute to spontaneous emission. The two-level model considered can be applied only when the rotational relaxation is slower than the vibrational relaxation, for example, when the length of the molecular free path is of the same order as the light-beam diameter.

Let us consider the influence of molecular rotational relaxation in the simplest model when the $\underline{v}$ velocity and the $\underline{J}$ angular momentum of a molecule after collision do not depend on $\underline{v}'$ and $\underline{J}'$ before collision. Taking into consideration the change of molecular distribution with velocities and rotational state $n_i(v,J)$, on condition that the standing light wave acts on a one-rotational-vibrational transition, $v_1J_1 \to v_2,J_2$ for the total population of the excited vibrational state $N_2 = \sum_J \int dvn_2(v,J)$, we have instead of (3.64) the relation [3.104],

$$N_2 = N_2^0 + (q_1N_1^0 - q_2N_2^0)\left(1 + \frac{\tau_v}{\tau_r}\right)\frac{G}{2}\, f(G,\Omega) \ , \tag{3.68}$$

where $q_i(J)$ is the portion of molecules at the rotational sublevel J of the ith vibration state, which is the factor that determines the population $n_1(v,J)$

$$n_i(v,J) = q_i(J)W(v)N_i \ , \tag{3.69}$$

τ_v and τ_r denote the times of vibrational and rotational relaxations; the saturation parameter G determines the degree of saturation of the considered rotational-vibrational transition with vibrational and rotational relaxations,

$$G = 2\sigma_0 P\left(\frac{1}{\tau_v} + \frac{1}{\tau_r}\right)^{-1} \ . \tag{3.70}$$

For rotational sublevels with $J \gg 1$, we assume that $q_1 = q_2 = q$. The population factor q for one rotational molecular sublevel usually ranges between 10^{-2} and 10^{-3}. For example, for a molecule of CO_2

$$q(J) = (2J+1)\frac{B}{kT}\exp\left[-\frac{BJ(J+1)}{kT}\right] \qquad (3.71)$$

and at the transition P(20) with $T = 300$ K $q = 0.03$.

For weak rotational relaxation, (3.68) is transformed to (3.64), provided that the population difference of the levels equals $(N_1^0 q_1 - N_2^0 q_2)$ and that the constants of population relaxation $\gamma_1 = \gamma_2 = 1/\tau_v$. As the rate of rotational relaxation increases, the number of molecules excited by field from the lower vibrational level to the upper grows by $(1 + \tau_v/\tau_r)$ times. This is conditioned by an additional "inflow" of molecules to the lower resonant rotational sublevel from other sublevels, due to rotational relaxation. The amplitude of a resonant dip and its field broadening depend on the relation between saturation parameter (3.70) of rotational-vibrational transition G and that of the total vibration band,

$$G_v = G\left(1 + \frac{\tau_v}{\tau_r}\right) = 2\sigma_0 P\tau_v \ . \qquad (3.72)$$

If the rotational relaxation is appreciable $(\tau_r \ll \tau_v)$, $G_v \gg G$. Thus, strong saturation of the whole band $(G_v \gg 1)$ and transfer of a considerable part of the molecules to the excited state can be realized at comparatively weak saturation of the resonant rotational-vibrational transition $(G \ll 1)$. This means that considerable values of peak contrast of excited-molecule density can be obtained with a small power broadening of the rotational-vibrational resonant transition.

The dependence of dip depth and width in the density of excited particles on the parameters G_v and τ_v/τ_r may be derived from (3.68). Figure 3.27 illustrates the shape of a resonant dip in the density of excited particles (or fluorescence intensity) at varying limiting values of G_v and τ_v/τ_r.

The case of weak rotational relaxation ($\tau_v \ll \tau_r$, and hence $G \simeq G_v$) is fully analogous to that of a two-level transition. In particular, the dip contrast h is maximum $(h^{max} \simeq 0.3)$ with $G_v \gg 1$ (curve 2), and $G_v \simeq 1$, when the dip is broadened slightly by field (by $\sqrt{2}$ times) and contrast is large enough ($h \simeq 0.1$, curve 1), can be considered an optimum value of the saturation parameter G_v. Because of strong rotational relaxation $(\tau_v \gg \tau_r)$, the absolute dip value with an optimum value of saturation parameter $G \simeq 1$ $(G_v \simeq \tau_v/\tau_r \gg 1)$ increases τ_v/τ_r times (curve 4). The maximum amplitude of the dip that can be attained with $G_v \simeq 2.8\,(\tau_v/\tau_r)$ increases by the same factor (curve 3).

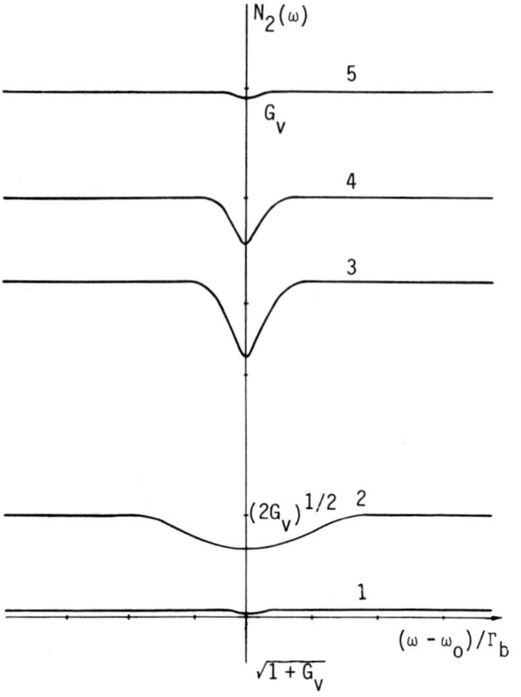

Fig.3.27 Shape of narrow dip in density of excited molecules $N_2(\omega)$ (in units $(\sqrt{\pi}/2)(N_1^0 - N_2^0)(\Gamma/ku)q$) for various saturation parameters of total band G_v and resonant rotational-vibrational transition G and various ratios of vibrational relaxation time τ_v and rotational relaxation time τ_r: 1 and 2 - $\tau_v/\tau_r < 1$, $G \simeq G_v$, $G_v \simeq 1$ and $G_v \gg 1$; 3, 4 and 5 - τ_v/τ_r, $G_v \gg 1$, $\tau_v/\tau_r \simeq (1/3)G_v$, $\tau_v/\tau_r \simeq G_v$, $\tau_v/\tau_r \gg G_v$

During very strong rotational relaxation ($\tau_v/\tau_r \gg G_v$), dip broadening by a strong field disappears, but in this case the dip depth decreases $\sim(\tau_r/\tau_v)$ (curve 5).

3.3.2 Fluorescence-Cell Method and Connected Methods

As is noted previously, the most important feature of the method of fluorescent dip is that it enables us to study transitions with extremely low absorption. That is why such a method was put forward to detect narrow resonances inside the Doppler-broadened line of an atomic or molecular beam (BASOV and LETOKHOV [3.5]). Lately, this possibility has been experimentally realized by SNYDER and HALL [3.105]. They measured, in a new way, the relativistic Doppler effect in a beam of accelerated Na atoms. The narrow dip at the center of the Doppler-broadened absorption line $1s_5 - 2p_2$ (5882 Å) of fast Na atoms in the beam was registered by the dip of fluorescence intensity at the upper level at 6599 Å (transition $2p_2 \rightarrow 1s_2$). The method of fluorescent dip was also successfully employed by Javan and Freed to investigate molecular transitions with weak absorption. The sensitivity of the method is so high that it permits using the transition between the excited levels of the CO_2 molecule, which at 300 K has absorption coefficient $1.5 \cdot 10^{-6}$ cm^{-1}. When highly sensitive ir photodetectors (Ge:Cu) with a large receiving area are used, it is possible to detect resonances in CO_2 and N_2O molecules (with CO_2 and N_2O lasers, respectively) at a pressure of 10^{-3}-10^{-4} torr [3.87].

It is clear that, because the photodetector receives no radiation of the strong-wave saturating absorption, the noise intensity in the photodetector must drop off, and hence the sensitivity of the nonlinear fluorescence method must be higher than that of the nonlinear absorption method when the resolutions are the same. SHIMODA [3.106] calculated the sensitivity of narrow-resonance recording by these two methods and found that the minimum number of molecules detected by nonlinear fluorescence is approximately 10^{-4} as much as that detected by nonlinear absorption.

To increase the sensitivity of recording resonances of fluorescence intensity, it is advisable to saturate absorption by independent, strong, counter-running waves whose intensities are modulated with different frequencies F_1 and F_2. A fluorescence signal modulated at the sum and difference frequencies contains information on the dip value. Such a method overcomes permanent fluorescence and parasitic-light backgrounds. In [3.58] this technique was used to resolve the hyperfine structure of the lines P(13), R(45) (43-0) of the I_2 molecule. An argon laser at 5145 Å was used. Figure 3.28 shows the fluorescence intensity nonmodulated and modulated at the sum frequency ($F_1 + F_2 = 2000$ Hz) at an iodine pressure of 10^{-3} torr, which was obtained in [3.58]. The unmodulated signal describes the contour of the total absorption line and the modulated signal describes the hyperfine structure of the absorption line of the iodine molecule.

Resonant changes of the total number of particles at the upper level can be revealed by other methods [3.107], in addition to recording the spontaneous radiation of excited particles (Fig.3.29). For example, it is possible to measure the coefficient of absorption of related transitions between the upper level of the saturated transition and higher atomic and molecular states, including transitions to the continuous spectrum (Fig.3.29b). The absorbing cell may also contain another gas that has a level with a short radiative decay rate, close to the excited level of the transition that is being saturated (Fig.3.29c). In this case, a narrow resonance can be observed in the fluorescence intensity of the added molecules. Such a method

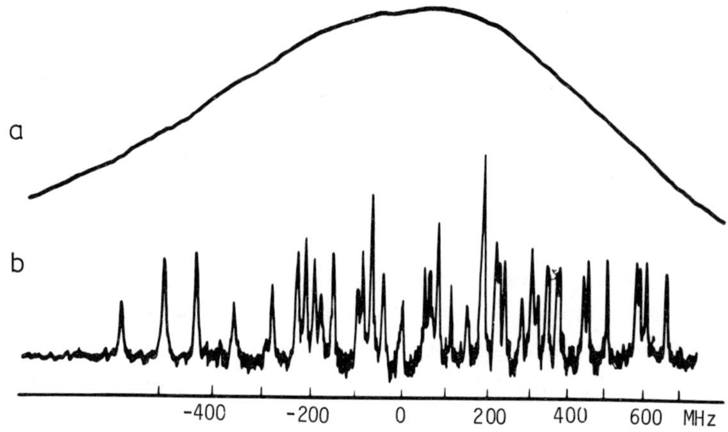

Fig.3.28 Unmodulated (a) fluorescence intensity and the fluorescence modulated (b) at the sum frequency $F_S = F_1 + F_2$, as a function of frequency, of argon-ion laser at $\lambda = 514.53$ nm which saturates the transitions P(13), R(15), 43-0 I_2. Unmodulated signal gives the contour of the total absorption line, and the modulated signal gives the hyperfine structure of the absorption lines (SOREM and SCHAWLOW [3.58])

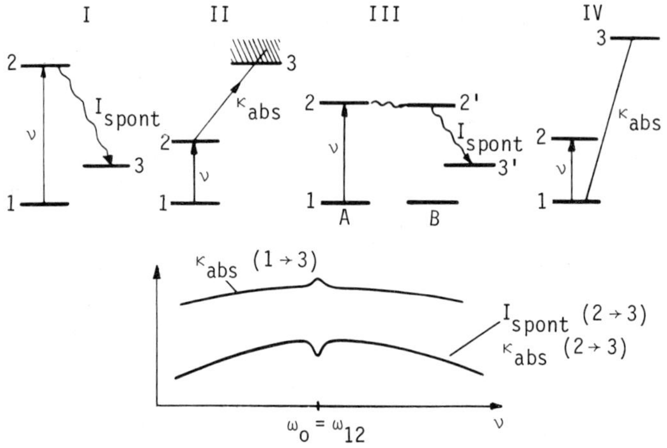

Fig.3.29 Various methods for detection of the resonant change of the total population of the upper (2) and lower (1) levels of the saturated transition (LETOKHOV [3.107])

may be useful for absorbing molecules that have a low quantum yield of luminescence. Resonant changes of the number of particles at the lower level of the transition can be detected by changes of the absorption coefficient at an upward transition (Fig.3.29d). Instead of a resonant minimum of excited particle density, in this case, a resonant peak may be observed in the density of excited particles in the ground state, because the populations of the levels of a saturated transition are related by

$$(N_1 - N_1^o) = -(N_2 - N_2^o) \quad . \tag{3.73}$$

Use of different ways of observation (Fig.3.29) of resonances in the total population of lower and upper levels of a saturated transition is of interest for identification of molecular transitions. By modulation of the standing-wave intensity and observing modulated fluorescence at different spectral lines,molecular transitions from the upper state may be detected. In much the same way, by observation of modulated absorption from the lower state, the transitions that terminate at the lowest level may be identified. In all of these cases it is possible to obtain resolution inside the Doppler contour of spectral lines. Not long ago, this important advantage of narrow density resonances was demonstrated in [3.108] to identify 113 lines of the Na_2 molecule.

Nonlinear fluorescence can be used to study absorption lines only 10^{-4} as intense as the weakest that can be studied by the method of nonlinear absorption, in particular, absorption lines at low-absorption molecular transitions between excited levels, vibrational-overtone transitions, and quadrupolar vibrational transitions of homonuclear molecules forbidden to a dipole approximation. It is advisable to use resonant dips in the density of excited molecules also to detect the narrow resonances of two-photon absorption considered in Chapter 4. To detect changes of the number of excited atoms and molecules, all methods for observing excited particles known in experimental physics can be used. In principle, it is possible to attain very high sensitivity of narrow-resonance detection at any electron-vibrational-rotational or vibrational-rotational transitions. In order to detect the density of excited molecules, we may, for example, employ subsequent photoionization of excited molecules by vuv laser radiation (selective two-step photoionization of molecules [3.109]) as well as detection of molecular ions. This method may lead to the development of novel quantum-laser detectors for trace amounts of complex molecules (LETOKHOV [3.110]) that will embody very high resolution (Doppler-free spectroscopy) and very high sensitivity (counting of individual molecules).

3.4 Resonance Effects at Mode Interaction

In Section 3.1 the output radiation properties of single-frequency lasers with saturated absorption were considered. Experimentally, such operation was obtained with a small excess of amplification over threshold. The conditions for single-frequency oscillation were therewith automatically met. With increase of amplification, when laser action may occur at several modes, in lasers with nonlinear absorption, some new effects arise that are consequences of resonant interaction of fields that result from inhomogeneous broadening of amplification and absorption lines. The oscillation regime of a laser with nonlinear absorption, and its spectrum depend on the amplification-absorption ratio, the positions of the modes with respect to the absorption and amplification lines, the homogeneous widths in the amplifying and absorbing media, and the frequency spacing between the modes. Of the whole variety of effects in the nonlinear absorption laser, let us consider just those directly related to the production of dips in absorption and amplification lines, and production of narrow resonances of output power owing to interaction of several modes. Resonance effects that lead to laser-mode selection are considered in Chapter 9.

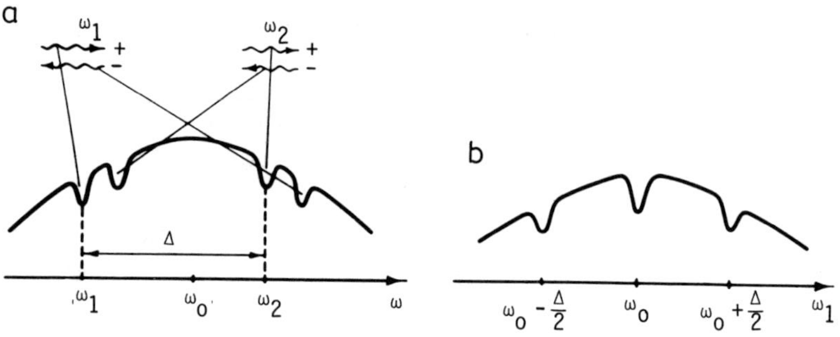

Fig.3.30a,b Interaction of two axial modes at the frequencies ω_1 and ω_2: (a) the shape of the Doppler-broadened line in the field of two axial modes and (b) the coefficient of saturated amplification (absorption) of the mode at the frequency ω_1 in the presence of the mode at the frequency ω_2

3.4.1 Interaction of Two Axial Modes

Let us consider the interaction of two axial modes at frequencies ω_1 and ω_2 within the Doppler-broadened amplification (absorption) line. Each mode is the sum of two counter-running waves and burns two holes in the Doppler profile. Figures 3.30a shows the locations of the holes and the running waves that burn these holes. When frequencies ω_1 and ω_2 are symmetric about the Doppler contour, two holes formed by the waves ω_1^+ and ω_2^- as well as those formed by ω_1^- and ω_2^+ coincide. These are due to the interactions between the counter-running waves at the different frequencies. The coefficient of saturated amplification (absorption) of a mode as a function of its frequency ω_1 has the form shown in Fig.3.30b. In addition to the Lamb dip, there are two symmetric dips at the frequencies $\omega_0 \pm (\Delta/2)$ at the center of line. This effect, termed "mode-crossing" can be observed in an amplifying (absorbing) cell located in the field of two axial laser modes outside the cavity. It is difficult to observe this effect directly in a gas laser, because the interaction of the waves ω_1^+ and ω_2^- as well as ω_1^- and ω_2^+ results in strong competition of axial modes whose frequencies are symmetric. Because of competition, the oscillation in one of the axial modes is suppressed and the laser action of two axial modes in symmetric positions becomes unstable.

The theory of two-mode oscillation in gas lasers was first developed by LAMB [3.111]. The output power of each mode is given by formulas of the type of (3.6), which are added to terms that describe amplification saturation at the frequency of one mode under the action of the field in the other mode. To a weak-saturation approximation, the equations for field intensities of each mode have the form,

$$\frac{dP_1}{dt} = 2\alpha_1 P_1 - 2\beta_1 P_1^2 - 2\theta_{12} P_1 P_2 \ , \quad \frac{dP_2}{dt} = 2\alpha_2 P_2 - 2\beta_2 P_2^2 - 2\theta_{21} P_1 P_2 \ , \quad (3.74)$$

where Lamb's notation is used: α_i is the coefficient that specifies the "net" linear amplification of the ith mode, β_i is the coefficient that specifies the "self-saturation" of amplification of each mode, θ_{12} and θ_{21} are coefficients that specify the "cross-saturation" of amplification in one mode by the field of the other mode. These coefficients are related to the notations introduced in Section 3.1 by

$$\alpha_i = \frac{1}{2}\left(c\kappa_{ao} - \frac{\omega}{Q_i}\right) = \frac{1}{2}\,c(\kappa_{ao} - \gamma_{oi}) \ ,$$

$$\beta_i = \frac{1}{8}\,c\kappa_{ao}\sigma_o\left[1 + L\left(\frac{\omega_i - \omega_a}{\Gamma_a}\right)\right] \ , \qquad (3.75)$$

$$\theta_{12} = \theta_{21} \simeq \frac{1}{8}\,c\kappa_{ao}\sigma_a L\left[\frac{1}{\Gamma_a}\left(\frac{\omega_1 + \omega_2}{2} - \omega_a\right)\right] \ ,$$

where the expression for the cross-saturation coefficients is given for the case $\Delta = c/2L \gg 2\Gamma_a$. If the modes are detuned far from the symmetric position, the coupling coefficient $\theta_{ij} \ll \beta_i$ and the condition of so-called soft coupling of modes is met,

$$\beta_1\beta_2 \gg \theta_{12}\theta_{21} \ . \tag{3.76}$$

In this case, the axial modes oscillate simultaneously and practically independently of one another. In a symmetric position, $\beta_1\beta_2 \simeq (1/4)\theta_{12}\theta_{21}$, and the coupling of modes becomes "hard". As a result, the simultaneous oscillation of two modes becomes unstable. Figure 3.31 illustrates the frequency dependence of output power of each gas-laser mode under oscillation of two axial modes. The experiments reported in [3.113,114] confirm this picture.

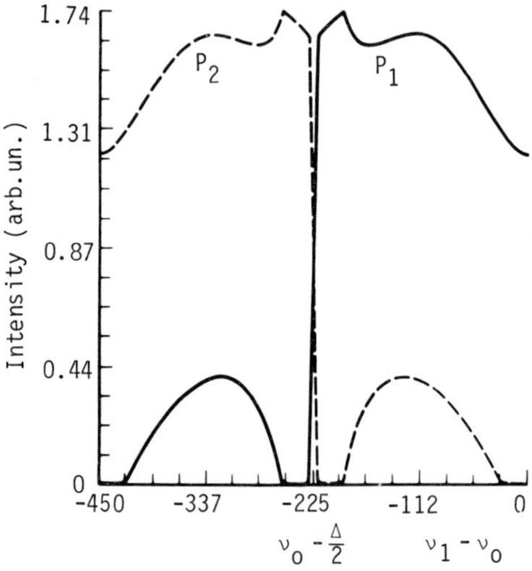

Fig.3.31 Variation of mode intensity when the mode frequency ν_1 scans through the region $\nu_0 - \Delta/2$ in a gas laser without saturated absorption ($\Delta = c/2L = 450$ MHz, $\Delta\nu_D = 1000$ MHz, $\gamma_1/2\pi = 40$ MHz, $\gamma_2/2\pi = 20$ MHz, $\Gamma_a/\pi = 100$ MHz, the excess of pumping power over threshold is 1.2). The dotted line shows the behavior of the second mode intensity at the frequency $\nu_2 = \nu_1 + \Delta$ (SARGENT and SCULLY [3.112])

In saturated-absorption gas lasers, the axial-mode interaction brings about contrary effects. If ω_b is close to ω_a, so that the areas of mode interaction due to nonlinear absorption $\omega_b \pm (\Delta/2)$ are coincident with the areas of their competition $\omega_a \pm (\Delta/2)$, owing to nonlinear interaction in an amplifying medium, nonlinear absorption can suppress the competition of laser modes. As a result, simultaneous stable oscillation of two symmetric axial modes becomes possible. When the frequencies of two modes ω_1 and ω_2 are scanned along the Doppler contour at the same time (the mode-frequency difference of them should be constant), which is easily achieved by scanning the cavity length L, the total laser-output power, as well as the power in each mode, has an additional peak when the mode frequencies are symmetric about ω_b, apart from those that arise when the frequency of each mode coincides with the center of the absorption line (Fig.3.32). Additional peaks of output power were experimentally observed in a He-Ne laser at 3.39 μm with a CH_4 nonlinear-absorption cell, as reported in [3.115]. The peak width was 300 kHz; the contrast was about 3%. In that work the frequency of two He-Ne lasers was stabilized with power peaks at $\omega_b \pm (\Delta_1/2)$ and $\omega_b \pm (\Delta_2/2)$ frequencies, where Δ_1 and Δ_2 was the difference between the frequencies of the modes in the two lasers.

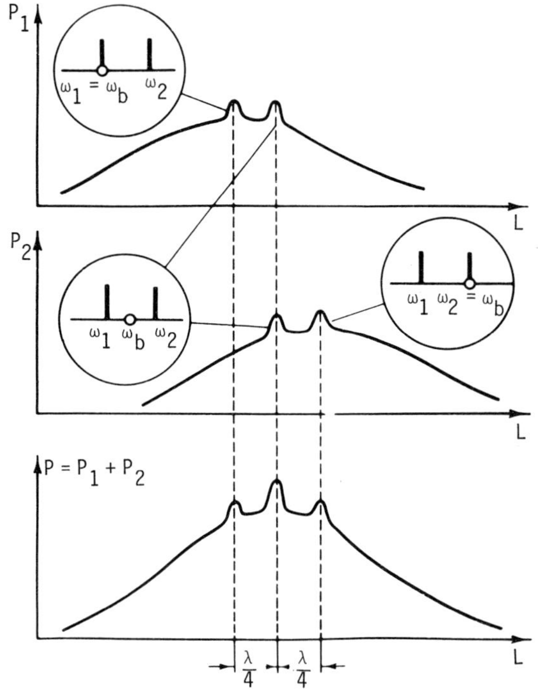

Fig.3.32 Variations of the output power of each axial mode (P_1 and P_2) and of the total mode power $P = P_1 + P_2$ when their frequencies ω_1 and ω_2 are scanned along the Doppler profile of an absorption line by changing the cavity length L

Power peaks caused by mode crossing appear; these are due to particles whose velocity projections onto wave direction comply with

$$kv = \pm \frac{\Delta}{2} = \pm\pi \frac{c}{2L} \; . \tag{3.77}$$

Because of this, there is an important difference between such resonances and Lamb dips, which occur with $kv = 0$. This characteristic property can be used to measure the ratio between collisional broadening of resonances and broadening due to particle velocities (see Subsec.3.2.2).

Nonlinear absorption also affects the ratios of mode intensities. When the intensity of a mode increases with decrease of intensity of another mode in a narrow frequency range, competitive resonances may occur. Such power resonances, owing to strong nonlinear coupling of modes, may be much narrower than the homogeneous width 2Γ. In [3.116] power resonances were observed experimentally in a He-Ne/CH_4 laser. The competitive power peak was 1/4 to 1/5 as wide as the ordinary power peak. The common feature of competitive power resonances, aside from additional narrowing, is their high contrast. Any small losses in the region of strong competition of modes always result in a considerable increase of peak power. To do this, for example, it is sufficient that the resonance dip in the absorption line at the frequency ω_b should fall within the region of mode competition near the $\omega_a \pm (\Delta/2)$ frequencies.

Competition power resonances are of interest for stabilizing laser frequency by use of a nonlinear absorption cell, because peak narrowing and associated contrast enhancement facilitate the work of the servosystem that stabilizes the oscillation frequency on the power peak. The first experiments of this kind were carried out with He-Ne/CH_4 lasers at 3.39 μm by BASOV et al. [3.117].

3.4.2 Interaction of Two Counter-Running Waves in a Ring Laser

In a ring-cavity laser, counter-running waves are not coupled rigidly together, as compared with Fabry-Perot cavity lasers. Therefore, when the oscillation frequency is far from the center of an inhomogeneously broadened amplification line ($|\omega - \omega_a| \gg \Gamma_a$), the two counter-running waves are not coupled by the active medium. In this case, the laser operates on two counter-running waves, and the total laser field represents a standing wave. When laser oscillation takes place near the center of the amplification line ($|\omega - \omega_a| \ll \Gamma_a$), the counter-running waves interact with the same particles. The competition of the counter-running waves causes one of the waves to be suppressed and the intensity of the other to rise. Thus, under the running-wave regime, oscillation occurs in a very narrow frequency range $\delta\omega_a$ near the center of Doppler line ω_a, whereas far from an inhomogeneously broadened amplification line, oscillation occurs in a standing wave. The change between these conditions occurs with a constant total intensity of the waves. The counter-running wave amplitudes change sharply during such a transition;

the frequency dependence of the output power of each wave contains resonances with half-width $\delta\omega_a \ll \Gamma_a$ called competitive resonances. Narrow resonances of of competition in ring lasers were first observed by LISITSYN and TROSHIN [3.118] in a He-Ne laser at 6328 Å. Saturation conditions being achieved, one of the counter-running waves was seen to be suppressed when the oscillation frequency passed through the center of Doppler amplification line, and the intensity of the other wave increased (Fig.3.33). The total intensity of the two waves had no such resonant rises, that is the energy was transferred from one direction to another. The widths of the competitive resonances in the power of the running waves amounted to several MHz, i.e., it was one tenth or less as wide as the homogeneous width of the amplification line.

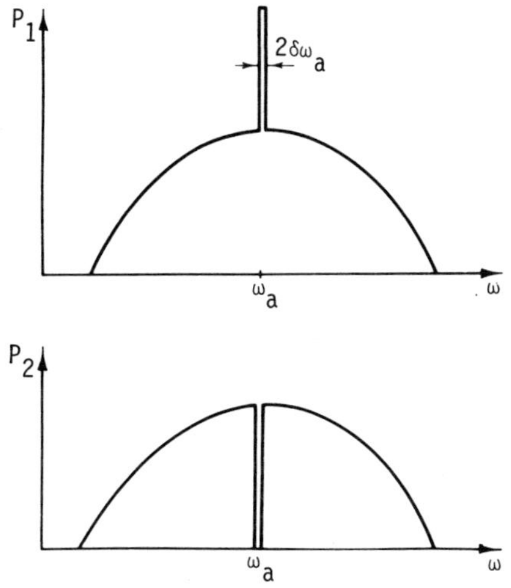

Fig.3.33 Dependence of the intensity of the counter-running waves P_1 and P_2 in a ring gas laser on the oscillation frequency ω

The change of generation conditions in a ring laser can be illustrated. To an approximation that allows only for the effect of spectral burning of Bennett holes in the Doppler profile, the total coefficient of saturated amplification of two counter-running waves with frequency ω and amplitude E that form a standing wave will be

$$\kappa_a^{st} = \kappa_{ao}\left\{1 - \frac{1}{2} G_a\left[1 + L\left(\frac{\omega - \omega_a}{\Gamma_a}\right)\right]\right\} \quad , \tag{3.78}$$

where G_a is the saturation of amplification by one running wave. With $\omega \neq \omega_a$ κ_a^{st} is always larger than the amplification coefficient for a travelling wave with the same frequency but with double power (Fig.3.34a)

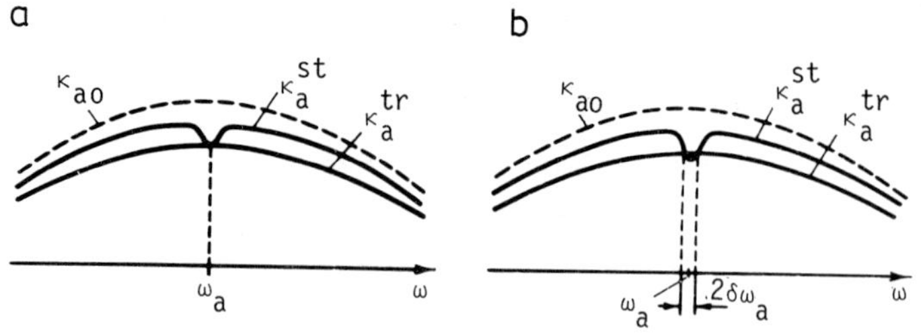

Fig.3.34a,b Frequency dependence of the amplification of the travelling wave κ_a^{tr} and the standing wave κ_a^{st} for a Doppler-broadened line (a) taking no account of, and (b) taking account of, the spatial burning effect

$$\kappa_a^{tr} = \kappa_{ao}(1 - G_a) \ . \tag{3.79}$$

However, if we consider the spatial effect in amplification saturation caused by travelling and standing waves, which arises because a travelling wave "burns" amplification uniformly over space, whereas a standing wave "burns" it mainly in field loops, (3.78) and (3.79) take the forms [3.119],

$$\kappa_a^{st} = \kappa_{ao}\left\{1 - \frac{G_a}{2}\left[1 + L\left(\frac{\omega - \omega_a}{\Gamma_a}\right) + \frac{\Gamma_a\gamma_a - 2\Gamma_a^2}{(ku_a)^2}\right]\right\} \tag{3.80}$$

or

$$\kappa_a^{tr} = \kappa_{ao}\left\{1 - G_a\left[1 - \left(\frac{\Gamma_a}{ku}\right)^2\right]\right\} \ , \tag{3.81}$$

where $2/\gamma_a = [(1/\gamma_{1a}) + (1/\gamma_{2a})]$. It follows from (3.80) and (3.81) that in the region of small detuning of the oscillation frequency ω with respect to the center of amplification line ω_a,

$$|\omega - \omega_a| < \delta\omega_a \frac{\sqrt{\Gamma_a\gamma_a}}{ku_a} \Gamma_a \ , \tag{3.82}$$

the inequality $\kappa_a^{tr} > \kappa_a^{st}$ exists (Fig.3.34b), and hence the regime of standing wave is changed by those of running wave.

The change of lasing conditions in a ring gas laser with nonlinear absorption considered by BASOV et al. [3.120] may be explained in a like manner. The nonlinear-absorption medium is described by (3.80) and (3.81) with index "a" replaced by "b", which as before corresponds to the absorbing medium.

Then a ring laser with a nonlinear-absorption cell will operate on a standing wave, provided that

$$(\kappa_a^{st} - \kappa_b^{st}) > (\kappa_a^{tr} - \kappa_b^{tr}) \tag{3.83}$$

and on a travelling wave otherwise. Condition (3.83) can be written in the form

$$\phi_a(\omega) > \beta\phi_b(\omega) \ , \quad \beta = \frac{\kappa_{bo}g_b}{\kappa_{ao}g_a} \ . \tag{3.84}$$

The functions $\phi_f(\omega)$ $(f = a,b)$ have the form

$$\phi_f(\omega) = 1 - L\left(\frac{\omega - \omega_f}{\Gamma_f}\right) - \frac{\Gamma_f\gamma_f}{(ku_f)^2} \tag{3.85}$$

and describe Lamb dips in amplification and absorption lines, with a correction for the effect of spatial burning of the gas medium (see Fig.3.34). Condition (3.84) is a generalization of the condition of standing-wave stability in ring gas lasers $\phi_a(\omega) > 0$ defined in [3.121]. The fulfillment of condition (3.84) depends completely on the relation between Lamb dips ϕ_a and ϕ_b (Fig.3.35). Note that the correction for the effect of spatial burning of the absorbing medium is not essential here because the predominant effect of one or another dip depends on the spacing between dips, i.e., on the distance between the center of the amplification and absorption lines.

Thus, with nonlinear absorption introduced into a ring laser, the two counter-running waves saturate absorption together, and the running-wave competition is suppressed. As a result, if the oscillation frequency lies in the region of standing-wave instability, the lasing conditions in a narrow frequency range near the center of the absorption line (Fig.3.35) become again stable. The width of the resultant narrow competitive resonance, that is the region of detuning $\delta\omega_b$ where the standing-wave regime is stable, is given by (3.83) and (3.84)

$$|\omega - \omega_b| < \delta\omega_b = \frac{\Gamma_b}{\Gamma_a}(\Delta_o^2 - \delta\omega_a^2)^{1/2} \ , \tag{3.86}$$

where $\Delta_o = |\omega_a - \omega_b|$, the value of $\delta\omega_a$ is determined by (3.82), and it is assumed that $\Gamma_b \ll \Gamma_a \ll ku$. When $\Delta_o \ll \delta\omega_a$, close to the center of amplification line the regime of standing-wave operation in the cavity does not exist in

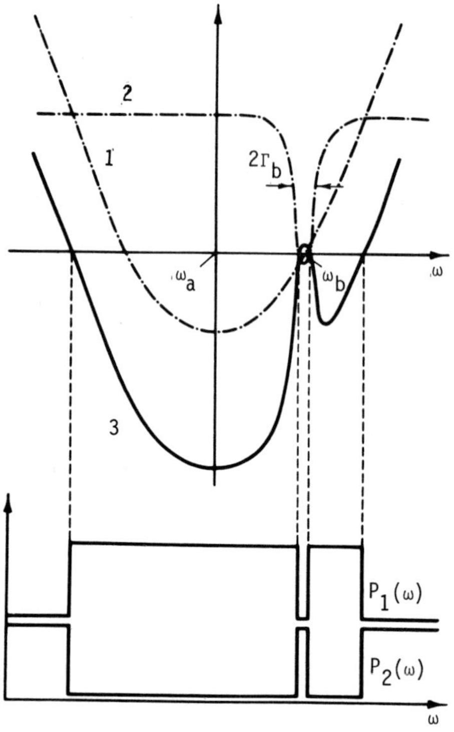

Fig.3.35 Occurrence of a competing resonance in a ring laser with a saturable gas absorber: 1 and 2 are Lamb dips in the difference between the amplification coefficients of a two-component medium, under conditions that generate standing and travelling waves with equal intensities; P_i are the intensities of the counter-running waves as functions of oscillation frequency

the model under consideration. With $\Delta_o = \delta\omega_a$, the dip effect in the amplification line compensates exactly for that in the absorption line (Fig.3.35) or, which is the same, the curve $\phi_a(\omega)$ comes in contact with $\beta\phi_b(\omega)$ at the point $\omega = \omega_b$ and the competitive-resonance width is zero. As Δ_o increases, the effect of the transmission peak in the absorption line prevails, and the competitive-resonance width increases according to (3.86).

Competitive-power resonances in ring lasers with nonlinear absorption have been studied in some detail by BASOV et al. [3.120,122-125] using a He-Ne/CH_4 laser. Figure 3.36 illustrates variations of intensity of counter-running waves when the cavity frequency is scanned along the Doppler contour. Production of resonances with width $\delta\omega_a \ll 2\Gamma_b$ has been observed [3.124]. The competitive resonances have considerable absolute amplitude and high

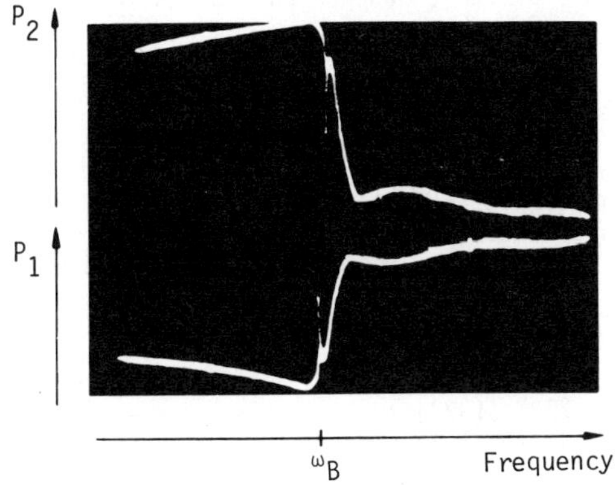

Fig.3.36 Intensity of the counter-running waves in the ring He-Ne laser with a CH_4 nonlinearly absorbing cell, as a function of the cavity frequency which was scanned over a section of the Doppler contour of the amplification line near the frequency ω_b. The width of the competing resonance equals 600 kHz (BASOV et al. [3.122])

contrast, which is of interest in stabilizing the frequency of gas lasers through narrow resonances.

Splitting of counter-running wave frequencies is another interesting resonance effect that occurs in ring gas lasers with nonlinear-absorption cells. The effect has been studied theoretically in [3.126] and should be verified by experiments.

References

3.1 V.S. Letokhov: Pis'ma Zh.Eksp.i Teor.Fiz. 6, 597 (1967)
3.2 V.N. Lisitsyn, V.P. Chebotayev: Zh.Eksp.i Teor.Fiz. 54, 419 (1968)
3.3 P.H. Lee, M.L. Skolnick: Appl.Phys.Lett. 10, 303 (1967)
3.4 V.S. Letokhov, V.P. Chebotayev: Pis'ma Zh.Eksp.i Teor.Fiz. 9, 364 (1969)
3.5 N.G. Basov, V.S. Letokhov: Report on URSI Conference "Laser Measurements", Sept. 1968, Warsaw, Poland; Electron Technology 2, 2/3, 15 (1969)
3.6 C. Freed, A. Javan: Appl.Phys.Lett. 14, 53 (1970)
3.7 B.L. Borovich, V.S. Zuev, V.A. Scheglov: Zh.Eksp.i Teor.Fiz. 49, 1031 (1965)
3.8 A.P. Kazantzev, S.G. Rautian, G.I. Surdutovich: Zh.Eksp.i Teor.Fiz. 54, 1409 (1968)
3.9 H. Greenstein: J.Appl.Phys. 43, 1732 (1972)
3.10 O.N. Kompanetz, V.S. Letokhov, V.V. Nikitin: Preprint Lebedev Physical Institute, No. 66, Moscow, USSR (1968)
3.11 V.P. Chebotayev, I.M. Beterov, V.N. Lisitsyn: IEEE J.QE-4, 788 (1968)
3.12 S.N. Bagaev, Yu.D. Kolomnikov, V.N. Lisitsyn, V.P. Chebotayev: IEEE J.QE-4, 868 (1968)

3.13 V.N. Lisitsyn: Thesis, Institute of Semiconductor Physics, Novosibirsk, USSR (1969)
3.14 V.M. Tatarenkov, A.N. Titov, A.V. Uspenskii: Optika i Spectroscopia 28, 572 (1970)
3.15 A.N. Titov: Thesis, VNIIFTRI-MFTI, Moscow, USSR (1970)
3.16 V.M. Tatarenkov, A.N. Titov: Optika i Spectroscopia 30, 803 (1971)
3.17 G.R. Hanes, K.M. Baird: Metrologia 5, 32 (1969)
3.18 G.R. Hanes, C.E. Dahlstrom: Appl.Phys.Lett. 14, 363 (1969)
3.19 J.D. Knox, Y.-H. Pao: Appl.Phys.Lett. 16, 129 (1970)
3.20 G.R. Hanes, J. Lapierre, P.R. Bunker, K.S. Shotton: J.Molec.Spectros. 39, 506 (1971)
3.21 G.R. Hanes, K.M. Baird, J. DeRemigis: Appl.Optics 12, 1600 (1973)
3.22 A.J. Wallard: J.Phys.E.Sci.Instr. 5, 926 (1972)
3.23 W.G. Schweitzer, Jr., E.G. Kessler, Jr., R.D. Deslattes, H.P. Layer, J.R. Wetstone: Appl.Optics 12, 2927 (1973)
3.24 J.T. LaTourette, R.E. Eng: Digest 7th Intern.Quant.Electr.Conf., May 1972, Montreal, Canada, p.43
3.25 W.G. Schweitzer, Jr.: Appl.Phys.Lett. 13, 367 (1968)
3.26 V.N. Lisitsyn, V.P. Chebotayev: Optika i Spectroscopia 26, 856 (1969)
3.27 R.L. Barger, J.L. Hall: Phys.Rev.Lett. 22, 4 (1969)
3.28 N.G. Basov, M.V. Danileiko, V.V. Nikitin: Zh.Prikl.Spectroscopii 11, 543 (1969)
3.29 J.L. Hall: In *Lectures in Theor. Phys.*, XII, ed. by K.T. Manauthega, W.E. Botin (Gordon Brendi and Co., New York 1970), p.161
3.30 S.N. Bagaev, V.P. Chebotayev: Pis'ma Zh.Eksp.i Teor.Fiz. 16, 614 (1972)
3.31 S.N. Bagaev, E.V. Baklanov, V.P. Chebotayev: Pis'ma Zh.Eksp.i Teor. Fiz. 16, 15 (1972)
3.32 S.N. Bagaev, E.V. Baklanov, V.P. Chebotayev: Pis'ma Zh.Eksp.i Teor. Fiz. 16, 344 (1972)
3.33 J.L. Hall, C. Borde: Phys.Rev.Lett. 30, 1101 (1973)
3.34 R.L. Barger, J.L. Hall: Appl.Phys.Lett. 22, 196 (1973)
3.35 J.L. Hall: *Proc. Conf. "Methodes de Spectroscopie sans largeur Doppler de niveaux excited de systems moleculaires simples"*, May 1973 (Publ. No. 217 CNRS, Paris 1974), p.105
3.36 S.N. Bagaev, E.V. Baklanov, E. Titov, V.P. Chebotayev: Pis'ma Zh.Eksp. i Teor.Fiz. 20, 292 (1974)
3.37 A.C. Luntz, R.G. Brewer: J.Chem.Phys. 53, 3380 (1970)
3.38 A.C. Luntz, R.G. Brewer, H.L. Foster, J.D. Swalen: Phys.Rev.Lett. 23, 951 (1969)
3.39 A.C. Luntz, R.G. Brewer: J.Chem.Phys. 54, 3641 (1971)
3.40 K. Uehara: J.Phys.Soc.Japan 34, 777 (1973)
3.41 Yu.M. Malushev, V.M. Tatarenkov, A.N. Titov: Pis'ma Zh.Eksp.i Teor. Fiz. 13, 592 (1971)
3.42 J.A. Maguar, J.L. Hall: Digest 7th Intern.Quant.Electr.Conf., May 1972, Montreal, Canada, p.44
3.43 W. Radloff, E. Below: Optics Comm. 13, 160 (1975)
3.44 A.C. Luntz, J.D. Swalen, R.G. Brewer: Chem.Phys.Lett. 14, 512 (1972)
3.45 M. Takami, K.S. Shimoda: Japan.J.Appl.Phys. 11, 1648 (1972)
3.46 Yu.V. Brzhazovsky, V.P. Chebotayev, L.S. Vasilenko: IEEE J.QE-5, 146 (1969); Zh.Eksp.i Teor.Fiz. 55, 2096 (1968)
3.47 V.S. Vasilenko, V.P. Chebotayev, G.I. Shershneva: Optika i Spectroscopia 19, 204 (1970)
3.48 T. Kan, G.J. Volga: IEEE J.QE-7, 141 (1971)
3.49 L.S. Vasilenko, M.I. Skvortzov, V.P. Chebotayev, G.I. Shershneva, A.V. Shishaev: Optika i Spectroscopia 32, 1123 (1972)
3.50 S.N. Bagaev, et al.: Rept. 2nd Nat.Symp.Gas Laser Physics, Novosibirsk (June 1975)
3.51 V.I. Bobrik, Yu.D. Kolomnikov, V.P. Chebotayev: Optika i Spectroscopia 35, 1179 (1975);
V.I. Bobrik, Yu.D. Kolomnikov, B.S. Mogil'nitzkii: Rept. 2nd Nat.Symp. Gas Laser Physics, Novosibirsk (June 1975)

3.52 V.S. Letokhov: Zh.Eksp.i Teor.Fiz. 54, 1248 (1968)
3.53 V.S. Letokhov, B.D. Pavlik: Serial edition "Kvantovai Electronika", N1, 53 (1971)
3.54 V.P. Chebotayev: Rept. 2nd Nat.Symp.Gas Laser Physics, Novosibirsk (June 1975); "Kvantovai Electronika 3, 694 (1976)
3.55 V.N. Lisitsyn, V.P. Chebotayev: Pis'ma Zh.Eksp.i Teor.Fiz. 7, 3 (1968)
3.56 S.N. Bagayev, A.K. Dmitriev: Optika i Spectroscopia 34, 337 (1973)
3.57 M.D. Levenson, A.L. Schawlow: Phys.Rev. A6, 10 (1972)
3.58 M.S. Sorem, A.L. Schawlow: Optics Comm. 5, 148 (1972)
3.59 T.W. Hänsch, M.D. Levenson, A.L. Schawlow: Phys.Rev.Lett. 26, 946 (1971)
3.60 T.W. Hänsch, I.S. Shahin, A.L. Schawlow: Phys.Rev.Lett. 27, 707 (1971)
3.61 Yu.A. Matiugin, B.I. Troshin, V.P. Chebotayev: Optika i Spectroscopia 31, 111 (1971); Digest Gas Laser Physics Symp., Novosibirsk, USSR (June 1969)
3.62 C.V. Shank, S.E. Schwarz: Appl.Phys.Lett. 13, 113 (1968)
3.63 P.W. Smith, T. Hänsch: Phys.Lett. 26, 740 (1971)
3.64 Yu.A. Matiugin, A.S. Provorov, V.P. Chebotayev: Zh.Eksp.i Teor.Fiz. 63, 2043 (1972)
3.65 T.W. Hänsch, I.S. Shahin, A.L. Schawlow: Nature 235, 63 (1972)
3.66 T.W. Hänsch, M.H. Nayfeh, S.A. Lee, S.M. Curry, I.S. Shahin: Phys. Rev.Lett. 32, 1336 (1974)
3.67 E.E. Uzgiris, J.L. Hall, R.L. Barger: Phys.Rev.Lett. 26, 289 (1971)
3.68 C.K.N. Patel: Appl.Phys.Lett. 25, 112 (1974)
3.69 J.W.C. Johns, A.R.W. McKellar, T. Oka, M. Römheld: J.Chem.Phys. 62, 1488 (1975)
3.70 C. Borde: C.R.Acad.Sci.Paris 271, 371 (1970)
3.71 R.G. Brewer: Phys.Rev.Lett. 25, 1639 (1970)
3.72 R.G. Brewer, R.L. Shoemaker, S. Stenholm: Phys.Rev.Lett. 33, 63 (1974)
3.73 S.M. Freund, G. Duxbury, M. Römfeld, J.T. Tiedje, T. Oka: J.Molec. Spectr. 52, 38 (1974)
3.74 N.G. Basov, I.N. Kompanetz, O.N. Kompanetz, V.S. Letokhov, V.V. Nikitin: Pis'ma Zh.Eksp.i Teor.Fiz. 9, 568 (1969)
3.75 P. Rabinowitz, R. Keller, J.T. LaTourette: Appl.Phys.Lett. 14, 376 (1969)
3.76 F. Shimizu: Appl.Phys.Lett. 14, 378 (1969)
3.77 N.G. Basov, O.N. Kompanetz, V.S. Letokhov, V.V. Nikitin: Zh.Eksp.i Teor.Fiz. 59, 394 (1970)
3.78 R.G. Brewer, M.J. Kelly, A. Javan: Phys.Rev.Lett. 23, 559 (1969)
3.79 R.G. Brewer, J.D. Swalen: J.Chem.Phys. 52, 2774 (1970)
3.80 M.J. Kelly, R.E. Francke, M.S. Feld: J. Chem.Phys. 53, 2979 (1970)
3.81 E.R. Petersen, B.L. Danielson: Bull.Amer.Phys.Soc. 15, N11 (1970)
3.82 I.M. Beterov, L.S. Vasilenko, B. Gangardt, V.P. Chebotayev: Kvantovai Electronika 1, 970 (1974)
3.83 Yu.A. Gorokhov, O.N. Kompanetz, V.S. Letokhov, G.A. Gerasimov, Yu.J. Posudin: Optics Comm. 7, 320 (1973)
3.84 O.N. Kompanetz, A.R. Kookoodjanov, V.S. Letokhov, V.G. Minogin, E.L. Mikhailov: Zh.Eksp.i Teor.Fiz. 69, 32 (1975)
3.85 T.W. Meyer, J.F. Brilando, C.K. Rhodes: Chem.Phys.Lett. 18, 382 (1973)
3.86 A.T. Mattick, A. Sanchez, N.A. Kurnit, A. Javan: Appl.Phys.Lett. 23, 675 (1973)
3.87 A. Javan: *Laser Spectroscopy*, Proc. 2nd Intern.Conf. 23-27 June 1975, Megeve, France (Springer-Verlag, Berlin, Heidelberg, New York 1975), p.439
3.88 C. Costain: Can.J.Phys. 47, 2431 (1969)
3.89 R.S. Winton, W. Gordy: Phys.Lett. 32A, 219 (1970)
3.90 O.N. Kompanetz, V.S. Letokhov: Zh.Eksp.i Teor.Fiz. 62, 1302 (1972)
3.91 V.S. Letokhov: In *Proc. Esfahan Symp.Fundamental and Applied Laser Physics*, Sept. 1971, ed. by M.S. Feld, A. Javan, N. Kurnit (Wiley-Interscience Publishers 1973), p.335

3.92 I.S. Shahin, T.W. Hänsch: Optics Comm. 8, 312 (1973)
3.93 T.W. Hänsch, P. Toschek: IEEE J.QE-4, 467 (1968)
3.94 S.M. Hamadani, A.T. Mattick, N.A. Kurnit, A. Javan: Appl.Phys.Lett. 87, 21 (1975)
3.95 A. Abragam: *The Principles of Nuclear Magnetism* (Oxford University Press 1961)
3.96 Yu.A. Vdovin, V.M. Ermachenko, A.L. Popov, E.D. Protzenko: Pis'ma Zh. Eksp.i Teor.Fiz. 25, 401 (1972)
3.97 Im-Tkhek-De, S.G. Rautian, E.G. Saprikin, R.I. Smirnov, A.M. Shalagin: Zh.Eksp.i Teor.Fiz. 62, 1661 (1972)
3.98 N. Ramsey: *Molecular Beams* (Clarendon, Oxford 1956)
3.99 V.S. Letokhov, B.D. Pavlik: Optika i Spectroscopia 32, 856 (1972)
3.100 V.S. Letokhov, B.D. Pavlik: Optica i Spectroscopia 32, 1057 (1972)
3.101 O.N. Kompanetz, V.S. Letokhov: Pis'ma Zh.Eksp.i Teor.Fiz. 14, 20 (1971)
3.102 V.S. Letokhov, B.D. Pavlik: Radiotekhnika i Electronika 17, 1030 (1972)
3.103 E.V. Baklanov, B.Y. Dubetzky, V.P. Chebotayev: Appl.Phys. 9, 177 (1976)
3.104 V.S. Letokhov, B.D. Pavlik: Zh.Eksp.i Teor.Fiz. 64, 804 (1973)
3.105 J.J. Snyder, J.L. Hall: *Laser Spectroscopy*, Proc. 2nd Intern.Conf. 23-27 June 1975, Megeve, France (Springer-Verlag, Berlin, Heidelberg, New York 1975), p.6
3.106 K. Shimoda: Appl.Phys. 1, 77 (1973)
3.107 V.S. Letokhov: Comments on Atomic and Molecular Physics 2, 181 (1971)
3.108 M.E. Kaminsky, R.T. Hawkins, F.V. Kowalski, A.L. Schawlow: Phys.Rev. Lett. 36, 671 (1976)
3.109 R.V. Ambartzumian, V.S. Letokhov: Appl.Optics 11, 354 (1972)
3.110 V.S. Letokhov: Upsekhi Fiz.Nauk 118, 199 (1976)
3.111 W.E. Lamb, Jr.: Phys.Rev. 134A, 1429 (1964)
3.112 M. Sargent, III, M.O. Scully: *Laser Handbook*, Vol.1, ed. by F.T. Areechi, E.O. Schulz-DuBois (North-Holland, Amsterdam 1972), p.45
3.113 R.L. Fork, M.A. Pollack: Phys.Rev. 139A, 1408 (1965)
3.114 S.A. Gonchukov, I.O. Leipunskii, E.D. Protzenko, A.Yu. Rumiantzev: Optika i Spectroscopia 27, 813 (1969)
3.115 S.N. Bagaev, A.K. Dmitriev, V.P. Chebotayev: Pis'ma Zh.Eksp.i Teor. Fiz. 15, 91 (1972)
3.116 M.A. Gubin, A.I. Popov, E.D. Protzenko: Serial edition "Kvantovai Electronika", Moscow, N3, 99 (1971)
3.117 N.G. Basov, M.A. Gubin, V.V. Nikitin, E.D. Protzenko, V.A. Spepanov: Pis'ma Zh.Eksp.i Teor.Fiz. 15, 525 (1972)
3.118 V.N. Lisitsyn, B.I. Troshin: Optika i Spectroscopia 22, 666 (1967)
3.119 Yu.L. Klimontovich, P.S. Landa, E.G. Lariontzev: Zh.Eksp.i Teor.Fiz. 52, 1616 (1967)
3.120 N.G. Basov, E.M. Belenov, M.V. Danileiko, V.V. Nikitin: Zh.Eksp.i Teor.Fiz. 57, 1991 (1969)
3.121 S.G. Zeiger, E.E. Fradkin: Optika i Spectroscopia 21, 386 (1966)
3.122 N.G. Basov, E.M. Belenov, M.V. Danileiko, V.V. Nikitin, O.N. Oraevskii: Pis'ma Zk.Eksp.i Teor.Fiz. 12, 145 (1970)
3.123 N.G. Basov, E.M. Belenov, M.V. Danileiko, V.V. Nikitin: Kratkie Soobschenia po Fizike, FIAN N10, 48 (1970)
3.124 N.G. Basov, E.M. Belenov, M.V. Danileiko, V.V. Nikitin: Zh.Eksp.i Teor.Fiz. 60, 117 (1971)
3.125 N.G. Basov, E.M. Belenov, M.V. Danileiko, V.V. Nikitin: Serial edition "Kvantovai Electronika" N1, 42 (1971)
3.126 V.S. Letokhov, B.D. Pavlik: Izv.VUZov, Radiofizika 14, 244 (1971)

4. Narrow Resonances of Two-Photon Transitions Without Doppler Broadening

Spectroscopy of two-quantum transitions without Doppler broadening considered in this chapter is based on quite a different principle than spectroscopy of absorption saturation discussed in two foregoing chapters. The method of two-photon absorption spectroscopy inside the Doppler contour was qualitatively considered in the Introduction (Sec.4.1). In this case the elimination of Doppler shift of the two-quantum transition frequency consists of mutual compensation for Doppler shifts during simultaneous absorption of two photons of the same frequency but oppositely directed. This is rather a common effect, which shows itself in some other two-quantum transitions that differ from two-quantum absorption. Figure 4.1 illustrates Doppler-effect compensation for a number of such cases. During inelastic scattering of photons by molecular vibrations, called the Raman effect [4.1,2], in the case of unidirectional photons of laser and scattered light, Doppler shifts are partially compensated for. It is evident that, during simultaneous absorption and emission of photons without real occupation of the upper level (resonance fluorescence), the Doppler effect also disappears if both photons are unidirectional. It should be noted that in the latter case the line width of the resonance fluorescence is free not only of Doppler broadening but also of natural broadening, owing to decay of the upper (virtual in this case) state [4.3,4]. In the present chapter, we are considering narrow resonances, free of Doppler broadening, for all of these three cases. Primary attention is placed, of course, on two-photon resonances in standing light waves as the most efficient and universal of the mentioned cases. The resonance effects considered below do not necessitate, in principle, saturation of two-quantum transition, that is excitation of a significant portion of the atoms or molecules to the upper state. This is a basic difference of the spectroscopy of two-quantum transitions from spectroscopy of absorption saturation in which narrow resonances can be obtained only at marked saturation of transition absorption. At high enough intensity, along with two-quantum resonances, those of two-quantum transition saturation may arise, of course. But, by proper choice of laser radiation intensity, it is always possible to eliminate saturation effects in two-quantum transitions and to study two-photon

$$\Delta\omega_D = (\underline{K}_L + \underline{K}'_L)\underline{v} \qquad \Delta\omega_D = (\underline{K}_L - \underline{K}_S)\underline{v} \qquad \Delta\omega_D = (\underline{K}_L - \underline{K}_S)\underline{v}$$

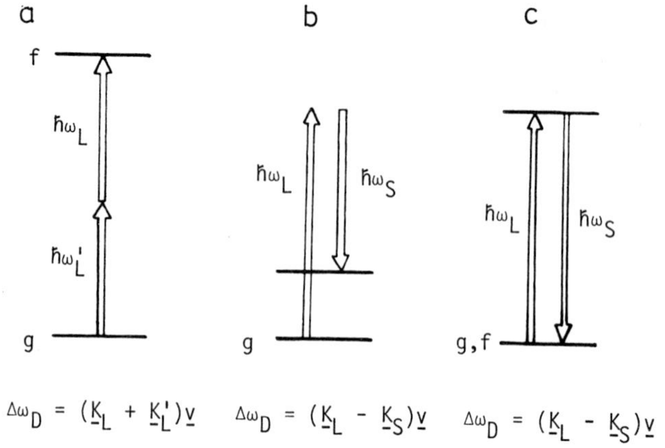

<u>Fig.4.1a-c</u> Three types of two-quantum transitions for which, with proper orientation of wave vectors of two quanta (laser and scattered), complete or partial compensation for linear Doppler effect is possible

resonances only. A quite-different phenomenon takes place in interaction of two laser fields with a three-level quantum system. In this case, resonance effects caused by saturation of absorption and two-quantum transitions can appear simultaneously and sometimes do not differ from one another. Reso-nance effects in three-level systems are comprehensively discussed in Chap-ter 5.

4.1 Two-Photon Absorption

Two-quantum absorption was theoretically studied almost half a century ago [4.5,6] but was experimentally revealed only after the laser was introduced [4.7]. Two-quantum transitions have the following features that are impor-tant for spectroscopy (see reviews [4.8,9]). First, the selection rules for two-quantum transitions differ from those for single-quantum transitions. For instance, to an electric-dipole approximation, two-photon transitions are allowed only between states of the same parity, whereas single-photon transitions are allowed between states of different parity. Besides, from two-photon spectra we can obtain more information than from single-photon spectra because the absorption coefficient is measured as a function of two frequencies ω_1 and ω_2. Second, during two-quantum excitation we may excite a particle state that in the single-photon process requires a shorter-wave photon absorbed, for example, by the matrix or particle environment (or even by the cell windows). So, owing to two-quantum transitions, selective excitation of particles inside the absorption medium is possible. The third important feature of two-quantum absorption is compensation of the Doppler effect when two photons from counter-running waves are absorbed by a moving particle. This effect was predicted not long ago [4.10]. From the theoretical point of view we can consider also a more-general situation in which, for example, three photons are absorbed at the same time [4.11] provided that their total momentum $\sum_i \hbar \underline{K}_i$ is equal to zero (Fig.4.2). But, in practice, only the method of two-photon absorption in the field of two counter-running waves is applied, which is discussed in detail below.

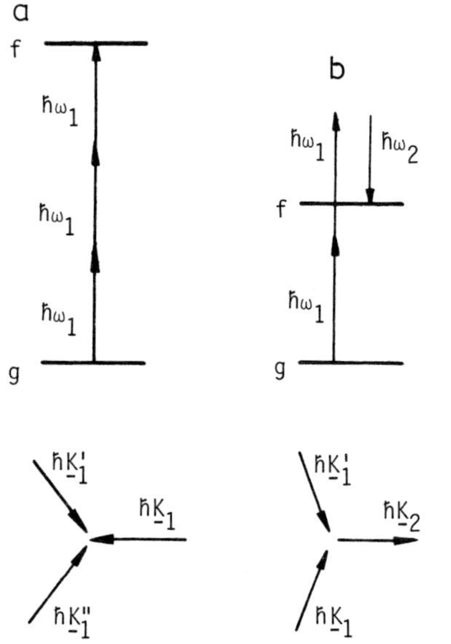

Fig.4.2a,b Three-photon processes of absorption (a) and scattering (b) in which Doppler-effect compensation is obtained because of proper orientation of wave vectors. These processes are analogous to processes (a) and (b) in Fig.4.1

4.1.1 Transition Probability and Selection Rules

Transition Probability

We consider a fixed atom in a two-frequency field

$$\underline{E} = \frac{1}{2}\left|\underline{e}_1 E_1\, e^{i(\underline{k}_1\underline{r}-\omega_1 t)} + \underline{e}_2 E_2\, e^{i(\underline{k}_2\underline{r}-\omega_2 t)}\right| + \text{c.c.}\ . \qquad (4.1)$$

Assume that the energy difference between two states $|g\rangle$ (ground state) and $|f\rangle$ (final state) $E_f - E_g$ is exactly equal to the total energy of two photons

$\hbar(\omega_1+\omega_2)$. In this case, two-quantum atomic transitions $g\to f$ are possible, and the probability of transition per unit time is

$$W_{gf} = (2\hbar^4\Gamma)^{-1}|A_{gf}|^2 E_1^2 E_2^2 \ . \qquad (4.2)$$

A_{gf} is the so-called composite matrix element of two-quantum transition,

$$A_{gf} = \sum_n \left[\frac{(\underline{p}\underline{e}_1)_{gn}(\underline{p}\underline{e}_2)_{nf}}{\omega_{ng}-\omega_1} + \frac{(\underline{p}\underline{e}_2)_{gn}(\underline{p}\underline{e}_1)_{nf}}{\omega_{ng}-\omega_2}\right] , \qquad (4.3)$$

where $\underline{p}$ is the electric-dipole operator, and the summation is over all intermediate quantum states of atom $|n\rangle$ that are connected with the states $|g\rangle$ and $|f\rangle$ by the electric-dipole transition; $2\Gamma = 2\Gamma_{gf}$ is total width at half height for the $g\to f$ transition.

Expressions (4.2) and (4.3) can be derived directly either by calculating the probability of two-quantum transition by perturbation theory [4.6,11] or from the expression for the imaginary part of the nonlinear third-order susceptibility $\chi''^{(3)}(-\omega_1, \omega_1, \omega_2, -\omega_2)$ for an atom under the action of the light field (4.1) [4.12,13]. It must be kept in mind that expressions (4.2) and (4.3) are written for a simple case in which the states $|g\rangle$ and $|f\rangle$ are nondegenerate and all the intermediate states $|n\rangle$ are far from resonance, that is

$$|\omega_{ng}-\omega_1| \gg \Gamma_{gf} \ .$$

Besides, it is assumed that the rate of two-photon transitions is much less than the relaxation rate of the population of the final state "f", so the effects of two-photon transition saturation may be neglected.

It is convenient sometimes to introduce the cross-section of two-photon absorption of radiation at the frequency ω_1 in the presence of the ω_2-field with the intensity $I_2 = (c/8\pi)E_2^2$,

$$\sigma^{(2)}(\omega_1) = (32\pi^3\omega_1/\hbar^3c^2)f(\omega_1+\omega_2)|A_{gf}|^2 I_2 \ , \qquad (4.4)$$

where $f(\omega_1+\omega_2)$ is the normalized function giving the line shape of two-photon absorption. For numerical estimates, this expression may be written

$$\sigma^{(2)}(\omega_1) = 1.15 \cdot 10^{-34}\omega_1 f(\omega_1+\omega_2) I_2 \cdot$$

$$\cdot \left| \sum_n \langle f|z|n\rangle\langle n|z|g\rangle \left(\frac{1}{E_n - \hbar\omega_1} + \frac{1}{E_n - \hbar\omega_2} \right) \right|^2 [\mathrm{cm}^2] \tag{4.5}$$

where the energy of levels and photons is given in Rydbergs $R_o = 2\pi^2 me^4/h^3$; the matrix elements of dipole transitions are expressed in terms of the Bohr radius $a_o = h^2/4\pi^2 me^2$; the function $\omega_1 f(\omega_1+\omega_2)$, accurately to the factor $\omega_1/(\omega_1+\omega_2)$, is a dimensionless quantity that can be termed "quality" of two-photon absorption line; I_2 is expressed in W/cm^2; E = 0.

Selection Rules

The composite matrix element A_{gf} may be written in a compact form by the use of the operator [4.11]

$$Q_{12} = \underline{e}_1 \underline{p} \frac{1}{\omega_1 - H_o} \underline{e}_2 \underline{p} + \underline{e}_2 \underline{p} \frac{1}{\omega_2 - H_o} \underline{e}_1 \underline{p} \ , \tag{4.6}$$

i.e.,

$$A_{gf} = \langle f|Q_{12}|g\rangle \ , \tag{4.7}$$

where H_o is the hamiltonian of the isolated atom with no light field present (in units of $\hbar$), $\underline{p}$ is the vector operator and $1/(\omega_i - H_o)$ are scalar operators. Q_{12} is a combination of second-order tensor operators. As Q_{12} is a symmetrical operator, it can be represented in the form of the sum of irreducible tensor second- and zero-rank operators,

$$Q_{12} = aT^{(2)} + bT^{(o)} \ . \tag{4.8}$$

Let the polarization vectors $\underline{e}_1$ and $\underline{e}_2$ of light waves correspond to σ^+, σ^- and π-polarizations. In this case, the expression (4.6) consists of the standard components p_{q1} and p_{q2} of the vector operator $\underline{p}$, where the indexes q_1 and q_2 take on the values +1, -1, and 0 for σ^+, σ^- and π polarizations, respectively. The operators $T_q^{(2)}$ and $T_q^{(o)}$ are composed of p_{q1} and p_{q2} components if the condition

$$q = q_1 + q_2 \tag{4.9}$$

is met.

Let us consider a simple case of an atom with LS coupling. The rank of the tensor operator Q_{12} automatically produces the general selection rule for atomic orbital angular momentum

$$\Delta L \leq 2 \ . \tag{4.10}$$

Selection rule (4.10) can be produced from simpler qualitative considerations. Indeed, in an electric-dipole approximation, in which we consider the connection of $|g\rangle$ and $|f\rangle$ states with the intermediate states $|n\rangle$, the hamiltonian of interaction between the atom and light field acts on the orbital electron only. The operator components of dipole interaction $\underline{e}_i\underline{p}$ are transformed as components of the irreducible tensor $T_q^{(1)}$ with $q = -1$ for σ^-, $q = +1$ for σ^+ and $q = 0$ for π-polarization. Each time when an odd-parity operator is used, the orbital angular momentum changes by ± 1. By using this operator twice in the composite matrix element A_{gf}, we find that the orbital angular momentum in a two-photon transition must change by

$$\Delta L = 0 \quad \text{or} \quad \Delta L = \pm 2 \ ; \tag{4.11}$$

apart from this, the condition

$$\Delta m_L = q_1 + q_2 \tag{4.12}$$

must be complied with.

By using the properties of the scalar operator $T^{(0)}$ and the irreducible second-order-tensor operator $T^{(2)}$ as well as those of Clebsch-Gordan coefficients, we can obtain the selection rules for the total angular momentum F of the atom (with allowance for electron and nuclear spins) and its projection onto the chosen direction for all different cases of orbital-momentum change, according to selection rule (4.10). These selection rules, obtained in [4.11], are listed in Table 4.1.

In a strong external magnetic field, when the energy of electron-field interaction is much greater than that of the spin-orbit interaction $\lambda\underline{L}\underline{S}$, the selection rules for the magnetic quantum numbers of electron and nuclear spin are

$$\Delta M_S = \Delta M_I = 0 \ . \tag{4.13}$$

They follow because the hamiltonian of electric-dipole interaction does not act on spin variables.

Table 4.1 Selection rules for two-photon transitions (LS-coupling case) [4.11]

Type of transitions	Allowed transitions	Forbidden transitions
$\Delta L = 2$	$\Delta F \leq 2$	$F = 0 \rightarrow F' = 0$
$\Delta L = 1$	$\Delta M_F = q_1 + q_2$	$F = 0 \rightarrow F' = 1$
$\Delta L = 0$, $\Delta M_F \neq 0$		$F = 1/2 \rightarrow F' = 1/2$
		$F, M_F = 0 \rightarrow F' = F + 1, M_{F'} = 0$ (F-integral)
		$F, M_F = -(1/2) \rightarrow F' = F, M_{F'} = 1/2$ (F-half integral)
$L_1 = L_2 = 0$	$\Delta F = 0$ $\Delta M_F = q_1 + q_2 = 0$	

The review by BLOEMBERGEN and LEVENSON [4.13] contains useful discussion of the validity of the above-given simple selection rules and considers important particular cases. In particular, the selection rules considered depend only on the properties of ground and finite states. This holds true as long as the energy in the denominators of the compound matrix element (4.3) can be the same for all of the terms in the multiplet of the intermediate state $|n\rangle$. It is necessary, in this case, that the values $\hbar(\omega_{ng} - \omega_1)$ and $\hbar(\omega_{ng} - \omega_2)$ should considerably exceed the constant λ of spin-orbit interaction $\lambda \underline{L}\underline{S}$ and the constant A of hyperfine interaction $A\underline{I}\underline{S}$. Then the summation over all substates of the intermediate multiplet that have equal weights is a scalar operation that cannot change the selection rules [4.13].

Systematic discussion of selection rules for two-photon transitions based on theory group approach is given also in [4.8,11,14].

Resonant Increase of Cross-Section

If one of two frequencies of a laser field approaches resonance with the frequency of transition to any intermediate state $|n\rangle$, according to (4.3) a sharp increase of two-photon absorption probability is expected. This was clearly demonstrated experimentally [4.15]. The two-photon transition N_aI 3S - 4D was under study, and the frequency of one of the lasers was tuned in the vicinity of the intermediate state 3P. When $E_{3P} - \hbar\omega_2 = 0.1\ cm^{-1}$ was detuned, the two-photon absorption cross-section $\delta^2(\omega_1)$ according to (4.5) may be as great as $(5 \cdot 10^{-14}) I_2\ cm^2$. Even when the laser power $I_2 \simeq 10^3\ W/cm^2$, the cross-section of two-quantum absorption of photon $\hbar\omega_1$ approximates the cross-section typical of single-photon transitions. Figure 4.3 shows experimental dependences of the rate of the two-photon transitions 3S (F = 2) → $4D_{5/2}$ and 3S (F = 2) → $4D_{3/2}$ on the wavelength λ_2 of the laser. Both two-photon transitions have high dispersion; then the frequency ν_2 approximates 3P intermediate states. For the 3S (F = 2) → $4D_{5/2}$ transition there is only one near-resonance intermediate state $3P_{3/2}$ because the single-photon transition $3P_{1/2} \to 4D_{5/2}$ is forbidden. At the same time, both $3P_{3/2}$ and $3P_{1/2}$ may be intermediate states for the transition 3S (F = 2) → $4D_{3/2}$. It is interesting that in this case the cross-section $\sigma^{(2)}$ has a sharp minimum when the laser frequency ω_2 lies exactly midway between two intermediate states. This can be explained by the opposite signs of the resonance contributions made by each intermediate state, that is by mutual compensation of their contributions to the probability of two-photon transition. In the region of resonances with intermediate states, the probability of two-photon absorption may be increased by 7 orders of magnitude. The points nearest to resonance

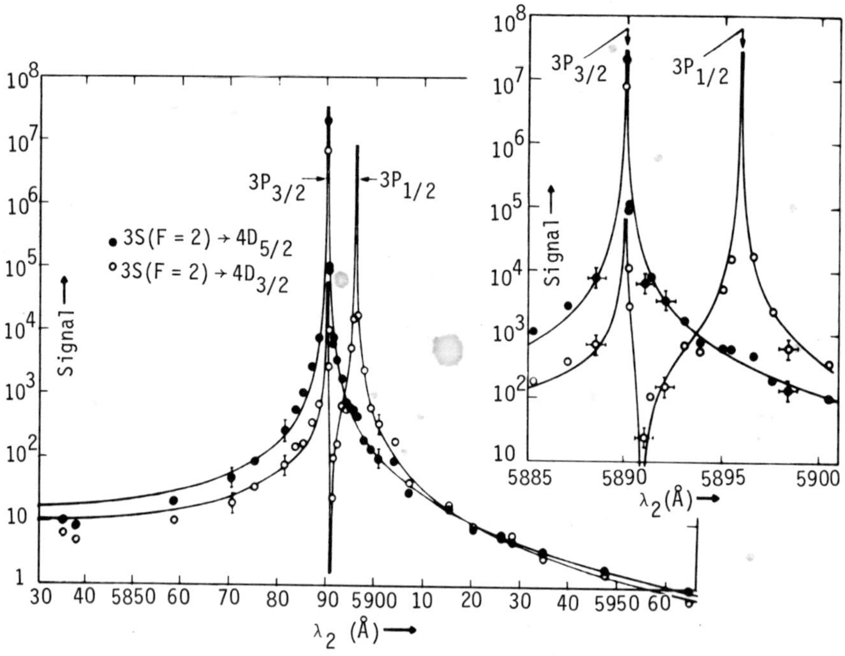

Fig.4.3 Normalized rates of two-photon transitions 3S (F = 2) → $4D_{5/2}$ and 3S (F = 2) → $4D_{3/2}$ as functions of laser wavelength λ_2. In the corner (right top) there is a region of near-resonance wavelengths with the intermediate states $^3P_{3/2}$ and $^3P_{1/2}$ shown to a large scale (from [4.15])

have such a high cross-section of two-quantum absorption that saturation of two-quantum transition occurs. The measurements near resonance therefore were taken with the laser-light intensity decreased by three orders of magnitude. The solid curves in Fig.4.3 are estimated, obtained from (4.5) with only two intermediate states allowed for. It can be seen that the simplest estimations of two-photon transition probability are in good agreement with the experimental results. Figure 4.3 is a good illustration of resonance behavior and selection rules for two-photon absorption.

4.1.2 Two-Photon Resonance Shape for Gas in Counter-Running Light Waves

Linear Doppler Effect

Let us consider the case when two light waves propagate in opposite directions; from simplest qualitative considerations, we expect a two-photon Doppler-free resonance to appear. Let the field wave vectors be $\underline{K}_1 = K_1\underline{n}$ and $\underline{K}_2 = -K_2\underline{n}$, where $\underline{n}$ is the unit vector in the direction of the Z axis. In this case, for an atom moving with velocity $\underline{v}$, the light waves have the frequencies $\tilde{\omega}_1 = \omega_1 - k_1 v$ and $\tilde{\omega}_2 = \omega_2 + k_2 v$, respectively, where $v_z = v$. To obtain the probability of two-photon transition, it is sufficient to substitute these new frequencies $\tilde{\omega}_1$ and $\tilde{\omega}_2$ into (4.3) for the composite matrix element. To allow for detuning of the transition frequency $\omega_{gf} = \omega_0$ of a fixed atom with respect to the total-field frequencies $[\omega_1 + \omega_2 - (k_1 - k_2)v]$, the corresponding lorentzian contour with half-width Γ should be introduced. As a result, the rate of two-photon transitions in the field of two counter-running waves is

$$W_{gf} = (2\hbar^4\Gamma)^{-1}E_1^2E_2^2 \frac{\Gamma^2}{[\omega_0 + (k_1 - k_2)v - \omega_1 - \omega_2]^2 + \Gamma^2} \cdot$$

$$\cdot \left|\sum_n \frac{(\underline{p}\underline{e}_1)_{gn}(\underline{p}\underline{e}_2)_{nf}}{\omega_{ng} + k_1 v - \omega_1} + \frac{(\underline{p}\underline{e}_2)_{gn}(\underline{p}\underline{e}_1)_{nf}}{\omega_{ng} - k_2 v - \omega_2}\right|^2 . \qquad (4.14)$$

Averaging $W_{gf}(v)$ over the velocities of atoms in a gas produces a contour with the reduced Doppler width $(k_1 - k_2)u$ because of the lorentzian resonant factor in (4.14). In case of very close wave frequencies, this residual Doppler broadening may prove to be even less than the uniform width of transition, 2Γ.

The case $\omega_1 = \omega_2 = \omega$ requires special calculation because, apart from simultaneous absorption of photons from the counter-running waves, simultaneous absorption of two photons from each wave may occur. Expression (4.14) for two-photon transition rate may be written in this case as the sum of three terms,

$$W_{gf} = (2\hbar^4\Gamma)^{-1}E_1^2E_2^2 L\left(\frac{\omega_0 - 2\omega}{\Gamma}\right)\left|\sum_n \frac{(p e_1)_{gn}(p e_2)_{nf} + (p e_2)_{gn}(p e_1)_{nf}}{\omega_{ng} - \omega}\right|^2 +$$

$$+ (2\hbar^4\Gamma)^{-1}E_1^4 L\left(\frac{\omega_0 - 2\omega + 2kv}{\Gamma}\right)\left|\sum_n \frac{(p e_1)_{gn}(p e_1)_{nf}}{\omega_{ng} - \omega}\right|^2 +$$

$$+ (2\hbar^4\Gamma)^{-1}E_2^4 L\left(\frac{\omega_0 - 2\omega - 2kv}{\Gamma}\right)\left|\sum_n \frac{(p e_2)_{gn}(p e_2)_{nf}}{\omega_{ng} - \omega}\right|^2 . \tag{4.15}$$

The first term corresponds to absorption of two photons from running light waves. It describes a two-photon Doppler-free resonance. The second and third terms, respectively, correspond to two-photon absorption from any of the waves. These terms describe the apparently Doppler-broadened profiles of two-photon gas absorption.

In the simplest case of running waves with the same polarization ($e_1 = e_2 = e$) and amplitude ($E_1 = E_2 = E$), (4.15) can be reduced to

$$W_{gf} = (2\hbar^4\Gamma)^{-1}E^4\left|\sum_n \frac{(p e)_{gn}(p e)_{nf}}{\omega_{ng} - \omega}\right|^2 \cdot$$

$$\left[4L\left(\frac{\omega_0 - 2\omega}{\Gamma}\right) + 2L\left(\frac{\omega_0 - 2\omega \pm 2kv}{\Gamma}\right)\right] . \tag{4.16}$$

After velocity averaging, we get the expression for the line shape of two-photon absorption of atoms in gases, first derived in [4.10]

$$\kappa^{(2)}(\omega) = \kappa_{tr}^{(2)}(\omega)\left\{\exp-[(\omega_0 - 2\omega)/2ku]^2 + \frac{4ku}{\sqrt{\pi}\Gamma} L\left(\frac{\omega_0 - 2\omega}{\Gamma}\right)\right\} , \tag{4.17}$$

where $\kappa_{tr}^{(2)}(\omega)$ is the coefficient of two-photon absorption for one running wave at the Doppler-broadened transition

$$\kappa_{tr}^{(2)}(\omega) = (N_g^0 - N_f^0)\,\frac{\sqrt{\pi}E^4}{2\hbar^4 ku}\left|\sum_n \frac{(p e)_{gn}(p e)_{nf}}{\omega_{ng} - \omega}\right|^2 . \tag{4.18}$$

The line shape of two-photon absorption in a standing wave is shown in Fig.1.16 (Chap.1). The width of narrow two-photon resonance is equal to the homogeneous width of two-photon transition, which is determined by the broadening of the initial $|g\rangle$ and final $|f\rangle$ transition states. The ratio of two-photon narrow resonance intensities to continuous Doppler background is $4ku/\sqrt{\pi}\Gamma$. Physically, this is because, when tuned to exact resonance, all atoms, regardless of their velocity $\underline{v}$, may take part in two-photon absorption. When the detuning about the center of the transition is much greater than Γ, the field resonates with a small portion of atoms whose resonant velocity satisfies the condition $2\underline{k}\underline{v} = 2\omega - \omega_0$.

The selection rules for two-photon transitions permit choosing the type of transition and polarizations of counter-running light waves $\underline{e}_1$ and $\underline{e}_2$ so that the Doppler-broadened background disappears completely. Let us consider the case of two-photon transition between two atomic S states. Because in the initial and finite states the orbital angular momentum equals zero ($L_1 = L_2 = 0$), it is evident that $\Delta M_L = 0$. If the light waves have the same circular

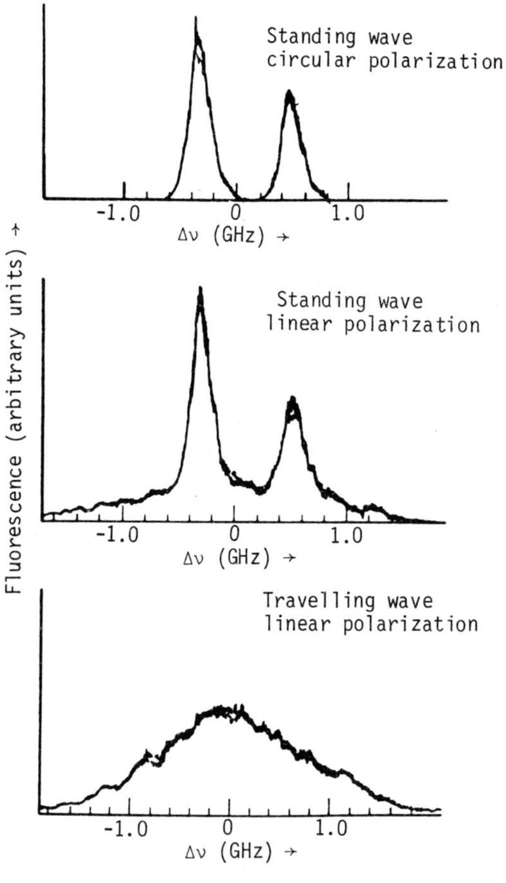

Fig.4.4a-c Line shape of two-photon absorption at the transition 3S - 5S of Na atom: a) standing wave with circular polarization of opposite signs for counter-running waves; b) standing light wave with linear polarization; c) running light wave with linear polarization (from [4.13])

polarization, that is $|q_1 + q_2| = 2$, all two-photon transitions are forbidden (see Table 4.1). However, if the waves have circular polarizations of opposite signs $q_1 + q_2 = 0$, and two-photon absorption may occur. Thus, in the field of two counter-running waves with different circular polarizations two-photon absorption is possible only during simultaneous absorption of photons from the counter-running waves, that is within a narrow resonance at the center of line, and quite impossible from one running wave. This possibility of eliminating the Doppler background of two-photon absorption is considered in [4.11]. Experimentally, this effect was clearly observed at the transition 3S - 5S of Na atom in [4.16]. Figure 4.4 shows the experimental dependence of two-photon-absorption line shape at this transition for the case of two counter-running waves with circular polarizations of opposite signs, and for linear polarization. In the first case, complete elimination of Doppler background can be seen clearly. For comparison, the spectrum of two-photon absorption for one running wave that is Doppler broadened in the ordinary way is shown below. As seen, inside the Doppler contour two two-photon resonances are formed, conditioned by two allowed transitions $F = I \rightarrow F' = I$ and $F = 2 \rightarrow F' = 2$ between the hyperfine-structure components of 3S- and 5S-states.

Second-Order Doppler Effect

In calculating the line shape of two-photon resonance, we restricted ourselves to the linear Doppler effect only. It is apparent that the line shape of two-photon resonance will be correct as long as the Doppler shift of the sum of the frequencies ω_Σ of the two counter-running waves with respect to double-wave frequency is considerably less than the homogeneous half-width of the transition,

$$\omega_\Sigma - 2\omega = 2\omega\left(1 - \frac{V^2}{c^2}\right)^{-1/2} - 2\omega << \Gamma \ , \qquad (4.19)$$

where $V = |\underline{v}|$ is the absolute atomic velocity. If this is not the case, we should allow for the contribution of the next higher-order terms in (V/c). By expanding the left-hand term of (4.19) in powers of (V/c), we can easily determine that the quadratic Doppler effect is the first nonzero approximation. The parameter that is responsible for two-photon resonance shape in this approximation is the ratio of the quadratic Doppler shift $\Delta\omega_q$ of the sum of the frequencies for atoms with the mean thermal velocity u, to the homogeneous half-width of the transition,

$$\varepsilon = \frac{\Delta\omega_q}{\Gamma} = \frac{\omega_0}{\Gamma}\frac{u^2}{c^2} \simeq \frac{\Delta\omega_D}{2\Gamma}\frac{u}{c} \ , \qquad (4.20)$$

where $\omega_0 \simeq 2\omega$ is the frequency of two-photon atomic transition.

To allow for the quadratic Doppler effect, let us substitute the value ω_Σ, which is equal, according to (4.19), to $2\omega + \omega(V^2/c^2)$ for the value 2ω in the first term of (4.16). The resultant expression must be averaged over the Maxwell distribution of the absolute velocities,

$$\langle W_{gf}\rangle = \frac{4}{\sqrt{\pi}u^3}\int_0^\infty W_{gf}(V)e^{-(V^2/u^2)}V^2dV =$$

$$= A\frac{4}{\sqrt{\pi}u^3}\int_0^\infty L\left(\frac{\omega_\Sigma - \omega_o}{\Gamma}\right)e^{-(V^2/c^2)}V^2dV \ , \qquad (4.21)$$

where A is the constant coefficient at the lorentzian contour of two-photon resonance. The appropriate integral can be precisely calculated and determined by [4.18]

$$\langle W_{gf}\rangle = A\frac{\Gamma}{\Delta\omega_q}\,\mathrm{Re}\{iZ^{1/2}\psi(3/2,3/2,Z)\} \ , \qquad (4.22)$$

where $Z = [(2\omega - \omega_o)/\Delta\omega_q] + i(\Gamma/\Delta\omega_q)$, $\psi(\alpha,\beta,\gamma)$ is the degenerate hypergeometric function of the second type.

The shape of two-photon resonance from (4.22) is fully determined by the magnitude of the parameter $\varepsilon = \Delta\omega_q/\Gamma$. Typical values of optical transitions in atoms are $\omega_o \simeq 10^{14}$-10^{15} Hz, $2\Gamma \simeq 10^6$-10^8 Hz, $u/c \simeq 10^{-6}$ and the parameter $\varepsilon \simeq 10^{-3}$-$10^{-6} \ll I$. But for very-light atoms, including abnormally light ones (positronium atoms) and/or very narrow absorption lines that arise from transitions from ground to metastable or highly excited (Rydberg) states, the parameter ε may exceed unity. So, to describe all situations, we want to consider two limiting cases, $\varepsilon \ll 1$ and $\varepsilon \gg 1$.

With $\varepsilon \ll 1$, paying attention on well-known asymptotic behavior of the function $\psi(\alpha,\beta,\gamma)$ at $|\gamma| \gg 1$, we derive [4.16] from (4.22) the approximate (to an accuracy of terms linear in ε) expression that holds for any detuning $(2\omega - \omega_o)$,

$$\langle W_{gf}\rangle = AL\left(\frac{2\omega - \omega_o}{\Gamma}\right)\left[1 - 3\varepsilon\frac{2\omega - \omega_o}{\Gamma}L\left(\frac{2\omega - \omega_o}{\Gamma}\right)\right] \ . \qquad (4.23)$$

This expression describes a slightly asymmetrical lorentzian profile, the maximum of which is red shifted and has the frequency

$$2\omega_{max} = \omega_o - \frac{3}{2}\Delta\omega_q \quad (\text{for } \Delta\omega_q \ll \Gamma) \ . \qquad (4.24)$$

This expression can be obtained also by direct calculation of the mean quadratic shift $\langle\omega(V^2/c^2)\rangle$, as has been done in [4.19,20]. Figure 4.5a shows the line shape of two-photon resonance at the transition 1S - 3S for hydrogen atom calculated with $2\Gamma \simeq 1.0$ MHz (radiative width of transition), $u/c = 7.5 \cdot 10^{-6}$ (T = 300 K) and $\varepsilon = 0.16$.

When $\varepsilon \gg 1$, by use of the asymptotic behavior of the $\psi(\alpha,\beta,\gamma)$ function with $|\gamma| \ll 1$, we can obtain the approximate expression for detuning [4.17,19] $|2\omega - \omega_0| \ll \Delta\omega_q$

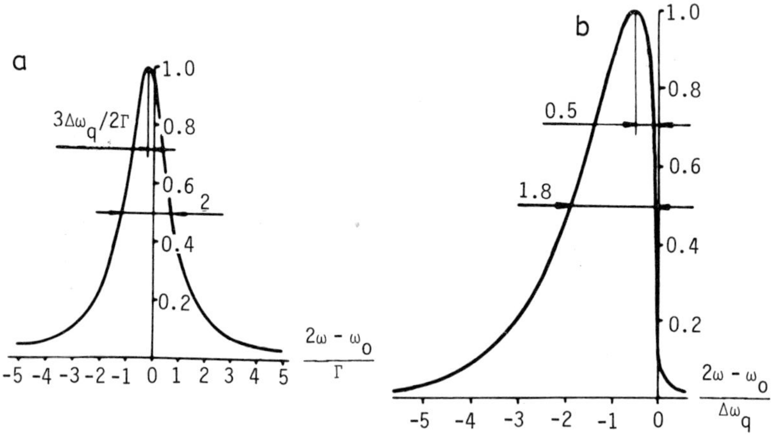

Fig.4.5a,b Narrow resonance shape of two-photon absorption with allowance made for quadratic Doppler effect: a) 1S - 3S transition of hydrogen atom with $\Delta\omega_q \ll \Gamma$; b) 1S - 2S transition of hydrogen atom with $\Delta\omega_q \gg \Gamma$ (from [4.18])

$$\langle W_{gf}\rangle = A\sqrt{2\pi}\,\frac{\Gamma}{\Delta\omega_q}\,e^{\frac{2\omega-\omega_0}{\Delta\omega_q}}\left\{\left[\left(\frac{2\omega-\omega_0}{\Delta\omega_q}\right)^2 + \left(\frac{\Gamma}{\Delta\omega_q}\right)^2\right]^{1/2} - \frac{(2\omega-\omega_0)}{\Delta\omega_q}\right\}^{1/2} . \quad (4.25)$$

Detailed analysis [4.18] shows that (4.25) is also valid with detuning $|2\omega - \omega_0| \simeq \Delta\omega_q$. For large detuning ($|2\omega - \omega_0| \gg \Delta\omega_q$) (4.23) applies. Equation (4.25) defines a sharply asymmetrical profile the maximum of which is red shifted and corresponds, unlike that from (4.24), to the frequency

$$2\omega_{max} = \omega_0 - \frac{1}{2}\,\Delta\omega_q \quad (\Delta\omega_q \gg \Gamma) ; \quad (4.26)$$

its half-height width equals $1.8\,\Delta\omega_q$. It is evident that with $\varepsilon \gg 1$ considerable resonance broadening occurs, due to the quadratic Doppler effect. In this case, the ratio of two-photon absorption intensity at a resonance maximum to the intensity of Doppler-broadened background of two-quantum absorption decreases from $4ku/\sqrt{\pi}\Gamma$ (with $\varepsilon \ll 1$) down to $4\sqrt{2}ku/\sqrt{e}\Delta\omega_q = 4\sqrt{2}c/\sqrt{e}u$ (with $\varepsilon \gg 1$). The resonance of two-photon absorption at the transition 1S - 2S of the hydrogen atom is a good example of this case. Figure 4.5b presents the calculated line shape of two-photon resonance for $2\Gamma = 1$ kHz (the homogeneous width of transition at a pressure of 10^{-4} torr) and $T = 300$ K when $\varepsilon = 70$. The total resonance half-height width is 130 kHz and resonance-maximum shift is -36 kHz.

Thus, the shape of two-photon resonance at narrow atomic transitions in a gas is the very rare case in optics when inclusion of the quadratic Doppler effect is essential in principle.

4.1.3 Power Shift and Power Broadening of Two-Photon Resonances

Two-photon transitions between levels are possible only because of perturbation of atomic or molecular quantum states by a light field. This perturbation inevitably acts on the ground and final atomic states; hence it results in a shift and broadening of two-photon resonance which is in proportion to laser radiation power. Frequency shift is inherent in the two-photon resonance because it is necessary that intermediate virtual quantum states should be involved in two-quantum transition. So the elimination of the Doppler effect in two-photon spectroscopy is realized at the expense of irreparable meddling in the "internal life" of a quantum system. This is a basic difference between two-photon nonlinear spectroscopy and saturation spectroscopy, in which the problem of frequency shift with power is almost absent because for an allowed single-photon transition between a pair of levels there is no need to disturb the atomic states and to employ the virtual intermediate quantum states.

In the general case, the shift of the atomic level "m" in the external nonresonant light field $\underline{E} = \underline{e}E \cos(\omega t - kz)$ can be easily evaluated according to the theory of perturbation [4.21,22],

$$\Delta E_m = \frac{E^2}{4} \sum_n \left[\frac{(\underline{p}_{nm}\underline{e})^2}{E_m - E_n - \hbar\omega} + \frac{(\underline{p}_{nm}\underline{e})^2}{E_m - E_n + \hbar\omega} \right] . \tag{4.27}$$

This formula is also valid in case of resonance, but the term Γ_{mn}, which gives the finite width of the resonance transition, must be added to the resonance denominators in this case. For example, in the simplest case of two-level system (4.27) is reduced to

$$\delta\nu = \frac{1}{4} E^2 \frac{(\underline{p}_{12}\underline{e})^2}{h^2(\nu_0 - \nu)} , \qquad (4.28)$$

where $h\nu_0 = E_2 - E_1$, $\nu = \omega/2\pi$ is the light-field frequency. Both levels have equal absolute shifts, but of opposite signs. With $\nu < \nu_0$ the perturbation caused by the light field increases the distance between the levels; vice versa, with $\nu > \nu_0$ it decreases the distance between them (Fig.4.6). At exact resonance ($\nu = \nu_0$), the shifts compensate for one another; because of this,

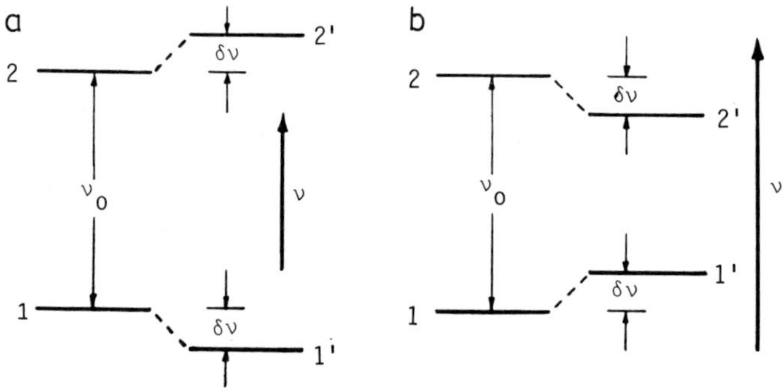

Fig.4.6a,b Level shift of a two-level system in a strong light field, with $\nu < \nu_0$ (a) and $\nu > \nu_0$ (b)

there are no light shifts in saturation spectroscopy of resonant absorption (accurately to the contribution of other nonresonance virtual states). Note that the shift of the level "m" in a linearly polarized field with the frequency $\omega \to 0$ (or rather with $\hbar\omega \ll |E_m - E_n|$) agrees with quadratic shifts in a dc electric field that has the same energy density.

Expression (4.27) holds for two-photon transitions as well. For example, the initial-state shift is expressed by

$$\Delta E_g = - \frac{E_1^2}{4\hbar} \sum_n \frac{(\underline{p}\underline{e}_1)_{gn}(\underline{p}\underline{e}_2)_{ng}}{\omega_{ng} - \omega_1} . \qquad (4.29)$$

If the frequency ω_1 approaches the resonance with ω_{ng}, the energy shift may be sufficiently high. As in the case of a two-level system, the sign of the shift changes when the frequency passes through the resonance. Near resonance, $\omega_1 \simeq \omega_{ng}$, the mixing of the states $|g\rangle$ and $|n\rangle$ becomes important. In the case of exact resonance, the term $-i\Gamma_{ng}$, which makes allowance for the finite width of the transition "g" → "n", should be added to the resonance denominator in (4.29). In this case, the real and imaginary parts of the resultant expression describe the shift that is dependent on power, and the broadening of the ground state $|g\rangle$ that is dependent on the frequency ω_1 of the first light field [4.13],

$$\Delta E_g = -\frac{E_1^2}{4\hbar}\,|(\underline{p}\underline{e}_1)_{gn}|^2\,\frac{\omega_{ng}+k_1v-\omega_1}{(\omega_{ng}+k_1v-\omega_1)^2+\Gamma_{ng}^2}, \tag{4.30}$$

$$\Delta\Gamma_g = +\frac{E_1^2}{4\hbar^2}\,|(\underline{p}\underline{e}_1)_{gn}|^2\,\frac{\Gamma_{ng}}{(\omega_{ng}+k_1v-\omega_1)^2+\Gamma_{ng}^2}, \tag{4.31}$$

where the Doppler frequency shift caused by atom motion is also allowed for. The shift with power and broadening of the levels caused by the second field ω_2 should also be allowed for. If $\omega_1 \simeq \omega_{ng}$ and $\omega_1+\omega_2 \simeq \omega_{fg}$, the second field becomes automatically set near the resonance with the frequency ω_{fn}. So, for the shift with power and the broadening of the level $|f\rangle$, we may write expressions similar to (4.30) and (4.31).

The two-photon resonance shift depends on the sum of the shifts of the ground and final levels of the two-photon transition [4.13],

$$\Delta\omega_{fg} = -\frac{E_1^2}{4\hbar^2}\,|(\underline{p}\underline{e}_1)_{ng}|^2\,\frac{\omega_{ng}+k_1v-\omega_1}{(\omega_{ng}+k_1v-\omega_1)^2+\Gamma_{ng}^2}$$

$$-\frac{E_2^2}{4\hbar^2}\,|(\underline{p}\underline{e}_2)_{fn}|^2\,\frac{\omega_{fn}-k_2v-\omega_2}{(\omega_{fn}-k_2v-\omega_2)^2+\Gamma_{fn}^2}. \tag{4.32}$$

When $\omega_1+\omega_2=\omega_{fg}=\omega_0$, the detuning $(\omega_{ng}+k_1v-\omega_1)$ and $(\omega_{fn}-k_2v-\omega_2)$ have equal values, with opposite signs. In this case, by fitting the field amplitudes E_1 and E_2 and the matrix-element values $(\underline{p}\underline{e}_1)_{ng}$ and $(\underline{p}\underline{e}_2)_{fn}$, we may make the level shifts so that the two-photon resonance frequency remains unchangeable.

Similarly, two-photon resonance broadening (far from saturation of two-photon transition) is determined by the total broadening of the ground and final states [4.13],

$$\Delta\Gamma_{fg} = \frac{E_1^2}{4\hbar^2}\,|(\underline{p}\underline{e}_1)_{ng}|^2\,\frac{\Gamma_{ng}}{(\omega_{ng}+k_1v-\omega_1)^2+\Gamma_{ng}^2}$$

$$+\frac{E_2^2}{4\hbar^2}\,|(\underline{p}\underline{e}_2)_{fn}|^2\,\frac{\Gamma_{fn}}{(\omega_{fn}-k_2v-\omega_2)^2+\Gamma_{fn}^2}\ . \qquad (4.33)$$

If we compare (4.15) for the two-photon transition rate and (4.30) for the frequency shift, in the case of predominant contribution of only one intermediate state to the process, we deduce the relationship between them [4.23]

$$W_{gf} \sim \Delta E_g \cdot \Delta E_f\ . \qquad (4.34)$$

This shows that the two-photon transition rate can be increased equally in any way that increases the level shift: either by increasing the field intensity or by bringing it closer to resonance with any intermediate level. At the same time, if we ensure a small shift of ground state ΔE_g, it is possible to produce a large shift of two-photon resonance frequency, for example, at the expense of the final level shift ΔE_f with the probability that the two-photon process is small. So it is quite possible to eliminate saturation effects of two-photon absorption and to observe therewith large shifts of the two-photon resonance frequency.

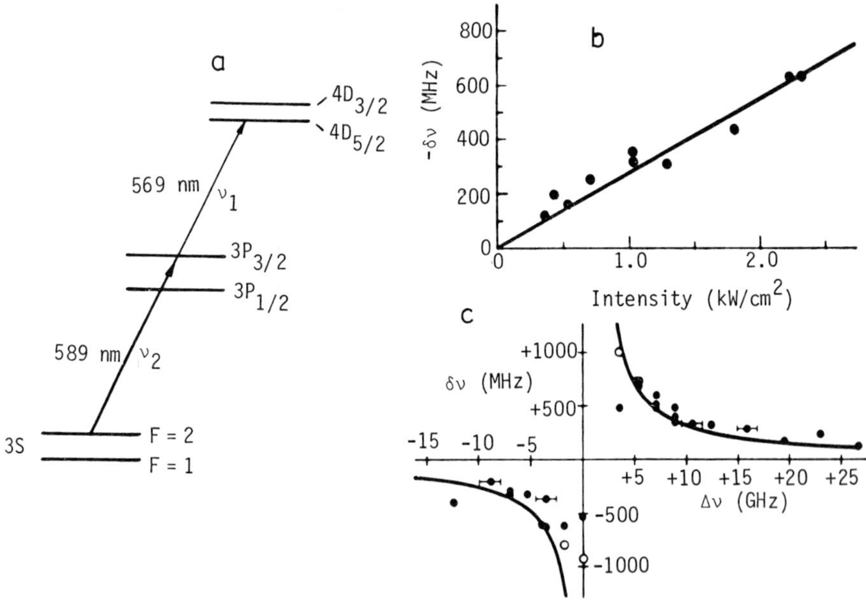

Fig.4.7a-c Power shift of two-photon resonance frequency at the transition 3S→4D of Na atom: a) scheme of the levels used; b) dependence of the shift of the initial level 3S (F = 2) on radiation intensity at 589 nm; c) dependence of the shift in the level 3S (F = 2) on detuning of frequency about exact resonance with the intermediate level 3S (F = 2) → $^3P_{3/2}$ (from [4.22])

Power shifts and broadening of two-photon resonance were experimentally studied in detail by LIAO and BJORKHOLM [4.23,24]. They used the transition 3S - 4D of Na (Fig.4.7a), already mentioned in Subsection 4.1.1. The frequencies of two dye cw lasers were tuned so that the levels $3P_{3/2}$ and $3P_{1/2}$ could serve as intermediate virtual ones. The two-photon transition 3S - 4D was controlled by the intensity of fluorescence at the transition $4P \rightarrow 3S$ (330 nm), which resulted from cascade during decay of the state 4D. By tuning the sum of the laser frequencies $(\nu_1 + \nu_2)$ within the frequency of two-photon resonance $\nu_0 = \nu(4D - 3S)$ they found the maximum of transition probability, that is the frequency of two-photon resonance. It was easier to investigate shifts and broadening, because the laser beams ν_1 and ν_2 were directed to meet one another, which decreased the Doppler-broadening value from 3.6 GHz down to the residual value $(k_1 - k_2)u = 62$ MHz. Against the background of such a small width it was convenient, of course, to control over negligible power shifts and broadening values that are usually concealed by Doppler broadening. Figure 4.7b shows the shift of the level 3S (F = 2), which depends mainly on the laser intensity I_2 at $\lambda_2 = 589$ nm. Figure 4.7c shows the relation between the shift of the same level and the detuning of λ_2 at the intermediate transition 3S $(F = 2) \rightarrow 3P_{3/2}$.

Similar effects arise in two-photon absorption at molecular transitions. Appropriate evaluations for power shift and broadening of two-photon resonances at rotational-vibrational molecular transitions are presented in [4.25, 26]. This effect has been studied experimentally in [4.27,28] where the two-photon resonance without Doppler broadening was observed at the vibrational-rotational transition $(\nu_3 = 0,\ J = 1) \rightarrow (\nu_3 = 2,\ J = 3)$ of the CH_3F molecule and at $(\nu_2 = 0,\ J = 5) \rightarrow (\nu_2 = 2,\ J = 5)$ of the NH_3 molecule. An increase of the two-photon transition cross-section was produced by adjusting the frequency of one of the CO_2 lasers near resonance with an intermediate rotational level of the vibrational state $\nu_i = 1$. When the radiation power of the cw CO_2 lasers was of the order of 10^3 W/cm^2 and the frequency of one of the lasers was detuned near exact resonance with an intermediate level of 5 GHz, the frequency shift of two-photon resonance in NH_3 observed was of the order of 300 kHz.

Thus, the frequency and width of narrow resonances of two-photon absorption inside the Doppler contour depend greatly on the intensity of fields that give rise to two-photon transitions. This must be taken into account in experiments on high-resolution spectroscopy. According to (4.34), this is of special importance in experiments in which a high rate of two-photon transitions is attained.

4.1.4 Two-Photon Resonances at Atomic Transitions

Two-photon resonances inside a Doppler-broadened absorption line were first observed in experiments on Na atoms [4.16,29]. Two counter-running pulses of a dye, narrow-line laser excited the transition 3S - 5S, which is forbidden for single-quantum transitions. The energy-level diagram of Na used in these

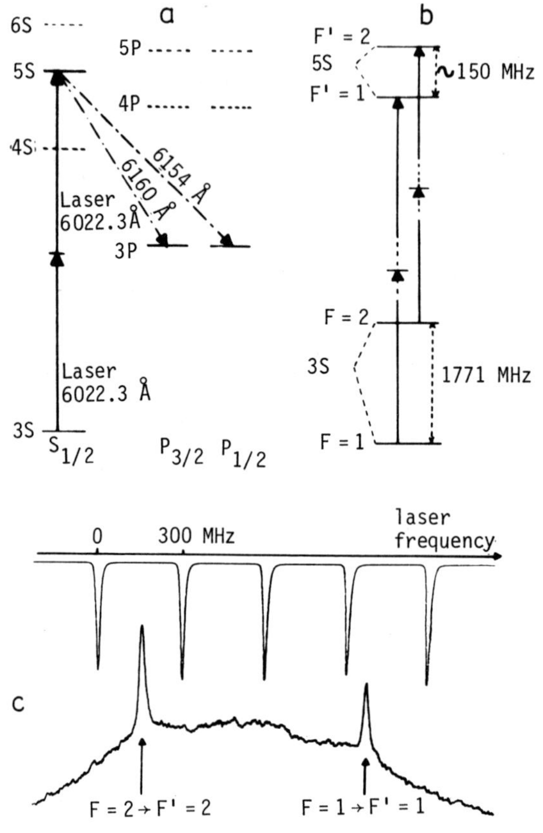

Fig.4.8a-c Two-photon spectroscopy of the transition 3S - 5S of Na atom: a) level scheme for excitation and fluorescence; b) hyperfine structure of 3S and 5S levels; c) two-photon absorption spectrum in the field of two counter-running waves of opposite circular polarization (from [4.30])

experiment is drawn in Fig.4.8a,b. The Na atom passes from the ground state 3S to the excited state 5S, absorbing two photons with the wavelength of 602.23 nm. The excited atoms were detected by subsequent decay of the 5S state with emission by successive transitions (5S - 3P or 4P - 3S). The nucleus of the ^{23}Na atom has spin $I = 3/2$, which causes hyperfine splitting of the levels of 3S and 5S states into two sublevels. For the two-photon transition $L = 0 \rightarrow L' = 0$ the selection rules demand that $\Delta F = 0$ and $\Delta M_F = 0$.

The line of two-photon absorption, therefore, is a superposition of two transitions: $F=1 \rightarrow F'=1$ and $F=2 \rightarrow F'=2$. The structure of two-photon resonances for counter-running waves, linearly polarized and circularly polarized with different signs, is shown in Fig.4.4. In the first experiments [4.16, 29] pulsed dye lasers were employed, but in next experiments [4.30,31] cw dye lasers were used. Figure 4.8c shows the shape of two-photon excitation rate of the 5S-state for Na recorded in the field of a cw-laser standing wave [4.30]. At the top there is a frequency scale obtained by means of a Fabry-Perot interferometer. The splitting of a two-photon resonance produces at once the difference of hyperfine interaction constants for two states,

$$\Delta\nu = A_{3S} - A_{5S} = 811 \pm 5 \text{ MHz} .$$

Because $A_{3S} = 886$ MHz, we can obtain a rather true value for the constant of hyperfine interaction, $A_{5S} = 75 \pm 5$ MHz. The resolution in these experiments is about 10 MHz and can be even better.

The method of two-photon resonances may be and has been applied to study all effects that require high resolution inside the Doppler contour, in particular, Zeeman and Stark effects, collision broadening, excitation transfer, isotope shift. Table 4.2 lists these experiments, which are first demonstrations of the scope of two-photon spectroscopy. The spectroscopic information obtained in these experiments is discussed in Chapter 7; the application of two-photon resonances to precision measurement of fundamental physical constants by means of positronium and hydrogen atoms is discussed in Chapter 10.

Table 4.2 Experiments with atomic two-photon resonances

Atom	Two-photon transition	Spectroscopic effect	Reference
^{23}Na	3S - 5S	Hyperfine splitting	[4.16,29,30]
	3S - 5S	Zeeman effect	[4.32]
	3S - 4D	Hyperfine splitting	[4.31]
	3S - 4D	Stark effect	[4.33]
	3S - 4D	Zeeman effect	[4.34]
	3S - 4D	Broadening and shift by Ne atom collisions	[4.35]
	3S - 4D	Transfer of excitation to Ne atoms	[4.36]
^{20}Ne, ^{22}Ne	3S, $J=2 \rightarrow$ 4d'(5/2), $J=3$	Isotopic shifts and hfs	[4.37,38]
H	1S - 2S	Lamb shift of 1S level	[4.39]
H and D	1S - 2S	Isotopic shift and Lamb shift	[4.40]

4.1.5 Two-Photon Resonances on Molecular Transitions

The first successful experiments on observing two-photon resonances at vibrational-rotational molecular transitions were reported in [4.27,28] on CH_3F and NH_3 molecules. In these experiments, the coincidence between several lines of CO_2 laser and the rotational-vibrational line of the ν_3 band was used. Figure 4.9a shows an example of agreement between the P(14) line of a CO_2 laser and the transition R(1,1) of the main band $0 \rightarrow \nu_3$ and between the P(30) line and the transition R(2,1) of the $\nu_3 \rightarrow 2\nu_3$ (so-called hot band). The difference between the sum of the frequencies of the two laser lines and the frequency of this two-photon transition is 139 MHz. By applying a dc electric field across the cell with CH_3F it is easy to obtain an exact resonance, due to Stark effect. The molecule of $^{12}CH_3F$ is a symmetrical top; in calculating the composite matrix element A_{gf} the selection rules for transitions of a parallel band should be met, $\Delta J = \pm 1$, $\Delta K = 0$, $\Delta M = 0, \pm 1$ (depending on the laser wave polarization and the dc electric-field direction). Figure 4.9b shows the shape of two-photon resonance (the signal of first derivative with respect to frequency) in CH_3F (20 mtorr) when the intensity of

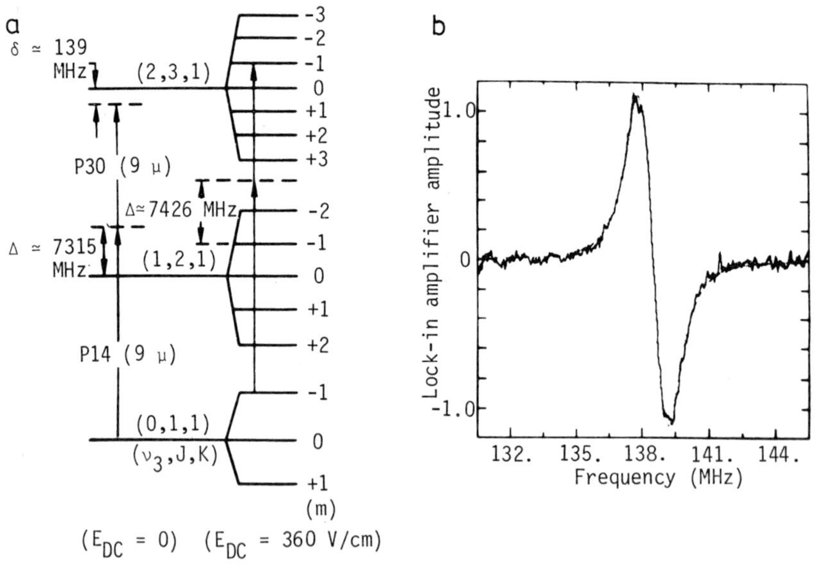

Fig.4.9a,b Two-photon spectroscopy of the rotational-vibrational transition $\nu_3 = 0 \rightarrow \nu_3 = 2$ of CH_3F molecules: a) scheme of energy levels in the external dc electric field for two-photon transition; b) resonance shape of two-photon absorption inside Doppler contour (from [4.27])

one of the fields (line P(14) 9.4 μm) is 400 W/cm^2 and the second field (line P(30) 9.4 μm) is used as a probe. The accuracy of measurement of the two-photon resonance parameters was sufficient to determine the pressure broadening (41.3 ± 1 MHz/torr) and pressure shift (2.1 ± 0.1 MHz/torr) of resonance [4.27].

The first experiment on two-photon Doppler-free spectroscopy at electronic-vibrational-rotational molecular transitions was reported in [4.41]. In this experiment, the transitions $A^2\Sigma_1^+(\nu = 0) \leftarrow X^2\Pi_{1/2}(\nu = 0)$ of the NO molecule and ${}^1B_{2u} \leftarrow {}^1A_{1g}$, 14_0^1 of benzene molecule were studied. Excitation was by radiation at $\lambda = 453$ nm (for NO) and 504 nm (for benzene) from a pulsed dye laser. The spectral resonances obtained, the width of the laser spectrum being 0.2 GHz, were much narrower than the Doppler broadening of the NO transition (3.0 GHz). Application of two-photon resonance technique to electron transitions of complex molecules is of great importance for it enables us to resolve the structure in the spectrum of a great number of overlapping Doppler lines, which often form a continuous spectrum. The successful experiment reported in [4.41] should be considered only as the first step in this important direction.

4.2 Raman Scattering in Gases Without Doppler Broadening

The frequency shift of Raman scattering ω_S with respect to the laser frequency ω_L depends on the scattering-particle velocity $\underline{v}$

$$\omega_L - \omega_S = \omega_{12} + (\underline{K}_L - \underline{K}_S)\underline{v} \ , \tag{4.35}$$

where $\underline{K}_L$ and $\underline{K}_S$ are the wave vectors of the laser and scattered photons, $\omega_0 = \omega_{12}$ is the frequency of transition for a stationary particle. Therefore, the line width of spontaneous Raman forward scattering (unidirectional running waves $\underline{K}_L$ and $\underline{K}_S$) is determined by the homogeneous width of transition 2Γ and a small residual Doppler broadening,

$$\Delta\omega_{SC}^{+} = 2\Gamma + (\omega_L - \omega_S)\,\frac{u}{c} = 2\Gamma + \frac{\omega_L - \omega_S}{\omega_L}\,\Delta\omega_D \ , \tag{4.36}$$

where $\Delta\omega_D$ is the ordinary Doppler width at the frequency of observation ω_S. The line width of backward scattering contains the almost-doubled Doppler broadening,

$$\Delta\omega_{SC}^{-} = 2\Gamma + (\omega_L + \omega_S)\,\frac{u}{c} = 2\Gamma + \frac{\omega_L + \omega_S}{\omega_L}\,\Delta\omega_D . \tag{4.37}$$

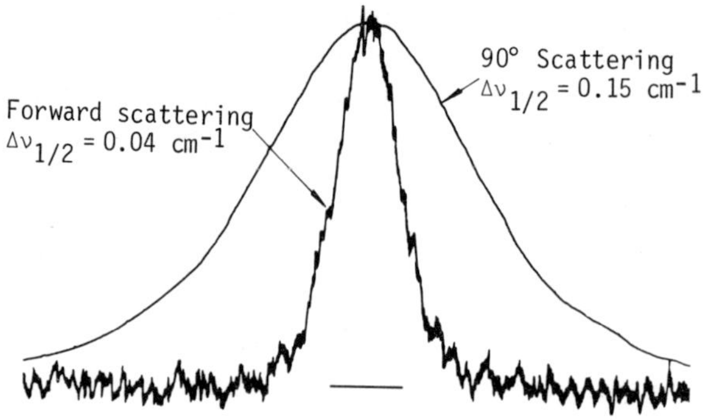

Fig.4.10 Profiles of the rotational line S(1) of Raman scattering at H_2 at 2 atm. in a forward direction and at an angle of 90° (from [4.42])

The effect of considerable reduction of Doppler broadening during collinear observation of scattered radiation in a narrow space angle was reported as early as in [4.42]. In this experiment, spontaneous Raman scattering was observed in rotations of H_2 molecules. Figure 4.10 shows the profiles of a rotational line of H_2 in forward scattering and at an angle of 90° to the incident radiation [4.42]. For 90° scattering, the total half-maximum width equals 0.15 cm^{-1}, which corresponds to the Doppler width at 20°C. In a forward direction, the observable line width is reduced to 0.04 cm^{-1}. It is evident that this width is determined by the total contribution of collisional broadening (0.006 cm^{-1} at 2 atm.), residual Doppler broadening (for the line S(1) located 608 cm^{-1} from the laser line 6328 Å, this width equals 0.015 cm^{-1}) and instrumental width (the resolution of Fabry-Perot interferometer and the line width of laser radiation). In this experiment, the resolution attained was an order of magnitude greater than that in all other experiments on H_2 and differed from the limiting value by only the factor of 2.

Attention should be drawn to the contribution of the residual Doppler effect, because of the inequality of the ω_L and ω_S frequencies, which limits the narrowing of the spectral line of Raman scattering. In [4.42] this effect was not allowed for; therefore, the author of [4.43] drew a wrong conclusion concerning the possibility of reducing the line width to 0.001 cm^{-1} by corresponding reduction of pressure. Actually, in experiment on H_2, the collisional broadening at 2 atm. is only 40 percent as much as the residual Doppler broadening; further reduction of pressure is not advisable. Nevertheless, further narrowing is possible in experiments on scattering at rotations of heavier molecules. In such cases, reduction of gas pressure is needed.

Increase of resolution in observing forward scattering requires that the scattered radiation should be detected in a narrow solid angle. It is possible to improve this technique by measuring the amplification line shape at the Raman scattering frequency by means of an additional tunable laser beam collinear with the exciting beam. The difficulty of recording weak amplification lines at the level 10^{-5} to 10^{-7} cm^{-1} can be overcome by using the optical-acoustic method to detect the energy absorbed by a gas due to Raman scattering. In such experiments, optical-acoustic signals in a molecular gas must be measured as the difference of frequencies of the two lasers is tuned to the frequency of molecular vibrations.

4.3 Spectral Shape of Resonance Fluorescence

Here we consider resonance scattering of a strong field by atoms in ground state. The parameter

$$G = \frac{1}{2}\left(\frac{P_{10}E}{\hbar\Gamma}\right)^2 , \tag{4.38}$$

is characteristic of the problem, where P_{10} is the dipole matrix element of the transition between the ground 0 and excited 1 states, $\gamma_1 = 2\Gamma$ is the rate of radiative decay of level 1. When $G \ll 1$, the incident field can be considered to be weak; to this case, we may apply WEISSKOPF's theory of resonance scattering [4.3]. Scattering, in this case, occurs at the incident-radiation frequency (coherent scattering).

In [4.44,45] the problem of light scattering by ground-state atoms was solved with no limitations on the incident-field intensity. The spectral distribution of scattered radiation consists of coherent and incoherent parts.

According to [4.45], the differential cross-section of incident-wave scattering by a stationary atom in the frequency interval $d\Delta$ may be written

$$\sigma(\theta,\Delta)dOd\Delta = [\sigma_{coh}\delta(\Delta) + \sigma_{inc}I_{inc}(\Delta)]\,\frac{3}{8}\sin\theta dOd\Delta , \tag{4.39}$$

where $dO = \sin\theta d\theta d\phi$, θ and ϕ are polar and azimuthal angles in the spherical-coordinate frame, $\Delta = \omega' - \omega$, ω is the incident-field frequency, ω' is the scattered-field frequency, and σ_{coh} and σ_{inc} are the cross-sections of coherent and incoherent scattering,

$$\sigma_{coh} = \lambda^2 \frac{3}{2\pi}\Gamma^2 \frac{\Omega^2 + \Gamma^2}{[\Omega^2 + \Gamma^2(1+G)]^2} , \tag{4.40}$$

$$\sigma_{inc} = \lambda^2 \frac{3}{2\pi} \Gamma^2 \frac{\Gamma^2 G}{[\Omega^2 + \Gamma^2(1+G)]^2} , \tag{4.41}$$

$2\Gamma = \gamma_1$, $\Omega = \omega - \omega_{10}$, λ is the wavelength. The spectral distribution of incoherent-scattering intensity is

$$I_{inc}(\Delta) = \frac{1}{\pi\left(1 - \frac{\sigma_{coh}}{\sigma}\right)} \mathrm{Re}\left\{\frac{1}{\Gamma - i(\Delta + \Omega)} - \right.$$

$$\left. - \frac{\Gamma^2 G[\Gamma - i(\Delta - \Omega)]}{(2\Gamma - i\Delta)[\Gamma - i(\Delta + \Omega)](\Omega^2 + \beta^2)} + i\,\frac{(\Gamma - i\Omega)[\Gamma - i(\Delta - \Omega)]}{\Delta(\Omega^2 + \beta^2)}\right\} , \tag{4.42}$$

where

$$\beta^2 = (\Gamma - i\Delta)(\Gamma - i\Delta + 2\Gamma^2 G)/(2\Gamma - i\Delta) .$$

At weak saturation ($G \ll 1$), we have for $I_{inc}(\Delta)$

$$I_{inc}(\Delta) = \frac{1}{\pi} \frac{2\Gamma(\Omega^2 + \Gamma^2)}{[(\Delta + \Omega)^2 + \Gamma^2][(\Delta - \Omega)^2 + \Gamma^2]} . \tag{4.43}$$

The spectral distribution of incoherently scattered photons has two maxima at the frequencies $\omega \pm \Omega$, that is there are two broadened side components at the frequency ω, along with a coherently scattered monochromatic component at frequency ω. The expression can be simplified when $G \gg 1$

$$I_{inc}(\Delta) = \frac{\Gamma_V}{2\pi\Gamma}\left[\frac{2V^2}{\Omega^2 + 2V^2}\,\frac{\Gamma_V}{\Delta^2 + \Gamma_V^2} + \frac{1}{2}\,\frac{b}{(\Delta - c)^2 + b^2} + \frac{1}{2}\,\frac{b}{(\Delta + c)^2 + b^2}\right] , \tag{4.44}$$

where $c = (\Omega^2 + 4V^2)^{1/2}$, $\Gamma_V = 2\Gamma(\Omega^2 + 2V^2)/(\Omega^2 + 4V^2)$, $b = 2\Gamma(\Omega^2 + 6V^2)/(2\Omega^2 + 8V^2)$, $V = P_{10}E/2\hbar$. The scattered field in this case consists of three lines spaced apart by a value that depends on the incident-light power. The result obtained may be interpreted as a consequence of level splitting due to dynamical Stark effect, as was done in Section 2.8.

To turn to the discussion of resonance fluorescence of gas, we should take into account the motion of the atoms. The scattering cross-section for atoms moving at the velocity $\underline{v}$ is given by expressions from (4.39) to (4.42) with the substitutions

$$\Omega \to \Omega - \underline{k}\underline{v} , \quad \Delta \to \Delta + \underline{k}\underline{v} - \underline{k}_s\underline{v} , \tag{4.45}$$

where $\underline{k}$ and $\underline{k}_s$ are the wave vectors of incident and scattered waves. If the axis z is directed along $\underline{k}$ and x lies on the plane of $\underline{k}$ and $\underline{k}_s$,

$$\Omega \to \Omega - kv \ , \quad \Delta \to \Delta + kv_z - kv_z \cos\theta - kv_x \sin\theta \ , \tag{4.46}$$

where v_z and v_x are the projections of atom velocity onto z and x axis, respectively.

For the cross-section averaged over atom velocities, we have

$$Q(\theta,\Delta)dOd\Delta = dOd\Delta \int_{-\infty}^{\infty} dv_z \int_{-\infty}^{\infty} dv_x W(v_z,v_x)\sigma(\theta,\Delta + kv_z - kv_z \cos\theta - kv_x \sin\theta) \ , \tag{4.47}$$

where $W(v_x,v_z)$ is the function of velocity distribution of atoms, which in the equilibrium case equals

$$W(v_x,v_z) = \frac{1}{\pi u^2} \exp\left(-\frac{v_x^2 + v_z^2}{u^2}\right) \ . \tag{4.48}$$

Our further calculations we carry out for equilibrium distribution (4.48) assuming that $ku \gg \Gamma$. For $G \ll 1$, accurate to G, we may write for cross-section of the total scattering

$$Q = \lambda^2 \frac{3}{8\sqrt{\pi}} \frac{\Gamma}{ku} e^{-(\Omega/ku)^2} (1 - \frac{G}{2}) \ , \tag{4.49}$$

where

$$Q(\theta,\Delta)dOd\Delta = QI(\theta,\Delta) \frac{3}{8\pi} \sin^2\theta dOd\Delta \tag{4.50}$$

denotes the differential scattering cross-section, $I(\theta,\Delta)$ is the spectral distribution of the scattered radiation.

For the angles $ku \sin\theta \gg \Gamma$ the spectral distribution equals

$$I(\theta,\Delta) = \frac{1}{ku\sqrt{\pi} \sin\theta} \exp\left\{-\left[\frac{\Delta + \Omega(1 - \cos\theta)}{ku \sin\theta}\right]^2\right\} \ . \tag{4.51}$$

Expression (4.51) is not true for forward and backward scattering, that is with $ku \sin\theta \ll \Gamma$. After integrating, we have, in this case,

$$I(0,\Delta) = \delta(\Delta)(1 - \frac{G}{2}) + \frac{3}{\pi} G \frac{\Gamma^3}{(\Delta^2 + 4\Gamma^2)(\Delta^2 + \Gamma^2)} \ , \tag{4.52}$$

$$I(\pi,\Delta) = \frac{1}{\pi}\frac{2\Gamma}{(\Delta+2\Omega)^2+4\Gamma^2} + \frac{G}{2\pi}\left\{\frac{2\Gamma}{(\Delta+2\Omega)^2+4\Gamma^2} - \right.$$

$$\left. - \frac{3(2\Gamma)^2}{[(\Delta+2\Omega)^2+(2\Gamma)^2]^2} + \frac{3(2\Gamma)^3}{[(\Delta+2\Omega)^2+(2\Gamma)^2][(\Delta+2\Omega)^2+(4\Gamma)^2]}\right\} . \quad (4.53)$$

The width of incoherently forward- and backward-scattering radiation is determined by the line width 2Γ, whereas for other angles of scattering it depends on the Doppler width. Expressions (4.52) and (4.53) describe narrow resonances without Doppler broadening in the fluorescence line during forward and backward scattering.

As follows from (4.39), in addition to incoherent scattering, there is always coherent scattering, the spectral width of which is determined by the spectrum of the incident laser radiation; hence, it may be much smaller than the line width of incoherent scattering. This has been proved experimentally recently [4.46]. This effect is usually used to explain the two-quantum process of coherent atom-photon interaction [4.4] and is sometimes called the "Heitler experiment".

References

4.1 G. Landsberg, L. Mandel'stam: Naturwissenschaften 16, 557,772 (1928)
4.2 C.V. Raman: Indian J.Phys. 2, 387 (1928)
4.3 V. Weisskopf: Ann.der Physik 9, 23 (1931)
4.4 W. Heitler: *The Quantum Theory of Radiation* (Oxford University Press, Oxford 1954), pp.348-353
4.5 M. Goeppert-Mayer: Ann.der Physik 9, 273 (1931)
4.6 P.A.M. Dirac: *The principles of Quantum Mechanics*, 1st edition (1930); 2nd edition (1958) (Clarendon Press, Oxford)
4.7 W. Kaiser, C.G.B. Garrett: Phys.Rev.Lett. 7, 229 (1961)
4.8 J.M. Worlock: In *Laser Handbook*, Vol.2, ed. by F.T. Arecchi, E.O. Schulz-Dubois (North Holland Publishing Co., Amsterdam 1972), p.1323
4.9 V.I. Bredikhin, M.D. Galanin, V.N. Genkin: Uspekhi Fiz.Nauk 110, 3 (1973)
4.10 L.S. Vasilenko, V.P. Chebotayev, A.V. Shishaev: Pis'ma Zh.Eksp.i Teor. Fiz. 12, 161 (1970)
4.11 B. Cagnac, G. Grynberg, F. Biraben: J.Phys. (Paris) 34, 56 (1973)
4.12 N. Bloembergen: *Nonlinear Optics* (W.A. Benjamin, Inc., New York 1965)
4.13 N. Bloembergen, M.D. Levenson: *High Resolution Laser Spectroscopy*, "Topics in Applied Physics", Vol. 13, ed. by K.Shimoda (Springer-Verlag, Berlin, Heidelberg, New York 1976), p.315.
4.14 M. Inoue, Y. Toyozawa: J.Phys.Soc.Japan 20, 363 (1965)
4.15 J.E. Bjorkholm, P.F. Liao: Phys.Rev.Lett. 33. 128 (1974)
4.16 F. Biraben, B. Cagnac, G. Grynberg: Phys.Rev.Lett. 32, 643 (1974)
4.17 V.S. Letokhov, V.G. Minogin: Z.Eksp.i Teor.Fiz. 71, 135 (1976)
4.18 V.G. Minogin: Kvantovaya Elektronika 3, 2061 (1976)
4.19 E.V. Baklanov, V.P. Chebotayev: Opt.Comm. 12, 312 (1974)
4.20 E.V. Baklanov, V.P. Chebotayev: Kvantovaya Elektronika 2, 606 (1975)

4.21 M. Mizushima: Phys.Rev. A133, 414 (1964)
4.22 A.M. Bonch-Bruevich, V.A. Khodovoi: Uspekhi Fiz.Nauk 93, 1967 (1967) (Sov.Phys.Usp. 10, 637 (1968)
4.23 P.F. Liao, J.E. Bjorkholm: Phys.Rev.Lett. 34, 1 (1975)
4.24 J.E. Bjorkholm, P.F. Liao: In *Laser Spectroscopy*, Proceedings of 2nd Intern. Conf., 23-27 June 1975, Megeve, France (Springer-Verlag, Berlin, Heidelberg, New York 1975), p.176
4.25 P.L. Kelley, H. Kildal, H.R. Schlossberg: Chem.Phys.Lett. 27, 62 (1974)
4.26 A.L. Golger, V.S. Letokhov, S.P. Fedoseev: Kvantovaya Elektronika 3, 1457 (1976)
4.27 W.K. Bischel, P.J. Kelley, C.K. Rhodes: Phys.Rev.Lett. 34, 300 (1975)
4.28 W.K. Bischel, P.J. Kelley, C.K. Rhodes: Phys.Rev. A 13, 1817 (1976)
4.29 M.D. Levenson, M. Bloembergen: Phys.Rev.Lett. 32. 645 (1974)
4.30 F. Biraben, B. Cagnac, G. Grynberg: Phys.Lett. 49A, 71 (1974)
4.31 T.W. Hansch, K. Harvey, G. Meisel, A.L. Schawlow: Opt.Comm. 11, 50 (1974)
4.32 N. Bloembergen, M.D. Levenson, M.M. Salour: Phys.Rev.Lett. 32, 867 (1974)
4.33 K.C. Harvey, R.T. Hawkins, G. Meisel, A.L. Schawlow: Phys.Rev.Lett. 34, 1073 (1975)
4.34 F. Biraben, B. Cagnac, G. Grynberg: C.R.Acad.Sci.Paris, 279, Ser.B, 51 (1974)
4.35 F. Biraben, B. Cagnac, G. Grynberg: J.de Physique 36, 41 (1975)
4.36 F. Biraben, G. Cagnac, G. Grynberg: C.R.Acad.Sci.Paris 280, Ser.B, 235 (1975)
4.37 F. Biraben, E. Giacobino, G. Grynberg: Phys.Rev. A 12, 2444 (1975)
4.38 B. Cagnac: In *Laser Spectroscopy*, Proceedings of 2nd Intern.Conf., 23-27 June 1975, Megeve, France (Springer-Verlag, Berlin, Heidelberg, New York 1975), p.165
4.39 T.W. Hansch, S.A. Lee, R. Wallenstein, C. Wieman: Phys.Rev.Lett. 34, 307 (1975)
4.40 S.A. Lee, R. Wallenstein, T.W. Hansch: Phys.Rev.Lett. 35, 1262 (1975)
4.41 J.A. Gelbwachs, P.F. Jones, J.E. Wessel: Appl.Phys.Lett. 27, 551 (1975)
4.42 W.R.L. Clements, B.P. Stoicheff: J.Molec.Spectros. 33, 183 (1970)
4.43 Yu.S. Bobovich: Uspecki Fiz.Nauk 108, 401 (1972)
4.44 B.R. Mollow: Phys.Rev. 188, 1969 (1969)
4.45 E.V. Baklanov: Zh.Eksp.i Teor.Fiz. 65, 2203 (1973)
4.46 H.M. Gibbs, T.N.C. Venkatesan: Phys.Rev.Lett. (in press)

5. Nonlinear Resonances on Coupled Doppler-Broadened Transitions

A new type of resonant phenomena that are of interest for laser spectroscopy of superhigh resolution and quantum electronics may occur by interaction of several light fields that are resonant to coupled transitions. Of great interest are the phenomena that occur in a gas of three-level atoms when the system is affected by two fields, each of which is resonant to transitions $1\to 0$, $0\to 2$, one of the levels being common to both transitions (Fig.5.1). The effect of a field on the transition $1\to 0$ influences considerably the shape of the radiated line of a probe signal of the adjacent transition $0\to 2$. The complex of methods based on the use of absorption or emission resonances resulting from nonlinear interaction of two or several fields with a gas of three-level atoms or molecules is the subject of so-called three-level laser spectroscopy of superhigh resolution (TLS). A great number of studies have been devoted to investigations of the TLS methods. References to most of those can be found in the surveys [5.1-4]. As compared with the resonances described previously, the ones that occur in three-level systems have new properties that are of interest for spectroscopy. The present chapter is devoted to discussion of these properties.

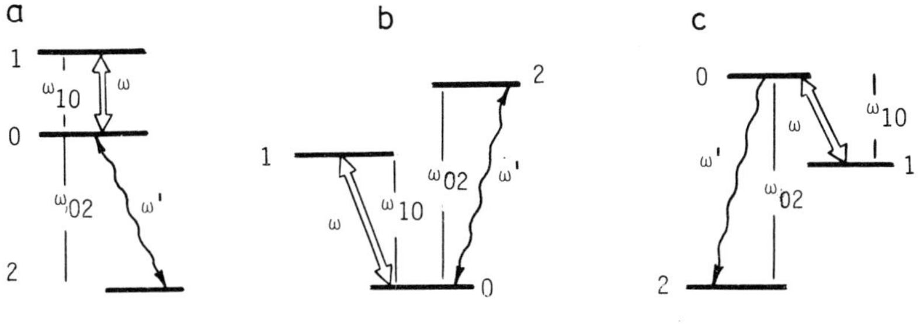

Fig.5.1a-c Energy-level configurations for two coupled transitions: a - cascade configuration; b,c - bent configurations

Historically, the investigations of three-level systems have played an outstanding part in forming the concept of interaction of radiation with quantum systems. The first theoretical investigations of interaction of three-level systems with a radiation field were associated with the discovery of Raman scattering in liquids and crystals [5.5,6]. Such effects as two-photon absorption and two-photon emission were treated together with Raman scattering [5.7]. WEISSKOPF used a three-level approximation in the theory of resonance fluorescence with decay of levels [5.8]. The general problems of light scattering were treated in the first years of origin of the radiation theory. EINSTEIN and EHRENFEST [5.9] assumed that the process of scattering could be considered as independent processes of absorption and emission following each other. We shall identify such processes with one-quantum, step-by-step ones. They lead to population of an intermediate level.

WEISSKOPF [5.8] and PLACZEK [5.10] showed that the processes of scattering of monochromatic light could not be considered as sequential processes of absorption and emission. Their treatment is adequate to the two-photon processes of radiation. The latter, as will be shown, play an important part in the processes of radiation in three-level systems under resonance conditions.

The first experimental studies of interaction of monochromatic radiation with quantum systems were made in a microwave range, because of the relative simplicity of production of intense sources of coherent radiation. Radiospectroscopic methods are practiced on a large scale in study of the hyperfine structure of molecular rotational spectra [5.11]. The high intensity of monochromatic fields of the microwave range permits observation of such phenomena as multiphoton transitions and splitting of levels in an electromagnetic field (dynamic Stark effect) [5.12]. Of great significance were the investigations of double radio frequency and optical resonances and of the methods of optical pumping [5.13]. The first method is based on the optical methods of recording low-frequency resonances; the second is pumping by optical radiation for creation of a nonequilibrium distribution between sublevels of a hyperfine structure.

Three-level systems of pumping attracted particular attention as one of the methods of production of active media for masers [5.14,15]. The theory of this method of pumping was treated in [5.16,17]. The interaction of strong fields of the radio range with a three-level system was treated in a series of surveys and monographs [5.18-20]. In connection with the problem of production of a three-level amplifier and generator, much consideration was given to investigation of the gain line shape of a weak signal in the presence of a strong signal on the coupled transition. For calculations, the atoms and molecules were assumed to be motionless (Doppler broadening of a line in the microwave range is usually much less than homogeneous width caused by collisions), and the relaxation constants of all levels were assumed to be identical. As a result of the calculations and experiments, the gain line shape of the probe signal on the coupled transition with a sufficiently strong external field was found to be greatly different from that which could be accounted for on the basis of only the change of population. The great changes of line shape were associated with two-quantum transitions or with the dynamic Stark effect.

The production of sources of coherent radiation in the optical and infrared ranges has opened up new chances to study the interaction of three-level systems with an electromagnetic field. The character of phenomena in three-level systems in the optical range as compared with those observed in the microwave range is related to the inhomogeneous Doppler-line broadening and the difference in level relaxation constants. Owing to this, the line shape of absorption (gain) of a weak field contains narrow resonances that may be used for creation of lasers and amplifiers with narrow gain lines, for spectroscopy of superhigh resolution under the conditions of dominant Doppler broadening.

Many studies have recently been published that are devoted to investigations of three-level systems as applied to the optical and near IR bands, for instance studies have been made of double infrared-microwave resonances, when one of the transitions is in the microwave region and another in the infrared region. The double-resonance method known previously is combined in these studies with the methods of nonlinear laser high-resolution spectroscopy of the optical band. Laser techniques have enriched these methods and opened up new possibilities for studies [5.21,22].

The nonequilibrium velocity distribution of particles, which arises on each level of a two-level system under the action of the field, can be found in the usual way when resonances are observed in the radiation line shape of the coupled transition (Fig.5.1). In this way, the processes of relaxation can be studied on each level. In one of the first studies of this kind, BENNETT et al. [5.23] found the nonequilibrium distribution of population of one of the levels when observing spontaneous radiation of the coupled transition and made spectroscopic investigations of collision broadening of a neon spectral line. CORDOVER et al. [5.24] used resonances in a three-level system to measure isotope shifts of neon lines, and SCHWEIZER et al. [5.25] studied broadening of narrow resonances in neon.

The line shape of the coupled transition is determined not only by effects of the level population changes in a strong field. The coherent phenomena lead to a number of peculiarities that manifest themselves, first of all, in the radiation line shape. SCHLOSSBERG and JAVAN [5.26] showed that, in the case of closely spaced levels, the width of the coupled transition is determined by the width of the initial and final levels. They have associated this phenomenon with two-quantum transitions. NOTKIN et al. [5.27] predicted an anisotropy of the radiation line of the coupled transition, which has been observed in spontaneous radiation of a gas-laser cavity [5.28], in the gain of a three-level gas laser [5.29], in the line shape of stimulated radiation [5.30,31], and in a level-crossing study [5.32,33].

A great variety of different scheme is encountered in three-level systems. Consequently, the authors of most studies centered attention on various aspects of the peculiarities of a three-level gas system, using various terminology and interpretations of the theory-predicted phenomena.

In their papers, SCHLOSSBERG and JAVAN [5.26], FELD and JAVAN [5.33], and FELD and FELDMAN [5.34] associated the effects of the radiation line narrowing under action of the laser field with two-quantum transitions as well as with change of the probability of one-quantum transitions in the presence of the field on the coupled transition. The authors of [5.26,33] treated the transition of an atom from level 1 to level 2 as a two-quantum transition. In this case, no distinction was made between two-quantum and step-by-step one-quantum transitions.

NOTKIN et al. [5.27] interpreted the changes of the line shape of spontaneous radiation by nonlinear interference effects that arise from mixing of stationary states of an isolated atom by the external field. HOLT [5.28] explained the change of the line shape by frequency correlation in the two-photon transition.

Reference [5.35] introduced a classification of the effects that lead to the change of absorption or emission spectra of a gas placed in the external monochromatic field resonant to the coupled transition. The first effect is formation of a nonequilibrium velocity distribution of atoms; the second is splitting of atomic levels; the third is a nonlinear interference effect that results from the coherence that is caused by the influence of the strong field upon an atom. HÄNSCH and TOSCHEK [5.36] analyzed two effects: an effect of frequency correlation and an effect of splitting, restricting themselves to consideration of polarization to the third order.

Finally, in [5.37] the phenomena observed in the framework of the perturbation theory are interpreted by two-quantum and one-quantum step-by-step processes and by their interference. The role of the processes in a resonance depends largely on the relation between the level relaxation constants. As usual, the interference of two- and one-quantum processes is more essential when the level relaxation constants are equal. The explanation based on the two- and one-quantum step-by-step processes directly connects the latest theories with the classical theories of scattering and describes simultaneously all of the phenomena in the framework of the perturbation theory. From here on, we shall use the treatment given in [5.37].

When describing the phenomena that occur in three-level systems, we assume that

a) detunings of the field frequencies from the centers of the lines of the transitions $0 \to 1$ and $2 \to 0$ $\Omega = \omega - \omega_{01}$, $\Omega' = \omega' - \omega_{02}$ are within the Doppler line width of the transitions, i.e., $\Omega \lesssim ku$ and $\Omega' \lesssim k'u$ where u is a thermal velocity, k and k' are wavenumbers of the transitions $1 \to 0$ and $2 \to 0$,

b) the homogeneous widths of the transitions $0 \to 1$ and $0 \to 2$ Γ_{01} and Γ_{02} are far less than the Doppler widths ku and k'u,

c) the energy of the atom-field interaction $P_{01}E/2\hbar$, $P_{02}E'/2\hbar$ (in units of $\hbar$) is far less than the Doppler width, P_{01} and P_{02} are matrix elements of dipole moments of the transitions $0 \to 1$ and $0 \to 2$, respectively, E and E' are amplitudes of the fields with the frequencies ω and ω',

d) the decay constants of the levels are different.

The conditions a, b, c suggest an inhomogeneous character of saturation of a gas of two-level atoms on the coupled transitions.

5.1 Two-Quantum and Step-by-Step Transitions

The processes that take place in a gas of three-level atoms are closely connected with those in two-level systems. An inhomogeneity of saturation leads to the formation of Bennett "holes" in the velocity distribution of atoms on levels 0 and 1 under the action of the field on the transition. A monochromatic field interacts effectively with the atoms whose velocities meet the resonance condition: the field frequency that is perceived by a moving atom is equal to the frequency of the transition $\omega_{01} = \omega - kv$.

One of the first experimental investigations of a three-level system in the optical band [5.23] was aimed at finding the nonequilibrium velocity distribution that arises under the action of a strong field. Indeed, a dip or

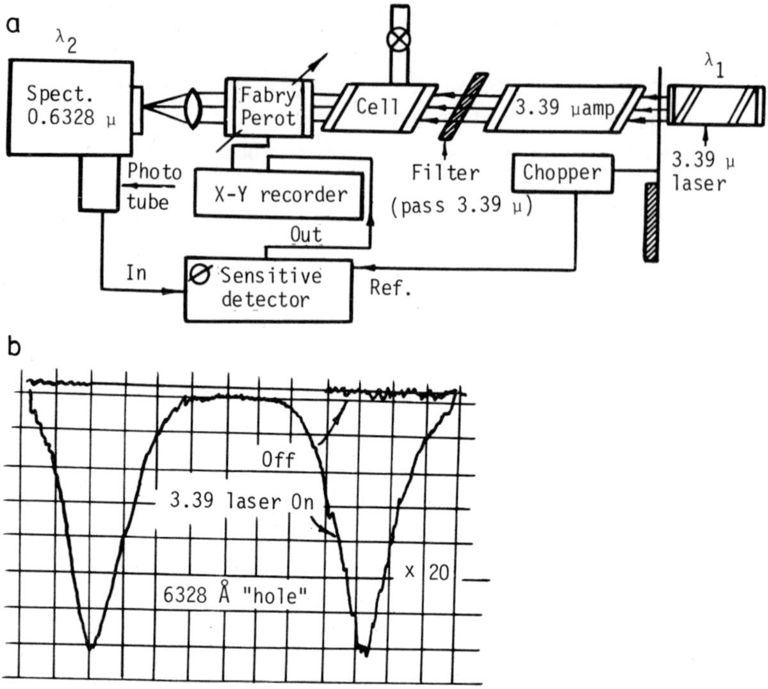

Fig.5.2a,b Experimental design for observing the variation of the line shape of radiation at $\lambda = 0.63$ μm under the action of the field at $\lambda = 3.39$ μm

a peak in the velocity distribution of excited particles can be detected by the change of the line shape of spontaneous radiation of the coupled transitions. The experiment [5.23] showed the change of the radiation spectrum of the Ne line at $\lambda = 0.63$ μm (transition $3s_2$-$2p_4$) under the action of the field at $\lambda = 3.39$ μm (transition $3s_2$-$3p_4$) (Fig.5.2). The stimulated transitions from the level $3s_2$ to the level $3p_4$ under the action of the field at $\lambda = 3.39$ μm reduce the population of the level $3s_2$, which is observed in the change of the spontaneous radiation spectrum at $\lambda = 0.63$ μm. Observation of a narrower spectrum as compared with the Doppler line width permits study of the influence of collisions on the homogeneous width of the transition $3s_2$-$2p_4$. The technique of observation of the line shape of spontaneous radiation does not possess the necessary sensitivity. Most subsequent experiments were therefore devoted to observation of the line shape of stimulated radiation of the coupled transition. We shall analyze that case.

These features of the line shape of radiation were explained by the change of particle distribution at the common level under the action of the field. The influence of the field on the line shape at the coupled transition is restricted to only the population effects. Let the field E be in resonance with the Doppler-broadened transition $0 \rightarrow 1$ and propagate along axis z. As a result of the interaction of the field E with particles of the level 1, on the level 0 a peak arises in the distribution of velocity projections of excited particles onto axis z. The field E', which is in resonance with the transition $0 \rightarrow 2$, and the particles on the level 0 produce an induced dipole moment on the transition $0 \rightarrow 2$, which is responsible for the energy absorption (emission) at the frequency ω'. It should be particularly emphasized that the energy absorption (emission) at the frequency of the field E' is due to the population of the level 0. We might obtain the same result if we excited the particles to the level 0 by collisions of electrons with a specially prepared beam of particles. Because the distribution of particles on the level 0 is narrower than the Maxwell distribution on the transition $0 \rightarrow 2$, we observe a narrow emission line. Knowing the velocity distribution of atoms on the level 0, the line shape of stimulated emission on the transition $0 \rightarrow 2$ can be found,

$$\kappa(\Omega') = \sigma_0 \int [N_o(v) - N_2(v)] \frac{\Gamma_{02}^2}{(\Omega' - k'v)^2 + \Gamma_{02}^2} dv \ , \tag{5.1}$$

where $2\Gamma_{02}$ is a homogeneous width of the transition $0 \rightarrow 2$, Ω' is a frequency detuning from the line center on the transition $0 \rightarrow 2$, k' is the wavenumber that corresponds to the transition $0 \rightarrow 2$. In accordance with the results of Chapter 2, the distribution of the population difference takes the form

$$N_o(v) - N_2(v) = \left[(N_o^0 - N_2^0) + (N_1^0 - N_o^0) \frac{\gamma_1}{2} \frac{\Gamma_{01} G}{(\Omega - kv)^2 + \Gamma_{01}^2 (1 + G)} \right] W(v) \ . \tag{5.2}$$

The change of the difference of population depends on the lifetimes of levels 0 and 1. The greater the lifetime of the common level 0 than that of level 1, the greater the change of the difference of population. It is easy to understand this, physically. The excitation rate of level 0, all other conditions being equal, is large when the rate of decay of the level 1 is increased. At a given excitation rate, the change of population of level 0 is large with a small rate of decay of level 0. Inserting (5.2) into (5.1), we can find the gain (absorption) line shape. Here and below we shall be interested in the most important case of collinear wave propagation, when $[\underline{k}\underline{k}'] = 0$. In this case, we have

$$\kappa(\Omega') = \kappa_o e^{-(\frac{\Omega'}{k'u})^2} \left\{ 1 + \frac{N_1^o - N_o^o}{N_o^o - N_2^o} \quad \frac{(\frac{k'}{2k})\Gamma_o \gamma_1 G}{[(\Omega' \pm \frac{k'}{k}\Omega)^2 + \Gamma_o^2]\sqrt{1+G}} \right\} , \tag{5.3}$$

where κ_0 is the absorption coefficient of a weak signal, for the transition $0 \to 2$ in the absence of an external field; the signs $\pm$ correspond to the cases of opposite and unidirectional propagations of two waves; $\Gamma_o = \Gamma_{02} + (k'/k)\Gamma_{01} \cdot \sqrt{1+G}$

The line shape (5.3) is given by the resultant of two dispersion contours, the Lorentz contour of radiation with a width $2\Gamma_{02}$ and the velocity distribution of atoms (in units of k) with a width $2(k'/k)\Gamma_{01}\sqrt{1+G}$. With increase of the field intensity, the resonance amplitude approaches a constant value, which depends on the relaxation constants and the initial difference of the level population. Increase of width of the inhomogeneous distribution leads to a corresponding resonance broadening in the radiation line of the coupled transition. Depending on the value and sign of the difference $N_1^o - N_o^o$ and $N_o^o - N_2^o$, the absorption coefficient of the transition may change sign. In the presence of a very strong external field, the condition of inversion of the population between the levels 0 and 2 has the form,

$$(N_o^o - N_2^o) > (N_1^o - N_o^o)\frac{\gamma_1}{2\Gamma_{01}} . \tag{5.4}$$

In this treatment, we ignore coherent processes that may occur in the nonlinear interaction of waves with a three-level system and greatly influence the radiation line shape of the transition $0 \to 2$. We give a qualitative explanation of coherent processes. A three-level system can be represented as a nonlinear oscillator with resonant frequencies ω_{01}, ω_{02} and ω_{12}. The dipole oscillations with corresponding frequencies arise in such a system by action of fields E and E'. Owing to nonlinearity of the system, the oscillations at frequency ω and the field E' at frequency ω' lead to the dipole oscillations at the frequencies $\omega' - \omega$ and $\omega' + \omega$ (the oscillation amplitude depends largely on how far the frequency $[\omega' - \omega]$ or $\omega' + \omega$ is from the frequency ω_{12}). The nonlinear interaction of oscillations at the frequency $[\omega' - \omega]$ or $\omega' + \omega$ with the field E leads to the appearance of the dipole moment at the frequency ω' of the transition $0 \to 2$. Note that the dipole moment of interest at the frequency ω' has arisen from nonlinear features of the oscillating system and coherent features of the fields E and E'.

The theoretical analysis of the influence of an external signal on a line shape, and experiments have shown that coherent processes can greatly influence the line shape of radiation of the coupled transition. When the fields are weak, these processes can be considered to be two-quantum, i.e., of the Raman-scattering type. Indeed, the three-level scheme in Fig.5.1 corresponds to the classical scheme of Raman scattering if a frequency of incident scattering ω is far away from a resonance frequency ω_{01}. The scattering of a photon by an electron system is absorption of an initial photon with simultaneous emission of another one. A scattering effect (or two-quantum absorption) can appear only in the second order of the perturbation theory [5.38]. The sum

$$V_{21} = \sum_n{}' \left(\frac{V'_{2n}V_{n1}}{\varepsilon_1 - \varepsilon'_n} + \frac{V_{2n}V'_{n1}}{\varepsilon_1 - \varepsilon''_n} \right)$$

is used as a matrix element for the process under consideration. Here ε_1 is an initial energy of the system "atom + photons"; $\varepsilon_1 = E_1 + \omega\hbar$, $\varepsilon'_n = E_n$; $\varepsilon''_n = \hbar\omega + \hbar\omega' + E_n$ are energies of intermediate states, V,V' are matrix elements of photon absorption and emission. Far from resonance the scattering cross-section is small; it can be observed in gases only at very high pressures. As the frequency approaches the resonance frequency ω_{01}, the scattering cross-section increases as does also the resonance-fluorescence cross-section. Near a resonance, the cross-section increases, so that it may be observed on the excited levels at moderate intensities of incident light in low-pressure gases.

At resonance, the interpretation of the processes of radiation becomes complicated. Indeed, far from a resonance the scattering line is far away from the line of the resonant transition at the frequency ω_{02}. Also, the radiation with the transition frequency ω_{02} arises in the second order of the perturbation theory and is a consequence of excitation (appearance of population) of the real level 0. The matrix element of the transition from the level 0 to the level 2 at the transition frequency ω_{02} as in the case of scattering will also contain the product of the matrix elements of photon emission and absorption. The real-level excitation leads to a finite probability of photon emission at the resonance frequency. The second process can be considered to be step-by-step, one-quantum transitions. The contribution of the second process to the scattering at a frequency that is far away from the resonance may be neglected. In this case, the interpretation of the processes is trivial. Thus, at resonance the two processes contribute to the radiation line, which cannot be interpreted unambiguously as a Raman-scattering line. The interpretation of the processes determines the resulting line shape of scattering in a resonance.

When analyzing the line shape of scattering near a resonance [5.37] has showed that the contribution of these two processes to the scattering line depends on the relation between the level relaxation constants. When the rate of the decay of the common level γ_0 is much more than that of the initial level γ_1 the scattering line in a resonance is connected with the two-quantum transitions and, consequently, can be treated as a "pure" line of the resonant Raman scattering. For $\gamma_0 \ll \gamma_1$ the main contribution to the scattering line is provided by the step-by-step, one-quantum transitions. With equal relaxation constants, the line shape observed is the result of the composition of the probability amplitudes of the two processes.[4] Thus, the differences of the level relaxation constants lead to the interesting peculiarities of the scattering line in a gas, which are characteristic of the optical band.

5.2 Fundamental Equations of TLS

We shall be interested in the line shape of absorption of a probe wave at a frequency ω' that is close to ω_{02}. When the probability of transition from level 1 to level 2 or the medium polarization at frequency ω' is known the power absorbed can be found. The second approach is more general and permits consideration of various relaxation processes. In this case, it is convenient to use the equations for a density matrix of particles. The transition probability of a particle can also be found with the aid of Schrödinger equations for probability amplitudes. Such an approach enables us to analyze elementary microscopic processes of interaction. We will discuss both approaches because the same elementary processes of radiation can manifest themselves differently in macroscopic properties of a medium; the transition probability depends on the direction of wave propagation at the frequencies ω and ω', on the relation between wave numbers k and k', and on the level scheme. It is obvious that the physical processes that occur in various configurations of three-level schemes are similar. Therefore, in order to be more concrete, we shall consider and analyze the three-level system that corresponds to the Raman-scattering scheme that is affected by an electric field of two frequencies

[4] Note that, in some cases, the second-order effect of the perturbation theory results from the amplitude composition (interference) of the processes of the second and third orders of the perturbation theory. It is essential at the initial excitation of a common level.

$$\underline{E}(\underline{r},t) = \underline{E}\cos(\omega t - \underline{k}\underline{r}) + \underline{E}'\cos(\omega' t - \underline{k}'\underline{r}) . \tag{5.5}$$

If, at the starting moment, a particle is on level 1 it can be also in states 0 and 2 under the action of fields. The wave function of an atom, describing its state, is

$$\psi = a_1(t)\psi_1 e^{-\frac{iE_1}{\hbar}t} + a_o(t)\psi_o e^{-\frac{iE_o}{\hbar}t} + a_2(t)\psi_2 e^{-\frac{iE_2}{\hbar}t} , \tag{5.6}$$

where ψ_i, E_i are the wave function and the energy of the stationary state, and a_i is the probability amplitude of the ith state.

The magnitudes $|a_i|^2$ define the probability of finding a particle on the ith level. Then $\gamma_i|a_i|^2dt$ is the probability of decay of the ith level in the time dt. Because, in the absence of fields E' and E, $|a_2|^2 = 0$, the total probability of decay of level 2 for an extremely long time is obviously equal to the transition probability of a particle from level 1 to the level 2 under the action of fields E and E'. So the transition probability of a particle from level 1 to level 2 is

$$W_{1\to2} = \gamma_2 \int_0^\infty |a_2|^2 dt . \tag{5.7}$$

The energy emitted by an atom under the action of the field at frequency ω' is

$$\Delta E = \hbar\omega' \cdot W_{1\to2} . \tag{5.8}$$

Sometimes the energy absorbed (emitted) by an atom at field frequency ω' is determined by the dipole moment of an optical electron in the atom.

The power absorbed (emitted) by an atom at frequency ω' is

$$\langle P\rangle = \langle \underline{j}\rangle \underline{E}'_{(t)} = e\,\frac{\partial\langle\underline{r}\rangle}{\partial t}\,\underline{E}'(t) , \tag{5.9}$$

where $\underline{j}$ is a current operator, e is the electron charge, $\underline{r}$ are the electron coordinates, and $\underline{E}'(t) = \underline{E}'\cos\omega t$.

For the mean value of $\langle\underline{r}\rangle$, we have

$$\langle\underline{r}\rangle = \int\psi^*\underline{r}\psi dq .$$

We are interested in the mean value of $<\underline{r}>$ at the frequency of the transition $0 \to 2$, which is equal to

$$<\underline{r}> = \underline{r}_{02}\, e^{i\omega_{02}t} a_0^*(t) a_2(t) + \text{c.c.}\ , \tag{5.10}$$

where $\underline{r}_{02} = \int \psi_0^*(r)\underline{r}\psi_2(r)dq$.

The energy absorbed (emitted) by an atom at frequency ω' is

$$E = \int_0^\infty <P>dt\ . \tag{5.11}$$

Hence, we have at last

$$E = \underline{P}_{02}\underline{E}' \cdot \text{Re}\left\{ i\omega_{02} \int_0^\infty a_0^*(t) a_2(t)\, e^{-i\Omega' t} dt \right\}\ . \tag{5.12}$$

Here $P_{02} = er_{02}$ is the matrix element of the transition dipole moment.

5.2.1 Equations for Probability Amplitude

When the frequencies ω and ω' are close to the resonant frequency ω_{01} and ω_{02}, the equations for the probability amplitudes are[5]

$$\begin{aligned} &1) \quad \dot{a}_0 + \frac{\gamma_0}{2} a_0 = iV\, e^{-i\Omega t} a_1 + iV'\, e^{-i\Omega' t} a_2 \\ &2) \quad \dot{a}_1 + \frac{\gamma_1}{2} a_1 = iV\, e^{i\Omega t} a_0 \\ &3) \quad \dot{a}_2 + \frac{\gamma_2}{2} a_2 = iV'\, e^{i\Omega' t} a_0\ . \end{aligned} \tag{5.13}$$

In (5.13) γ_i is the rate of decay of the corresponding levels, $\Omega = \omega - \omega_{01}$, $\Omega' = \omega' - \omega_{02}$, $V = P_{01}E/2\hbar$, $V' = P_{02}E'/2\hbar$.

We note some features of the system of (5.13). Write down two evident equalities

$$\dot{a}_2 a_2^* + \frac{\gamma_2}{2} |a_2|^2 = iV'\, e^{i\Omega' t} a_0 a_2^* \tag{5.14a}$$

[5] Here we usually assume that $\omega_{12} \gg \Omega, \Omega'$. This means that the fields E and E' do not interact with the transitions $0 \to 2$ and $0 \to 1$, respectively.

$$a_2\dot{a}_2^* + \frac{\gamma_2}{2}|a_2|^2 = -iV' e^{-i\Omega' t} a_0^* a_2 . \tag{5.14b}$$

Summing them, we obtain

$$\frac{d}{dt}|a_2|^2 + \gamma_2|a_2|^2 = -2\mathrm{Re}\{iV' e^{-i\Omega' t} a_0^* a_2\} . \tag{5.15}$$

Integrating (5.15) from 0 to ∞ over t, we have

$$|a_2|^2 \Big|_0^\infty + \gamma_2 \int_0^\infty |a_2|^2 dt = -2\mathrm{Re}\left\{iV' \int_0^\infty e^{-i\Omega' t} a_0^* a_2 dt\right\}$$

because, at $t=0$ and $t=\infty$, $a_2 = 0$. Then

$$W_{1\to 2} = \gamma_2 \int_0^\infty |a_2|^2 dt = -2\mathrm{Re}\left\{iV' \int_0^\infty e^{-i\Omega' t} a_0^* a_2 dt\right\} . \tag{5.16}$$

Comparing (5.8), (5.16), and (5.12), we are convinced of their identity. Writing the equalities similar to (5.14) for the first two equations and summing them, we obtain

$$\gamma_1 \int_0^\infty |a_1|^2 dt + \gamma_2 \int_0^\infty |a_2|^2 dt + \gamma_0 \int_0^\infty |a_0|^2 dt = 1 . \tag{5.17}$$

We give the solution of the system (5.13) in the perturbation theory for weak fields.

In a zero-order approximation (in the absence of fields)

$$a_0 = 0 ; \quad a_1 = e^{-\frac{\gamma_1 t}{2}} ; \quad a_2 = 0 . \tag{5.18}$$

Inserting $a_1(t) = \exp(-\gamma_1 t/2)$ into (5.13), we have

$$\dot{a}_0 + \frac{\gamma_0}{2} a_0 = iV e^{-i\Omega t - \frac{\gamma_1}{2} t} . \tag{5.19}$$

With the initial conditions $a_0 = 0$ at $t=0$, we obtain

$$a_0(t) = iV \frac{e^{-i\Omega t - \frac{\gamma_1 t}{2}} - e^{-\frac{\gamma_0 t}{2}}}{(\gamma_0 - \gamma_1)/2 - i\Omega} . \tag{5.20}$$

The amplitude $a_0(t)$ contains two exponential terms. The first comprises information on the frequency of the incident radiation and on the lifetime of the initial level 1. The second represents damping, according to the lifetime of level 0.

The behavior of the probability amplitude $a_0(t)$ as a function of time differs considerably from the case of incoherent excitation of a level, for instance, by electron impact. The dipole moment of an atom involves two frequencies: the transition and the field frequencies. Each component damps according to the reciprocal lifetimes γ_1 and γ_0, respectively.

At large frequency detuning Ω ($\Omega \gg \gamma_0, \gamma_1$) the power absorbed by an atom is related only to the induced dipole moment. But in resonance both dipole moments contribute to absorption. (Due to the damping of the dipole moment at the frequency ω_{02} there is always a Fourier component at the field frequency). At precise resonance and equal relaxation constants $\gamma_0 = \gamma_1$ the dipole moments coincide and, accordingly, their contributions are the same.

The amplitude a_2 is found by integrating the third of (5.13) after substitution of $a_0(t)$,

$$\dot{a}_2 + \frac{\gamma_2}{2} a_2 = -V'V \frac{e^{i(\Omega'-\Omega)t-\frac{\gamma_1}{2}t} - e^{i\Omega't-\frac{\gamma_0}{2}t}}{\frac{\gamma_0 - \gamma_1}{2} - i\Omega} . \tag{5.21}$$

Hence,

$$a_2 = -V'V \frac{1}{\frac{\gamma_0 - \gamma_1}{2} - i\Omega} \left\{ \frac{e^{i(\Omega'-\Omega)t-\frac{\gamma_1}{2}t} - e^{-\frac{\gamma_2}{2}t}}{\frac{\gamma_2 - \gamma_1}{2} + i(\Omega' - \Omega)} - \frac{e^{i\Omega't-\frac{\gamma_0}{2}t} - e^{-\frac{\gamma_2}{2}t}}{\frac{\gamma_2 - \gamma_0}{2} + i\Omega'} \right\} . \tag{5.22}$$

The amplitude $a_2(t)$ contains some oscillating and damping terms with the appropriate constants.[6] The oscillating term with frequency $\Omega' - \Omega$ is responsible for the appearance of the dipole moment of the transition $1 \rightarrow 2$ with the frequency $\omega' - \omega$ ($d_{12} = P_{12} a_2 a_1^* e^{-i\omega_{21}t}$). The dipole moment of the transition $0 \rightarrow 2$ contains the frequencies ω_{02}, ω', as in the two-level scheme.

---

[6] For the scheme of cascade transitions, the system of equations for the probability amplitudes is obtained from (5.13) by replacing Ω' with $-\Omega'$. Therefore, all of the results obtained for the scheme can be automatically extended to the cascade scheme by corresponding replacements in the final formulas.

Under the influence of the field on the coupled transition an induced dipole moment arises at the frequencies $\omega_{02}+\Omega$ at $\omega_{02}+(\Omega'+\Omega)$ and $\omega_{02}+(\Omega'-\Omega)$ with different damping coefficients. Fourier components of the dipole moment at the frequency ω_{02} determine the absorption properties at this frequency.

5.2.2 Equation for Density Matrix of Three-Level Atoms

The equations for the density matrix permit simplification of calculations to include the phenomenological relaxation constants due to collisions. The interpretation of phenomena on the basis of the equations for the density matrix also becomes simpler. We give the equations averaged over excitation moments. With the notation used previously the equations take the forms,

$$\left(\frac{d}{dt}+\gamma_o\right)\rho_{00} = iV(\underline{r},t)\rho_{10} - iV(\underline{r},t)\rho_{01} + iV'(\underline{r},t)\rho_{20} - iV'(\underline{r},t)\rho_{02}$$

$$\left(\frac{d}{dt}+\gamma_1\right)\rho_{11} = -i[V(\underline{r},t)\rho_{10} - V(\underline{r},t)\rho_{01}] + \gamma_1\rho_{11}^{(o)}$$

$$\left(\frac{d}{dt}+\gamma_2\right)\rho_{22} = -i[V'(\underline{r},t)\rho_{20} - V'(\underline{r},t)\rho_{02}]$$

$$\left(\frac{d}{dt}+i\omega_{02}+\Gamma_{02}\right)\rho_{02} = -iV'(\underline{r},t)(\rho_{00}-\rho_{22}) + iV(\underline{r},t)\rho_{12} \qquad (5.23)$$

$$\left(\frac{d}{dt}+i\omega_{01}+\Gamma_{01}\right)\rho_{01} = -iV(\underline{r},t)(\rho_{00}-\rho_{11}) + iV'(\underline{r},t)\rho_{21}$$

$$\left(\frac{d}{dt}+i\omega_{12}+\Gamma_{12}\right)\rho_{12} = iV(\underline{r},t)\rho_{02} - iV'(\underline{r},t)\rho_{10}$$

$$\rho_{ik} = \rho_{ki}^{*}\ ; \quad V(\underline{r},t) = \frac{\underline{P}_{01}\underline{E}(\underline{r},t)}{\hbar}\ ; \quad V'(\underline{r},t) = \frac{\underline{P}_{02}\underline{E}'(\underline{r},t)}{\hbar}\ ;$$

$d/dt = \partial/\partial t + \underline{v}(\partial/\partial\underline{r})$, $\rho_{11}^{(o)}$ is the solution of the equations in the absence of fields.

Here we assume that pumping is to level 1. An atom interacts with two fields $\underline{E}(\underline{r},t) = \underline{E}\cos(\omega t - \underline{k}\underline{r})$, $\underline{E}'(\underline{r},t) = E'(\cos\omega' t - \underline{k}'\underline{r})$. The frequencies ω and ω' are close to the frequencies ω_{01} and ω_{02}, respectively. The solution of the system of equations (5.21) makes it possible to determine both the emitted or absorbed power and the absorption coefficient of the field at the frequency ω' of the transition $0\rightarrow 2$.

The linear absorption coefficient of the field at frequency ω' is determined by

$$\kappa' = -4\pi k' \operatorname{Im}\{x'\} , \tag{5.24}$$

where x' is the polarizability of the medium at the field frequency, which is connected with the polarization of the medium by the relation

$$P(\underline{r},t) = x'E' \, e^{-i\omega' t + i\underline{k}\underline{r}} . \tag{5.25}$$

The polarization of the medium is determined by an off-diagonal element,

$$P(\underline{r},t) = P_{02}(\rho_{02} + \rho_{20}) N_o . \tag{5.26}$$

In the stationary case, (5.23) for stationary atoms, after the substitutions

$$\rho_{01} = iV \, e^{-i\omega t + i\underline{k}\underline{r}} r_{01} ,$$

$$\rho_{02} = iV' \, e^{-i\omega' t + i\underline{k}'\underline{r}} r_{02} , \tag{5.27}$$

$$\rho_{12} = V'V \, e^{i(\omega-\omega')t - i(\underline{k}-\underline{k}')\underline{r}} r_{12} ,$$

where $V = \underline{P}_{01}\underline{E}/2\hbar$, $V' = \underline{P}_{02}\underline{E}'/2\hbar$, are reduced to

$$\begin{aligned}
\gamma_o \rho_{00} &= V^2(r_{01} + r_{10}) + V'^2(r_{02} + r_{20}) \\
\gamma_1 \rho_{11} &= -V^2(r_{01} + r_{10}) + \gamma_1 \rho_{11}^{(o)} \\
\gamma_2 \rho_{22} &= -V'^2(r_{02} + r_{20}) \\
[\Gamma_{01} - i\Omega] r_{01} &= (\rho_{11} - \rho_{00}) + V'^2 r_{21} \\
[\Gamma_{02} - i\Omega'] r_{02} &= (\rho_{22} - \rho_{00}) + V^2 r_{12} \\
[\Gamma_{12} - i(\Omega' - \Omega)] r_{12} &= -(r_{02} + r_{10}) .
\end{aligned} \tag{5.28}$$

The solution of (5.28) can be obtained in the general case of arbitrary fields. But the expressions are rather unwieldy. Of greatest importance for spectroscopy is the case in which the field E' of the transition $0 \rightarrow 2$, which we shall call a probe field, is weak, and the field E is arbitrary.

Omitting the term V'^2 in the first three equations, we find

$$\rho_{11} - \rho_{00} = \frac{\Omega^2 + \Gamma_{01}^2}{\Omega^2 + \Gamma_{01}^2(1+G)} \cdot \rho_{11}^{(o)}$$

$$\rho_{00} = \frac{G^2}{\gamma_o} \cdot \frac{2\Gamma_{01}}{\Omega^2 + \Gamma_{01}^2(1+G)} \cdot \rho_{11}^{(o)} \tag{5.29}$$

$$r_{01} = \frac{\Gamma_{01} + i\Omega}{\Omega^2 + \Gamma_{01}^2(1+G)} \cdot \rho_{11}^{(o)} \ ,$$

where $G = (2V^2/\Gamma_{01})[(1/\gamma_o) + (1/\gamma_1)]$ is a parameter of saturation of the transition $0 \to 1$ by the field E.

Inserting ρ_{00} and r_{01} in the fourth and fifth equations of (5.28), we obtain two equations for r_{02} and r_{12},

$$r_{02} = - \frac{V^2 \left\{ \frac{2\Gamma_{01}}{\gamma_o} [\Gamma_{12} - i(\Omega' - \Omega)] + (\Gamma_{01} - i\Omega) \right\} \rho_{11}^o}{\{(\Gamma_{02} - i\Omega')[\Gamma_{12} - i(\Omega' - \Omega)] + V^2\}[\Omega^2 + \Gamma_{01}^2(1+G)]} \ . \tag{5.30}$$

At $G \ll 1$ we have

$$r_{02} = -V^2 \left\{ \frac{2\Gamma_{01}}{\gamma_o(\Gamma_{02} - i\Omega')(\Omega^2 + \Gamma_{01}^2)} + \frac{1}{(\Gamma_{01} + i\Omega)[\Gamma_{12} - i(\Omega' - \Omega)](\Gamma_{02} - i\Omega')} \right\} \rho_{11}^o \tag{5.31}$$

By use of (5.31) we can obtain expressions for the transition probability, for the absorption coefficient, and so on.

In accordance with (5.28) the dipole moment arises on the transition $0 \to 2$ at the frequency ω' due to the population difference between levels 0 and 2 and to the induced moment as well. The main features of the line shape are related to the latter. The appearance of the dipole moment at the frequency ω may be graphically represented by

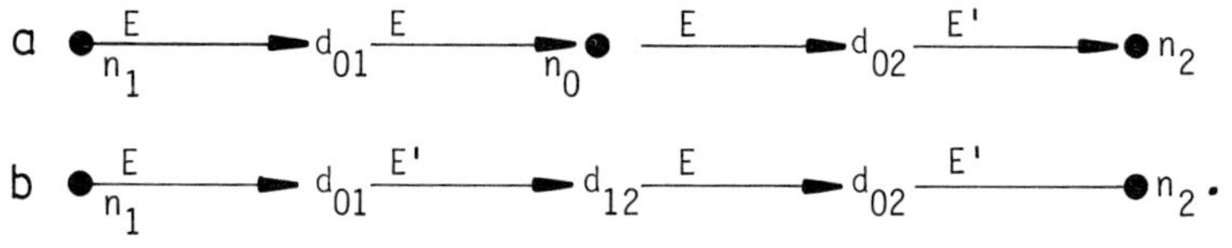

(5.32)

The first process corresponds to an ordinary step-by-step transition. The field E polarizes the transition $0 \to 1$. The interaction of the dipole moment induced by the field E with the field itself results in population of level 0. Then the process repeats itself but on the transition $0 \to 2$ and with the field E'.

The second process (b) is not accompanied by the change of population of an intermediate level. Interaction of the polarization P_{01} with the field E' polarizes the transition $2 \to 1$. The polarization P_{12} and the field E produce the polarization P_{01} of the transition $0 \to 2$ whose interaction with the field E' results in excitation of level 2.

When considering the interactions of the fields with excited levels, we should take into account the processes associated with excitation of the other levels. It leads to new processes in addition to processes (a) and (b). Within the framework of perturbation theory, when all of the levels are populated all of the main processes are

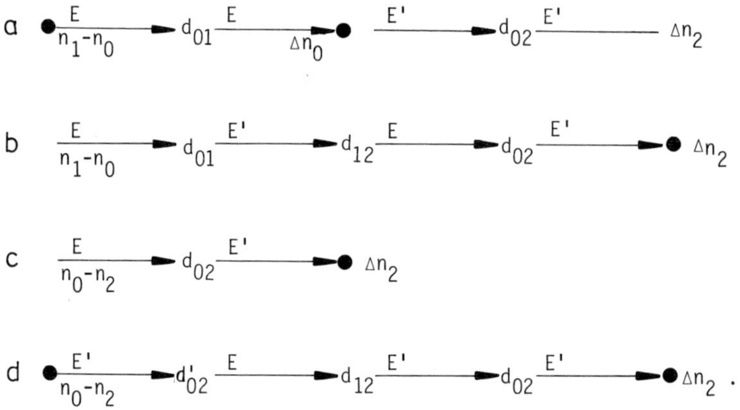

a $\quad \bullet \dfrac{E}{n_1-n_0} \blacktriangleright d_{01} \dfrac{E}{\Delta n_0} \bullet \dfrac{E'}{} \blacktriangleright d_{02} \dfrac{E'}{} \quad \Delta n_2$

b $\quad \dfrac{E}{n_1-n_0} \blacktriangleright d_{01} \dfrac{E'}{} \blacktriangleright d_{12} \dfrac{E}{} \blacktriangleright d_{02} \dfrac{E'}{} \blacktriangleright\bullet \Delta n_2$

$$(5.33)$$

c $\quad \dfrac{E}{n_0-n_2} \blacktriangleright d_{02} \dfrac{E'}{} \blacktriangleright\bullet \Delta n_2$

d $\quad \bullet \dfrac{E'}{n_0-n_2} \blacktriangleright d'_{02} \dfrac{E}{} \blacktriangleright d_{12} \dfrac{E'}{} \blacktriangleright d_{02} \dfrac{E'}{} \blacktriangleright\bullet \Delta n_2$.

The linear absorption coefficient of the transition $0 \to 2$ is related to process (c). Process (a) is connected with the change of population of level 0, owing to the difference of populations of levels 1 and 0. The coherent processes (b) and (d), which lead to the appearance of the dipole moment P_{02}, are not connected with the change of the populations of levels 0 and 2. If the interpretation of the process (b), within the framework of second-order perturbation theory corresponds to the two-quantum transition $1 \to 2$, then, with level 0 populated, coherent effects result from the interference of the amplitude $a_2^{(1)}$ in the first order and the amplitude $a_2^{(3)}$ in the third-order perturbation theory.

5.2.3 Transition Probability in a Three-Level Scheme

Inserting (5.20) and (5.22) for $a_0(t)$ and $a_2(t)$ into (5.16), we can obtain the expression for the transition probability $W_{1\to2}$. The expression found for the matrix element f_{12} (5.31) can also be used to find the probability $W_{1\to2}$ when the corresponding substitutions are made in (5.16). The expression for $W_{1\to2}$ has been obtained by many authors. Even for the case of weak fields it is rather unwieldy. Reference [5.39] provides the most convenient expression for $W_{1\to2}$, which we use here to analyze the role of various processes described in Section 5.1,

$$W_{1\to2} = \frac{2V^2V'^2}{\left(\frac{\gamma_1+\gamma_0}{2}\right)^2+\Omega^2}\,\mathrm{Re}\left\{\frac{1}{\gamma_1}\,\frac{1}{\left[\frac{\gamma_1+\gamma_2}{2}+i(\Omega'-\Omega)\right]}+\right.$$
$$\left.+\frac{1}{\gamma_0}\,\frac{1}{\frac{\gamma_0+\gamma_2}{2}+i\Omega'}+\frac{1}{\left(\frac{\gamma_0+\gamma_2}{2}+i\Omega'\right)\left[\frac{\gamma_1+\gamma_2}{2}+i(\Omega'-\Omega)\right]}\right\}\,. \tag{5.34}$$

Equation (5.34) describes the behavior of the transition probability both in the resonance ($\Omega\sim\gamma_1+\gamma_0$) and outside it ($\Omega\gg\gamma_1+\gamma_0$). Therefore, when using (5.34) we can see the way the probability $W_{1\to2}$ when the conditions that correspond to the classical Raman scattering $\Omega\gg\gamma_1+\gamma_0$ are gradually changed to those of resonance ($\Omega\sim\gamma_1+\gamma_0$).

When (5.34) is changed it is not difficult to make such an analysis. The probability $W_{1\to2}$ contains three terms; each of them describes the characteristic features of the behavior of $W_{1\to2}$.

The first term describes the line that has the maximum at the frequency $\omega'-\omega_0$, i.e., when $\Omega'-\Omega=0$. The line has the width of the forbidden transition $1\to2$. So it may be ascribed to the two-quantum transition $1\to2$.

The second term describes the line that is inherent in the one-quantum transition $0\to2$. Its maximum is at the frequency of the transition $0\to2$ ($\Omega'=0$) and its width is equal to the width of this transition.

The absolute values of both terms depend equally on the frequency detuning Ω.

The third term is associated with the interference of the two processes. It is appreciable at resonance only under certain conditions that we shall discuss. The third term, like the interference term, contains information concerning the two processes.

With large frequency detunings Ω from the resonance $\Omega \gg \gamma_1 + \gamma_0$, two maxima, with widths $\gamma_1 + \gamma_0$ and $\gamma_1 + \gamma_2$, are observed at the frequencies $\Omega' = 0$ and $\Omega' = \Omega$ in the stimulated-radiation line; their relative intensities depend on the relation between the relaxation constants of levels 1 and 0. The ratio of the line intensities of their maxima is similar to $(\gamma_0 + \gamma_1)\gamma_0/(\gamma_1 + \gamma_2)\gamma_1$. That is why, when the Raman scattering was studied, attention was concentrated on one component only; it was related to the Raman-scattering line situated at the frequency $\omega' = \omega_{12} + \omega$. When Raman scattering from the ground state $\gamma_0 \gg \gamma_1/2$ is studied, therefore, the line intensity at frequency ω_{02} is very small, compared with the Raman-scattering line. When Raman scattering near the frequency $\Omega' \simeq \Omega$ is studied, the influence of the second and third terms may be neglected. In this case, the expression describes ordinary Raman scattering,

$$W_{1\to 2} = \frac{2V^2V'^2}{\Omega^2} \operatorname{Re}\left\{\frac{1}{\gamma_1} \cdot \frac{1}{\frac{\gamma_1 + \gamma_2}{2} + i(\Omega' - \Omega)}\right\}, \quad \Omega' \sim \Omega . \tag{5.35}$$

Near $\Omega' \sim 0$, the probability is described by the second term, which we associate with one-quantum, step-by-step transitions,

$$W_{1\to 2} = \frac{2V^2V'^2}{\Omega^2} \operatorname{Re}\left\{\frac{1}{\gamma_0} \cdot \frac{1}{\left(\frac{\gamma_0 + \gamma_2}{2} + i\Omega'\right)}\right\} . \tag{5.36}$$

Thus, far from resonance, the lines of Raman scattering and of one-quantum, step-by-step transition are separated. The contributions of both processes depend on the frequency Ω' and the relaxation-constant relation.

When resonance is approached, the two first terms increase. The third term begins to play a, relatively, greater part and, at last, both lines coincide in the resonance. It is impossible to separate the processes in the general case. However, we have already noted that the intensities of the step-by-step and two-quantum transitions depend on the relation between the relaxation constants. It is the most important feature at the resonance conditions. It is easy to notice that if $\gamma_0 \gg \gamma_1\gamma_2$ the second and third terms can be neglected as well and the probability $W_{1\to2}$ is to be connected only with the two-quantum process in the resonance ($\Omega \sim \gamma_0$).

In this case, the probability is described by

$$W_{1\to2} = \frac{2V^2V'^2}{\Omega^2 + \gamma_0^2/4} \,\mathrm{Re}\left\{\frac{1}{\gamma_1} \cdot \frac{1}{\frac{\gamma_1+\gamma_2}{2} + i(\Omega' - \Omega)}\right\} . \tag{5.37}$$

Thus, the condition $\gamma_0 >> \gamma_1, \gamma_2$ indicates that at resonance the stimulated radiation line can be considered a resonance-Raman-scattering line.

At $\gamma_0 << \gamma_1$ the contribution of the first and third terms to $W_{1\to2}$ may be neglected. The one-quantum, step-by-step transition becomes the main process.

The probability $W_{1\to2}$ is given by

$$W_{1\to2} = \frac{2V^2V'^2}{\Omega^2 + \gamma_1^2/4} \,\mathrm{Re}\left\{\frac{1}{\gamma_0} \cdot \frac{1}{\frac{\gamma_0+\gamma_2}{2} + i\Omega'}\right\} . \tag{5.38}$$

When the relaxation constants are equal, the stimulated-radiation line in the resonance can be ascribed to no process. It is determined by their interference. Figure 5.3 shows the change of the probability $W_{1\to2}$ as a function of the frequency Ω' with the change of the frequency Ω.

Up to now, we have considered the case in which a particle is excited to level 1. Similarly, we can consider the influence of the field at the transition $1\to0$ upon the probability of the transition $0\to2$ when a particle is excited to level 0, and upon that of the transition $2\to0$ when a particle is excited to level 2. Leaving the derivation of the probability W_{02} and W_{20} to the reader, we give the expression for $W_{0\to2}$ and $W_{2\to0}$ when at the initial moment a particle is on level 0 and 2, respectively,

$$W_{0\to2} = \frac{V'^2(\gamma_0+\gamma_2)}{\gamma_0} \cdot \frac{1}{\left[\Omega'^2 + \left(\frac{\gamma_0+\gamma_2}{2}\right)^2\right]} -$$

$$- \left\{\frac{4V'^4\left(\frac{\gamma_0+\gamma_2}{2}\right)^2}{\gamma_0}\left(\frac{1}{\gamma_2} + \frac{1}{\gamma_0}\right) \cdot \frac{1}{\left[\Omega'^2 + \left(\frac{\gamma_0+\gamma_2}{2}\right)^2\right]^2} + \right.$$

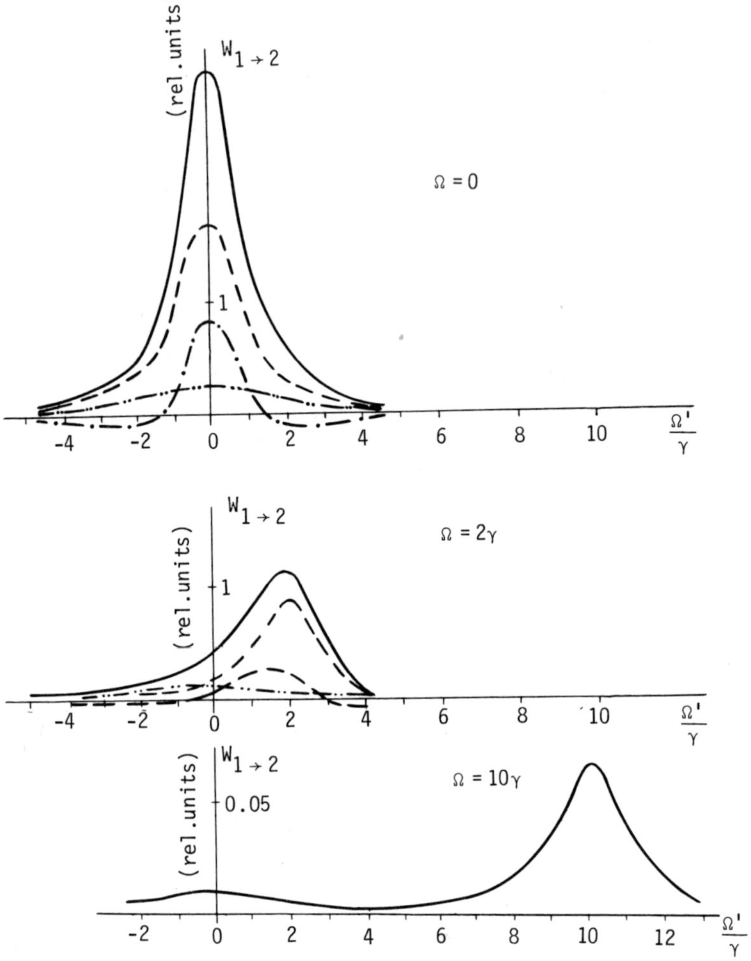

Fig.5.3 Dependence of the transition probability $W_{1\to 2}$ on the detuning frequency Ω' when changing the frequency Ω

$$+\frac{2V^2V'^2}{\gamma_0}\,\mathrm{Re}\Bigg[\frac{\gamma_0+\gamma_2}{\gamma_0\left(\frac{\gamma_0+\gamma_1}{2}+i\Omega\right)\left[\Omega'^2+\left(\frac{\gamma_0+\gamma_2}{2}\right)^2\right]}+$$

$$+\frac{1}{\left(\frac{\gamma_0+\gamma_2}{2}-i\Omega'\right)^2\left[\frac{\gamma_1+\gamma_2}{2}+i(\Omega-\Omega')\right]}+$$

$$+\frac{1}{\left(\frac{\gamma_0+\gamma_2}{2}-i\Omega'\right)\left(\frac{\gamma_0+\gamma_1}{2}+i\Omega\right)\left[\frac{\gamma_1+\gamma_2}{2}+i(\Omega-\Omega')\right]}\Bigg]\Bigg\}\ , \tag{5.39}$$

$$W_{2\to 0} = \frac{\gamma_0}{\gamma_2} W_{0\to 2} \ . \tag{5.40}$$

5.3 Line Shape of the Coupled Transitions in a Gas

Equations (5.34), (5.39), and (5.40) can be used to find the line shape of absorption in a gas. In the system of the center of mass, a moving atom perceives the field frequencies: $\omega \to \omega - \underline{k}\underline{v}$ and $\omega' \to \omega' - \underline{k}'\underline{v}$. Then the corresponding detunings Ω and Ω' should be replaced with $\Omega - \underline{k}\underline{v}$ and $\Omega' - \underline{k}'\underline{v}$, respectively. The case of collinear propagation of waves E and E' is of interest for spectroscopy. Let us choose axis z along the direction of propagation of wave E. Then the corresponding Doppler corrections to the frequency are equal to $-kv$ and $\pm k'v$.

The sign (-) corresponds to the propagation of waves E and E' in the same direction, the sign (+) in the opposite direction. For brevity, we shall distinguish these cases by calling the line shape of absorption (radiation) of the wave E' in the presence of the wave E of the same direction κ_- the line shape of forward scattering; conversely, κ_+ is the line shape of backward scattering.

The linear absorption coefficient of the wave E' is equal to the ratio of the energy absorbed per unit volume to the density of energy flow of the incident radiation,

$$\kappa(\omega') = \hbar\omega' R_1 \langle W_{1\to 2}\rangle \left(\frac{cE'^2}{8\pi}\right)^{-1} + \hbar\omega' R_0 \langle W_{0\to 2}\rangle \left(\frac{cE'^2}{8\pi}\right)^{-1} -$$

$$- \hbar\omega' R_2 \langle W_{2\to 1}\rangle \left(\frac{cE'^2}{8\pi}\right)^{-1} - \hbar\omega' R_2 \left(\frac{cE'^2}{8\pi}\right)^{-1} \langle W_{2\to 0}\rangle \ , \tag{5.41}$$

where R_1, R_0, and R_2 are the excitation rates of levels 1, 0, and 2. The first term in (5.41) is associated with the two-photon transition of a particle from level 1 to level 2. The second term arises from the initial excitation of level 0. Naturally, it includes a linear part of the absorption and the influence of the field E at the transition $0 \to 1$ on the probability of the transition $0 \to 2$. The use of the equations for the density matrix permits us to obtain the expression for absorption of the wave E' when we know the difference of level populations. It allows us to simplify the calculations. However, for particular cases that are of great interest for spectroscopy it is advisable to emphasize the contribution of various levels to absorption. We shall therefore analyze in detail the case that corresponds to Raman scattering or two-photon absorption far from resonance. Under the conditions of resonance, we shall give general expressions that account for populations of all levels.

The transition probability $W_{1\to2}$ is of the form (for Λ-scheme of levels)

$$W_{1\to2}^{(\mp)}(v) = \frac{2V^2V'^2}{(\Omega - kv)^2 + \left(\frac{\gamma_0+\gamma_1}{2}\right)^2}\,\mathrm{Re}\left\{\frac{1}{\gamma_1\left[\left(\frac{\gamma_1+\gamma_2}{2}\right) + i(\Omega' - \Omega + kv \mp k'v)\right]} + \right.$$

$$+ \frac{1}{\gamma_0\left[\frac{\gamma_0+\gamma_2}{2} + i(\Omega' \mp k'v)\right]} +$$

$$\left. + \frac{1}{\left[\frac{\gamma_1+\gamma_2}{2} + i(\Omega' - \Omega + kv \mp k'v)\right]\left[\frac{\gamma_0+\gamma_2}{2} + i(\Omega' \mp k'v)\right]}\right\} . \qquad (5.42)$$

Depending on the magnitude of the detuning Ω as compared with a Doppler-transition width we shall discuss two cases: 1) when the detuning frequency $\Omega \gg ku$, i.e., it is outside the Doppler-broadened transition $1\to0$; 2) the detunings Ω and Ω' are within the Doppler-broadened transitions $1\to0$ and $0\to2$, respectively.

5.3.1 Line Shape of Radiation Far from Resonance ($\Omega \gg \gamma_0, \gamma$)

This case corresponds to the classical stimulated Raman scattering. Near the detuning frequencies Ω', the second and third terms in (5.42) may be neglected. Independent of velocity, all atoms emit the line of the same shape but with Doppler shifts depending on velocity. The line shape is given by

$$W_{1\to2}^{(\mp)}(v) = \frac{2V^2V'^2}{\Omega^2}\,\mathrm{Re}\left\{\frac{1}{\gamma_1}\,\frac{1}{\sqrt{\pi}\cdot u}\int_{-\infty}^{\infty}\frac{e^{(-v/u)^2}\,dv}{\frac{\gamma_1+\gamma_2}{2} + i(\Omega' - \Omega \mp k'v + kv)}\right\} . \qquad (5.43)$$

In the integrand, the characteristic Doppler shift of the line is determined by the sum or difference of wave numbers k and k'. It is the mathematical description of the elimination of a Doppler shift by coherent processes that we discussed in Chapter 4. The elimination of the Doppler shift occurs with unidirectional wave propagation. With oppositely travelling waves, the picture is reversed: Doppler frequency shifts are added that lead to an increase of a Doppler line width of scattering, as compared with a Doppler width of the one-quantum transition.

Integrating (5.43), we obtain expressions for the transition probability $<W^{(-)}_{1\to 2}(v)>$ for unidirectional waves,

$$<W^{(-)}_{1\to 2}(v)> = \frac{2V^2 V'^2}{\Omega^2} \mathrm{Re}\left\{\frac{1}{\gamma_1}\cdot\frac{1}{\frac{\gamma_1+\gamma_2}{2}+i(\Omega'-\Omega)}\right\}$$

$$\gamma_1+\gamma_2 >> |k'-k|u \ , \quad \Omega' \sim \Omega \tag{5.44}$$

$$<W^{(-)}_{1\to 2}(v)> = \frac{2V^2 V'^2}{\Omega^2}\frac{\sqrt{\pi}}{|k'-k|u}\, e^{-\left[\frac{\Omega'-\Omega}{(k'-k)u}\right]^2}\cdot\frac{1}{\gamma_1}$$

$$(\gamma_1+\gamma_2) << |k'-k|u \ , \quad \Omega' \sim \Omega \ . \tag{5.45}$$

When $(\gamma_1+\gamma_2) >> |k'-k|u$ the integration over velocities means summation of a number of particles. It is connected with the fact that all atoms scatter the line of the same shape at the same frequency. When $|k'-k|u >> \gamma_1+\gamma_2$, a SRS line is the totality of Doppler-shifted lines; each of them has a width $\gamma_1+\gamma_2$. Thus, the resulting line is an inhomogeneously broadened Doppler line with an effective width $(|k'-k|u)$.. In the intermediate case $(\gamma_1+\gamma_2) \approx |k'-k|u$, (5.43) cannot be integrated in the analytical form. The scattering line is the resultant of dispersion and gaussian contours.

For oppositely travelling waves, we have

$$<W^{(+)}_{1\to 2}(v)> = \frac{2V^2 V'^2}{\Omega^2}\cdot\frac{\sqrt{\pi}}{(k'+k)u}\frac{1}{\gamma_1}\, e^{-\left[\frac{\Omega'-\Omega}{(k'+k)u}\right]^2} \ . \tag{5.46}$$

Near the frequency detuning $\Omega'=0$ the expression for $<W^{(-)}_{1\to 2}(v)>$ is obtained by integrating the second term in (5.42) over velocities. As should be expected, the line shape of radiation on the transition $2\to 0$ does not depend on the direction of wave propagation, because it is connected with the processes of level population only

$$<W^{(\pm)}_{1\to 2}(v)> = \frac{2V^2 V'^2}{\Omega^2}\frac{\sqrt{\pi}}{\gamma_0}\frac{e^{-(\Omega'/k'u)^2}}{k'u} \ , \quad (\Omega'\approx 0) \ . \tag{5.47}$$

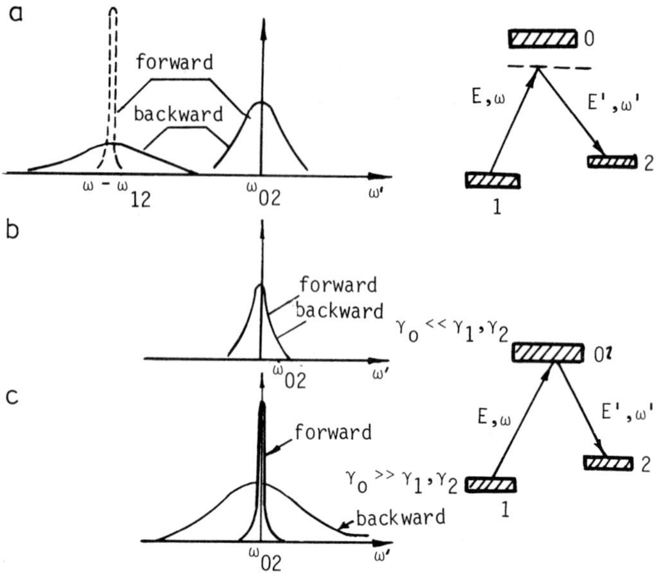

Fig.5.4a-c Radiative processes in three-level system near resonance: a) Doppler-broadened line shape of Raman scattering and of single-quantum stepwise transition with the field strongly off resonance ($\Omega \gg kv$); b) Doppler broadened line of single-quantum stepwise transition at exact resonance ($\Omega \approx 0$) c) Doppler-broadened shape of resonant Raman scattering ($\Omega \approx 0$)

The line shape coincides with the Doppler-broadened line on the transition $2 \to 0$.

Figure 5.4 shows the line shape of radiation of the transition $2 \to 0$ for unidirectional and oppositely travelling waves.

Similarly, we can obtain the expression for the transition probability $W_{1\to 2}$ in a gas for the cascade level scheme. As we have already noted, the transition probability for the cascade level scheme can be obtained from (5.34) by replacing Ω' with $-\Omega'$. Unlike the scheme of Raman scattering, the Doppler shift is compensated for the oppositely travelling waves. In this case, the two-photon resonance, which is described in detail in Chapter 4, occurs. By replacing Ω' with $-\Omega'$ we obtain the probability of the two-photon transition from (5.42)

$$\langle W_{1\to 2}(v)\rangle = \frac{2V^2V'^2}{\Omega^2}\,\mathrm{Re}\left\{\frac{1}{\gamma_1}\,\frac{1}{\sqrt{\pi}\cdot u}\int_{-\infty}^{\infty}\frac{e^{-(v/u)^2}dv}{\frac{\gamma_1+\gamma_2}{2} - i(\Omega'+\Omega+k'v-kv)}\right\} \ . \qquad (5.48)$$

For oppositely travelling waves of the same frequency near a resonance $2\omega \simeq \omega_{12}$, $k' \simeq k$. We note that all of the atoms have the same line shape of two-photon absorption, independently of their velocities, and the integration over velocities means summation over a number of particles. The transition probability $W_{1\to2}$ gives the resonance which is similar to that described in Chapter 4[7]

$$W_{1\to2} = \frac{8V^2V'^2}{\Omega^2} \cdot \frac{1}{\gamma_1} \mathrm{Re}\left\{ \left[\frac{\gamma_1+\gamma_2}{2} - i(2\omega - \omega_{12})\right]^{-1} \right\} . \tag{5.49}$$

5.3.2 Line Shape of Absorption in Weak Fields under Resonance

Both fields are in resonance with the Doppler-broadened transition. It is the case that was studied in detail. The condition of resonance with the Doppler-broadened transition means that there are always atoms that are in exact resonance with the fields E and E'. Detailed theoretical and experimental investigations of phenomena in three-level schemes under resonance conditions have been reported by a number of authors [5.33,35,36]. The line shape of radiation in a gas is dependent on the relations between transition frequencies, the direction of wave propagation, level population and so forth.

The main theoretical results obtained in many papers enable us to consider any situation that can be encountered in spectroscopy of three-level gas systems.

We give the main results for some cases:

Weak Field $G \ll 1$

a) $k' > k$

$$\kappa_+ = \kappa_0\, e^{-\left(\frac{\Omega'}{k'u}\right)^2} \left\{ 1 + V^2 \frac{N_1^0 - N_0^0}{N_0^0 - N_2^0} \cdot \frac{2\Gamma_+}{\gamma_0} \cdot \frac{k'/k}{\left(\Omega' + \frac{k'}{k}\Omega\right)^2 + \Gamma_+^2} \right\}$$

$$\kappa_- = \kappa_0\, e^{-\left(\frac{\Omega'}{k'u}\right)^2} \left\{ 1 + V^2 \frac{N_1^0 - N_0^0}{N_0^0 - N_2^0} \cdot \frac{2\Gamma_-}{\gamma_0} \cdot \frac{k'/k}{\left(\Omega' - \frac{k'}{k}\Omega\right)^2 + \Gamma_-^2} \right\} . \tag{5.50}$$

κ_0 is the linear absorption coefficient of the transition $0\to2$; N_1^0, N_0^0, and N_2^0 are the populations of levels 1, 0, and 2, respectively, in the absence of fields.

[7] The appearance of the factor 8 instead of 2 in (5.49) is connected with the fact that fields E' and E can interact also with the transitions $0\to1$ and $0\to2$, respectively.

$$2\Gamma_+ = \gamma_0 + \gamma_2 + \frac{k'}{k}(\gamma_1 + \gamma_0)$$

$$2\Gamma_- = \gamma_1 + \gamma_2 + (\frac{k'}{k} - 1)(\gamma_1 + \gamma_0) \ . \qquad (5.51)$$

Γ_- and Γ_+ are the line halfwidths of forward and backward scattering.

b) $k' < k$

$$\kappa_+ = \kappa_0 \, e^{-(\frac{\Omega'}{k'u})^2} \left\{ 1 + \frac{N_1^0 - N_0^0}{N_0^0 - N_2^0} \; \frac{2\Gamma_+}{\gamma_0} \cdot \frac{V^2 k'/k}{(\Omega' + \frac{k'}{k}\Omega)^2 + \Gamma_+^2} \right\}$$

$$\kappa_- = \kappa_0 \, e^{-(\frac{\Omega'}{k'u})^2} \left\{ 1 + \frac{2k'}{k}(1 - \frac{k'}{k})V^2 \cdot \frac{\Gamma_-^2 - (\Omega' - \frac{k'}{k}\Omega)^2}{[\Gamma_-^2 + (\Omega' - \frac{k'}{k}\Omega)^2]^2} + \right.$$

$$\left. + \frac{N_1^0 - N_0^0}{N_0^0 - N_2^0} V^2 \cdot \frac{2\Gamma_- \frac{k'}{k}}{\gamma_0[(\Omega' - \frac{k'}{k}\Omega)^2 + \Gamma_-^2]} \right\} \qquad (5.52)$$

$$2\Gamma_+ = \gamma_0 + \gamma_2 + \frac{k'}{k}(\gamma_1 + \gamma_0)$$

$$2\Gamma_- = \frac{k'}{k}(\gamma_1 + \gamma_2) + (1 - \frac{k'}{k})(\gamma_0 + \gamma_2) \ .$$

The main feature of the line of stimulated radiation in a gas is the difference between κ_+ and κ_-. The greater the width of the common level γ_0 as compared with γ_1 and γ_2, the greater is the difference. When $\gamma_0 \gg \gamma_1, \gamma_2$, the line width of forward radiation is determined by the width of the two-quantum transition $1 \to 2$, which is Doppler-broadened due to the difference between k' and k. The backward-radiation line has the shape that is characteristic of a step-by-step transition. Its width is due to the width of the transition $0 \to 2$ and to the velocity distribution of particles on the level 0.

It is necessary to pay attention to the sharp difference of the line shape of radiation of an ensemble of atoms from that of an isolated particle. At $k' > k$, the ensemble of atoms emits a pure line of resonant SRS independently of the relation between relaxation constants of individual levels. Recall that individual particles emit a line that cannot be unambiguously ascribed to the two-quantum process. The integration of (5.34) over velocities has the following remarkable feature. The expression corresponding to the first term is equal to the sum of the first member plus the second member = K (the third member), where K is the constant value. Thus, the sum of the contributions of the processes that correspond to the step-by-step transition (second term in (5.34)) and to the interference of the step-by-step and two-quantum transitions (third term in (5.34)) yields the line width of the two-quantum transition. So a gas of atoms makes it possible to observe two-quantum processes in a pure form. The explanation of the appearance of a backward-radiation line that coincides with a step-by-step transition line is similar. For $k' < k$ and the population of a common level, a forward-radiation line is not lorentzian. As we have pointed out, the additional changes of the line shape are due to the interference process when the amplitudes $a_2^{(1)}$ and $a_2^{(3)}$ are added. The interference processes are essential when the oscillating terms in the probability amplitude $e^{-i\Omega' t}$ have close frequencies. When passing to Doppler-shifted frequencies, we obtain the condition for which the interference gives the maximum contribution to a radiation line,

$$k'v = (k - k')v \ . \tag{5.53}$$

Hence $k' = k/2$. In this case, total compensation of the Doppler frequency shift occurs and, as we have already shown, the radiation line of an ensemble coincides with that of an individual particle. Thus, in forward scattering within the ensemble of atoms, the case $k' = k/2$ makes it possible to observe directly quantum-mechanical effects of the interference that take place at transitions in an isolated atom. The change of the line shape of the transition $0 \to 2$ occurs when $N_1 = N_0$. This case has been studied in [5.31,47] and is, without justification, called a dynamic Stark effect.

Broadening of a SRS Line

With greatly different k' and k, the line width of forward and backward scattering by the ensemble of atoms proves to be much more than the width of radiation of an isolated atom. The resonant SRS line is inhomogeneously broadened. The degree of broadening inhomogeneity depends on the direction of observation and the ratio of the relaxation constants. In forward resonant SRS the degree of broadening inhomogeneity is determined by the ratio k'/k (when $k' > k$). In backward scattering a SRS line is always inhomogeneously broadened. The character of the inhomogeneous broadening can be manifest when a saturation effect is considered, e.g., in SRS of a gas laser. We shall dwell on these phenomena, later.

In accordance with the nature of broadening the nonequilibrium velocity distribution that arises in a three-level system has its peculiarities as well. Equation (5.42) can be used to determine the velocity distribution of atoms on level 2,

$$N_2(v) = RW_{1\to 2}(v)\,\frac{1}{\gamma_2}\;. \qquad (5.54)$$

From (5.50), we can see that, depending on the ratio of the level relaxation constants, peaks with widths $\sim\Gamma_{12}/|k' - k|$ and $\Gamma_{12}/k' + k$ arise in the velocity distribution on level 2; these peaks may be much narrower than those that arise in a two-level system $0 \to 2$ under the action of the field E'. At $k'/K \gg 1$ and $\gamma_o \gg \gamma_1, \gamma_2$, the atoms in the narrow velocity interval participate in the two-quantum process of the region, whereas each field E and E' interacts independently with atoms in a larger velocity interval.

5.3.3 Strong Field

a) $k' > k$

The line shape of radiation of the transition $0 \to 2$ in the arbitrary field E was considered in [5.48] for this case

$$\kappa_+ = \kappa_o\, e^{-\left(\frac{\Omega'}{k'u}\right)^2}\left[1 + \frac{N_1^o - N_o^o}{N_o^o - N_2^o}\cdot\frac{Q^2 - 1}{Q}\cdot\frac{k'}{2k}\cdot\frac{\gamma_1\Gamma_+}{\Gamma_+^2 + (\Omega' + \frac{k'}{k}\,\Omega)^2}\right]$$

$$\kappa_- = \kappa_o\, e^{-\left(\frac{\Omega'}{k'u}\right)^2}\left[1 + \frac{N_1^o - N_o^o}{N_o^o - N_2^o}\cdot\frac{Q^2 - 1}{Q}\cdot\frac{k'}{2k}\cdot\frac{\gamma_1\Gamma_-}{\Gamma_-^2 + (\Omega' - \frac{k'}{k}\,\Omega)^2}\right], \qquad (5.55)$$

$$2\Gamma_- = \gamma_2 + \gamma_1 Q + \left(\frac{k'}{k} - 1\right)(\gamma_1 + \gamma_o)Q$$

$$2\Gamma_+ = \gamma_2 + \gamma_o Q + \frac{k'}{k}(\gamma_1 + \gamma_o)Q\;, \quad Q = \sqrt{1 + G}\;. \qquad (5.56)$$

b) $k' < k$

This case was discussed in [5.49]. The line shape differs strongly from that of the case when $k' > k$ for unidirectional waves. The changes of the line shape can be of interest for spectroscopy. For strong saturation, the line shape is described by

$$\kappa'_-(\Omega') = \kappa_o\, e^{-\left(\frac{\Omega'}{k'u}\right)}\cdot F\;,$$

where

$$F \sim \frac{1}{\sqrt{2S\Omega_-}}\left[\frac{1}{(\Omega_- - S)^2 + (\gamma_-/2)^2} - \frac{1}{(\Omega_- + S)^2 + (\gamma_-/2)^2}\right]^{1/2}$$

$$\gamma_- = \gamma_2 + \gamma_0 + \frac{k'}{k}(\gamma_1 - \gamma_0) \qquad (5.57)$$

$$\Omega_- = \Omega' - \frac{k'}{k}\Omega$$

$$S^2 = 4V^2\,\frac{k'(k-k')}{k^2}\;.$$

The line shape is described by two resonances with widths $\sqrt{3}\gamma_-$ located at frequencies S and -S, respectively. Because $S \gg \gamma_-$, the distance between peaks is great compared with their widths. For sufficiently large S, the distance between the maxima of the peaks is given to good accuracy by

$$2S = 4V\sqrt{\frac{k'(k-k')}{k^2}}\;. \qquad (5.58)$$

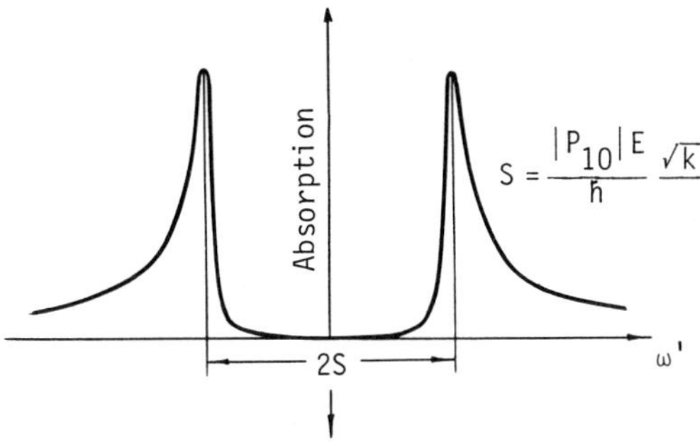

Fig.5.5 Absorption line shape on the transition $0 \to 2$ in the strong field on the transition $1 \to 0$ at $k' = k/2$

The authors of [5.49] attribute a fine structure that occurs in strong fields (Fig.5.5) to a high-frequency Stark effect [5.12], well known in the microwave region, where Doppler broadening is negligible.

As is known, in a strong field, for the transition $0 \to 1$, the probability of finding a particle on level 0 oscillates with a low frequency dependent on the magnitude of the field. This oscillation results in modulation of the dipole moment of the transition $0 \to 2$. As a result, additional resonances arise in the line shape of radiation from the transition $0 \to 2$; they can be interpreted as an effect of splitting of level 0.

In the strong field, for the transition $0 \to 1$, with the initial conditions $a_1 = 1$, $a_o = 0$ the expression for $a_1(t)$ has the form

$$a_o(t) = -\frac{iV}{\alpha_1 - \alpha_2}\left(e^{\alpha_1 t} - e^{\alpha_2 t}\right) , \tag{5.59}$$

where

$$\alpha_{1,2} = -\frac{i\Omega + \Gamma}{2} \pm \sqrt{\left(\frac{i\Omega + \gamma}{2}\right)^2 - V^2} , \quad \Gamma = \frac{\gamma_1 + \gamma_o}{2} , \quad \gamma = \frac{\gamma_1 - \gamma_o}{2} .$$

The wave function of the atom is

$$\psi = a_1\psi_1 e^{-iE_1 t/\hbar} + a_o\psi_o e^{-iE_o t/\hbar} . \tag{5.60}$$

It will have oscillating terms that correspond to new quasi-stationary states in the system "atom + field".

The condition under which we can use the splitting model is $V \gg \Gamma$. When the decay of levels is neglected, the energies that correspond to new states of an atom are

$$\frac{E_o^{(1)}}{\hbar} = \frac{E_o}{\hbar} + \frac{\Omega}{2} + \sqrt{\left(\frac{\Omega}{2}\right)^2 + V^2} , \quad \frac{E_o^{(2)}}{\hbar} = \frac{E_o}{\hbar} + \frac{\Omega}{2} - \sqrt{\left(\frac{\Omega}{2}\right)^2 + V^2} . \tag{5.61}$$

The resonance frequencies of the transition $0 \to 2$ are

$$\omega_{02}^{(1,2)} = \omega_{02} + \frac{\Omega}{2} \pm \sqrt{\left(\frac{\Omega}{2}\right)^2 + V^2} . \tag{5.62}$$

Equation (5.58) can be obtained for qualitative considerations of the model of splitting of levels in a strong electromagnetic field. The resonance frequencies of the transition $2 \to 0$ in a strong field, for the transition $1 \to 0$, taking into account a Doppler shift are (for unidirectional waves)

$$\omega' - k'v = \omega_{02}^{(1,2)} . \tag{5.63}$$

For simplicity, we assume $\Omega = 0$. Then from (5.62) it is easy to find the velocities of the atoms that interact resonantly with the fields,

$$v_p = \frac{\Omega'(k' - \frac{k}{2}) \pm \sqrt{(\frac{k\Omega'}{2})^2 + (k'^2 - k'k)V^2}}{k'^2 - k'k} . \tag{5.64}$$

For oppositely travelling waves, by similar calculations we obtain

$$v_p = \frac{-\Omega'(k' + \frac{k}{2}) \pm \sqrt{(\frac{k\Omega'}{2})^2 + (k'^2 + k'k)V^2}}{k'^2 + k'k} . \tag{5.65}$$

When analyzing (5.64), we note that for the case of unidirectional waves there is the detuning Ω' for which no atoms interact resonantly with the probe wave.

The magnitude of the detuning Ω' at which there are resonantly interacting atoms is defined by the condition that the expression under the radical in (5.64) must be positive,

$$\Omega' \gtrsim 2V\sqrt{\frac{k'k - k'^2}{k^2}} = S . \tag{5.66}$$

The absence of resonant atoms means that absorption of the probe wave is zero over a wide range of frequencies $\Omega' < S$. For oppositely travelling waves there are always resonant atoms, and no similar peculiarities appear. The result obtained agreed with the precise calculation. We may conclude that the dynamic Stark effect on Doppler-broadened transitions consists not in the splitting of lines but in the appearance of the frequency range in which absorption is zero. The dynamic Stark effect manifests itself in two-level systems in a similar way (see Chap.3).

5.4 Methods of Research of Resonance Phenomena in Three-Level Systems

Studies concerned with a radiation line shape can be classified into three groups:

1) Experiments for observing resonances of spontaneous emission of a gas in the presence of an external monochromatic field. The simplest experimental design is one in which spontaneous emission is observed directly from the gas-laser cavity (Fig.5.6a).

2) Experiments for investigating the line shape of absorption or gain of a weak monochromatic probe signal scanning an appropriate frequency range in the presence of a strong field (Fig.5.6b). Unlike the experiments with spontaneous emission, these require two monochromatic fields that have different frequencies.

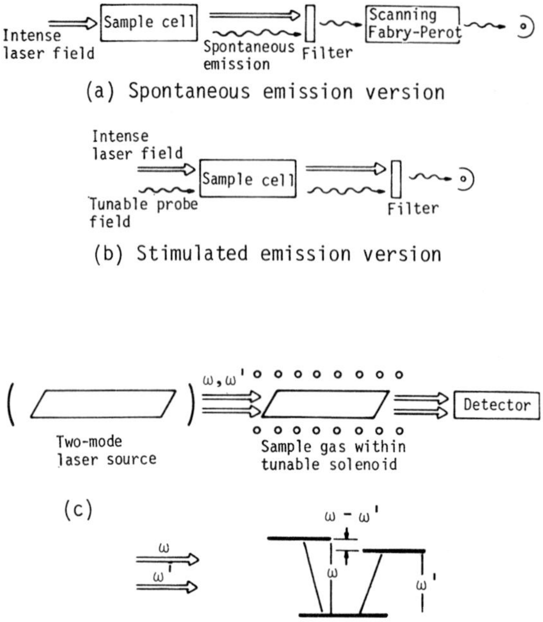

Fig.5.6a-c Experimental design for observing narrow resonances on the coupled transition a) at the spontaneous radiation and b) at the stimulated radiation. 1 - cell, 2 - filter, 3 - scanning Fabry-Perot interferometer, 4 - photodetector. c) Experimental design for observing narrow resonances at the expense of the mode-crossing effect (a) and the scheme of levels and transitions (b). 1 - two-mode laser, 2 - gas cell with solenoid, 3 - photodetector

3) Experiments of the "mode-crossing" type, in which the resonances in the absorption of two-frequency radiation are studied when the magnitude of the electric or magnetic field applied to the external absorption cell is changed (Fig.5.6c). In these experiments, the frequency of the field is not varied but the distance between the levels of the atomic or molecular system is changed.

5.4.1 Resonances in Spontaneous Emission

In the experiments on spontaneous emission it was demonstrated for the first time that there is a fine structure in the Doppler emission line in the presence of a monochromatic field on the coupled transition. BENNETT et al. [5.23] investigated the line shape of the spontaneous radiation from the transition $3s_2 - 2p_4$ $(2 \rightarrow 0)$ of neon ($\lambda = 0.63$ μm) in the presence of a field, for the coupled transition $3s_2 - 3p_4$ $(0 \rightarrow 2)$ ($\lambda = 3.39$ μm) (see Fig.5.7). This experiment was described in Section 5.1.

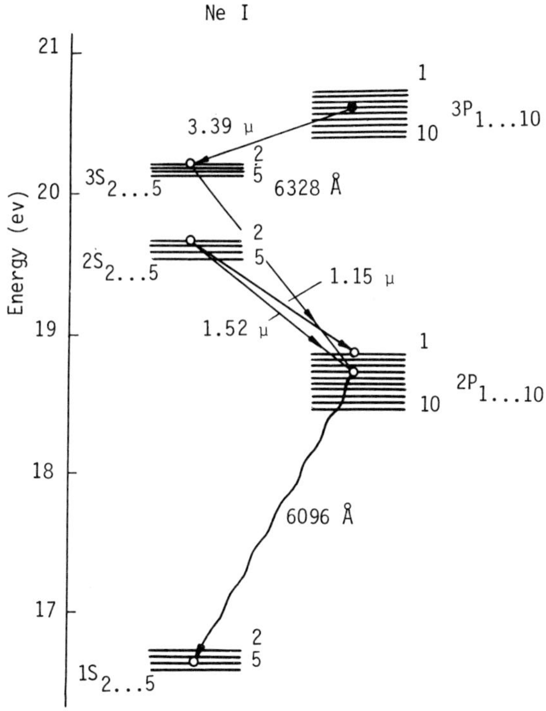

Fig.5.7 Diagram of levels and transitions of the atom Ne I frequently used in experiments

A number of studies have explored spontaneous radiation that is observed along the axis of propagation of laser radiation. The authors of [5.24,25, 28] demonstrated the appearance of a fine structure in the spontaneous emission line, as well. In these papers, a short single-mode He-Ne laser operating at $\lambda = 1.15$ μm (transition $2s_2 - 2p_4$) was used. The spontaneous emission at the 0.6096 μm line ($2p_4 - 1s_4$), which begins at the lower laser level, was observed through one of the mirrors, which was transparent in the 0.6 μm region. The spontaneous-emission spectrum was analyzed with the aid of a pneumatically scanned Fabry-Perot interferometer and recorded with a photomultiplier. A characteristic record of the spontaneous-emission spectrum is shown below in Fig.7.14a. The upper trace corresponds to the usual Doppler-broadened profile of spontaneous emission at 0.6096 Å for the isotope ^{20}Ne. A record of the fine structure that arises in the 1.15 μm line in the presence of field is shown on lower traces. The middle record corresponds to the case with the laser frequency tuned to the center of the gain line, and the lower record to the case of the laser frequency detuned from the line center. The presence of two peaks is due to the fact that the field in the gas-laser cavity is a standing wave. When the field frequency is detuned from the resonance with the atomic transition at saturation, there appear in the velocity distribution two groups of atoms that correspond to two oppositely travelling waves. Consequently, the spontaneous-emission line contains two peaks that are determined by the radiation of these two groups of atoms. When the laser is tuned to the center of the line, as was already indicated, the two components of the standing wave interact with the same atoms and a fine structure, in the shape of a lorentzian peak, is observed.

In these experiments, the interpretation of the fine structure of the spontaneous-emission spectrum that arises in the gas-laser field has invoked only the effect of the generation of a nonequilibrium velocity distribution of the atoms. Coherent effects were observed, qualitatively, by HOLT [5.28]. She explored the fine structure of spontaneous emission in the system of transitions that was considered in [5.24,25]. By improvement of techniques for recording the fine structure, she managed to increase the signal-to-noise ratio and to observe the radiation-line asymmetry in the presence of a standing wave field. Accuracy of experiments with spontaneous emission is greatly limited by the shot noise of the photodetector, laser-frequency drift and the resolution of the Fabry-Perot interferometer. In the first experiments, in order to provide single-frequency operation, the laser worked near threshold, which led to a small signal amplitude. Apart from this, it was necessary to use spontaneous radiation into a very small angle along the laser axis. According to the data of [5.28], a beam spread of 0.025 radian gave a broadening of the peaks of 17 ± 4 MHz. Because the laser was to be detuned from the center of the gain line in order to get the separation of the peaks, it was not stabilized. As a result, the frequency drift of the laser determined the lower limit of the scanning rate of the Fabry-Perot interferometer. In the experiment, the scanning rate was 6 minutes/order. For the data that were used in the analysis, the laser frequency was varied by not more than 5 MHz per order. The instrument line width determined by the Fabry-Perot interferometer and the monochromator was 95 ± 10 MHz. The length of the interferometer base was 6.6 cm, in order to get a sufficiently large spectral range (2275 MHz).

The records of the spontaneous-emission spectrum are shown in Fig.5.8. Within the errors of measurements, each peak is described by a lorentzian profile. This means that the spontaneous-emission line caused by the external field has a lorentzian form, because the instrument function of the interferometer is dispersional. In this case, the width of the fine structure that is of interest can be obtained by a simple calculation of the width of

a

b

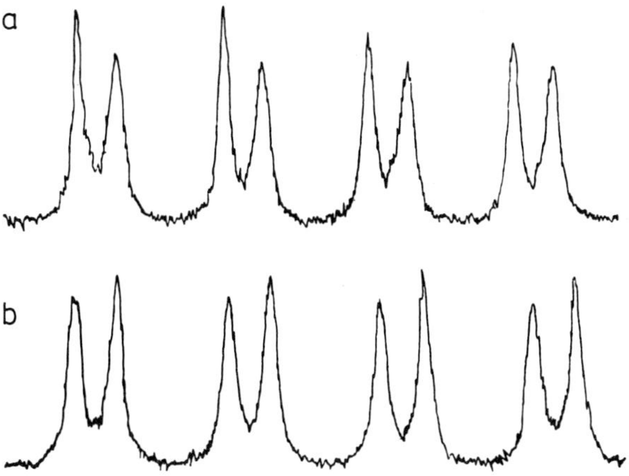

Fig.5.8a,b Spontaneous-radiation spectrum at $\lambda = 0.6096$ μm (transition $2p_4 - 1s_4$) from the He-Ne laser cavity at $\lambda = 1.15$ μm (transition $2s_2 - 2p_4$) [5.28] Ω - the field frequency detuning relative to the Lamb-dip center. a) $\Omega = 160$ MHz; b) $\Omega = -170$ MHz

the peak of the instrument function of the interferometer. The fine structure observed on the spontaneous-radiation profile, which arises in the standing-wave field is asymmetric; the nature of the asymmetry depends on the sign of the detuning frequency with respect to the gain-line center. The asymmetry is due to the dependence of the spontaneous-emission line shape on the direction of observation. The experiment shows that the fine-structure line caused by the wave that propagates away from the observer is sharper than that associated with the wave that propagates towards the observer. In the presence of an external monochromatic field at the coupled transition for a cascade scheme of transitions, the observed behavior of spontaneous radiation is in full agreement with the predictions of theory. However, analysis of the absolute magnitudes of the widths and an attempt to determine the constants of the transitions and levels led to a negative magnitude of the width of one of the levels. This was most probably caused by insufficient accuracy of the experiment and by influence of collisions, which were essential in the experiment described.

Investigations of the interaction of monochromatic radiation with matter were important for progress, because they diverted attention from investigations of the active media of gas lasers to investigations of the line shapes of stimulated and spontaneous radiation in external absorption cells. This was made possible by the attainment of single-frequency laser action by use of mode selection in gas lasers. The most effective and simple method is based on the introduction of nonlinear absorption (see Chap.9). The introduction of external cells permitted considerable expansion of the range of pressures investigated, and made it possible to observe directly the dependence of the spontaneous-emission line shape on the direction of observation.

Reference [5.40] explored spontaneous radiation by the $1s_1 - 2p_4$ transitions of Ne in the presence of a monochromatic field from the transition $3s_2 - 2p_4$. The volume investigated was a cell containing a discharge in pure Ne; much attention was centered on comparison of theory with the experimental results. Though a considerable anisotropy of the spontaneous-emission line was observed in the directions along and opposite to the direction of propagation of the laser radiation, quantitative agreement with the theory was not obtained. The dependence of the line shape of the spontaneous emission on direction was reported in [5.41] when the hyperfine structures of the transitions $2s_2 - 2p_4$ and $2p_4 - 1s_4$ were investigated. This experiment will be considered in more detail in Chapter 7.

The experiments mentioned are practically the only ones that have investigated the line shape of spontaneous emission in the presence of an intense monochromatic field for a coupled transition. Most probably this is due to the difficulty of recording the weak signals of spontaneous radiation, which naturally arise in spectral instruments of superhigh resolution. This difficulty is, to a considerable degree, avoided in experiments with a probe wave, i.e., on induced transitions.

5.4.2 Resonances on Induced Transitions

The first experiments on the line shape of stimulated emission [5.29-31] also discovered line-emission anisotropy. In these experiments, the transitions $3s_2 - 2p_4$ and $2s_2 - 2p_4$, with the common level $2p_4$, were investigated, but under different conditions. In [5.29] the line shape of a three-level gas laser was explored, i.e., of a laser in which the gain is achieved by pumping with a monochromatic field. The properties of three-level lasers will be discussed in detail in Chapter 9. In [5.29], the external field was a travelling wave, and the weak field was a standing wave. In the first approximation, the dependence of the generated power on frequency is determined by the gain line shape of the coupled transition. Moreover, near threshold, this dependence proves to be rather sensitive to small changes of the gain line shape of the transition $0 \to 2$. Therefore, the experiments for observing generation in three-level lasers are suitable for observing, first of all, small but qualitative changes of the radiation line shape of the coupled transition. The attainment of exact quantitative data concerning the parameter of the radiation line is difficult, because it is necessary to allow for generation characteristics and saturation effects in a laser. The experiment demonstrated a considerable asymmetry of the generation line (Fig. 5.9).

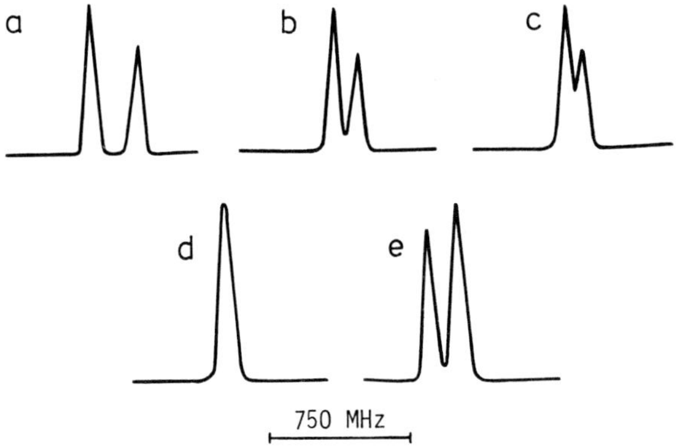

Fig.5.9a-e Line shape of generation of a three-level gas laser at $\lambda = 1.15$ µm ($2s_2 - 2p_4$ of Ne) at the optical pumping of discharge in pure Ne by a monochromatic field at $\lambda = 0.63$ µm (transition $3s_2 - 2p_4$) with fixed detuning of the external field frequencies relative to the center of an absorption line. a) $\Omega = -194$ MHz, b) 100 MHz, c) 67 MHz, d) 0 MHz, e) 119 MHz

When the external-field frequency is detuned from the line center, the travelling-wave field interacts with the atoms whose velocity projections onto the axis of wave propagation are equal to $v = \Omega/k$. The standing-wave field of the coupled transition ($\lambda = 1.15$ µm) will interact with the same atoms if its frequency detuning satisfies the condition $\Omega' = \pm(k'/k)\Omega$. Same signs of detunings Ω' and Ω correspond to unidirectional waves in a resonator at $\lambda = 1.15$ µm and $\lambda = 0.63$ µm; different signs correspond to oppositely travelling waves. Therefore when the laser-resonator frequency at $\lambda = 1.15$ µm is scanned, generation will occur at two frequencies in the neighborhood of $\Omega' = \pm(k'/k)\Omega$. Because, at small excess of gain over the threshold, the generated power is proportional to unsaturated gain, the gain-line anisotropy causes the appropriate asymmetry of the frequency dependence of the power generated by a three-level laser.

The line shape of absorption of a probe signal at $\lambda = 1.15$ µm in the presence of the strong field at $\lambda = 0.63$ µm was investigated in [5.31]. An additional discharge cell was introduced in the cavity of the 0.63 µm laser. By

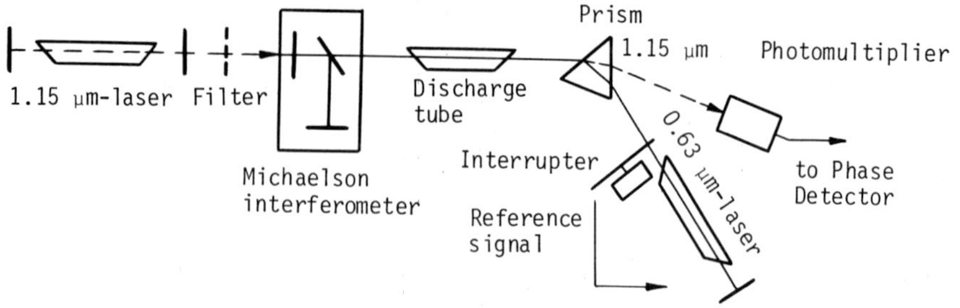

Fig.5.10 Experimental design for observing a splitting effect by the method of probe-signal absorption at $\lambda = 1.15$ μm in the He-Ne laser cavity. 1 - laser at $\lambda = 1.15$ μm, 2 - filter, 3 - Michaelson interferometer, 4 - discharge tube, 5 - prism, 6 - photomultiplier, 7 - laser at $\lambda = 0.6328$ μm, 8 - chopper, 9 - signal to the phase detector

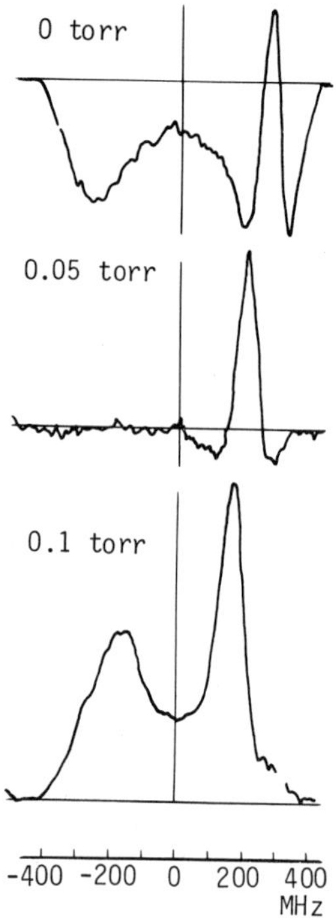

Fig.5.11 Splitting effect of the gas-radiation spectrum on the transition $2s_2 - 2p_4$ in Ne (λ = 1.15 μm) in the presence of the standing-wave field on the transition $3s_2 - 2p_4$ in Ne ($\lambda = 0.63$ μm) in the region of full transmittance of a He-Ne discharge for $\lambda = 0.63$ μm [5.31] (the numbers on the left indicate He pressure)

matching the partial and total pressures of He and Ne and adjusting the discharge current, a condition could be obtained at which the absorption coefficient of the strong field was zero. A weak probing signal at 1.15 μm was passed through this cell. The layout of the experiment is shown in Fig.5.10. Figure 5.11 shows three spectra of the weak probe signal at 1.15 μm caused by the standing-wave field at 0.63 μm when the strong field was detuned from the central transition frequency. The structure on the right and left corresponds to the nonlinear interaction of parallel and oppositely travelling waves, respectively. At 1.15 μm, gain occurs over the full range of pressures of He while the difference of populations in the transition $3s_2 - 2p_4$ changes its sign; the upper record in Fig.5.11 corresponds to absorption; the lower record corresponds to gain. The change from absorption to gain was achieved by a slight change of He pressure. These experiments confirmed that, in the case $k' < k$, the fine-structure shape for unidirectional waves is not lorentzian; a change of the absorption line shape of the probe signal is observed even when the external field passes through the gas cell without being absorbed. The characteristic shape of the structure reported in [5.31] is in full agreement with theoretical predictions (see Sec.5.3).

Theory and experiments were compared quantitatively in [5.37,42] for the case $k' > K$ when the line shape of radiation caused by the external field is lorentzian both for oppositely travelling and unidirectional waves. Investigations of resonance SRS in neon are discussed in Section 5.5.

5.4.3 Mode-Crossing Resonances

The idea of mode-crossing experiments [5.32,33] is shown in a simplified manner in Fig.5.6c. Laser emission containing two monochromatic components at frequencies ω_2 and ω_1 traverses a gas absorption cell on which a weak magnetic field is acting. The intensity of the passing light is recorded as a function of the change of the magnetic-field strength H. When the magnitude of the Zeeman splitting is approximately equal to the frequency difference between the modes, a resonance occurs. This effect has been called the mode-crossing effect. Its theory was given in [5.26] in the analysis of saturation behavior of the Doppler-broadened transition that contains levels with a closely spaced structure.

In a longitudinal magnetic field, the atomic levels with $I \neq 0$ split, because of the Zeeman effect. The corresponding circular components of the linearly polarized laser modes are in resonance with two coupled Doppler-broadened transitions of the gas that fills the cell. When the field intensities are small and it is possible to ignore saturation effects, the intensity of the light that passes through the cell varies slowly with the magnetic field strength because of the shift of the Doppler-broadened absorption line. Large changes of the absorption arise only at magnetic fields at which the Zeeman splitting is comparable with the Doppler line width. But when the field intensity increases, saturation of the transition that is in resonance with the field occurs. As a result, an absorption change arises in the coupled transition. It is assumed that the Zeeman splitting is greater than the level width, but considerably less than the Doppler width. The mode-crossing effect corresponds to the case of the interaction of two unidirectional waves with closely spaced frequencies in a three-level system. The difference is that, in these experiments, the field frequency is not changed, but the levels of the three-level system are shifted.

Taking into account the above considerations, we can easily obtain an expression for the resonance shape,

$$\kappa(H) = \kappa_0 \left[1 - \frac{V^2}{\gamma_2} \frac{2\gamma_1}{(\omega_2 - \omega_1 - \Delta)^2 + \gamma_1^2} \right] , \qquad (5.67)$$

where $\Delta = \mu_0 gH$ is Zeeman splitting of levels, μ_0 is the Bohr magneton, g is the Landé factor, and H is the magnetic-field strength. We assume that $G \ll 1$. The resonance has a width equal to twice the width of the upper level that undergoes Zeeman splitting.

The condition that modes cross depends only on the distance between the laser modes; it is insensitive to their individual frequencies. The resonance width is the sum of the widths of the "crossing" levels (including the effects of broadening by the field $\gamma_1 \to \gamma_1 + \gamma_1 Q$) and does not depend on the width of the lower level.

Experimentally, the phenomenon of mode-crossing in a magnetic field was observed in investigations of the hyperfine structure of xenon [5.26] and the g factors in atomic oxygen [5.33]. Figure 5.12 illustrates a mode-crossing signal observed in Xe at $H = 20$ gauss [5.33]. The experimental design was like that shown in Fig.5.6c. In order to increase the signal-to-noise ratio, a small modulation of the stationary longitudinal magnetic field was introduced. The detected signal is fed to a phase-sensitive amplifier tuned to the modulation frequency, and the output signal is recorded as a function of the magnetic field. As the radiation source, a 3.37 μm Xe laser was used with a mode spacing of 50 MHz. The crossing levels were a pair of Zeeman components of the upper laser level. The one-meter cell contained a discharge in Xe at pressure of about 10^{-2} torr so that it was possible to neglect the collision broadening. The observed g factor was 0.929. The observed width (0.5 MHz) of the upper level of the 3.37 μm transition was about one-thirtieth of the width of the lower level.

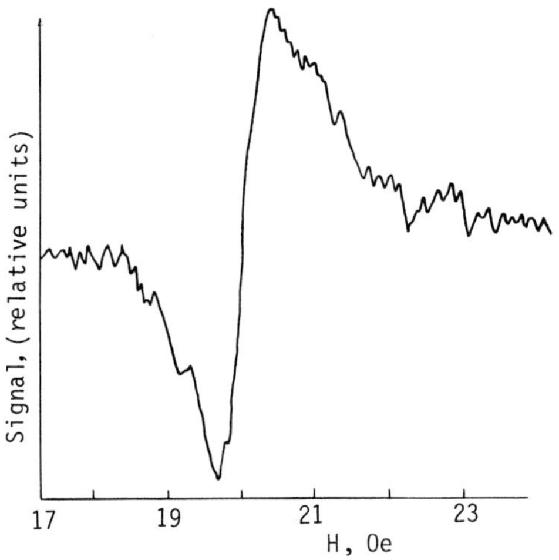

Fig.5.12 Experimentally observed signal of the mode crossing in the Xe discharge at $\lambda = 3.37$ μm [5.33]

The mode-crossing phenomenon is closely connected with the effect of a dip at the line center of a laser located in a magnetic field; this effect was predicted by DIAKONOV and PEREL [5.43] from the theory of a Zeeman laser. It was interpreted as a result of the overlapping of "own" and "foreign" holes in the line center. By "foreign" is meant a hole in an inhomogeneously broadened gain line that is produced, for example, by a right-handed polarized wave for a wave with left-circular polarization. The formation of a dip in the generation-line center is connected with the interaction of two oppositely travelling waves that have orthogonal circular polarizations; it is actually an effect of mode crossing in a zero magnetic field.

Spectroscopic applications of the mode-crossing effect and level crossing in the zero electric and magnetic fields are discussed in the next chapter.

5.5 Investigations of Stimulated Raman Scattering (SRS)

5.5.1 Selection of Transitions

As we have already pointed out, under the conditions for resonance, when the lifetime of the common level is much less than those of the initial and final states, radiation from the transition $0 \to 2$ in the presence of a field can be interpreted as resonance-stimulated Raman scattering. Two-quantum processes determine the most significant features of the radiation line shape, sharp-line anisotropy and narrowing of a resonance SRS line, as compared with a line of the one-quantum transition $0 \to 2$. Thus, under the conditions of resonance it is possible to study two-quantum transitions and to use them in superhigh resolution spectroscopy. Resonance SRS was first studied in neon, from whose coupled transitions many lines were generated. The most suitable and very convenient pair of transitions for investigating resonant SRS are the transitions $2s_2 - 2p_4$ ($\lambda = 1.15$ μm) and $2s_2 - 2p_1$ ($\lambda = 1.5$ μm). The first experiments on RSRS were on these transitions [5.30,37,42,44]. Laser action on the above transitions is observed in ordinary, so-called commercial gas lasers. Lifetimes of operating levels comply with the requirements for observing resonant SRS. A common level $2s_2$ is resonant, i.e., associated with a ground state by a strong optical transition. The lifetime of level $2s_2$ is far less than those of levels $2p_4$ and $2p_1$. The wavelengths of the transitions $2s_2 - 2p_4$ and $2s_2 - 2p_1$ are close enough so that cancellation of Doppler shifts in a two-quantum transition should be effective enough and the contribution of Doppler broadening should be small. As a result of numerous studies a great amount of information about the relaxation times of levels 2p and 2s of neon was obtained. With finite population of level 2s an interference effect of one- and two-quantum processes should not manifest itself in the approximation of weak fields, for the scheme under investigation. All of the above considerations enable us to consider an experiment as the direct observation of a line shape of stimulated resonant Raman scattering.

The lifetimes of levels 2p were measured in [5.45]. These measurements gave the values $\gamma 2p_1 = 6.95 \times 10^7$ s^{-1}, $\gamma 2p_4 = 5.24 \times 10^7$ s^{-1}. The rate of decay of level $2s_2$, according to the data of [5.46], has a value $\gamma 3s_2 = 1.6 \times 10^8$ s^{-1}. The width of the forbidden transition $2p_1 \to 2p_4$ is equal to 19 MHz. The width of a peak of the velocity distribution on level $2s_2$ is about 36.8 MHz. As stated previously, the line width of forward scattering is equal to that of two-quantum scattering, which is composed of widths of the forbidden transition $2p_1 \to 2p_4$ and a Doppler part $(k'/k - 1)$ $(\gamma 2p_1 + \gamma 2p_2)$. Thus, for $2\Gamma_-$ we have 30.8 MHz. The line width of backward scattering is equal to the width of a step-by-step transition. It is caused by the contribution of a nonequilibrium velocity distribution of atoms that is equal to k'/k $(\gamma 2p_1 + \gamma 2s_2)$ = 48.6 MHz and by the width of the transition $2s_2 - 2p_4$. For $2\Gamma_+$ we obtain 82.6 MHz. Note that the line width of two-photon backward scattering is 104 MHz.

5.5.2 Description of Experiment for Observing Resonant SRS

The experiment for observing the line shape of resonant SRS consisted of investigations of the line shape of gain of a signal at $\lambda = 1.15$ μm which arises under the action of a monochromatic signal at $\lambda = 1.15$ μm. The scheme of the experimental installation is shown in Fig.5.13. The radiation of a high-

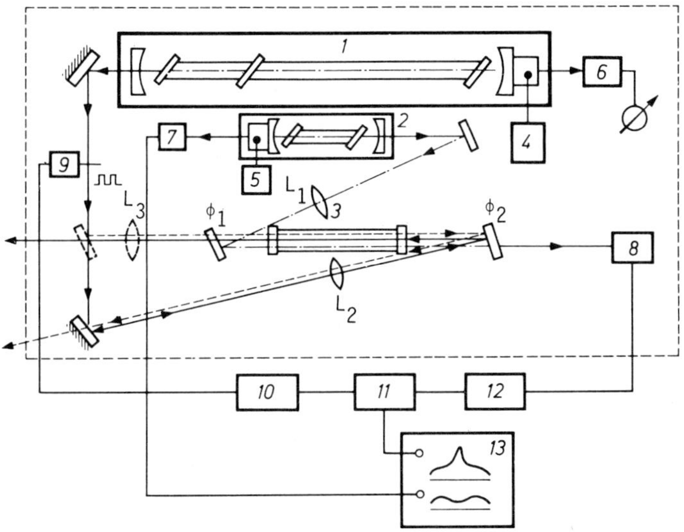

Fig.5.13 Experimental arrangement for observing the anisotropy of the fine structure of the SRS line in neon. 1 - He-Ne laser at 1.52 μm (strong field); 2 - He-Ne laser at 1.15 μm (weak field); 3 - discharge cell; 4,5 - units to control piezoceramics; 6,7,8 - photodiodes; 9 - mechanical modulator; 10 - audiofrequency oscillator; 11 - phase sensitive detector; 12 - selective amplifier; 13 - recorder

intensity single-frequency laser 1 at the 1.52 μm line was directed into a discharge tube 3 (length 50 cm, diameter 2 mm), which was filled with pure isotope Ne. In order to measure the line shape of gain or absorption on the transition $2s_2 - 2p_4$ the radiation at $\lambda = 1.15$ μm of a short scanned single-mode He-Ne laser 2 was introduced into cell 3 either in the same direction with the strong field (path of rays is dotted in this case), or in the opposite direction. In the case of parallel-travelling waves the strong field at 1.52 μm should traverse mirror Φ_1, with the aid of which a weak field is introduced into the cell. The wavefronts were matched by use of lenses, which at the same time focussed the field, in order to achieve great saturation in the cell. Nonmatching of wavefronts acts upon the signal value and the line shape. For the optical system used in the experiment, the influence of these factors can contribute not more than 1 MHz to the width.

The photodiode was used as photodetector 8. It recorded laser radiation at $\lambda = 1.15$ μm after passing through the absorbing cell 3. The strong field at $\lambda = 1.52$ μm was modulated by use of an electromechanical modulator 9 with a frequency of about 40 Hz. The modulation depth was nearly 100%. As a result of nonlinear resonant interaction of the field with the gas, the laser radiation at $\lambda = 1.15\ \mu m$, after passing through the cell, was found to be modulated at the 40 Hz frequency. The alternating signal from the photodetector was amplified by a selective amplifier 12 and supplied to an input of a synchronous detector 11.

The radiation frequency of the weak field was slowly scanned by supplying a sawtooth voltage to the piezoceramic on which one of the mirrors of the laser 2 cavity was mounted. Laser 2 had a discharge tube of 1 mm inner diameter and 7 cm length of the discharge section. The tube was filled with the mixture of neon and helium isotopes in the ratio 1:10 at a total pressure of 3 torr. The high He pressure in the amplifying tube produced an almost flat-top frequency dependence of the generated power, which provided a minimum change of the weak-field intensity in the range of 300-400 MHz near the center of the gain line. The distance between the mirrors was 12.3 cm (the weak-field frequency was about 1220 MHz). The radiation from laser 2 was fed to photodetector 7; the output of the photodetector was fed to the second channel of the automatic recorder 13. Thus, the second channel of the automatic recorder recorded the change of the weak-field amplitude by scanning its frequency. The alternating signal at the output of the synchronous detector was proportional to the magnitude of the weak field. It is therefore clear that the ratio of the first spectrogram of automatic recorder 13 to the second spectrum was proportional to the alternating part of the absorption coefficient for the 1.15 μm line of neon in tube 3; this part was caused by the external strong field at $\lambda = 1.52\ \mu m$.

The recording time for one order was about 40 s. High stability of the strong-field frequency was therefore required. The length of resonator 1 was held constant by placing the experimental setup on a massive steel plate that rested on an isolated concrete foundation. Massive steel mounts held the mirrors rigidly to the plate. The spaces between the windows that sealed the discharge tubes and between the mirrors were carefully isolated from the surrounding air. As a result of these arrangements, the error of determining the line width, due to the strong-field-frequency drift during the recording time of the narrow structure of the line shape amounted to less than 1 MHz. The relative short-term variation of the laser frequency was 10^{-9}.

Finally, we describe how the influence of a pedestal due to trapping of the resonant radiation was eliminated [5.46]. This was done in order to improve the accuracy of measurement and to facilitate processing of experimental data. The presence of the pedestal greatly complicates processing of results concerning narrow resonances (Fig.5.14). It is eliminated by using

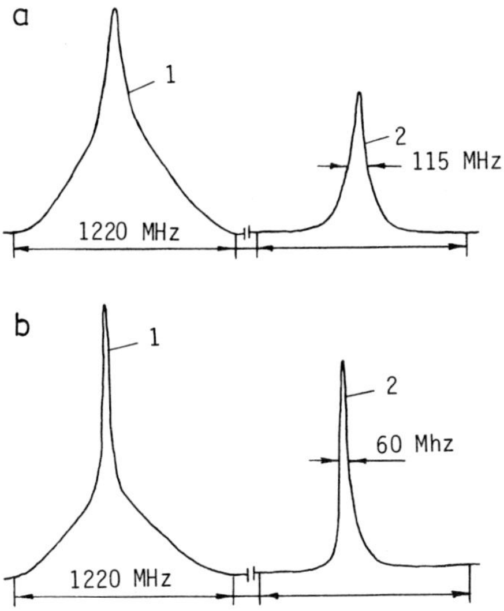

Fig.5.14a,b Line shapes of SRS in neon at $\lambda = 1.15$ μm ($2s_2 - 2p_4$) in the presence of the strong field 1.52 μm. a) forward scattering, b) backward scattering; 1 - without cancellation of the Doppler "underlining", 2 - pure-scattering line

the coincidence of the pedestal shape with the Doppler line shape of absorption. In order to cancel the pedestal, the discharge current in the absorbing cell was modulated out of phase with the strong-field modulation. Modulation of the discharge current led to modulation of the absorption coefficient of the weak signal. The modulation depth of the discharge current was such that the alternating-signal amplitude at the frequency of detuning from a narrow peak was zero. In this way, the influence of the Doppler lining could be practically completely removed.

5.5.3 Line Shape of Resonant SRS

Anisotropy of the Radiation Line Shape

Figure 5.14 shows the record of the line shape of stimulated radiation (absorption) for parallel-propagating waves ($\underline{k}\underline{k}' > 0$) and for oppositely travelling waves ($\underline{k}'\underline{k} < 0$) of the probe and strong field. At the same time, it shows the line shape of the probe field. The gain line has a rather complicated shape; it is a narrow peak on the background of rather considerable pedestal both for the same and for the opposite directions. The widths of the peaks are far less than the Doppler width of the 1.15 μm line and comparable with the radiative width of transition. We draw the reader's attention to the strong anisotropy of the radiated line. For unidirectional waves ($\underline{k}'\underline{k} > 0$), the peak is considerably narrower (about half) than for oppositely travelling waves ($\underline{k}'\underline{k} < 0$). The wide lining with a width that is equal to a Doppler width, associated with the equilibrium velocity distribution, can be explained by diffusion of excitation in a velocity distribution of atoms. The anisotropy agrees qualitatively with theoretical predictions. For a probe wave that propagates in the same direction as the strong field, the radiated line has a width of 58 MHz at a pressure of 0.9 torr, which is half as narrow as the width that results from population-saturation effects. For oppositely travelling waves, the width of the radiation peak was 118 MHz and was approximately equal to the sum of the width of the transition $2s_2 - 2p_1$ and of the width due to the velocity distribution of atoms on the level $2p_4$.

The resonance widths observed were close to those predicted by theory. The dependences of the resonance widths on field intensities and gas pressures were experimentally observed. The analysis of results is simple in the case of weak fields. In this case, there is no need to take into account transverse or longidutinal field inhomogeneities. But the fact that at very weak fields the signal magnitudes decrease must be considered. Therefore, the widths that correspond to very small field were obtained by extrapolating the dependences of the widths to zero value of the field. Because in experiments the saturation parameter G did not exceed about 0.5, extrapolation according to the linear law was possible. The widths $2\Gamma_+ = 86 \pm 3$ MHz and $2\Gamma_- = 31 \pm 2$ MHz extrapolated to zero field and pressure were in good agreement with those calculated from known values of level lifetimes. Therefore, the efficiency of this method for spectroscopic investigations can be judged.

5.5.4 Polarization Characteristics of Resonant SRS

Up to now we have not taken into account polarization characteristics of resonant SRS directly associated with degeneration of levels 1, 0, 2. Investigations reported in [5.44] indicated that the fine structure observed is extremely sensitive to polarization of fields E' and E.

Figure 5.15 shows the line shape of scattering for a circularly polarized wave at $\lambda = 1.52$ μm. The weak field at 1.15 μm has circular polarization, in case a) opposite to the direction of rotation of the plane of polarization of the strong field, and in case b) coincident with that direction. The intensity ratio of the maxima of the scattered line is 1:6, which coincides (within the accuracy of measurements) with the ratio of the Zeeman components calculated from GJ symbols for the transition $2s_2 - 2p_4$ $(J = 2)$.

The degree of polarization of the scattered line, as a function of the angle between the polarization planes of the weak and strong fields was investigated (Fig.5.16). The maximum degree of polarization of the scattered

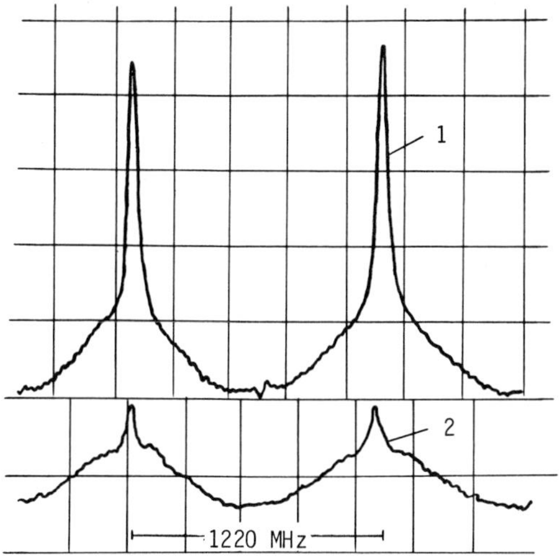

Fig.5.15 Line shape of RSRS for the right- (curve 1) and left-hand (curve 2) circular polarizations of probe and pumping waves

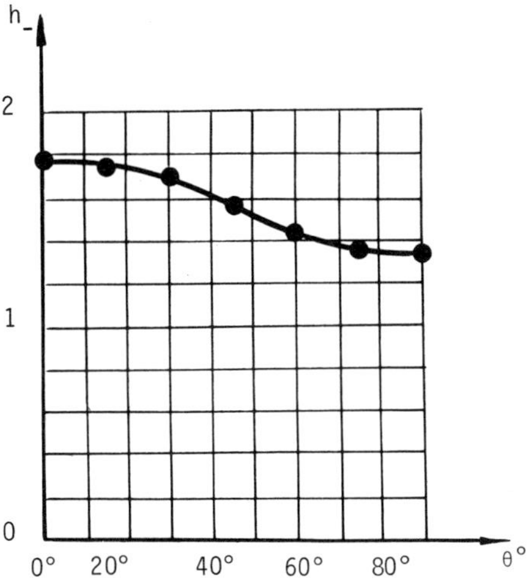

Fig.5.16 Dependence of the degree of polarization of RSRS lines as a function of an angle between the polarization planes of the probe and pumping waves

line observed was 25%. The polarization anisotropy of the fine structure of the line is described to a sufficiently high accuracy by

$$\frac{\kappa}{\kappa_0} = 1 - 0.25 \sin^2\theta \ , \qquad (5.68)$$

where κ_0 is the gain for coincident linear polarizations.

In the range of pressures under investigation (0.3 to 0.5 torr) polarization properties of the scattered line did not significantly change. This indicates that in a gas, under resonance exchange by excitation, reorientation of magnetic moment is accompanied by a simultaneous change of atomic velocity. In the case of "weak" collisions, orientation relaxation must lead to depolarization of the fine structure. If, as a probe wave, the field from the transition $2s_2 - 2p_4$ ($\lambda = 1.1767$ μm) is used, narrow resonances are not observed with linear polarizations of the external and probe fields. This is easy to understand. The transitions between sublevels of the fine structure of levels $2s_2$ ($J = 1$) and $2p_2$ ($J = 1$) are resolved for $\Delta m = \pm 1$. At the same time, the field at $\lambda = 1.52$ μm causes the transitions between the sublevels with $m = 0$. Interactions of waves with circular polarization lead to the fine structure of the line radiated at $\lambda = 1.1767$ μm in the presence of the field at $\lambda = 1.5$ μm.

5.6 Resonances at Saturation of Stimulated Raman Scattering in the Standing-Wave Field

The line shape in a linear-probe-field approximation was investigated in the previous paragraphs. It is interesting to consider saturation of absorption (gain) of a probe signal. Saturation effects at multiphoton transitions can result in new resonances, whose properties are of great interest for super-high-resolution spectroscopy. Investigation of saturation is of practical interest for the theory of three-level lasers, in which gain is caused by the action of the field on the coupled transition (see Chap.9). It is easier to explore saturation effects when the characteristics of a three-level laser are studied.

Consider the saturation effects in a resonant SRS laser when a SRS line is inhomogeneously broadened (widths Γ_+ and Γ_- are more than that of a two-photon transition). The gain line in such lasers with a Fabry-Perot cavity is sharply anisotropic. When the pumping frequency Ω of a travelling wave is detuned from the center of the transition line, the gain of the transition $0 \to 2$ has two maxima, at frequencies $\Omega' = \pm(k'/k)\Omega$ (see Subsec.5.4.2), which corresponds to SRS resonances on interaction of unidirectional and oppositely travelling waves on the transitions $0 \to 1$ and $0 \to 2$. As a result, the dependence of the generated power on the frequency proves to be very asymmetric. As has been shown in Section 5.5, the gain line at 1.15 μm in Ne at the transition $2s_2 - 2p_4$ when pumping by the field at $\lambda = 1.5$ μm is determined by two-quantum processes. A laser that uses these transitions of Ne may be considered to be a resonant Raman laser. It is convenient for investigations of nonlinear phenomena in SRS. Theoretical and experimental investigations [5.39] have detected saturation resonance of multiphoton transitions. It arises from saturation of the gain of the transition 0 - 2 by the standing-wave field E' when the difference of frequencies of incident and scattered light is equal to the frequency of a forbidden transition, i.e., when $\omega' = \omega + \omega_{12}$. A resonance dip in the SRS line was observed in the dependence of generated power on frequency. Unlike the resonance SRS line, it does not undergo a Doppler line broadening and has the width of the forbidden transition $1 \to 2$.

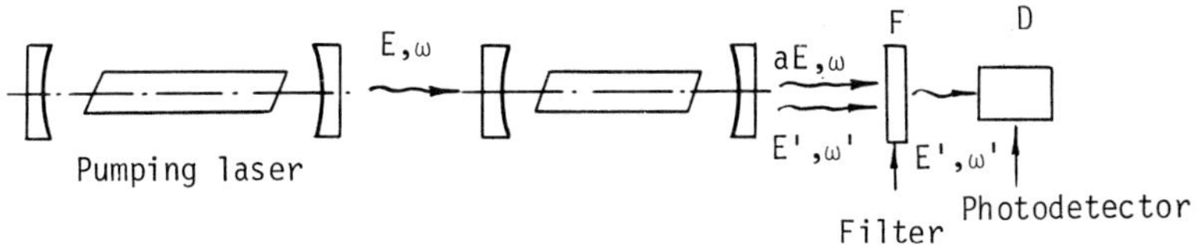

Fig.5.17 Experimental design for observing the frequency characteristics of the laser at $\lambda = 1.15$ μm pumped by the 1.5 μm field

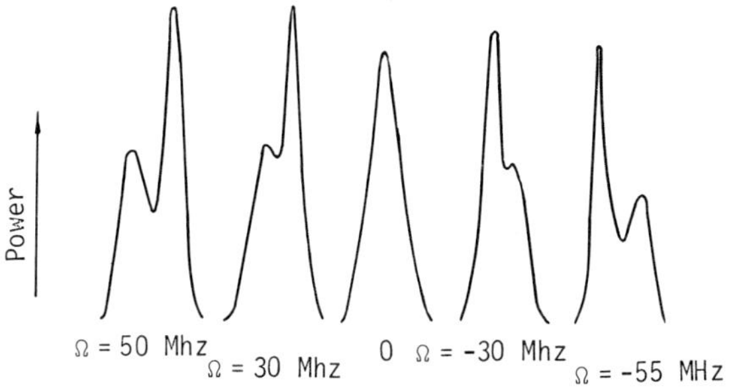

Fig.5.18a Dependence of the generated power on frequency of the field-frequency detuning at $\lambda = 1.5$ μm from the line center

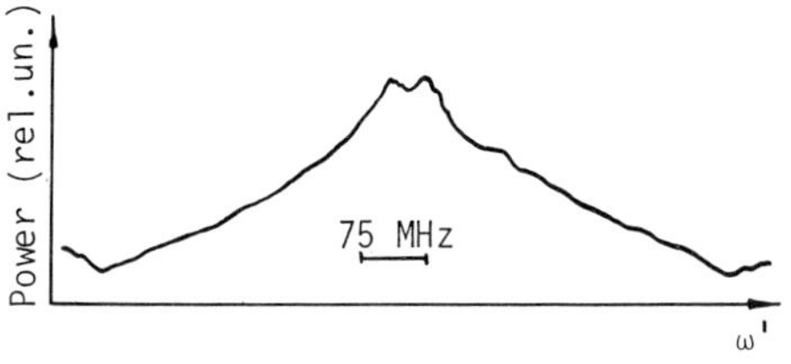

Fig.5.18b Resonance dip of the generated power of a three-level laser

The experimental design for observing the frequency characteristics of the 1.15 μm laser pumped by the field at $\lambda = 1.5$ μm is typical for similar experiments. It is shown in Fig.5.17. Radiation of a single-frequency He-Ne laser with mode selection is focused in the laser cavity at $\lambda = 1.15$ µm. Figure 5.18 shows the dependence of the generated power on the frequency when the frequency of the field at $\lambda = 1.5$ µm is detuned from the line center. A sharp generated peak with a large amplitude corresponds to interaction of one of the components of the standing wave travelling in the same direction as the pumping wave at $\lambda = 1.5$ µm. When the pumping intensity and the generated power at $\lambda = 1.15$ µm are increased, near the detuning frequency $\Omega \simeq 0$ a sharp dip of a small amplitude and a width of about 10 MHz arises in the dependence of the generated power on the frequency (Fig.5.18). The width of the dip is close to that of the two-quantum transition $2p_1 - 2p_4$.

Theoretical calculation of the saturated RSRS line is very unwieldy. Analysis of the processes responsible for saturation becomes difficult. On the assumption $\gamma_0 \gg \gamma_1, \gamma_2$, $(k' - k)/k \sim 1$ the saturated-gain line is [5.39]

$$\kappa = \kappa_0 \left\{ 1 - G \left[\phi(\Omega) + \frac{k'^2 - k^2}{4k'^2} \frac{\gamma_2 \Gamma[\gamma_{12}\Gamma + \Omega(\Omega' - \Omega)]}{(\Gamma^2 + \Omega^2)[\gamma_{12}^2 + (\Omega' - \Omega)^2]} \right] \right\} , \qquad (5.69)$$

where $\kappa_0 = 8\pi^{3/2} P_{02}^2 N V^2 k'^2 / \hbar u (k'^2 - k^2)(\Gamma^2 + \Omega^2)$ is the linear gain coefficient of the transition $0 \to 2$ when pumped by the field from the transition $0 \to 1$. N is the gas density; G is the saturation parameter of the transition $0 \to 2$; $\phi(\Omega)$ is a smooth function with width of the order of Γ $(\phi(0) = 1)$; $\Gamma = \gamma_0/2$; $\gamma_{12} = (\gamma_1 + \gamma_2)/2$. Correspondingly, the generated power P of a three-level laser at $\Omega = 0$ is described by (gain is equal to losses h)

$$P \sim \frac{\kappa_0 - h}{\kappa_0} \left\{ 1 + \frac{k'^2 - k^2}{4k'^2} \frac{\gamma_2}{\gamma_{12}} \frac{\gamma_{12}^2}{[\gamma_{12}^2 + (\Omega' - \Omega)^2]} \right\}^{-1} . \qquad (5.70)$$

Equation (5.70) describes the dip at the frequency ω' when $\omega' - \omega = \omega_{12}$. Its contrast is

$$K = \frac{P(\infty) - P(0)}{P(0)} = \frac{k'^2 - k^2}{4k'^2} \frac{\gamma_2}{\gamma_{12}} . \qquad (5.71)$$

The dip depth depends on relaxation constants and on the frequency detuning Ω. When $\Omega \gg \gamma_0$ a standing wave from the transition $0 \to 2$ interacts with two different groups of atoms and the described resonance dip vanishes.

As theoretical analysis has showed, the dip with the width of the forbidden transition is caused only by coherent processes. It is not associated with saturation of population of levels. The processes of interest can be represented graphically

$$\circ \xrightarrow[n_1]{E_1} d_{01} \xrightarrow{E'} d_{12} \xrightarrow{E'} d_{01} \xrightarrow{E'} d_{12} \xrightarrow{E} d_{02} \xrightarrow[\Delta n_2]{E'} \bullet \qquad (5.72)$$

The probability of transition to level 2, which is proportional to $E^2E'^4$ is not connected with the change of population of levels 0 and 2. The polarization of the transition $0 \to 2$ resulting from (5.72) can be represented by 1) the field resonant with transition $0 \to 2$ and polarization of transition $1 \to 0$ polarize transition $2 \to 1$ at frequency $\omega' - \omega$; 2) the field from transition $0 - 2$ and polarization of transition $2 \to 1$ polarize transition $1 \to 0$; 3) the field from transition $1 \to 0$ and polarization of transition $2 \to 1$ polarize transition $0 \to 2$; 4) polarization and field of transition $0 \to 2$ produce absorption (emission) at frequency ω.

The process of saturation caused by the change of population can be represented by

$$\circ \xrightarrow[n_1]{E} d_{01} \xrightarrow{E'} d_{12} \xrightarrow{E} d_{02} \xrightarrow[n_2]{E'} \bullet \quad \xrightarrow{E'} d_{02} \xrightarrow[\Delta n_2]{E'} \bullet \qquad (5.73)$$

The width of the saturation dip associated with (5.73) is equal to that of transition $2 \to 0$.

We shall explain qualitatively the formation of the dip. The probability of transition of an individual particle from level 1 to level 2 under the action of the fields E and E' is given by (5.34). In the case of a standing-wave field on the transition $0 \to 2$, the transition probability for the atom that has a projected velocity onto the field v direction has the form

$$W_{1\to 2}(v) = W^-_{1\to 2}(v) + W^+_{1\to 2}(v) \ , \qquad (5.74)$$

where $W^-_{1\to 2}$ and $W^+_{1\to 2}$ correspond to the interaction of unidirectional and oppositely travelling waves, respectively. $W^-_{1\to 2}$ is obtained from $W_{1\to 2}$ by replacing Ω with $\Omega - kv$ and Ω' with $\Omega' - k'v$, and $W^+_{1\to 2}$ by replacing Ω with $\Omega - kv$ and Ω' with $\Omega' + kv$. For the case $\gamma_0 >> \gamma_1, \gamma_2$, we have

$$W_{1\to 2}(v) = \frac{2V^2V'^2}{(\gamma_0/2)^2 + (\Omega - kv)^2}\,\frac{1}{\gamma_1}\left[\frac{\gamma_{12}}{\gamma_{12}^2 + (\Omega' - \Omega - k'v + kv)^2} + \right.$$

$$+ \frac{\gamma_{12}}{\gamma_{12}^2 + (\Omega' - \Omega + k'v + kv)^2} \Bigg] . \tag{5.75}$$

From (5.75) it follows that in the general case the standing wave from the transition $0 \rightarrow 2$ interacts with two groups of atoms. The main contribution to the radiation from the transition $0 \rightarrow 2$ in the standing-wave field is provided by atoms whose projected velocities satisfy the condition (at $k' > k$)

$$v^{\pm} = \pm \frac{\Omega' - \Omega}{|k' \pm k|} ; \tag{5.76}$$

and near the range of the projected velocities

$$\Delta v^{\pm} = \frac{\gamma_{12}}{|k' \pm k|} . \tag{5.77}$$

Therefore, when $\Omega' - \Omega = 0$ the standing wave from the transition $0 \rightarrow 2$ interacts simultaneously with two groups of atoms. As a result, owing to (5.72) the saturation effects become stronger and the dip appears.

According to the analysis given in [5.39], in the case of standing waves from the transitions $0 \rightarrow 1$ and $0 \rightarrow 2$ the resonance dip appears at any values of k' and k. Because of the progress achieved in the production of tunable lasers, the method described can be widely used in the spectroscopy of forbidden transitions for frequency stabilization. In addition to the two-photon resonance method, it can be used for precise measurements of frequencies of forbidden transitions, such as a frequency of the transition 1s - 2s of H, and frequencies of vibrational-rotational molecular transitions. Unlike the two-photon resonance method, it does not require great intensities of optical fields.

5.7 Lamb Dip of Coupled Closely Spaced Transitions. Crossing Resonances

In Section 3.2.5, we discussed the fact that the nature of the resonant interaction of two fields of similar frequencies with a gas depends largely on their polarization. According to the field polarizations and level degeneration we can consider their interaction both with a two-level and with a three-level system. In the general case, the interaction of fields has properties of two- and three-level systems. In this chapter, we shall analyze in more detail the nonlinear interaction of the standing-wave field with two closely spaced transitions that have a common level. The analysis is of great practical significance since it permits analysis of the influence of

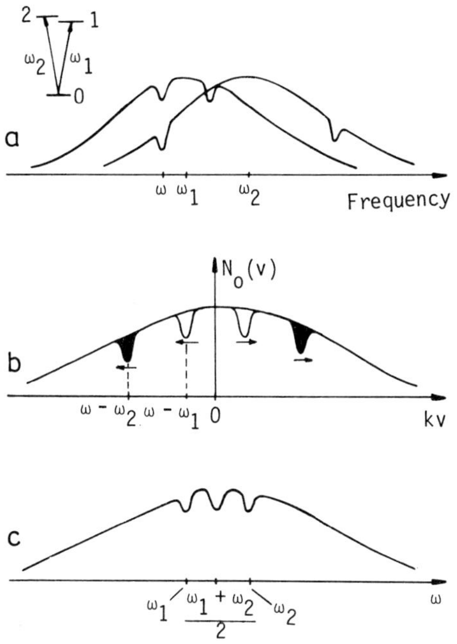

Fig.5.19a-c Occurrence of additional cross-resonances when the absorption of overlapping transitions is saturated with a common level, in the field of a standing light wave with a frequency ω. a) the Doppler contour shape of two lines; b) velocity distribution on common level 0; c) the shape of the saturated absorption spectrum

neighboring transitions on the reproducibility of frequency of lasers. The presence of a coupled transition influences the Lamb-dip shape of each of transitions $0 \to 1$ and $0 \to 2$ (see Fig.5.19). A two-level system has no resonances that arise in such a system. The interaction of resonances is connected with that of the field with two transitions. The resonances described below have been called crossing. Their nature is close to the phenomenon called mode crossing. Let the standing wave saturate absorption of a spectral line formed by overlapping of two Doppler-broadened lines with a common level (Fig.5.19). Two holes are burnt in each of these lines at the expense of the interaction with the particles that have the projected velocities $kv = \pm(\omega - \omega_{10})$ and $\pm(\omega - \omega_{20})$ ("light" and "dark" holes on Fig.5.19b), respectively. As a result, the velocity distribution of particles on the common level $N_0(v)$ acquires four holes. When the standing-wave frequency is scanned to determine the spectral-line contour, two dips appear at the frequencies ω_{10} and ω_{20} that are due to overlapping of holes in the line center for each of the transitions; "light" holes (due to field on ω_1) overlap each other, and "dark" holes (due to field on ω_2) overlap each other, too (Fig.5.19).

Besides this, a dip appears at the symmetric frequency $\omega_{10}+\omega_{20}/2$, which is due to overlapping of "light" with "dark" holes. This crossing dip is caused by a decrease of absorption of one of the travelling waves with the frequency ω_{10} at the expense of the interaction with the hole in the velocity distribution of particles on the common level formed by an oppositely travelling wave with the frequency ω_{20} and vice versa. The possibility of occurrence of crossing nonlinear resonances was first noted in [5.45](an effect of its own and another's dip in a gas laser in the magnetic field) and in [5.26]. The formation of a crossing resonance was experimentally explored in [5.50] at splitting of an absorption line in a He-Ne/Ne 1.5 μm laser, in [5.51] at Zeeman splitting of an absorption CH_4 line and in [5.52] when the magnetic hyperfine structure $F_2^{(2)}$ of a component of the line P(7) of the band V_3 of methane was resolved.

In order to determine the absorption coefficient of the standing-wave field

$$E(z,t) = E \cos kz \cos\omega t \tag{5.78}$$

in a gas of atoms whose transition frequency is split into two similar components with a common level, we shall use the system of equations for the density matrix (5.23), assuming that $\omega_{21} \ll \omega_{10},\omega_{20}$. The polarization of a medium saturated by the field (5.78) is equal to

$$P(z,t) = P_{01}\int_{-\infty}^{\infty} dv\rho_{10} + P_{02}\int_{-\infty}^{\infty} dv\rho_{20} + \text{k.c.} \ . \tag{5.79}$$

When averaging over velocities, we assume that a Doppler width is much more than all homogeneous widths of atomic levels. The absorption coefficient is found from (5.79) in the usual way; it is equal to [5.53]

$$\begin{aligned}\kappa = \kappa_0 \Bigg\{ 1 &- \frac{G_1}{2}\left(1 + \frac{\Gamma_{10}^2}{\Gamma_{10}^2 + \Omega_1^2}\right) - \frac{G_2}{2}\,\frac{\gamma_0\gamma_2\Gamma_{20}}{(\gamma_0+\gamma_2)(\Gamma_{21}^2+\omega_{21}^2)} \,\cdot \\ &\cdot \left[\frac{\Gamma_{21}\Gamma_+ - \omega_{21}^2/2}{\Gamma_+^2 + (\omega_{21}/2)^2} + \frac{\Gamma_{21}\Gamma_{10} - \omega_{21}\Omega_1}{\Gamma_{10}^2+\Omega_1^2}\right] - \\ &- \frac{G_2}{2}\,\frac{\gamma_2\Gamma_{20}}{(\gamma_0+\gamma_2)\Gamma_+}\left[\frac{\Gamma_+^2}{\Gamma_+^2+(\omega_{21}/2)^2} + \frac{\Gamma_+^2}{\Gamma_+^2+\tilde{\Omega}^2}\right]\Bigg\} + (1 \rightleftarrows 2) \ ,\end{aligned} \tag{5.80}$$

where $\kappa_{vo} = 4\pi^{3/2} n_{vo}^{(o)} P_{vo}^2 (\hbar u)^{-1}$, $V = 1,2$ are linear absorption coefficients of the transitions $0 \to 1$ and $0 \to 2$; P_{vo} is dipole matrix elements of the transitions $0 \to 1$ and $0 \to 2$; $n_{vo}^{(o)} = N_o - N_v$; N_o, N_v are populations of atomic levels per unit volume in the absence of the field, u is a thermal velocity; γ_2, γ_1, γ_o are inverse lifetimes of levels 2, 1, 0, respectively; $\Gamma_{10}, \Gamma_{20}, \Gamma_{21}$ are the halfwidths of the lines of the transitions $1 \to 0$, $2 \to 0$, $2 \to 1$, respectively; and $\Gamma_+ = (\Gamma_{10} + \Gamma_{20})/2$ $\Omega_v = \omega - \omega_{vo}$, $\tilde{\Omega} = \omega - (\omega_{10} + \omega_{20})/2$, $G_v = P_{vo}^2 E^2 / 2\Gamma_{vo} \hbar^2 (1/\gamma_o + 1/\gamma_v)$. Equation (5.80) describes three nonlinear resonances at frequencies ω_{10}, ω_{20}, and $(\omega_{10} + \omega_{20})/2$. The resonance at the frequency ω_{10} corresponds to the Lamb dip of the transition $1 \to 0$, the interaction of the field with the transition $2 \to 0$ that leads to an additional homogeneous saturation and renormalization of the parameter G_1. Moreover, this resonance is asymmetric. The center of the Lamb dip is shifted with respect to the frequency ω_{10}. The origin of the resonance at the frequency ω_{20} is similar, i.e., "repulsion" of the Lamb dips takes place at the transitions $1 \to 0$ and $2 \to 0$. Analysis has indicated that the resonance shifts are associated with a known Stark shift in optical fields (optical shift). The difference is that a nonlinear resonance arises in the same approximation that a Stark effect does. So the shift is independent of the field. The dependence on the field can manifest itself in the next order of the perturbation theory.

As has been already noted, the peculiarity of the scheme considered is the appearance of a resonance of a dispersion form at the frequency $(\omega_{10} + \omega_{20})/2$ which is the mean of ω_{10} and ω_{20}. The relative intensity of the resonance depends on the intensity of initial "parent" transitions and on the lifetimes of the particles on the levels. In the particular case of equal lifetimes of all three levels it is the geometric mean of relative intensities of "parent" resonances.

When the laser frequencies are stabilized, of great importance is the problem of the interference of three resonances at frequencies ω_{10}, ω_{20}, and $(\omega_{10} + \omega_{20})/2$. In the general case, it is difficult to analyze this problem. We consider the situation in which the intensities of the transitions $2 \to 0$ and $1 \to 0$ greatly differ, i.e., $P_{10}^2 / P_{20}^2 \ll 1$ and find the way the resonance maximum shifts at the frequency ω_{20}. This case is of importance for analyzing the influence of a methane HFS on frequency reproducibility of a He-Ne laser stabilized over the line P(7) of the band V_3 of CH_4. The position of the resonance maxima is determined by the roots of the equation $\partial\kappa/\partial\omega = 0$. The case $N_1 = N_2 = 0$, $\gamma_e = \Gamma_{ik} = \Gamma$ for all $i,k,e = 0,1,2$ will be analyzed further. Taking into account these simplifications, we obtain for the resonance maximum

ω_p corresponding to the frequency ω_{20} in the $P_{10}^2/P_{20}^2 \ll 1$ linear approximation,

$$\omega_p = \omega_{20} - \frac{P_{10}^2}{P_{20}^2}\frac{|\omega_{21}|}{2}F\ ,$$

$$F = \frac{1}{2(1+\delta^2)}\left[-1 + \frac{1}{1+\delta^2} + \frac{2(1-\delta^2)}{(1+\delta^2)^2}\right] + \frac{1}{(1+\delta^2/4)^2}\ . \tag{5.81}$$

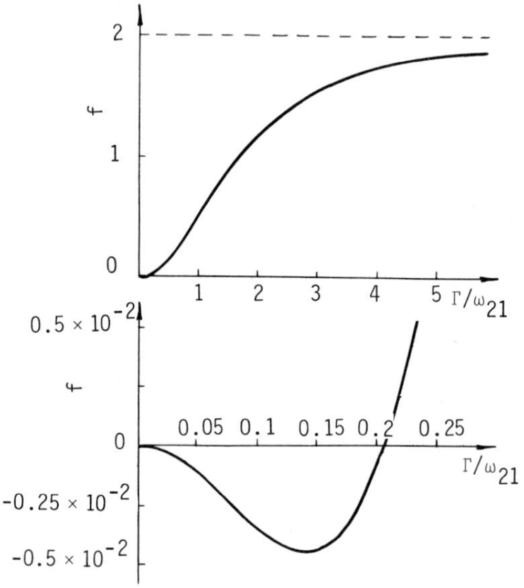

Fig.5.20 Shift of the maximum of the nonlinear power resonance formed by two components that are greatly different in intensity and have a common level

The function F plotted in Fig.5.20 indicates the change of the position of the resonance maximum as a function of $\delta^{-1} = \Gamma/|\omega_{21}|$, i.e., pressure. At low pressures ($\delta \gg 1$), $F \simeq -1/2\delta^2$ and $\omega_p - \omega_{20} = (P_{10}^2/P_{20}^2)(|\omega_{21}|/4\delta^2)$. This "repulsion" of resonances at frequencies ω_{10} and ω_{20} at low pressures occurs because of the asymmetry mentioned above. At $\Gamma/|\omega_{21}| > 0.2$ the sign of the shift changes. The influence of wings of the crossing resonance proves to be essential in this region. They "pull" the Lamb dips at frequencies ω_{10} and ω_{20} towards each other.

References

5.1 M.S. Feld: *Proc. of the Esfahan Symp. on Fundamental and Applied Laser Physics, August 29-September 5, 1971*, ed. by M.S. Feld, A. Javan, N. Kurnit, p.369
5.2 R.G. Brewer: Science 178, 247 (1972)
5.3 I.M. Beterov, V.P. Chebotayev: In *Progress in Quantum Electronics*, ed. by J.H. Sanders, S. Stenholm, Vol.3 (Pergamon Press 1974), p.1
5.4 V.P. Chebotayev: In *High Resolution Laser Spectroscopy*, Topics in Applied Physics, Vol.13, ed. by K. Shimoda (Springer-Verlag, Berlin, Heidelberg, New York 1976)
5.5 C.V. Raman, K.S. Krishnan: Nature 121, 501 (1928)
5.6 G.S. Landsberg, L.I. Mandelstam: Naturwiss. 16, 577, 772 (1928)
5.7 M. Goepert-Mayer: Ann.d.Phys. 9, 273 (1931)
5.8 V. Weiscopf: Zeit.f.Phys. 85, 451 (1933)
5.9 A. Einstein, P. Ehrenfest: Zeit.f.Phys. 19, 301 (1923)
5.10 G. Placzek: *Rayleigh-Strenung und Raman-Effect* (Leipzig 1934)
5.11 C.H. Townes, A.L. Schawlow: *Microwave Spectroscopy* (McGraw-Hill, New York 1955)
5.12 S.H. Autler, C.H. Townes: Phys.Rev. 100, 340 (1955)
5.13 A.C. Kastler: Nobel Lecture, Stockholm (1967)
5.14 N.G. Basov, A.M. Prokhorov: Zh.Eksp.i Teor.Phys. 27, 431 (1954)
5.15 N. Bloembergen: Phys.Rev. 104, 324 (1956)
5.16 V.M. Kantorovich, A.M. Prokhorov: Zh.Eksp.i Teor.Phys. 33, 1428 (1957)
5.17 A. Javan: Phys.Rev. 107, 1579 (1957)
5.18 A. Vuylsteke: *Elements of Maser Theory* (D. Van Nostrand Co., Inc., Princeton, N.J. 1960)
5.19 A. Sigman: *Microwave Solid-State Masers* (McGraw-Hill, New York 1964)
5.20 V.M. Fain, Ya.I. Khanin: *Quantum Radiophysics*, M; *Quantum Electronics*, Vol.1,2 (Pergamon Press, Oxford 1965, 1968)
5.21 K. Shimoda: In *Proc. of an Intern. Conf. on Laser Spectr., Vail, Colorado, June 25-29, 1973*, ed. by R.C. Brewer, A. Mooradian, p.29
5.22 S.B. Decomps, M. Dumont, M. Ducloy: In *Laser Spectroscopy*, Topics in Applied Physics, Vol.2, ed. by H. Walther (Springer-Verlag, Berlin, Heidelberg, New York 1976)
5.23 W.R. Bennett, Jr., V.P. Chebotayev, J.W. Knutson: In *Proceedings of V ICPEAC* (Leningrad 1967), p.521
5.24 R.H. Cordover, P.A. Bonczyk, A. Javan: Phys.Rev.Lett. 18, 730 (1967)
5.25 W.E. Schweitzer, Jr., M.M. Birky, J.A. White: J.Opt.Soc.Amer. 57, 1226 (1967)
5.26 H.R. Schlossberg, A. Javan: Phys.Rev.Lett. 17, 1242 (1966)
5.27 G.Ye. Notkin, S.R. Rautian, A.A. Feoktistov: Zh.Eksp.i Teor.Phys. 59, 1673 (1967)
5.28 H.K. Holt: Phys.Rev.Lett. 19, 1275 (1967); 20, 410 (1968)
5.29 I.M. Beterov, V.P. Chebotayev: Zh.Eksp.i Teor.Phys.Pis.Red. 9, 216 (1969)
5.30 I.M. Beterov, Yu.A. Matyugin, V.P. Chebotayev: Zh.ETP Pis.Red. 10, 296 (1969)
5.31 T. Hänsch, R. Keil, A. Schabert, Ch. Schmelzer, P. Toschek: Z.Phys. 226, 293 (1969)
5.32 G.W. Flynn, M.S. Feld, B.J. Feldman: Bull.Amer.Soc. 12, 669 (1967)
5.33 M.S. Feld, A. Javan: Phys.Rev. 177, 540 (1969)
5.34 B.J. Feldman, M.S. Feld: Phys.Rev. A5, 899 (1972)
5.35 T.Ya. Popova, A.K. Popov, S.G. Rautian, R.I. Sokolovsky: Zh.Eksp.i Teor.Phys. 57, 850 (1969)
5.36 T. Hänsch, P. Toschek: Z.Phys. 236, 213 (1970)
5.37 I.M. Beterov, Yu.A. Matyugin, V.P. Chebotayev: Zh.Eksp.i Teor.Phys. 64, 1495 (1975)

5.38 Ye.M. Lifshits, L.P. Pitayevsky: *Relativistic Quantum Theory*, Part 2 (Pergamon Press, Oxford 1973)
5.39 Ye.V. Baklanov, I.M. Beterov, B.Ya. Dubetsky, V.P. Chebotayev: Zh.Eksp. i Teor.Phys.Pis.Red. 22, 289 (1975)
5.40 S.N. Atutov, V.S. Kuznetsov, S.G. Rautian, E.G. Saprykin, R.I. Yudin: *Proceedings of the All-Union Symposium on Gas Laser Physics* (Novosibirsk 1969), p.43
5.41 T.W. Ducas, M.S. Feld, L.W. Ryan, N. Skribanowitz, A. Javan: Phys.Rev. A5, No.3, 1036 (1972)
5.42 I.M. Beterov, Yu.A. Matyugin, V.P. Chebotayev: Zh.E.T.P.Pis.Red. 12, 174 (1970)
5.43 M.I. D'yakonov, V.I. Perel: Zh.Eksp.i Teor.Phys. 50, 448 (1966)
5.44 I.M. Beterov, Yu.A. Matyugin, V.P. Chebotayev: Preprint of the Inst. of Semiconductor Physics of the Siberian Branch of the USSR Ac. of Sci., No.22 (1971)
5.45 W.R. Bennett, Jr., P.J. Kindlman: Phys.Rev. 149, 38 (1966)
5.46 I.M. Beterov, Yu.A. Matyugin, S.G. Rautian, V.P. Chebotayev: Zh.Eksp.i Teor.Phys. 58, 1243 (1970)
5.47 A.K. Popov: Zh.Eksp.i Teor.Phys. 58, 1623 (1970)
5.48 N. Skribanowitz, M.S. Feld, R.E. Franke, M.J. Kelly, A. Javan: Appl. Phys.Lett. 19, 161 (1971)
5.49 N. Skribanowitz, M.J. Kelly, M.S. Feld: Phys.Rev. A6, 2302 (1972)
5.50 V.N. Lisitsyn, V.P. Chebotayev: Optika i Spektroskopiya 26, 856 (1969)
5.51 E.E. Uzgiris, J.L. Hall, R.L. Barger: Phys.Rev.Lett. 26, 289 (1971)
5.52 C. Borde, J.L. Hall: Phys.Rev.Lett. 30, 1101 (1973)
5.53 Ye.V. Baklanov, Ye.A. Titov: Kvantovaya Elektronika 2, 1893 (1975)

6. Narrow Nonlinear Resonances in Spectroscopy

Classical methods of spectroscopy of atoms and molecules in a gas offer information concerning, for the most part, line shifts; they assume homogeneous width. Due to the fairly low resolution of previous methods, which on the one hand is caused by a screening effect of Doppler broadening, and on the other hand by very limited potentialities of spectral instruments, investigations were usually carried out with considerable gas pressures. The production of homogeneous line width was provided by perturbation of both levels, and information on the processes of relaxation of individual levels was lost. Lifetimes of level relaxation were measured by means of direct observation of damping of spontaneous radiation.

Narrow nonlinear resonances induced by laser radiation from saturated Doppler-broadened transitions have become the basis of a new trend in spectroscopy, which is frequently called nonlinear laser spectroscopy. The methods of nonlinear laser spectroscopy have extremely high resolution and make it possible to make measurements in low-pressure gases with very high resolution and high accuracy. Heterodyne methods permit spectroscopic investigations in the optical band to an absolute accuracy that is characteristic of radio-frequency ranges.

Relaxation lifetimes of individual levels, and differential cross-sections of elastic scattering of atoms and molecules on collisions can be determined by observing narrow resonance shapes. An experimental procedure with a relative accuracy of 10^{-11} to 10^{-13} required taking into account the influence of relativistic effects. Previously, these phenomena were not usually taken into consideration in optical spectroscopy.

The history of nonlinear laser spectroscopy is very short, and the choice of subjects for investigation is so far limited. In spite of this, however, effective methods have been developed and valuable information that could not be gained by use of previously known spectroscopic methods has been obtained. With the advent of tunable monochromatic lasers, the potentialities of laser spectroscopy will grow considerably. However it is the physical principles of the methods of nonlinear spectroscopy should remain permanent. Of great importance for spectroscopic applications is the problem of resolution and sensitivity of the methods of laser spectroscopy.

6.1 Resolution of Nonlinear Resonances

In previous chapters, we have considered different types of narrow nonlinear resonances induced by a coherent laser field, and have found their widths and amplitudes. On the basis of this, we now consider the problem of resolution of nonlinear resonances from the unified point of view.

6.1.1 Absorption Saturation Resonances

If a spectral absorption line consists of two independent lines that have no common level, spaced at a distance less than the Doppler width, then from the standpoint of the Rayleigh criterion they are unresolved by methods of conventional linear spectroscopy. At absorption saturation, a resonance minimum arises in the center of each of the lines that is confused by Doppler broadening (Fig.6.1).

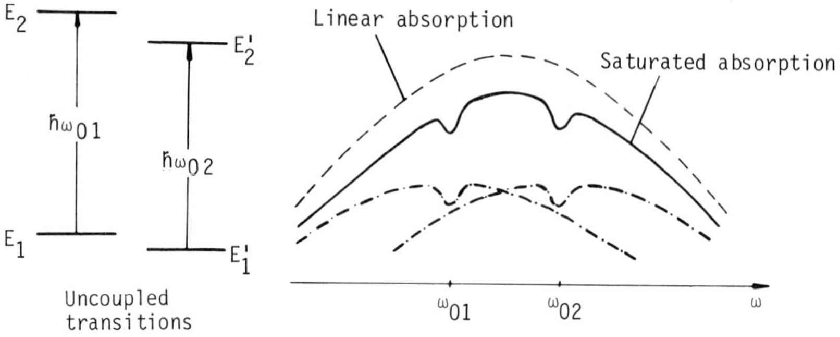

Fig.6.1 Shape of spectrum of saturated absorption for the case when a spectral line consists of two overlapping Doppler-broadened lines due to two uncoupled transitions

First, assume that the line intensities are identical. This means identity of reasonable intensities. Then the resolution limit is determined by the narrow dip width $\Delta\omega$. When absorption is saturated by a standing-wave field, the resonance width is

$$\Delta\omega = 2\Gamma f_p \ , \tag{6.1}$$

where Γ is the homogeneous halfwidth of the transition, f_p is a factor that characterizes the field broadening of the dip. To approximate the noncoherent field-medium interaction, the factor f_p is described by the dependence given in Fig.2.5. In the simplest approximation, it may be assumed that

$$f_p = \sqrt{1+G} \ . \tag{6.2}$$

When an oppositely travelling probe wave is used, the resonance width is also described by (6.1), but the factor f_p takes another form. To approximate the noncoherent interaction,

$$f_p = \frac{1}{2}\left(1+\sqrt{1+G}\right) \ . \tag{6.3}$$

At strong saturation ($G \gg 1$) the field broadening of the dip is half as much as in the case of a standing wave. This is important in absorption saturation in cells that have great optical density ($\kappa_0 \ell_b \gg 1$), when strong saturation is necessary for brightening. With allowance for coherence effects, the field-broadening factor is as shown in Fig.2.10. In the method of nonlinear fluorescence the resonance width is also determined by (6.1).

The following effects make contributions to the homogeneous width: 1) radiative decay of levels; 2) particle collisions; 3) geometric broadening; 4) finite flight time of molecules through the light beam. At sufficiently low pressure of the gas under investigation, when the free path length of particles Λ relative to the broadening collisions exceeds the light-beam diameter a, the contribution of collisions becomes negligible. The geometric (spatial) broadening of resonance arises due to sphericity of the beam wavefront or nonparallelism of oppositely travelling waves. In order for the broadening due to the sphericity of wavefronts to be less than the flight broadening, the radius of curvature r of the wavefront must satisfy the condition

$$r > \frac{a^2}{2\pi\lambda} \ . \tag{6.4}$$

According to (1.14), due to the wavefront sphericity the broadening decreases at increasing r proportionally to $\sqrt{r}$. Nonparallelism of wavefronts of oppositely travelling beams also results in resonance broadening. In order to reduce the broadening due to this effect to a value comparable with the flight width, the angle between oppositely travelling beams must satisfy the condition

$$\theta < \frac{\lambda}{a} \ . \tag{6.5}$$

If (6.4) and (6.5) are satisfied with a rather large margin, the geometric broadening can also be made negligible. In this case, the flight broadening is the only instrumental contribution to the resonance broadening. The magnitude of the flight broadening for molecules with a specified radial velocity, which fly through the center of the beam with a half-height diameter a, with a gaussian transverse profile of intensity is given by (3.21). If it is taken into account that the path length of molecules crossing the beam off center is somewhat less, the mean diameter $\langle a \rangle$ of the beam for molecules is $(\pi/4)a$. The mean value of the projected velocity v_r of molecules onto the wavefront plane is $v \ll u$ provided that the projected velocity $\langle v_r \rangle$ onto the beam direction is $(\sqrt{\pi/2})u$. Then the expression $\Delta\nu_{tr} \simeq u/2\pi a$ can be used to estimate the flight width. Calculation of the influence of the flight effects on the Lamb-dip width [6.1] yields, for the flight width in the region 0.2 $\Gamma\tau$, (see Subsec.6.2.3)

$$\Delta\nu_{tr} = \frac{1.16\sqrt{\ln 2}}{2\pi} \cdot \frac{u}{a} \; . \tag{6.6}$$

Experiments on absorption saturation yielded resonance widths that are close to that determined by the flight effects. Results of such experiments, as well as theoretical values of the flight width calculated from (6.6) are summarized in Table 6.1. In a number of experiments, the observed value of the width is almost twice as large as the limiting width. Such a difference appears to be attributable to the other contributions to broadening 2°, 3° and to broadening by the field saturating a transition. The highest value of resolution of nonlinear resonance $R = \nu_0/\Delta\nu$ was reported in [6.2,3] when a methane cell and a light beam of 30 and 20 cm diameter, respectively, were used.

In saturated absorption of long-lived molecules at a very low gas pressure, when the free-path length Λ considerably exceeds the beam diameter, the restriction of resolution imposed by (6.6) can be eliminated. In this case, the Lamb dip is sharpened at the top [6.4]; this is connected with anomalously large contributions to saturation of slow molecules having $v_r \ll u$. As the first experiments have shown [6.5], detection of this effect requires a recording system of high sensitivity because absorption at low pressures is small and the resonance intensity decreases.

With decrease of resonance width, absolute intensity falls, also. This is especially appreciable when the resonance width is of the order of 10^3 to 10^4 Hz. We give a simple, common estimate of the absolute intensity of the Lamb dip under optimal conditions. At optimal saturation of $G = 1.4$ the

Table 6.1 Resolving power of saturated-absorption resonances

Molecule	Wavelength (μm)	Diameter of beam (cm)	Temp. (K)	$\Delta\nu_{tr}$ calcul. (kHz)	$\Delta\nu$ exper. (kHz)	$R = \frac{\nu_0}{\Delta\nu}$	Reference
CH_4	3.39	0.83	300	12	60	1.5×10^9	[3.29]
CH_4	3.39	1.0	300	10	50	1.8×10^9	[3.30]
CH_4	3.39	5.0	77	1.0	5.8	1.5×10^{10}	[3.33]
OCS	8.2×10^3	60	300	0.08	3.4	1.1×10^7	[3.88]
SF_6	10.6	0.6	300	6	100	3×10^8	[3.90]
CH_4	3.39	20	300	0.9	1.5	6×10^{10}	[6.2]
CH_4	3.39	10	300	1.8	3	3×10^{10}	[6.22]

relative dip depth is equal to 0.13. Then the absorbed power per unit volume is

$$\Delta I = 0.13\kappa_0 \left(\frac{cE^2}{8\pi}\right) .$$

Recall that $\kappa_0 = 4\pi^{3/2}p^2N/\hbar u$. Because $G = 1.4 = p^2E^2/\hbar^2\Gamma^2$, we obtain

$$\Delta I_\Sigma[W] \sim 10^2 \frac{\left(\frac{\partial\Gamma}{\partial p}\right)^2\left[\frac{MHz}{torr}\right] \cdot p^3[torr] \cdot L[cm] \cdot S[cm^2] \cdot q}{\sqrt{\frac{T[K]}{M[at.un]}}} , \tag{6.7}$$

where q is the relative fraction of the particles on the absorbing level.

Here we assume that the collision broadening is proportional to pressure. It is interesting that the absolute resonance intensity is independent of the transition probability. This is connected with the fact that the resonance intensity is proportional to the product of the linear absorption and the incident power. Under optimal conditions, the first is proportional to the transition probability and the second is inversely proportional to the transition probability. When the transition probability is decreased it is necessary to increase the field intensity, in order to obtain optimal saturation. This suggests that the resonance intensity is not very dependent on the nature of the gas. The value q is the main factor that determines the absolute resonance intensity in molecular gases. For absorption in methane ($\lambda = 3.39$ μm), and in carbon dioxide, the values are 0.1 and 0.01, respectively.

These values of resonance intensities present upper limits. Under real conditions, the value is less. In lasers with an internal absorption cell, at small absorptions, the absolute resonance intensity should be multiplied by T/γ, where T is the transmittance of the mirror, and γ is the total relative loss in the mirrors.

Not infrequently, closely spaced transitions have different transition probabilities. For slight saturations, the resonance intensity is proportional to the square of the transition probability (i.e., the square of the linear absorption). This can make difficult the observation of weak transitions against stronger ones. Comparison of resonance intensities becomes difficult as well. The change of the field in wide transitions permits observation of very weak transitions, because resonance intensities of these transitions are comparable with those of strong transitions. Comparison of resonance intensities, under optimal conditions, permits qualitative estimates of values p_{ik} for different transitions $i \to k$. This may assist identification of transitions. Such investigations have not yet been performed.

If closely spaced spectral lines latent by Doppler broadening belong to transitions having a common level, a nonlinear absorption spectrum can contain additional resonances, which should be taken into account in analysis of line structures. Some properties of these resonances were studied in Section 5.7.

The relative intensity of crossing resonances depends on the intensity of "parent" transitions and particle lifetimes on the levels. In the particular case of equal lifetimes on all levels, in the approximation of weak saturation of transitions in the standing wave, the relative amplitude of the crossing resonance is the geometric mean of the relative amplitudes of the parent resonances. This facilitates observation of weak transitions. In the general case of unequal lifetimes, the intensity of crossing resonances depends strongly on the ratio of the lifetimes on the levels. For instance, if the common level 0 is short-lived compared to levels 1 and 2 then the hole in the distribution $N_0(v)$ is far less than the peaks in the distributions $N_1(v)$ and $N_2(v)$. Therefore, the amplitudes of the main resonances determined by the hole depths in the distributions of population differences $\Delta N_1(v) = N_0(v) - N_1(v)$ and $\Delta N_2(v) = N_0(v) - N_2(v)$ are much more than those of the crossing resonance determined only by the hole dimension in $N_0(v)$.

6.1.2 Competitive Resonances

In Section 3.4, we noted that the width of competitive resonances, for instance, in a ring gas laser with nonlinear absorption may be considerably less than the homogeneous width 2Γ, and the contrast of the resonances reaches 100%. However, of importance for nonlinear spectroscopy is the resolution of "competitive" resonances, i.e., the minimum frequency difference between two closely spaced absorption lines of the same intensity, at which its own competitive resonance arises for each line. Consider the potentialities of competitive resonances in this respect [6.6].

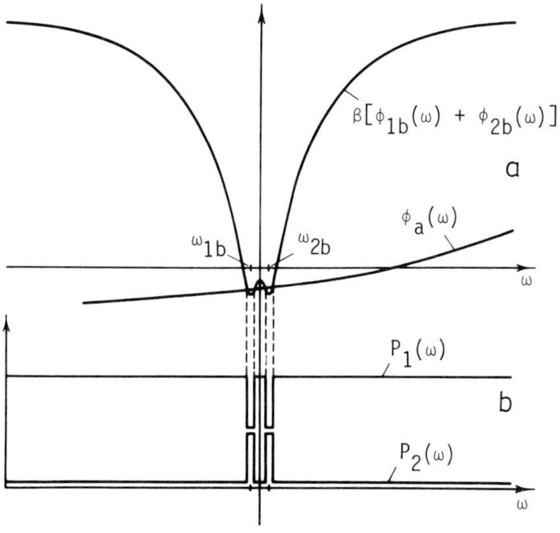

Fig.6.2a,b Resolution of competing resonances in a ring laser with nonlinear absorption. a) form of the function $\phi_b(\omega)$ in the case of two overlapping absorption lines; b) output power of oppositely travelling waves in a ring laser

The condition of occurrence of two separate competitive resonances is readily found from the relations given in Section 3.4. In (3.84) the function $\phi_b(\omega)$ that describes the dip in the absorption line should be replaced with the function $\phi_b(\omega) = \phi_{1b}(\omega) + \mu\phi_{2b}(\omega)$ that describes the dip in the absorption line that consists of two components. The curves in Fig.6.2a show that for appearance of two competitive resonances near the centers ω_{ib} of each spectral component, at least, it is necessary that the function $\phi_b(\omega)$ should have two minima and a maximum between them. For identical lorentzian lines ($\mu = 1$) this is possible only if the distance between the centers of lines ω_{1b} and ω_{2b} meets the condition,

$$\Delta\omega_{min} = |\omega_{1b} - \omega_{2b}| > \frac{2}{\sqrt{3}}\Gamma_b . \quad (6.8)$$

If the detuning of frequencies of gain and absorption lines $\Omega = \omega_a - (1/2)(\omega_{1b} + \omega_{2b})$ is such that

$$\beta\phi_b(\omega_{max}) > \phi_a(\omega_b) > \beta\phi_b(\omega_{min}) , \quad (6.9)$$

where ω_{min} and ω_{max} are frequencies of the minimum and the maximum, two separate competitive resonances arise (Fig.6.2b). The resonance widths can be much less than the homogeneous width $2\Gamma_b$ but the difference between the absorption lines at which these resonances arise cannot be less than $\Delta\omega_{min}$, i.e., the value of an order $2\Gamma_b$.

If (6.8) is not fulfilled, the dips on the curve $\phi_b(\omega)$ are not resolved. Nevertheless, the existence of two unresolved lines manifests itself in some rise of the function minimum $\phi_b(\omega)$. This results in the reduction of the region of stability of the two-wave conditions. In principle, the precise measurement of the width of an "unresolved" competitive resonance and the dependence of this width on pressure can provide information on the existence of two unresolved lines, even though (6.8) is not fulfilled [6.7]. Naturally, such a method is unsuitable for studying lines with more-complex structure.

It is not surprising that the limiting resolution is not determined by the competitive-resonance width. Some ways of narrowing of nonlinear resonances are known which nevertheless do not permit increased resolution. Examples are narrowing of the generated power peak in a laser with an internal nonlinear absorption cell with $\beta \approx 1$, which is described by (3.14). In particular, in a He-Ne/CH_4 laser operating near threshold and with a sufficient absorption value, power peaks with a width of 5 kHz can be obtained [6.8], i.e., about one tenth as much as the homogeneous width. When operating with an external nonlinear absorption cell having a large optical density ($\kappa_0 l_b \gg 1$) an effect of the extra broadening of the power peak is also observed [6.9]. However, all of these effects as well as the effect of competitive resonance narrowing only sharpen the top of the common unresolved resonance. As long as resonances themselves are not separated, none of the narrowing methods lead to the formation of two separate resonances. Nevertheless, all the methods of the resonance narrowing, even without increasing their resolution, are rather useful for facilitating laser frequency stabilization on them.

Thus, a valuable feature of competitive resonances, namely their sensitivity to the dip-peak steepness, is not essentially used with the lorentzian dip shape, because its steepness coincides with the width. Indeed, the dip steepness for the lorentzian shape is

$$|\Theta|^{-1/2} = \left|\frac{1}{\kappa}\frac{\partial^2\kappa}{\partial\omega^2}\right|^{-1/2} = \frac{\Gamma}{\sqrt{2}} . \qquad (6.10)$$

However, the sensitivity of competitive resonances to the dip-peak steepness can be used to reveal the effect of a "sharpened peak" by saturating the absorption of a very-low-pressure gas. LETOKHOV and PAVLIK [6.6,10] treated theoretically this possibility of increasing the resolution of competitive resonances and showed that, in this case, instead of (6.8),

$$|\omega_{1b} - \omega_{2b}| > 2\gamma_{rad} . \qquad (6.11)$$

Consequently, competitive resonances may be used in this particular case for nonlinear spectroscopy, with resolution that is not limited by flight time. In this case, two dips with sharpened peaks can be resolved (for instance, by the Rayleigh criterion) by the method of absorption saturation in the standing-wave field. However, the absolute value of the dip is negligible due to negligibility of the absorption coefficient of the gas at such low pressures (far less than 10^{-3} torr) that are necessary for making the collisional width of the dip far less than the radiative one. An effect of competition of travelling waves in a ring laser is extremely sensitive to the slightest resonance changes in the absorption coefficient and makes it possible to convert nonlinear absorption dips of negligible amplitudes into high-contrast competitive resonances of laser output power.

6.1.3 Spectroscopy Within a Radiative Width of Two-Quantum Transitions

As mentioned in Chapter 5, in some methods (based on the use of two-quantum transitions) the width of nonlinear optical resonances is less than a homogeneous width of the transition 2Γ. If the main contribution to the homogeneous transition width is provided by the radiative decay of levels, the resonance width may be less than the radiation (natural) width $\gamma_{rad} = \gamma_1 + \gamma_2$. It is therefore reasonable to discuss the potentialities of spectroscopy within the radiative width of a spectral line by the methods of nonlinear optical resonances. We consider in brief methods based on the use of resonantly stimulated scattering and two-quantum absorption.

When narrow resonances are observed by the method of unidirectional waves (Sec.3.2) resonances with widths equal to rates of decay of a datum and finite levels arise. If the travelling wave ω_1 at the frequency is strong, and ω_2 is a weak probe wave with the scanning frequency, the absorption coefficient of the weak wave at $\omega_2 \approx \omega_1$ has a resonance minimum with a complex structure. For example, if $\gamma_1 << \gamma_2$, the complex-resonance shape is the sum of three dips of dispersion form with half-widths $2\Gamma = \gamma_{rad}$ (with no collisions), γ_1 and γ_2, respectively, relative to the Doppler width. Thus, observation of narrow resonances by this method provides information on level lifetimes not by direct measurements of lifetimes but by the methods of nonlinear spectroscopy. The occurrence of a narrow resonance with width $\gamma_1 <<$ γ_{rad} by no means indicates the possibility of splitting spectral lines with that resolution. If the spectral line consists of overlapping lines, the resonance arises only at one frequency, which is equal to that of the strong field. The resonance-dip shape can become more complicated and, in principle, precise measurement and analysis of the shape can provide indirect information about the existence of structure within a line.

Resonances with width $\Delta\omega < \gamma_{rad}$ arise also by competition of oppositely travelling waves in a ring laser with nonlinear absorption (gain). As shown in Subsection 6.1.2, separate competitive resonances arise only provided that the distance between spectral lines is more than $(1/\sqrt{3})\gamma_{rad}$. As in the other methods of the narrowing by nonlinear resonances, here the resolution within a radiative width is not achieved, either. However, the precise measurement of the width of any nonlinear resonance (Lamb dip, dip of fluorescence, generated-power peak, competitive resonance and others) can yield indirect data on the structure within a line. In particular, the competitive-resonance shape in a ring laser may be more sensitive to a line structure and thus somewhat facilitate requirements for precision of measurements. Certainly, all of these remarks are applicable only to the simplest case of overlap of two spectral lines. If a priori the spectral-line structure is unknown, these methods are practically of no use.

The cases of spectroscopy with the other levels involved besides the levels of a saturated transition are worthy of special consideration. Examples are found in resonances obtained with light waves that have opposite circular polarizations (Sec.3.3). In this case, the waves interact with two coupled transitions. In the case of unidirectional waves, a narrow resonance width is determined by those of a datum and finite levels, and an intermediate level does not contribute to broadening. This is a common feature of nonlinear spectroscopy of three-level systems (Fig.6.3). If an intermediate-level width is $\gamma_2 >> \gamma_1, \gamma_3$ the width of the narrow resonance observed at the frequency $\omega_2 \approx \omega_{32}$ is $\gamma_1 + \gamma_3$, i.e., far less than the radiative width of the probe transition $\gamma_{rad} = \gamma_2 + \gamma_3$. Such a narrow resonance arises at the frequency $\omega_2 =$

$\omega_1 - \omega_{13}$, where ω_{13} is the frequency of the "forbidden" transition $1 \rightarrow 3$. If, for example, level 3 has a complicated structure, in which the distance between levels $\delta\omega$

$$\gamma_2 + \gamma_3 >> \delta\omega > \gamma_1 + \gamma_3 \, , \tag{6.12}$$

this structure will be resolved by scanning the frequency of a probe field ω_2.

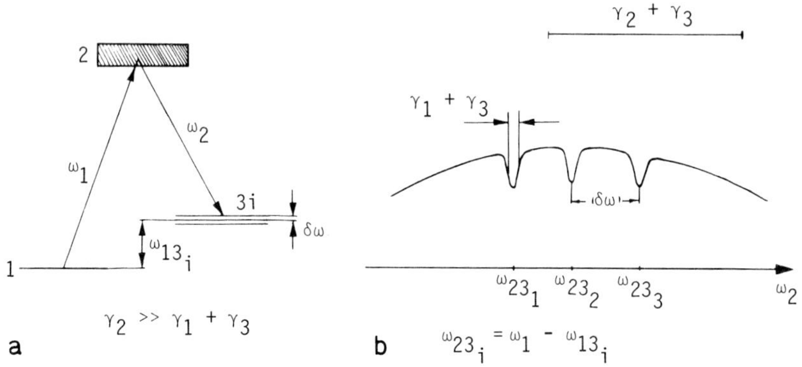

Fig.6.3a,b Potentialities of spectroscopy inside the radiation width of the transition 2 - 3 when the three-level system (a) lies in the field of two unidirectional light waves (ω_1 is a strong wave with a fixed frequency, ω_2 is a weak probe wave with a variable frequency). The absorption coefficient of the probe wave (b) has narrow resonances with a width less than the radiation width of the transition $\gamma_{rad} = \gamma_1 + \gamma_2$

6.1.4 Influence of Finite Line Width of Laser Radiation

Up to now we have considered laser radiation to be strictly monochromatic. Such an approximation is satisfactory as long as the line width of the laser radiation is far less than the resonance width.

Under real conditions, the line width of laser radiation is determined by frequency fluctuations caused by acoustical and mechanical perturbations of a cavity length. The radiated line width is usually of the order of 1 to 10 kHz in a highly stable laser when special precautions are taken to protect the cavity against perturbations. When a resonance with a width of about 1 to 100 kHz is recorded, the radiated line width is comparable with the resonance width, and its influence on the shape of the recorded resonance has to be taken into account. In the general case, the spectrum of perturbations of the cavity length that broaden the laser-radiation frequency can be rather complicated and consists of a discrete set of frequencies. The spectral density of fluctuations increases as the perturbation frequency decreases. The frequency fluctuations in a gas laser with nonlinear absorption were analyzed in [6.11]. Taking into account real noise requires the solution of a nonlinear statistical problem. For qualitative understanding and estimations we restrict ourselves to a more simple problem in which a laser is disturbed by one frequency. With this restriction, the problem amounts to finding nonlinear absorption in a gas when it interacts with a standing-wave field whose frequency is modulated by an external signal. We do not take into consideration the effects of nonlinear frequency pulling. For slow variations of frequency, the absorption coefficient depends on time and is

$$\kappa(\Omega)/\kappa_0 = 1 - \frac{G}{2}\left(1 + \frac{\Gamma^2}{\Gamma^2 + (\Omega + p\ \cos ft)^2}\right) , \tag{6.13}$$

where $\Omega = \omega - \omega_0$, p is the frequency deviation, and f is the frequency modulation. We are interested in the mean absorption coefficient during the time of observation $T \gg 1/f$; then

$$\bar{\kappa} = \frac{1}{T}\int_0^T \kappa(t)dt , \quad \frac{\bar{\kappa}}{\kappa_0} = 1 - \frac{G}{2}(1 + R) , \tag{6.14}$$

where

$$R(\delta) = \frac{\left[\sqrt{(1 + S^2 - \delta^2)^2 + 4\delta^2} + 1 + S^2 - \delta^2\right]^{1/2}}{\sqrt{2}\sqrt{(1 + S^2 - \delta^2)^2 + 4\delta^2}} , \tag{6.15}$$

$\delta = \Omega/\Gamma$, and $S = P/\Gamma$. When $S = 0$ $R(\delta) = 1/(1 + \delta^2)$, the resonance shape corresponds to an ordinary Lamb dip. The resonance amplitude is equal to

$$R(\delta = 0) = \frac{1}{\sqrt{1 + S^2}} ; \quad \text{at } S \ll 1 \quad R(\delta = 0) \simeq 1 - S^2/2 .$$

The resonance halfwidth is determined by

$$R\left(\frac{\gamma_{1/2}}{\Gamma}\right) = \frac{R(0) + R(\infty)}{2} .$$

When $S \ll 1$, $\gamma_{1/2}/\Gamma \simeq 1 + (3/4)S^2$. For arbitrary S the dependence of $\gamma_{1/2}/\Gamma$ on S is shown in Fig.6.4.

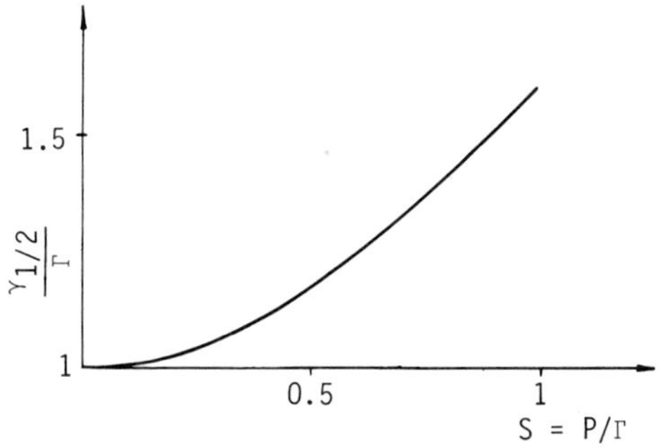

Fig.6.4 Dependence of the Lamb-dip halfwidth on the deviation amplitude of the cavity frequency

6.2 Limiting Resolution of Narrow Resonances

The most remarkable achievement of nonlinear laser spectroscopy is the production of extremely narrow resonances with relative widths less than 10^{11}. Various physical factors that influence the shape of narrow resonances will be considered in this section. These factors are: collisions of particles, flight effects, and secondary Doppler effect. We consider also some ways to obtain supernarrow resonances.

6.2.1 The Influence of Collisions

Collisions of excited atoms with surrounding particles randomize the oscillator phase, and scatter them at some angle; it increases the width of the nonequilibrium velocity distribution of atoms. The two phenomena occur simultaneously at the same collision. As compared to a Doppler contour, nonlinear resonances are sensitive to elastic scattering at small angles Θ, even in the case of the phase-randomizing collisions. Description of the influence of collisions under these conditions requires a new approach. Spectroscopic characteristics of a resonance at collisions must be connected directly with amplitudes of elastic and inelastic scattering. The latter can be introduced into the equations for the density matrix that describe the interaction of colliding particles with a field.

Inelastic collisions are those followed by the variation of an internal state of particles. In our case this means that, besides the radiative probability A, a probability P_n^{in} of quenching of a level at collisions appears

$$P_n^{in} = Nu\sigma_n^{in} \quad , \tag{6.16}$$

where σ_n^{in} is the inelastic-scattering cross-section of an atom in the state n, N is the density of scattering centers, which are considered to be stationary, for the sake of simplicity. Because the processes of emission and collision are statistically independent, for line width we have

$$\gamma = \gamma_1 + Nu\sigma_1^{in} + \gamma_2 + Nu\sigma_2^{in} \ . \tag{6.17}$$

Analysis of the influence of elastic scattering is more complicated. It will be considered strictly, later; now, we consider the physical picture of the phenomenon. In elastic collision, the scattering of an atom on an upper level occurs at potential U_2, and on a lower level at the potential U_1. The difference between U_2 and U_1 invalidates the classical treatment of the motion of an atom in a mixed state, because it is not clear what trajectory an atom moves along, either along the trajectory determined by the potential U_2 or by U_1.

The situation becomes clear in the quasi-classical treatment in which a mixed atomic state is considered to be a superposition of overlapping wave packets on the upper and lower levels, with probability amplitudes C_2 and C_1. If U_2 and U_1 are different, according to [6.12,13] the packets move along the corresponding classical trajectories with the probabilities $|C_1|^2$ and $|C_2|^2$, respectively. When the packets diverge for a distance of the order of the de Broglie particle wavelength (we can consider that the packets have such a dimension) the coherent mixed state may be considered to be disturbed, which corresponds to phase randomization of the oscillator by collision.

The probability

$$P_n^e = Nu\sigma_n^e \ , \tag{6.18}$$

of elastic scattering on the level n, where σ_n^e is the total elastic scattering cross-section of level n, has to be added to the probabilities γ_n and P_n^{in}. In this case, the line width is

$$\gamma = \gamma_1 + Nu(\sigma_1^{in} + \sigma_1^e) + \gamma_2 + Nu(\sigma_2^{in} + \sigma_2^e) \ . \tag{6.19}$$

If the potential of scattering on both levels is the same, $U_2 = U_1$, then no phase randomization occurs at collisions and, consequently, such collisions do not contribute to broadening. When the particle deviation at scattering is neglected, i.e., when Doppler shift of emission of an atom after scattering is not taken into account, $\Delta\omega_v = \underline{k}(\underline{v} - \underline{v}')$, where $\underline{v}'$ is the atomic velocity after scattering. In this case, line broadening depends on the location of the observer who records the emission.

The line broadening and shift can be connected more precisely with the characteristics that describe the process of scattering, namely with the amplitudes of scattering on both levels.

In the quasi-classical approximation for scattering at small angles, particle motions in the potential field are considered rectilinear, and the action of the potential is taken into account by introducing an extra phase on the level n (see [6.13])

$$\Delta\phi_n(t) = -\frac{i}{\hbar}\int_{-\infty}^{t} U_n dt \ . \qquad (6.20)$$

The collision leads to the oscillator phase gain

$$\Delta\phi = \phi_2 - \phi_1 = -\frac{i}{\hbar}\int_{-\infty}^{\infty} (U_2 - U_1) dt \ . \qquad (6.21)$$

Reference [6.14] provides the calculation of a Fourier component that takes into account the phase gain (6.21) at collision and averages over impact parameters, which is equivalent to the averaging over the positions of the scattering centers. As a result, for the line shift and width expressions involving the scattering amplitudes $f_1(\Theta)$ and $f_2(\Theta)$ can be obtained [6.15],

$$\gamma = 2Nu\sigma' \ , \quad \Delta = Nu\sigma'' \ , \quad \sigma' = \frac{1}{2}(\sigma_2^e + \sigma_1^e) - \mathrm{Re}\{\sigma_{21}\}$$

$$\sigma'' = \frac{2\pi}{k}[\mathrm{Re}\{f_2(0)\} - \mathrm{Re}\{f_1(0)\}] - \mathrm{Im}\{\sigma_{21}\} \qquad (6.22)$$

$$\sigma_{21} = 2\pi\int_0^{\pi} d\Theta \ \sin\Theta f_2^*(\Theta) f_1(\Theta) \ .$$

For considerably different scattering amplitudes, the cross-section σ_{21} is very small, because, in the quasi-classical approximation, $f_1(\Theta)$ and $f_2(\Theta)$ are strongly oscillating functions of the angle Θ. The line width, as must be expected, is equal to

$$\gamma = Nu(\sigma_1^e + \sigma_2^e) \ , \qquad (6.23)$$

and the shift is equal to

$$\Delta = Nu\frac{2\pi}{k}[\mathrm{Re}\{f_2(0)\} - \mathrm{Re}\{f_1(0)\}] \ . \qquad (6.24)$$

For equal scattering amplitudes $f_2(\Theta) = f_1(\Theta) = f(\Theta)$, $\sigma_2^e = \sigma_1^e = \sigma$, $\mathrm{Re}\{\sigma_{21}\} = \sigma$, $\mathrm{Im}\{\sigma_{21}\} = 0$, and so $\sigma' = 0$, $\sigma'' = 0$, i.e., there is no shift or broadening.

Consider the influence of collisions on the broadening and shift of a narrow resonance of nonlinear absorption. In the model of relaxation constants, the Lamb-dip width coincides with a collision line width [6.16]. With low-pressure gases they cannot be always identified. In order to find the Lamb-dip width it is necessary to solve a kinetic equation for the density matrix in the standing-wave field [6.17]. A gas-kinetic approach developed in [6.17] includes recession and approach terms in this equation by use of precise scattering amplitudes on levels 1 and 2. In qualitative considerations [6.18], however, the fact that the Lamb-dip width in a gas is, in essence, an inverse time of coherent interaction with the field can be used. Physical analysis indicates that the main contribution to the creation of the Lamb dip is by the atoms that have approximately zero projected velocities $v_z \approx 0$ in the direction of the light beam, i.e., move almost perpendicular to the light beam (the Doppler width is much more than the homogeneous width γ).

The influence of collisions on the Lamb-dip shape will depend considerably on the specific character of the two-level system. On collision, two qualitatively different cases may be distinguished: a) coherence between levels 1 and 2 is completely lost at collision. The Lamb-dip width is given by (6.17) in this case. b) at scattering, the coherence between levels remains (there is no phase randomization of the oscillator). In this case, the elastic scattering amplitudes on levels 1 and 2 are equal. Case b differs, in principle, from case a because the effective duration of the coherent atom-field interaction depends considerably on the relation between the characteristic angle of scattering Θ and the magnitude γ/ku. If $\bar{\Theta} \gg \gamma/ku$, then when an atom in the region of interaction undergoes collision, the halfwidth of the Lamb dip is

$$\gamma = 2Nu\sigma + \gamma_0 , \tag{6.25}$$

where γ_0 is determined by quenching collisions, time-of-flight, and other conditions. For $\bar{\Theta} \sim \gamma/ku$, the atoms scattered at angles $\Theta < \gamma/ku$ do not go out of the region of coherent interaction with the field, i.e., we cannot distinguish between scattered and nonscattered atoms (there is no phase randomization). Hence we obtain

$$\gamma \sim 2Nu2\pi \int_{\gamma/ku}^{\pi} d\Theta \, \sin\Theta\sigma(\Theta) + \gamma_0 , \tag{6.26}$$

where $\sigma(\Theta)$ is the differential-elastic-scattering cross-section.

Equation (6.26) is, in essence, the equation relative to γ and describes the nonlinear dependence of γ on pressure. With an increase of pressure γ_0 increases, so $\gamma \to \gamma_0$ and the elastic scattering does not lead to the Lamb-dip broadening.

Equation (6.26) indicates the possibility of measurement of the elastic scattering cross-section in a gas by the Lamb-dip broadening. When (6.26) is used to measure the total cross-section, the situation is similar to that arising in experiments on scattering of atomic beams. The quantity γ/ku plays the role of instrumental resolution. Provided that the error of determination of the total cross-section does not exceed 10%, the angle Θ_m, the minimum angle that can be considered as a collision, is introduced as the quantity that determines the angular instrumental resolution. To estimate Θ_m, the formula given in [6.19] can be used,

$$\Theta_m \approx \frac{277}{a(AT)^{1/2}} \frac{\pi}{180} , \tag{6.27}$$

where a is the sum of the gas-kinetic radii of colliding particles in Å (a can be found from viscosity or some other data), A is the mass of the falling atom (in atomic units), T is its equivalent Kelvin temperature. The line width required is connected with Θ_m by the relation

$$\gamma = \Theta_m ku . \tag{6.28}$$

It is interesting to compare the conclusion obtained with the models of "strong" and "weak" collisions [6.16], which were used for phenomenological description of the influence of collisions. When $ku\Theta \gg Nu\sigma$ we have the case of "strong" collisions, because all scattered atoms go out of the interaction region. The case $ku\Theta \ll Nu\sigma$ corresponds to "weak" collisions. Varying the density of atoms N passes from one model of collisions to another. When $ku\Theta = Nu\sigma$, the above models cannot be used.

The shift of the center of the Lamb dip in low-pressure gases can also be a nonlinear function of pressure. When deriving the expression for the line shift (6.22) we neglected the change of velocity at collision. When the shift of the narrow Lamb dip Δ_L is calculated, the change of trajectory cannot be neglected, because atoms scattered at angles $\Theta > \Theta_L = \gamma/ku$ do not make any contribution to the dip shift. A decrease of γ results in a decrease of Θ_L and of the Lamb-dip shift.

6.2.2 Influence of Flight Effects on the Lamb-Dip Shape

Because the total elastic scattering cross-section of molecules is 10^{-14} to 10^{-15} cm^2, in order to obtain resonances with a width of $\sim 10^3$ Hz it is necessary to use pressure of $\sim 10^{-5}$ torr.

At such low gas pressure, when the free path length Λ exceeds considerably the beam radius, the Lamb-dip width is determined by the duration of the atom-field interaction $T = a/v_r$ (a is the beam radius, and v_r is the transverse atomic velocity). The saturation parameter of an atom with the velocity v_r is

$$G(v_r) = \left(\frac{pE}{\hbar}\right)^2 T^2 = \left(\frac{pE}{\hbar}\right)^2 \frac{a^2}{v_r^2} . \tag{6.29}$$

Equation (6.29) shows atoms moving with small transverse velocities give larger relative contributions to saturation. The nonlinear absorption coefficient of these atoms in the line center is

$$\frac{\kappa(v_r)}{\kappa_0} = 1 - G(v_r) = 1 - \left(\frac{pE}{\hbar}\right)^2 \frac{a^2}{v_r^2} . \tag{6.30}$$

The equilibrium distribution of the transverse velocities of the atoms is defined

$$W(v_r) = \frac{2v_r}{u^2} \exp\left(-\frac{v_r^2}{u^2}\right) . \tag{6.31}$$

By averaging the absorption coefficient (6.30) with the distribution (6.31), we obtain

$$\frac{\langle\kappa(v_r)\rangle}{\kappa_0} = 1 - 2\left(\frac{pE}{\hbar}\right)^2 \frac{a^2}{u^2} \int_0^\infty \frac{dv_r}{v_r} e^{-v_r^2/u^2} . \tag{6.32}$$

The integral in (6.32) diverges at the lower limit. Physically, this means that consideration of collisions is necessary for very small velocities, because for velocities $v_r \leq \Gamma a$ (Γ is the collisional-line halfwidth) the saturation parameter is $G = (pE/\hbar\Gamma)^2$. The lower limit of the integral in (6.32) is determined by these velocities. Taking this into account, we have with logarithmic accuracy,

$$\frac{\langle\kappa(v_r)\rangle}{\kappa_0} = 1 - 2\left(\frac{pE}{\hbar\Gamma}\right)^2 \frac{\Gamma^2 a^2}{u^2} \ln\Gamma \frac{a}{u} . \tag{6.33}$$

In this case, the characteristic parameter is $\beta = \Gamma\tau$, which determines the relation between the collisional width $\sim\Gamma$ and the width caused by the finite time of flight of an atom through a light beam $\sim 1/\tau$, $\tau = a/u$.

A width of the Lamb dip produced by atoms moving with velocities $v_r \lesssim \Gamma a$ is of the order of Γ. According to (6.33), the contribution of these atoms is not small, which leads to narrowing of the resulting resonance, as compared with the width caused by the mean time of flight.

RAUTIAN and SHALAGIN [6.4] were the first who paid attention to the role of the slow atoms. They showed that in the case $\Gamma\tau \ll 1$ the ordinary lorentzian shape of the Lamb dip is distorted; its top is sharpened and the wings are raised a little. Here we follow [6.1], the authors of which considered the more realistic case of a field with a gaussian profile.

In order to find the absorption coefficient of the standing-wave field with a gaussian profile we shall use the equation for the atom-density matrix in a cartesian system with volume-homogeneous pumping,

$$\left(\frac{d}{dt} + \gamma_2\right)\rho_{22} = i2(\rho_{21}\overset{*}{V} - \overset{*}{\rho}_{21}V)$$

$$\left(\frac{d}{dt} + \gamma_1\right)\rho_{11} = -i2(\rho_{21}\overset{*}{V} - \overset{*}{\rho}_{21}V) + \gamma_1 N(v_z,v_r) \qquad (6.34)$$

$$\left(\frac{d}{dt} + i\omega_{21} + \Gamma\right)\rho_{21} = -2iV(\rho_{11} - \rho_{22}) \ ,$$

where $d/dt = \partial/\partial t + v_z(\partial/\partial z) + v_x(\partial/\partial x) + v_y(\partial/\partial y)$, v_x, v_y, v_z are the atom velocity projections onto the coordinate axes, $N(v_z,v_r) = (N/\sqrt{\pi}u)\exp(-v_z^2/u^2)\cdot(2v_r/u^2)\exp(-v_r^2/u^2)$ is the velocity distribution of atoms on the lower level in the absence of field, $v_r = \sqrt{v_x^2+v_y^2}$ is the transverse atom velocity, u is thermal velocity, γ_2,γ_1 are the inverse lifetimes of the upper and lower levels, Γ is the homogeneous-line halfwidth, $V = -pE(x,y,z,t)/2\hbar$, and p is the dipole matrix element of the transition $2\to1$. The absorption coefficient of the standing-wave field

$$E(x,y,z,t) = E\cos kz\, e^{-\frac{(x^2+y^2)}{a^2}}\cos\omega t \qquad (6.35)$$

in the Lamb approximation is found in a usual way,

$$\kappa(\Omega) = \kappa_0\left[1 - \frac{G}{2}\beta^2\int_0^\infty\int_0^\infty d\xi d\eta\,\frac{e^{-2\xi-\eta}}{\beta^2+\xi^2+(\xi+\eta)^2}\left(1+\cos\frac{2\Omega}{\Gamma}\xi\right)\right] \ , \qquad (6.36)$$

where κ_0 is the nonsaturated absorption coefficient, $G = (pE)^2/\hbar^2\gamma\Gamma$ is the saturation parameter, $1/\gamma = (1/2)(\gamma_1^{-1} + \gamma_2^{-1})$ (we assume that $\gamma_1 = \gamma_2 = \Gamma$), $\Omega = \omega - \omega_{21}$, ω is the field frequency, and ω_{21} is the frequency of the transition $2 \to 1$.

When $\beta >> 1$, (6.36) takes the usual dispersion form with a halfwidth $\gamma_{1/2} \approx \Gamma(1 + 2.5/\beta^2)$. From (6.36), we obtain the equation for the resonance halfwidth, $\kappa(\gamma_{1/2}) = [\kappa(0) + \kappa(\infty)]/2 \cdot$

$$\int_0^\infty\int_0^\infty d\xi d\eta \frac{e^{-2\xi-\eta}}{\beta^2 + \xi^2 + (\xi + \eta)^2}\left(\cos \frac{2\gamma_{1/2}}{\Gamma}\xi - \frac{1}{2}\right) = 0 . \tag{6.37}$$

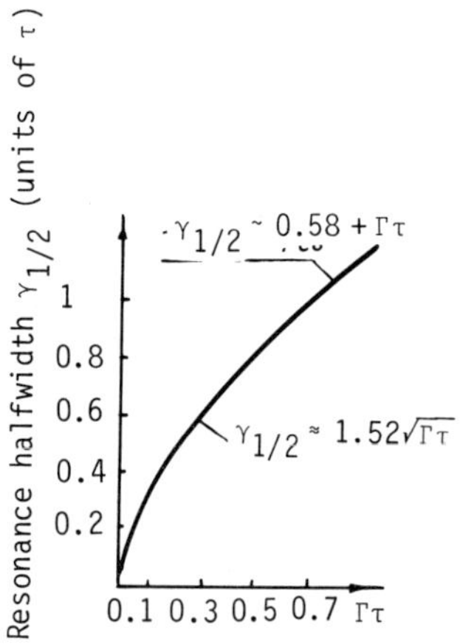

Fig.6.5 Dependence of the Lamb-dip halfwidth (in units of $1/\tau$) on parameter

The root $(\gamma_{1/2}\tau)$ of this equation has been found numerically and is shown in Fig.6.5 as a function of β. Figure 6.5 shows three characteristic regions of the variation of the halfwidth as a function of β. When $\beta < 0.5$, $\gamma_{1/2}\tau$ is an exponential function of β; it is the region of influence of slow atoms [6.1]. When $0.5 \leq \beta \leq 2$, $\gamma_{1/2} \simeq \Gamma + 0.58/\tau$, where $0.58/\tau$ can be interpreted as the flight resonance halfwidth. In this region of variation of the parameter β the collisional and flight widths are added. When $\beta >> 1$, $\gamma_{1/2} \to \Gamma$.

In the region $\beta \to 0$ an approximate expression for (6.37) can be written,

$$\ln \frac{1}{\beta} - \alpha = \frac{4}{\pi}\int_1^\infty \frac{dz}{z^2 + 1} \cdot \ln\left(1 + \frac{4(\gamma_{1/2}/\Gamma)^2}{(z+1)^2}\right) , \tag{6.38}$$

where

$$\alpha = \frac{4}{\pi}\int_1^\infty \frac{dz}{z^2 + 1} \ln\left(e^C \frac{z+1}{\sqrt{z^2+1}}\right) ,$$

c is the Eiler constant. For sufficiently small β, the solution of (6.38) should be sought in the region $\gamma_{1/2}/\Gamma \gg 1$. In that case, instead of (6.38), the approximation can be written

$$\ln \frac{1}{\beta S} = \ln \frac{4\gamma_{1/2}^2}{\Gamma^2} + \frac{2\Gamma}{\gamma_{1/2}} ,$$

hence

$$\gamma_{1/2}\tau \approx \frac{\sqrt{\beta}}{2\sqrt{S}} (1 - 2\sqrt{S}\sqrt{\beta}) , \tag{6.39}$$

where $2\sqrt{S} = 0.66$. A correction term makes it possible to estimate the region of approximation over β. The correction is less than 10% in the region $\beta \lesssim 0.02$. Actually, $1.52\sqrt{\beta}$ describes rather well the whole region of action of slow atoms, i.e., up to $\beta \lesssim 0.5$.

For applications connected with frequency stabilization and for spectroscopic investigations of the shape of a nonlinear power resonance it is interesting to find the halfwidth (6.36) from the maximum of the derivative of frequency ("dimension" of the derivative κ). The derivative maximum (6.36) is determined by

$$\int_0^\infty \int_0^\infty d\xi d\eta \frac{e^{-2\xi-\eta}}{\beta^2 + \xi^2 + (\xi+\eta)^2} \xi^2 \cos \frac{2\tilde{\gamma}}{\Gamma} \xi = 0 . \tag{6.40}$$

When $\beta \gg 1$, $\tilde{\gamma} \approx (\Gamma/\sqrt{3})(1 + 4/\beta^2)$, i.e., the collisional resonance halfwidth Γ and the derivative maximum halfwidth $\tilde{\gamma}$ in this region are connected by a simple relation. This circumstance is used in experiments for measuring resonance halfwidth. When $\beta \ll 1$ (6.40) transforms into

$$\int_1^\infty \frac{dz}{(z^2+1)[4(\frac{\tilde{\gamma}}{\Gamma})^2 + (z+1)^2]} = 8(\frac{\tilde{\gamma}}{\Gamma})^2 \int_1^\infty \frac{dz}{(z^2+1)[4(\frac{\tilde{\gamma}}{\Gamma})^2 + (z+1)^2]^2} , \tag{6.41}$$

from which it is readily seen that $\tilde{\gamma} \sim \Gamma$. This indicates the sharpening of the resonance top noted in [6.1]. Precise calculation yields $\tilde{\gamma} = (2.485/\sqrt{3})\Gamma$.

6.2.3 Methods of Reduction of the Average Flight Width

The obtaining of supernarrow resonances is based on methods that produce increase of the time of particle-field interaction. Two-photon resonances permit use of extended beams with long time of interaction with the field. The time of coherent interaction with the field can be increased by using a nonlinear-optical Ramsey resonance (see Chap.3). The two methods should be considered to be the most promising ones to obtain supernarrow resonances with relative widths of 10^{-12} to 10^{-13}. However, these methods have not been yet explored experimentally. We will review the methods that have been tried experimentally. First, are applications of telescopic expanders of a laser beam and of the selection effect of atoms with low velocities.

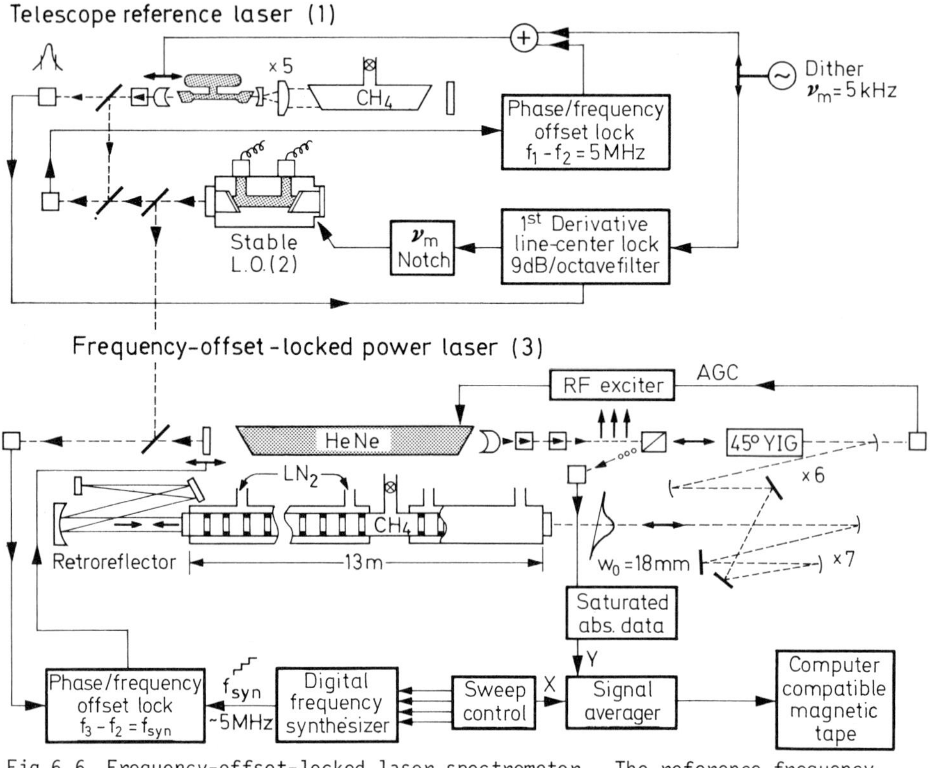

Fig.6.6 Frequency-offset-locked laser spectrometer. The reference frequency output is taken from the stable, offset, LO lasers. The wavefront errors of the beam in the absorption cell are estimated to be ≲0.1 waves (of 3.39 μm radiation). The retroreflector ensures parallelism of incident and returned beams

Telescopic Expanders of Beams

Telescopic expanders increase the time of particle-field interaction, decrease diffraction divergence of beam and increase resonance intensities.

The authors of [6.2,20,21] used an external absorption cell with a telescopic expander of the beam in order to obtain narrow resonances in methane at $\lambda = 3.39\ \mu m$. The scheme of the experimental arrangement [6.21] (frequency-offset-locked laser spectrometer) is shown in Fig.6.6. Three He-Ne lasers were used. A He-Ne laser with an external methane cell whose frequency was stabilized over a methane absorption line was used as a reference laser. The reference laser yielded high long-term frequency stability of the order of 3 Hz (in Allan variance) for averaging times of 10^2 to 10^4 s. The observed frequency drift was about 30 Hz/min; perhaps it was associated with the influence of a magnetic hyperfine structure in CH_4. A short stable laser was used in order to eliminate the influence of one laser on the frequency of the other and the detuning from the region of zero beat. This laser was rigidly constructed and insulated from mechanical and acoustical disturbances (vibrations); this made it possible to obtain high short-term frequency stability, which was transmitted to other lasers by electronic feedback systems. The spectrum width of the laser radiation was about 300 Hz.

The radiated frequencies of the reference and short stable lasers were offset-locked by detuning 5 MHz by use of frequency-phase feedback systems that permitted optimal use of the advantages of the two lasers. The short- and long-term frequency stability achieved was further transmitted to a high-power He-Ne laser whose radiation interacted with the methane absorption cell under investigation. The frequency reproducibility achieved in the laser system described was ±300 Hz. This value restricted the accuracy of finding of the position of the center of an absorption line.

An external cell 13 m long and 5 cm diameter was used in the experiments described in [6.20]. However, because the cell was not an ideally straight cell, the useful laser beam radius was limited to 18 mm. As can be seen from Fig.6.6, the system of cascade reflecting telescopes was used in order to expand the output beam of the high-power He-Ne laser to a transverse dimension of 36 mm. The cell operated under a methane pressure of 30 to 100 μtorr and temperature of 77 K or 300 K. The resolution of 6 kHz achieved in this system made it possible to resolve the magnetic hyperfine structure in methane ($F_2^{(2)}$ component of line P(7) of band ν_3). The results of measurement of the hyperfine resolution in methane will be presented in detail in Section 8.1. Note that the resonance intensity was very low. The time of measurement was 8 hours. Recently, the authors of [6.2] have obtained narrow resonances with widths of 1.5 kHz, in methane under a pressure of 20 μtorr by using a telescopic cell 30 cm in diameter (the laser-beam diameter was about 10 cm) and 13 m long. The scheme of the experimental arrangement was similar to that described previously. With the resolution achieved (of the order of 1.5 kHz) a duplet in methane could be partly resolved in the pressure range 20 to 75

The other method of using a telescopic system is in lasers with telescopic cavities. The authors of [6.3,22] obtained narrow resonances in methane with a width of 3 kHz by use of a four-mirror telescopic laser (TL). Use of a telescopic expander inside the cavity provides a high field density in the absorption cell, obtains high resonance contrast and permits a resonance to be used in the laser-radiation spectrum.

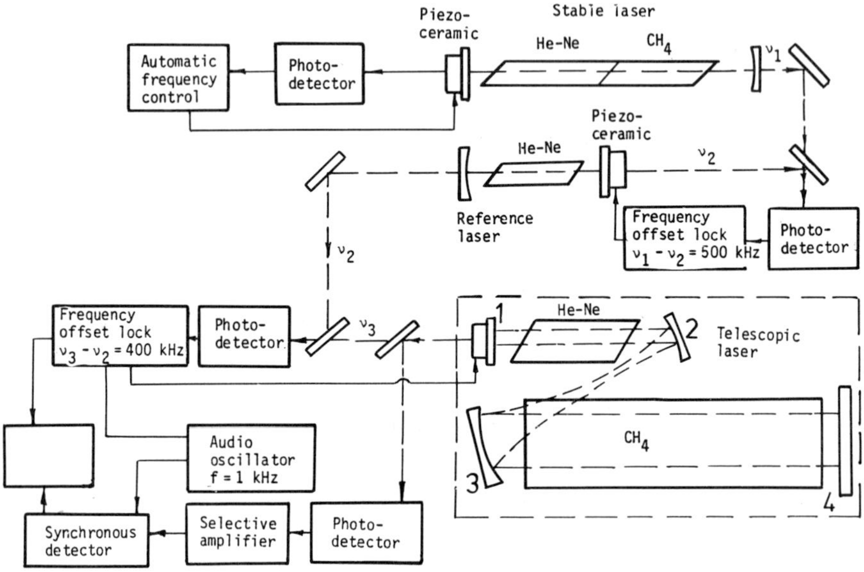

Fig.6.7 Experimental arrangement of the laser with a telescopic cavity

Figure 6.7 shows a laser whose cavity is formed by four mirrors: a plane or long-focus mirror (1), a telescopic beam expander (short- (2) and long-focus (3) mirrors) and a plane mirror (4). The angle of reflection of light waves from mirrors 2 and 3 was made small (of the order of several degrees) in order to reduce the influence of astigmatism on the laser-beam shape. The distance between the spherical mirrors 2 and 3 is approximately equal to the sum of focal distances of these mirrors. An amplifying tube is located in the narrow part of the light beam, an absorption cell in the wide part.

In the experiments reported in [6.22] with a four-mirror cavity, the parameters of the cavity were such that the transverse dimension of the laser beam in the methane absorption cell was 10 cm. The length of the absorption cell was 500 cm. The length of the amplifying part of the cavity was ≈2 m. The absorption cell was made of stainless steel and had a diameter of more than 20 cm. The methane pressure in the cell ranged from 10 to 100 μtorr.

Recording of narrow resonances with widths of the order of 1 kHz requires that the width of the laser-radiation spectrum be less than the resonance width. A number of arrangements of passive frequency stabilization were therefore used in order to reduce the width of the radiated spectrum of the laser. The construction was rigid, the whole optical path of the laser beam in the cavity outside the amplifying and absorbing tubes was carefully isolated from air movement, all of the main components of the installation were covered with sound-absorbent materials. The installation was placed on a 12 t metal plate on special multilayer shock absorbers. Besides the measures described for passive frequency stabilization of the telescopic laser, active frequency stabilization was used, by offset-locking the frequency with that of a stabilized reference laser.

A linear He-Ne laser with an internal methane absorption cell was used as a reference laser. A frequency of the reference laser was stabilized over the resonance maximum in methane by use of an electron system of automatic frequency control. The parameters of resonance obtained in methane (width 50 kHz, contrast 50% and amplitude 1 mw) made it possible to achieve long-term frequency stability of the reference laser of the order of 10^{-12} to 10^{-13} for an averaging time of 1 s [6.23]. The width of the laser-radiation spectrum was 1 kHz. In order to eliminate the interference of frequencies of the reference and telescopic lasers, the scheme of measurement used a heterodyne laser whose radiated frequency was offset locked with the frequency of a stable laser by use of frequency-offset lock (FOL) with a shift of 500 kHz; the frequency of the telescopic laser was offset locked with a shift of -500 kHz. The value of the shift could be adjusted relative to 500 kHz within large limits; in this way, the frequency of the telescopic laser could be tuned near the center of the methane absorption line.

The resonance in methane was registered by means of recording the dispersion curve of a power peak of the telescopic laser oscillation on an XY-recorder while tuning the oscillation frequency near the center of the absorption line. A signal with the frequency of 800 Hz was fed to a frequency discriminator of the FOL system. This signal caused the oscillation-frequency deviation of the telescopic laser to be 1 kHz. The recording lasted about 10 min. and had a speed of 100 Hz/s with an averaging time of 1 s.

The frequency resolution of 3 kHz obtained made it possible to register the hyperfine structure in methane on the transition $F_2^{(2)}$ of the line P(7) of the band ν_3. A characteristic record of the dispersion curve of resonance in methane at a pressure of 45 mtorr is shown in Fig.6.8. The nonlinear resonance in methane is seen to be formed by three strong components of the hyperfine structure. The magnitude of the field in the cavity, at which

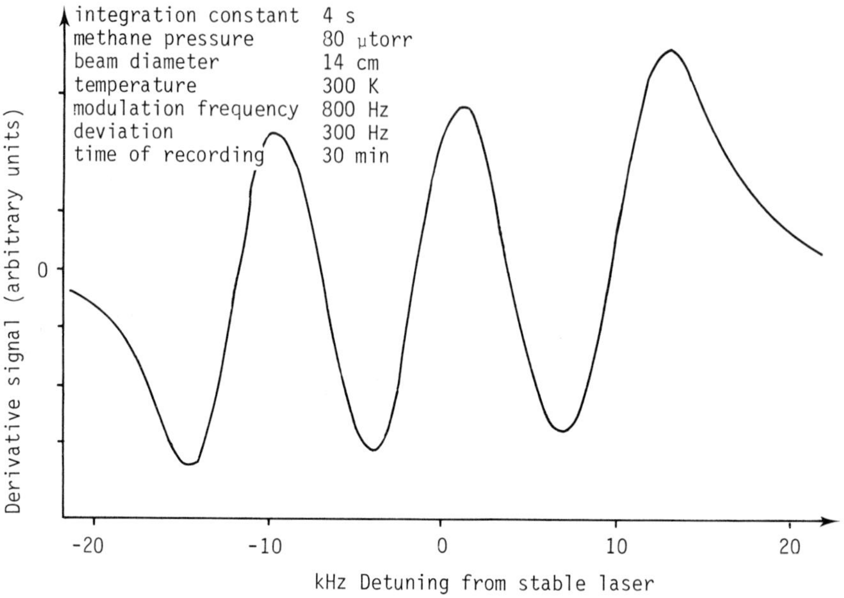

Fig.6.8 Record of the resonance in methane at $\lambda = 3.39$ μm by use of a laser with a telescopic cavity

parameters of the individual hyperfine-resonance components were measured, corresponded to a saturation parameter less than 0.1. The contrast of the resonance observed in the telescopic laser amounted to 1% and the amplitude was more than 10 μW which considerably exceeded the resonance parameters in an external cell.

Processing of a great number of records of the resonance dispersion curve, by use of a computer, made it possible to determine the main parameters of the hyperfine components of the structure. The resonance shape of each individual component of the MHFS was close to a Lorentz shape. The frequency difference between the central and two side components was $\Delta_1 = 11.5 \pm 0.5$ kHz and $\Delta_2 = -11.1 \pm 0.5$ kHz. The ratio of the intensity of the main resonance to that of a crossing resonance is 20 to 1.

The distortion of the shape of the recorded dispersion curve caused by the recoil duplet was observed on each component with methane pressures of the order of 30 μtorr and less. Analysis of experimental data by use of an electronic computer, with the least squares method, yielded a value of line splitting due to the recoil effect $\Delta = 2 \pm 0.5$ kHz that corresponded to the calculated value ($\Delta = 2.16$ kHz).

Effect of Slow Atoms

Another simple method, which increases the duration of interaction of the coherent field atoms, involves selection of atoms according to their velocities. As was shown in the previous paragraph, in the time-of-flight range $\Gamma\tau << 1$ the contribution of slow atoms to the formation of the Lamb dip is more important than the contributions of atoms that have average thermal velocities. An effect of narrowing the Lamb dip is appreciable when $\Gamma\tau < 0.7$ (see Fig.6.5). Experimental study of the Lamb-dip narrowing caused by slow atoms was reported in [6.5,24] in a 3.39 μm He-Ne/CH_4 laser. The scheme for observing the resonance was similar to that shown in Fig.6.7, with the difference that use of an ordinary laser with a nonlinear absorber was explored instead of one with a telescopic cavity. The cavity of the former was formed by mirrors that had radii of curvature $R_1 \simeq 10$ m and $R_2 \simeq \infty$. The cavity was 5 mm long and the absorption cell was 3 mm long. The light beam had a 4 mm

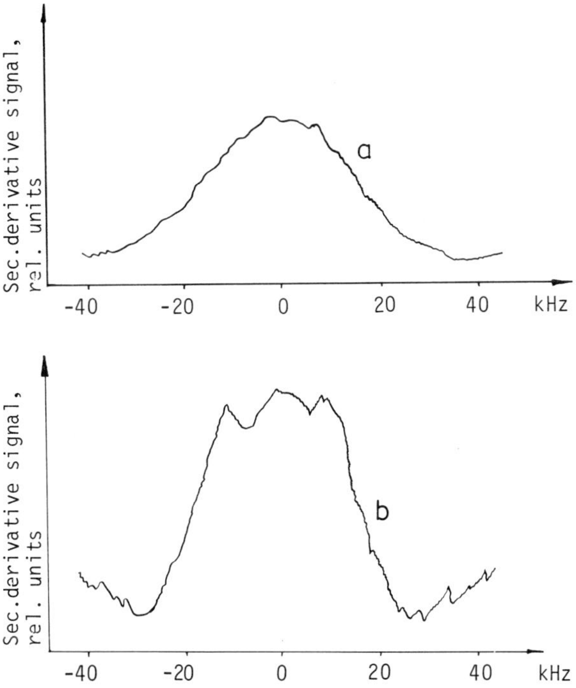

Fig.6.9a,b Shape of the Lamb dip at $\lambda = 3.39$ μm in CH_4 (registration on the second harmonic of the cavity modulation frequency) within the flight region at methane pressures, a) 360 μtorr; b) 60 μtorr

diameter waist. The resonance shape was recorded in the second harmonic of the cavity-modulation frequency. Figure 6.9 shows the results of experiments. At methane pressure of 360 μtorr, the resonance shape was similar to the Lorentz shape and the magnetic structure of the methane line F^2 was not resolved. At this pressure, the collision-resonance halfwidth was about 5 kHz. The parameter $\Gamma\tau$ was $\Gamma\tau \simeq 0.2$. In accordance with this, the dip halfwidth is $0.6(1/\tau_0) \cdot 1/2\pi = 25 \cdot 10^3$ Hz, i.e., it is caused by atoms with average thermal velocities. When the pressure was decreased to 60 μtorr, the parameter $\Gamma\tau$ became about 0.03, the Lamb-dip width was about 7 kHz, and the hyperfine structure of the methane line was resolved. Thus, the resonance narrowing caused by slow atoms makes it possible to obtain high resolution without using telescopic cavities.

6.2.4 Potentiality of Nonlinear Spectroscopy with Resolution that is not Restricted by the Finite Flight Time

The resolution of all methods of nonlinear spectroscopy is restricted by the finite duration of field-particle interaction, i.e., by the time of flight of a particle through the region of the field. With lateral dimensions of the light field from 1 to 10 cm the flight width of different particles ranges from 10^3 to 10^5 Hz. At the same time, the natural width of forbidden transitions of atoms and molecules in the visible band and of resolved rotational-vibrational transitions of molecules in the ir band is much less than these values. Therefore of great importance is a search for new methods of nonlinear spectroscopy whose resolution is not limited by the finite time of flight of a particle through the region of the light field. Such a method was proposed by LETOKHOV [6.25]. The method is based on trapping of particles with slow velocities in a standing light wave that is not in resonance with any transition of the particles and plays the role of only a space-periodic potential field. Trapping of the particles eliminates Doppler broadening.

Electrons in an atom or a molecule in a high-frequency electromagnetic field are acted upon by a force proportional to the averaged square of the electric field [6.26,27]. It is determined by the striction force [6.28] that acts upon the atom or a molecule

$$f = \frac{1}{2}\alpha_\omega \mathrm{grad}(\overline{E^2}) \ , \tag{6.42}$$

where α_ω is polarizability of the particle at the external-field frequency ω determined by the gas refractive index n_ω

$$\alpha_\omega = \frac{n_\omega - 1}{2\pi N} , \tag{6.43}$$

where N is the density of the gas particles.

In the case of the plane standing light wave

$$E(z,t) = \mathcal{E} \sin\omega t \sin kz , \tag{6.44}$$

the striction force is a periodic function

$$f = \frac{1}{4} \alpha_\omega \mathcal{E}^2 k \sin 2kz . \tag{6.45}$$

Moving particles are pulled into the standing-wave loop (at $\alpha_\omega > 0$) or pushed out of it (at $\alpha_\omega < 0$). The change of coordinate of the center of gravity of the particle is described by

$$\ddot{z} = \frac{k\alpha_\omega}{4M} \mathcal{E}^2 \sin 2kz , \tag{6.46}$$

or

$$\ddot{\xi} = \Omega^2 \sin\xi , \quad \Omega^2 = \frac{k^2 \alpha_\omega \mathcal{E}^2}{2M} , \tag{6.47}$$

where $\xi = 2kz$ is a dimensionless coordinate, and M is the mass of the particle. Equations (6.46) and (6.47) are the well-known equation of pendulum vibrations, the solution of which, under initial conditions

$$\xi|_{t=0} = \xi_0 = 2kz_0 ; \quad \dot{\xi}|_{t=0} = \xi_0' = 2kv_0z , \tag{6.48}$$

is

$$\dot{\xi} = \pm[\Omega^2(\cos\xi_0 - \cos\xi) + (\xi_0')^2]^{1/2} , \tag{6.49}$$

or

$$\dot{z} = \pm \frac{\Omega}{k} [(\cos 2kz_0 - \cos 2kz) + (\frac{k}{\Omega} v_{0z})^2]^{1/2} . \tag{6.50}$$

Let $\alpha_\omega > 0$. Consider particles that are first in minima of the potential field supplied by the standing light wave

$$|2kz_0 - \pi n| \ll 1 , \quad n \text{ is an integer} . \tag{6.51}$$

From (6.50) it follows that if such particles have an initial velocity in the direction of the OZ axis which is less than the critical velocity determined by

$$v_{cr} = \frac{\Omega}{k} = E\left(\frac{|\alpha_\omega|}{2M}\right)^{1/2} , \tag{6.52}$$

then instead of free motion they undergo vibrations in the direction of the OZ axis with frequency Ω and amplitude that does not exceed $\lambda/4$. If $|v_{oz}| \leq v_{cr}$, the particles undergo harmonic vibrations with frequency Ω. As v_{oz} approaches v_{cr}, the particle motion becomes essentially nonsinusoidal. If the initial position of the particle is far from the minima, (6.52), some of the particles with the velocity $|v_{oz}| < v_{cr}$ undergo vibrations of limited amplitude, and others are capable of unlimited motion. For $|v_{oz}| > v_{cr}$ the particles are free to cross the standing light wave, but z, a component of their velocities, is periodically modulated.

This effect of particle trapping between loops (at $\alpha_\omega < 0$) and nodes (at $\alpha_\omega > 0$) of the standing wave occurs, evidently, for particles whose energy of motion in the direction of the OZ axis is less than the height of the potential barrier produced by the light wave (Fig.6.10). It is plausible that the probability of particle trapping at $|v_{oz}| < v_{cr}$ depends on the initial position z_0, i.e., on the initial height with respect to the potential barrier.

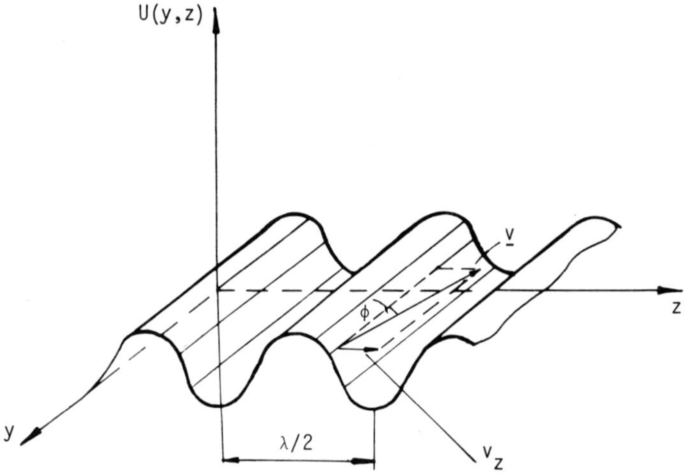

Fig.6.10 Potential field U(y,z) of the standing light wave E sinkz · sinωt

It should be noted that the value v_{cr} even at field intensities of several kW/cm^2 is several orders lower than the mean thermal velocity of the particles $u=(2\ kT/M)^{1/2}$. Such trapping is, therefore, possible only for the particles with very low absolute speed or for particles with velocity of the order of the mean thermal velocity that move at very narrow angles ϕ to a wave surface (Fig.6.10),

$$|\phi| < \phi_{cr} = \frac{v_{cr}}{u} = E\left(\frac{|\alpha_\omega|}{4kT}\right)^{1/2} . \tag{6.53}$$

Thus, the standing light wave changes considerably the velocity distribution of $W(v_z)$ in the neighborhood

$$-v_{cr} < v_z < v_{cr} . \tag{6.54}$$

The motion of the center of gravity of the particles with velocity that satisfies (6.54) becomes oscillatory instead of translatory. This must change considerably the Doppler-broadened line shapes of emission and absorption of particles that are observed along the OZ axis. The change of line shape in the center of Doppler-broadened lines, i.e., for particles with velocities that satisfy (6.54), is the most radical. The Doppler shift for trapped particles must vanish. It resembles the narrowing of lines in the microwave range by diffusion motion of particles in a buffer gas [6.29] or by wandering of particles in a storage bulb [6.30].

It is easy to consider a small neighborhood near the center of the Doppler-broadened line of some transition, i.e., particles with $|v_{oz}| << v_{cr}$ whose motion in the standing light wave is sinusoidal. The analysis of such a small neighborhood near the Doppler-line peak has physical significance for transitions that have very small natural (radiative) width Γ and for a light beam of very large diameter, so that the time of flight of the beam τ_{tr} is sufficiently great

$$\Gamma << k_o v_{cr} , \tag{6.55}$$

$$\frac{\pi}{\tau_{tr}} << k_o v_{cr} , \tag{6.56}$$

where $k_o = \omega_o/c$, ω_o is the frequency of the transition under consideration. Under these conditions a narrower peak with width $\Delta\omega_{res}$ determined by radiative broadening (at $\Gamma >> 1/\tau_{tr}$) or by the time of passage of a particle across the beam (at $\Gamma << 1/\tau_{tr}$) is formed in the center of the Doppler line, in a

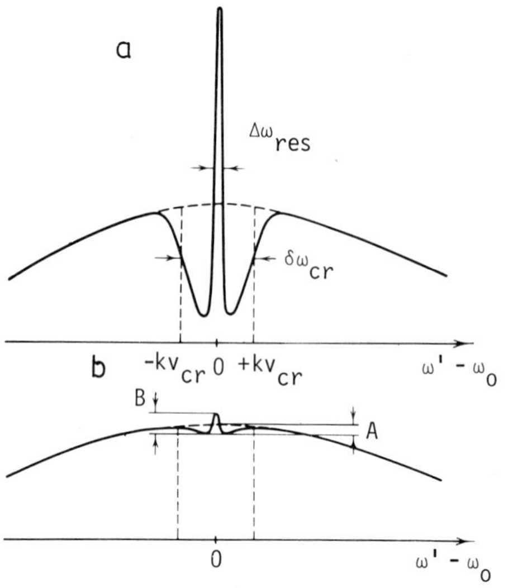

Fig.6.11a,b Shape of the Doppler contour with one-dimensional trapping of most particles (a) and with three-dimensional (b) trapping of a small fraction of the particles

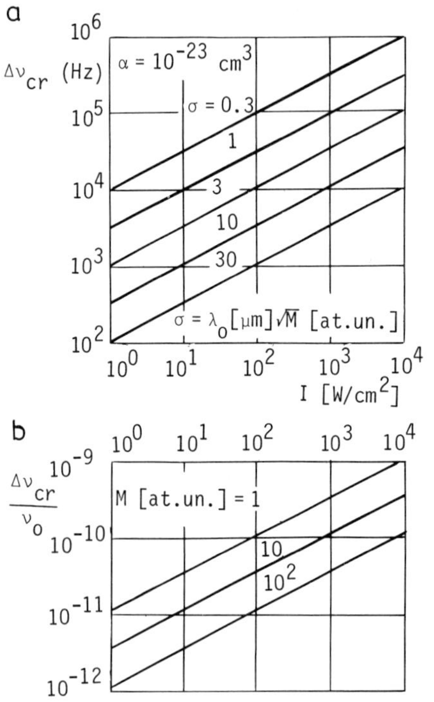

Fig.6.12a,b Dependence of the absolute (a) and relative (b) widths of the spectral interval $\delta\nu_{cr} = \delta\omega_{cr}/2\pi = k_0 v_{cr}/\pi$, in which the effect of particle trapping is significant, ($\alpha_\omega = 10^{-23}$ cm^3), on the standing-wave intensity and on the magnitude of the parameter $\sigma = \lambda_0[\mu m]\sqrt{M}$ [atomic units]

narrow dip, with width of the order of k_0v_{cr}. Of course collisions can also restrict the line narrowing if the pressure is low enough. The qualitative picture of the Doppler-contour shape for this case is given in Fig.6.11a. Note that the peak height may be rather high because the area of the peak must be equal to that of the dip. The background in the neighborhood of the peak is determined, first, by side components of frequency modulation of the spectral line produced by vibrational motion of molecules (increases with the ratio of the vibration amplitude to the wavelength λ_0 of the transition under observation) and, second, by nonsinusoidality of vibrations (for particles with velocity v_{oz} comparable with v_{cr}).

The width of the spectral region $\delta\omega_{cr} = 2k_0v_{cr}$ near the center of the Doppler line in which the trapping effect is essential can be estimated. Figure 6.12 shows the values of the absolute and relative widths of this spectral interval for the case when the strong standing-wave frequency ω lies outside resonances and absorption bands of molecules, and polarizability is $\alpha_\omega \approx 10^{-23}$ cm^2 (this corresponds to the refractive index $n_\omega - 1 = 2 \cdot 10^{-3}$ atm^{-1}). The mass of particles M in Fig.6.12 is expressed in atomic units. Even at the wave intensity of 10^4 W/cm^2 this region for the lightest particle M = 1 at. unit does not exceed 10^{-9}, i.e., 10^{-4} times the Doppler width.

In spite of the fact that the value $\delta\omega_{cr}$ is small, (6.55) can be satisfied by using transitions of molecules of small radiative width that are of most interest for spectroscopy with no Doppler broadening. It is much more difficult to fulfill (6.56), which imposes a severe restriction on the broadening $\Delta\omega_{tr} = 2\pi/\tau_{tr}$ due to the finite duration of field-particle interaction. The ratio of the width $\Delta\omega_{tr}$ to $\delta\omega_{cr}$ is

$$\mu = \frac{\Delta\omega_{tr}}{\delta\omega_{cr}} = \frac{\pi}{ak_0\phi_{cr}} = \frac{\lambda_0}{8}\left(\frac{2ckT}{\alpha_\omega P}\right)^{1/2} , \qquad (6.57)$$

where P = SI is the wave power, $S = (\pi/4)a^2$ is the beam cross-section, and I is the wave power density. In order to narrow a spectral line it is necessary that $\mu \ll 1$. At $\alpha(\omega) = 10^{-23}$ cm^2, T = 300 K, the standing-light-wave power necessary to obtain $\mu = 1$ amounts to P = 3.7 kW, to observe the transition at $\lambda = 1$ µm; it increases proportionally to λ_0^2. Thus, flight broadening makes difficult the observation of the narrowing of a Doppler line at the expense of the effect of particle trapping in the standing light wave. Nevertheless, this difficulty may be overcome when particles are trapped in a three-dimensional space, but not by one-dimensional trapping along one Z coordinate. In principle, it is possible when a three-dimensional standing wave, formed by intersection of three perpendicular standing waves, is used. In this case, particles with absolute velocities that satisfy the condition

$$|V| < v_{cr} \tag{6.58}$$

execute a finite motion in a volume of the order of $(\lambda/2)^3$. This must lead to the formation of a narrow resonance in the Doppler contour when the radiation is observed in any direction, if (6.55) is fulfilled.

Note that the intensity of the standing light field necessary to trap even very slow particles must be sufficiently great (about 10^3 W/cm^2). So we raise the question about possible broadening and shift of narrow spectral lines due to the Stark effect in the strong light field. The energy of particle interaction when it moves in the standing light wave varies from 0 to $1/2\ \alpha_\omega E^2$. This causes a Stark shift of levels of the observed supernarrow transition by about $\Delta E_{St} \simeq \hbar\delta\omega_{cr}$ and might be thought to eliminate the narrowing effect, completely. In fact, it does not do so when supernarrow lines of transitions whose initial and final levels have the same or almost the same value of Stark shift in a nonresonant light field. For example, this is characteristic of vibrational-rotational transitions of molecules within the limits of an electron state.

The main problem is to have sufficient sensitivity for resonance recording, because an extremely small portion N_{tr} of particles is trapped in the formation of a supernarrow resonance with a width 2Γ

$$N_{tr} \simeq N_0 \left(\frac{v_{cr}}{u}\right)^3 . \tag{6.59}$$

In the three-dimensional case, a relative dip depth A and the supernarrow peak height in the center of its B (see Fig.6.11b) are determined by the approximate relations,

$$A \approx \left(\frac{v_{cr}}{u}\right)^2 \qquad B \approx \left(\frac{kv_{cr}}{\Gamma}\right) A . \tag{6.60}$$

At low pressure of the gas under investigation (about 10^{-4} to 10^{-3} torr) which is necessary for the absence of collisions, the value of N_{tr} amounts to only several tens of particles per cm^3. This is many orders less than the maximum density of trapped particles achieved in the case when there is one particle in each elementary trapping volume of $(\lambda/2)^3$. This represents the evident fact that the portion of the particles whose velocities meet the condition of trapping (6.57) is extremely small in a gas at a thermal equilibrium. The density of trapped particles can be increased when the slow particles are classified and gradually stored in the volume occupied by the three-dimensional light field, up to the density

$$N_{max} \approx \left(\frac{2}{\lambda}\right)^3 , \qquad (6.61)$$

at which there is no interaction of the particles with each other. In principle, the density of trapped particles N_{max} can reach values of 10^9 to 10^{12} cm^{-3}, at which the supernarrow peak amplitude will be sufficient for recording.

In spite of considerable experimental difficulties of practical realization of the proposed method, it is worthy of serious attention because of its universal nonresonant character. When the strong light field is available at only one frequency, outside the resonant transitions of a particle, narrow resonances can be obtained in the center of any Doppler transition of a particle for which the homogeneous width is $2\Gamma << \delta\omega_{cr}$. A detailed discussion of this method is given in [6.38].

In order to obtain and to store very slow particles, an effect of laser-radiation cooling of atoms and molecules can be used. When atoms are illuminated by isotropic monochromatic radiation whose frequency ω_L lies on a long-wave wing of a Doppler-broadened line, then at each reradiation of an absorbed photon the atomic velocity decreases by $\Delta v = \hbar\omega/Mc$, owing to the recoil effect. After $v_0/\Delta v$ events of reradiation, i.e., after the time $t_{cool} = 2A_{21}^{-1}(v_0/c)(Mc^2/\hbar\omega)$ the atomic velocity should decrease to a value $v_{min} \simeq v_0 \cdot \Gamma/ku \simeq v_0(A_{21}/ku)$. This method of radiation cooling of particles was proposed recently in [6.31]. A more detailed calculation carried out in [6.32] showed that a further cooling of particles, down to a minimum velocity $v_{min} \simeq (\Gamma/k) \cdot (R_{rec}/\hbar\Gamma)^{1/2}$, $R_{rec} = (\hbar\omega)^2/Mc^2$ is recoil energy that is determined by a velocity jump at reradiation of one photon is possible in a standing wave. For such strong cooling, the frequency of the three-dimensional standing light wave must be tuned to a frequency $\omega_c = \omega_0 - \Gamma$ slightly shifted to the red with respect to the center of ω_0 of an allowed radiative transition. After such strong cooling of particles, it is necessary to apply an additional nonresonant field, in the form of the three-dimensional standing wave that must trap the slow particles. The intensity of this nonresonant field must satisfy the condition,

$$v_{cr} < v_{min} = \frac{\hbar\omega_L}{Mc} . \qquad (6.62)$$

It goes without saying that a strong field must also keep the particles in the region of trapping, in spite of the recoil effect in a probe light field of the frequency ω used for measuring the supernarrow absorption lines of the trapped particles. This requires fulfillment of the condition

$$\frac{\hbar\omega}{Mc} < v_{cr} \ . \qquad (6.63)$$

It is clear that the probe-field frequency ω must be less than the cooling field frequency ω_c. For the radiative cooling, allowed electronic transitions of atoms and molecules can be used and supernarrow absorption lines on forbidden transitions in the visible and ir bands can be observed.

Reference [6.32] showed that the required field intensities can be considerably decreased and the velocity range of the trapped particles can be significantly increased by use of both cooling and trapping. The forces are caused by recoil during spontaneous and/or induced transitions of the atom in the resonant field of a standing laser wave with frequency scanned by a certain law.

The idea of laser trapping of a particle, together with that of laser cooling to eliminate completely a linear and quadratic Doppler effect deserves attention in spite of evident difficulties in the practical realization of this complex method. For spectroscopy with no Doppler ion broadening, trapping of ions by a high-frequency electromagnetic field can be used. For instance, the authors of [6.33] have recently suggested that ions Tl^+ and Al^+ should be trapped in regions of the order of 10 μm in dimension and the spectroscopy of their forbidden transitions should be performed with the resolution $R \gtrsim 10^{14}$. In this case, the physics of elimination of Doppler broadening is the same as that in the method described for trapping neutral particles in a three-dimensional standing light wave, though the mechanisms of trapping are essentially different.

6.2.5 Influence of a Quadratic Doppler Effect on a Resonance Shape

When the relative width of resonance is less than 10^{-11}, owing to collisions, flight effects and so on, it is necessary to take into account the influence of the quadratic Doppler effect on the shape of nonlinear optical resonances. The quadratic Doppler effect is a relativistic effect connected with the reduction of duration in a moving system of reference. When a moving atom interacts with a field of a frequency ω, it is acted upon by the frequency

$$\omega' = \frac{\omega - \underline{\underline{k}}\underline{\underline{v}}}{\sqrt{1 - V^2/c^2}} \simeq \omega - \underline{\underline{k}}\underline{\underline{v}} + \frac{1}{2}\frac{V^2}{c^2} . \qquad (6.64)$$

The relative shift caused by the quadratic Doppler effect is of the order of $\pm\Delta\omega/\omega \sim 10^{-12}$ (V is the particle velocity, c is the light speed). The quadratic Doppler effect depends on the particle velocity and may be detected in the line shift by varying the temperature of a gas (temperature red shift). The temperature dependence of the shift of the center of the Lamb dip in methane measured by use of frequency-stable 3.39 μm He-Ne lasers [6.34] was linear with a slope of 0.5 Hz/degree and agreed, to a good accuracy, with the calculated value

$$\Delta\omega_0 = -\frac{kT}{Mc^2}\omega_0 , \qquad (6.65)$$

obtained in [6.34] (see Chap.10).

This line shift leads to an additional inhomogeneous broadening. We consider the influence of the quadratic Doppler shift on the Lamb-dip shift in a low-pressure gas. The resonance shape of two-photon absorption in a standing-wave field was considered in Chapter 4.

As has been shown, atoms whose velocities are less than the average thermal velocity play a leading role in the formation of the Lamb dip in a low-pressure gas. This must lead to a decrease of the influence of the quadratic Doppler effect and, hence, to a decrease of the Lamb-dip shift caused by this effect.

Reference [6.35] offers the solution of the problem of the shift of the Lamb-dip center caused by the quadratic Doppler effect when the finite dimension of the light beam is taken into account. Taking into account the quadratic Doppler shift amounts to replacement of ω with $\omega + V^2\omega/c^2$ in (6.34). Solution of (6.34) gives the dependence of the shift of the Lamb-dip center $\Delta\omega$ on the parameter $\beta = \Gamma\tau$ for a beam that has a gaussian profile. The function

$$\psi(\beta) = \frac{\Delta\omega}{\Delta\omega_0}$$

that determines this dependence is shown in Fig.6.13. When $\beta \ll 1$

$$\psi(\beta) = \beta^2\left(a_1 \ln\frac{1}{\beta} - a_2\right) , \qquad (6.66)$$

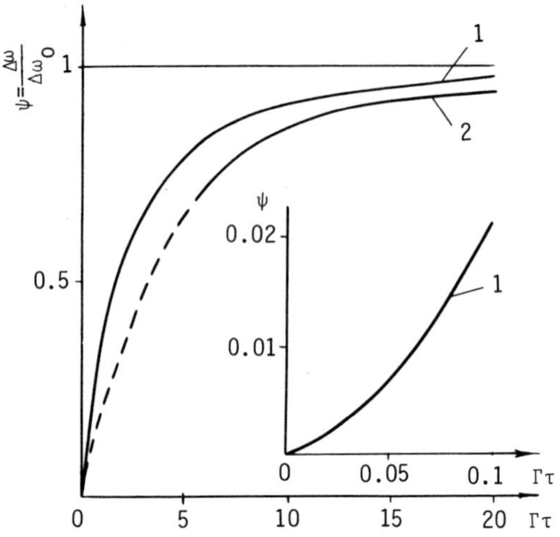

Fig.6.13 Dependence of the shift of the Lamb-dip center on the parameter $\beta = \Gamma\tau$. 1 - a gaussian beam; 2 - a beam with constant amplitude over its cross-section. Dotted curve was obtained by interpolation

where $a_1 = 1.86$, $a_2 = 2.58$. In the case $\beta >> 1$

$$\psi(\beta) = 1 - 11/\beta^2 \ . \tag{6.67}$$

When $\beta \to \infty$ we have a spatially homogeneous case.

6.3 Information Capacity of Nonlinear Spectroscopy Methods

A nonlinear laser spectrometer that performs spectroscopy of atoms and molecules without Doppler broadening, with resolution $R >> \nu_0/\Delta\nu_D$, where $\Delta\nu_D$ is a Doppler width of spectral lines in a gas, has, at least, two advantages important for applied spectroscopy. First, the information capacity of a spectral interval $\Delta\nu$ with such a spectrometer is

$$\rho = \rho_0 R \frac{\Delta\nu}{\nu_0} \ [\text{bit}] \ , \tag{6.68}$$

where ρ_0 is the number of information bits obtained in each resolved spectral interval ν_0/R, based on the accuracy of the value of the spectral intensity measured at each frequency. For instance, with the resolution that is normal for nonlinear spectroscopy $R = 10^8$ the spectral interval $\Delta\nu = 1\text{cm}^{-1}$ at $\nu_0 = 3 \cdot 10^3\ \text{cm}^{-1}$ contains $\rho_0 \times 3 \cdot 10^4$ bits, i.e., if $\rho_0 = 10$ the value $\rho = 3 \cdot 10^5$ bits. For the usual infrared spectrometer of very good quality with a resolution of 0.1 cm^{-1}, the same spectral interval contains about 10^2 bits only.

Second, overlap of a great number of rotational-vibrational lines, the distance between which is less than a Doppler width, is typical for complicated molecules; a spectral analysis may be performed only on vibrational bands. In this case, the maximum number of bits in one octave of the ir region is

$$\rho_{max} \simeq \rho_0 \frac{\nu_0}{\delta\nu_{vib}} , \tag{6.69}$$

where $\delta\nu_{vib}$ is the vibrational bandwidth. It is a rather serious restriction for possibilities of ir molecular spectroscopy. There is no such restriction for nonlinear spectroscopy; information on complicated molecular mixtures, which is sufficient for quantitative and qualitative spectral analysis can be gained from a rather narrow spectral interval, even inside overlapping bands of absorption of molecules in a mixture.

6.4 Sensitivity of Nonlinear Laser-Spectroscopy Methods

In a brief discussion of the sensitivity of detection of narrow resonances inside a Doppler contour of absorption lines, the main attention will be paid to a comparison of three methods: absorption-saturation spectroscopy, two-photon spectroscopy and spectroscopy of trapped particles. Then we will consider the limiting sensitivity of recording of narrow resonances by absorption and fluorescence methods, omitting all practical details.

6.4.1 Comparison of the Sensitivity of Methods

In the general case, the sensitivity of detection of a narrow resonance by a method is determined by the maximum power ΔP that can be absorbed by particles from a light wave from the transition 1 - 2. The particles responsible for the formation of a narrow resonance with a homogeneous halfwidth Γ cannot do stimulated quantum transitions between levels 1 and 2 under the action of a light field if they have a velocity greater than Γ. Hence, the maximum power absorbed per unit volume is

$$\Delta P \gtrsim \hbar\omega_{12}\Gamma N_{int} , \tag{6.70}$$

where N_{int} is the density of particles that resonantly interact with the light wave. In absorption spectroscopy, a small fraction of the particles with a certain projected velocity is resonant to the field, i.e., $N_{int} \simeq (\Gamma/ku)N_0$, where N_0 is the total density of particles on levels 1 and 2. Thus, the sensitivity of this method is inversely proportional to width of a narrow resonance.

In two-photon spectroscopy, in spite of their velocity, all of the particles are resonant with the standing light wave. So, in this method $N_{int} = N_0$, which makes an essential gain in the method sensitivity, of the order of $(ku/\Gamma) \simeq 10^3$ to 10^5 when very narrow resonances are observed. True, a much higher intensity of the standing light wave is required than in spectroscopy of saturation of one-quantum transitions, to obtain a rate of induced two-quantum transitions of the order of Γ.

In particle-trapping spectroscopy the number of interacting particles N_{int} is equal, in essence, to the number of trapped particles N_{trap}. With total (three-dimensional) particle trapping in the region of crossing of three mutually perpendicular standing light waves, i.e., with particle trapping in volumes $(\lambda/2)^3$, the sensitivity decreases proportionally to $(v_{cr}/v_0)^3$, where v_{cr} is the critical velocity. An exceptionally small fraction of particles in an equilibrium gas with an absolute velocity $v < v_{cr}$ requires the accumulation of slow particles for registration of resonances.

In order to form very narrow resonances in any of the described methods it is necessary to eliminate the influence of particle collisions, at least so that a broadening due to collisions $2\Gamma_{coll}$ can be less than the required resonance width 2Γ. This is achieved by decreasing the pressure to values that are far less than 10^{-3} torr; this causes a corresponding decrease of N_{int} and a decrease of the sensitivity of the method.

6.4.2 Limiting Sensitivity of Methods

The sensitivity of any method of nonlinear spectroscopy depends essentially on the manner of detection of the energy absorbed in a gas ΔP. Usually, the change of intensity of the wave P that detects the narrow resonance is measured after the wave has passed through a gas cell. With small absorption ($\Delta P \ll P$), such a change is negligible, and the sensitivity is restricted by the fluctuation noise of the detector, arising from the incidence of the principal wave with the intensity P on the detector.

We make a simple estimate of the limiting sensitivity of detection of a narrow resonance of saturated absorption, following [6.36]. When a beam P passes through a gas cell the initial power of the beam decreases by

$$\Delta P = \Delta P l S \ , \tag{6.71}$$

where ΔP is the power absorbed per unit volume, and S is the cross-sectional area of the light beam, l is the length of cell. In the optimal case of strong absorption saturation, within the limits of a homogeneous width 2Γ the value ΔP determined by (6.70) is approximately

$$\Delta P = \hbar\omega_{12}\Gamma(N_1^o - N_2^o)\,\frac{\Gamma}{ku} \ , \tag{6.72}$$

where $(N_1^o - N_2^o)$ is the difference of the population densities of the two levels of the transition.

Heterodyne detection is the most sensitive method for detecting coherent laser radiation. The sensitivity threshold of heterodyne detection determined by quantum fluctuations is

$$P_{min} = 4\sqrt{PP_n} \ , \tag{6.73}$$

where P_n is the noise power of the photodetector due to quantum fluctuations, determined by

$$P_n = \frac{\hbar\omega B}{\eta} \ , \tag{6.74}$$

where B is the detection band in Hz, and η is the quantum efficiency of the photoelectric detector.

Equating the maximum power ΔP absorbed in a cell with the minimum detected power P_{min}, we obtain, for the minimum number of particles necessary for detection of a narrow resonance of absorption saturation,

$$n_{min} = (N_1^o - N_2^o)_{min}(Al) = \frac{P_{min}}{\hbar\omega\Gamma}\left(\frac{ku}{\Gamma}\right) \ . \tag{6.75}$$

When narrow resonances of molecules are to be detected, the fact that a small fraction q of molecules interacts with the field, owing to the distribution of molecules over many rotational levels, must be taken into account. The value of q ranges from 10^{-1} for simple molecules to 10^{-3} for complicated ones. In this case, instead of (6.75), we have

$$n_{min} = \frac{P_{min}}{\hbar\omega\Gamma q}\left(\frac{ku}{\Gamma}\right) \ . \tag{6.76}$$

It is quite realistic to detect $P_{min} \simeq 10^{-9}$ W by use of either a photoelectric detector in the visible region or a photoconductor detector in the ir band. Detection of resonances with a width $\Gamma \simeq 10^5$ s^{-1} and $(\Gamma/ku) \simeq 10^{-3}$ requires about 10^8 atoms or $(1/q)10^8$ molecules.

It is more advantageous to record directly the value of absorbed intensity ΔP. This is possible, if, for example, the excited particles fluoresce. In this way, a narrow resonance on the change of fluorescence intensity can be recorded (see Sec.3.4). In this method, the principal radiation is not incident on the detector and the sensitivity for recording narrow resonances is greatly increased. For instance, the narrow resonances of fluorescence of the CO_2 molecule at absorption saturation of the radiation of a CO_2 laser is recorded at very low CO_2 pressure (down to 10^{-5} torr) [6.37]. The same method of detection of narrow resonances is quite applicable to the methods of two-photon spectroscopy and spectroscopy of particle trapping. For instance, when trapped particles are observed by the fluorescence method with a visible or uv, tunable-frequency laser, in principle, sensitivity sufficient to record an individual particle can be achieved. In particular, such a possibility has been considered for the Tl^+ ion trapped in an electromagnetic field [6.33]. Attention should be given to the possibility of detecting excited particles by the other methods, for instance, at induced transitions to upper-lying states or by direct detection of excited particles in experiments with atomic beams. These possibilities have not been realized yet but promise essential improvement of sensitivity of all of the mentioned methods of nonlinear spectroscopy, in experiments with very low densities of atoms and molecules.

References

6.1 Ye.V. Baklanov, B.Ya. Dubetsky, Ye.A. Titov, V.A. Semibalamut: Kvantovaya Elektron. 2, 11 (1975)
6.2 J.L. Hall, C.J. Borde, K. Uehara: Digest of the Second Symposium on Gas Laser Physics, Novosibirsk, June 1975, p.107
6.3 V.P. Chebotayev: In *Laser Spectroscopy*, Proceedings of 2nd Intern. Conf., 23-27 June 1975, Megeve, France (Springer-Verlag, Berlin, Heidelberg, New York 1975), p. 150
6.4 S.G. Rautian, A.M. Shalagin: Zh.Eksp.i Teor.Phys.Pis.Red. 10, 686 (1969); 58, 962 (1970)
6.5 S.N. Bagaev, L.S. Vasilenko, A.K. Dmitriev, M.N. Skvortzov, V.P. Chebotayev: Zh.Eksp.i Teor.Phys.Pis.Red. 23, 399 (1976)
6.6 V.S. Letokhov, B.D. Pavlik: Opt.Commun. 6, 202 (1972)

6.7 V.A. Alexeyev, N.G. Basov, E.M. Belenov, M.I. Vol'nov, M.A. Gubin, V.V. Nikitin, A.N. Nikolaenko: Zh.Eksp.i Teor.Phys. 66, 887 (1974)
6.8 S.N. Bagaev: Thesis, The Institute of Semiconductor Physics, Siberian Branch, The USSR Academy of Sciences (1975)
6.9 N.G. Basov, O.N. Kompanets, V.S. Letokhov, V.V. Nikitin: Zh.Eksp.i Teor.Phys. 49, 394 (1970)
6.10 V.S. Letokhov, B.D. Pavlik: Preprint of the Institute of Spectroscopy of the USSR Academy of Sciences, No. 140/21 (1973); Kvantovaya Elektronika 1, 2425 (1974)
6.11 V.S. Letokhov, B.D. Pavlik: Kvantovaja Electronika 10, 32 (1972)
6.12 L. Shiff: *Quantum Mechanics* (McGraw-Hill Book Co., Inc., New York 1955)
6.13 L.D. Landau, Ye.M. Lifshits: *Quantum Mechanics* (Moscow 1963)
6.14 I.I. Sobelman: *Introduction into the Theory of Atomic Spectra* (Pergamon Press 1970)
6.15 P.R. Berman, W.E. Lamb, Jr.: Phys.Rev. 2A, 2435 (1970)
6.16 S.G. Rautian: Proc.Phys.Inst.USSR Acad.Sci. 43, 3 (1968)
6.17 V.A. Alexeyev, T.L. Andreyeva, I.I. Sobelman: Zh.Eksp.i Teor.Phys. 62, 614 (1972)
6.18 S.N. Bagayev, Ye.V. Baklanov, V.P. Chebotayev: Zh.Eksp.i Teor.Phys. Pis.Red. 16, 15 (1972)
6.19 H.S.W. Massey, E.H.S. Burhop: *Electronic and Ionic Impact Phenomena* (Clarendon Press, Oxford 1952)
6.20 C. Borde, J.L. Hall: Phys.Rev.Lett. 30, 1101 (1973)
6.21 J.L. Hall: Colloques Internatioux du C.N.R.S. No. 217, p.105
6.22 S.N. Bagayev, L.S. Vasilenko, V.G. Gol'dort, A.K. Dmitriyev, M.N. Skvortsov, V.P. Chebotayev: Report at the Second Symposium on Gas Laser Physics, Novosibirsk, June 1975 (to be published)
6.23 S.N. Bagayev, Ye.V. Baklanov, V.P. Chebotayev: Zh.Eksp.i Teor.Phys. Pis.Red. 16, 314 (1972)
6.24 V.P. Chebotayev: Kvantovaja Electronika 3, 694 (1976)
6.25 V.S. Letokhov: Zh.Eksp.i Teor.Phys.Pis.Red. 7, 348 (1968)
6.26 A.V. Gaponov, M.A. Miller: Zh.Eksp.i Teor.Phys. 34, 242, 751 (1959)
6.27 G.A. Askar'yan: Zh.Eksp.i Teor.Phys. 42, 1567 (1962)
6.28 L.D. Landau, Ye.M. Lifshits: *Electrodynamics of Continuous Media* (Moscow 1975)
6.29 R.H. Dicke: Phys.Rev. 89, 471 (1953)
6.30 H.M. Goldenberg, D. Kleppner, N.F. Ramsey: Phys.Rev.Lett. 5, 361 (1960)
6.31 T.W. Hänsch, A.L. Schawlow: Opt.Commun. 13, 68 (1975)
6.32 V.S. Letokhov, V.G. Minogin, B.D. Pavlik: Opt.Commun. 19, 72 (1976)
6.33 H. Dehmelt: Abstracts submitted for New Haven Meeting of DEAP Amer. Phys.Soc. (Dec. 1973), and Anaheim Meeting of the Amer.Phys.Soc. (Jan.-Feb. 1975)
6.34 S.N. Bagayev, V.P. Chebotayev: Zh.Eksp.i Teor.Phys.Pis.Red. 16, 614 (1972)
6.35 Ye.V. Baklanov, B.Ya. Dubetsky: Quantum Elec. 2, 2041 (1975)
6.36 K. Shimoda: Appl.Phys. 1, 77 (1973)
6.37 A. Javan: In *Laser Spectroscopy*, Proceedings of 2nd Intern. Conf., 23-27 June 1975, Megeve, France (Springer-Verlag, Berlin, Heidelberg, New York 1975), p.439
6.38 V.S. Letokhov, B.D. Pavlik: Appl.Phys. 9, 229 (1976)

7. Nonlinear Atomic Laser Spectroscopy

Now we consider use of narrow nonlinear resonances in two- and three-level systems to measure natural widths of spectral lines and level lifetimes, investigation of collisions, study of hyperfine and isotopic structures of quantum transitions, and splitting and shift of spectral lines in external electric and magnetic fields. In all cases, we shall restrict ourselves, for the most part, to exposition of experimental results.

7.1 Measurement of a Natural Width and Lifetimes of Levels

In the optical spectral region, a natural line width caused by spontaneous decay of levels ranges usually from 10 to 100 MHz (Table 1.1). Also, collisional line broadenings at collisions range from 10 to 100 MHz/torr. At pressures less than 0.1 torr the collisional line broadening becomes less than the natural width γ_{rad} or comparable with it. So measurements of a homogeneous width at low pressures can give values of the natural line width immediately. Spectroscopic methods based on analysis of the shape of a Doppler contour yield absolute accuracy of measurements of a homogeneous width of about 10 MHz, even with the most careful experimental procedure, employing electronic computers. Therefore these methods, used to explore the collisional broadening of a Doppler contour, are of little use for measurements of the natural line width. It is different with the application of narrow resonances, whose widths are determined by a homogeneous line width. With them the accuracy of measurement of the natural line width can be very high.

Investigations with a He-Ne laser at 0.63 and 1.15 μm on the neon transitions $3s_2 - 2p_4$ and $2s_2 - 2p_4$, respectively, were among the first investigations of narrow resonance widths in a pressure range of about 1 torr. As experiments indicated, at operating gas pressures in the He-Ne laser the collision broadening of a Lamb dip exceeds the natural line width of 0.63 μm several times. In order to obtain its value it was therefore necessary to extrapolate the pressure dependence of the Lamb-dip width to zero. The extrapolation from a rather high pressure range can reduce the accuracy of determination of the natural width. Usually the extrapolation follows a linear law, which cannot always describe correctly the dependence of the dip width on pressure. The principal results of measurements of the natural width in the He-Ne laser are summarized in Table 7.1. As seen from it, in spite of the difficulties mentioned, the value obtained for the 0.63 μm line by analysis of the Lamb-dip shape in the He-Ne laser is in good agreement with data gained by use of other methods.

Table 7.1 Natural halfwidth of the ^{20}Ne 0.63 μm line

N_N	Papers	MHz	Method of measurement
1	Lisitsyn, Chebotayev [3.2]	14	Power peak in a laser with internal absorbing cell
2	Lee, Skolnick [3.3]	15	Power peak in a laser with internal absorbing cell
3	Letokhov, Nikitin, Kompanets [3.10]	13 ± 1	Power peak in a laser with internal absorbing cell
4	Tatarenkov, Titov [3.16]	9 ± 1.5	Power peak in a laser with internal absorbing cell
5	Cordover, Bonzyak, Javan [7.13]	13 ± 12	Lamb dip in the He-Ne laser
6	Dietel [7.15]	12	Lamb dip in the He-Ne laser
7	Provorov, Matyugin, Chebotayev [3.64]	11 ± 3	Opposing probe wave

In those few cases where lasers can operate at low gas pressures, measurements of the dip width yield directly γ_{rad}. Examples may be found in pure Ne lasers ($\lambda = 1.15$ μm), Xe lasers ($\lambda = 3.5$ μm), and Hg lasers ($\lambda = 1.52$ μm) operating at pressures less than 0.1 torr. In more detail, the Lamb dip was studied in a mercury laser. In a pure Ne laser, even at low pressures when the influence of collisions is small, it was necessary to take into account resonant radiation trapping. The Lamb-dip shape was analyzed by use of a formula that allowed for a saturation homogeneity due to resonant radiation trapping and collisions [7.1]

$$P(\Omega) \simeq \frac{\eta - \exp\left(\frac{\Omega}{ku}\right)^2}{1 + \frac{\Gamma^2}{\Gamma^2 + \Omega^2} + \tilde{\alpha}} \; ; \qquad (7.1)$$

where η is the excess of initial gain over the threshold value, $\tilde{\alpha}$ is a parameter that allows for saturation homogeneity and depends on level relaxation constants. Figure 7.1 shows the influence of saturation homogeneity on the dip shape in a Ne laser. This figure shows the experimental record of the

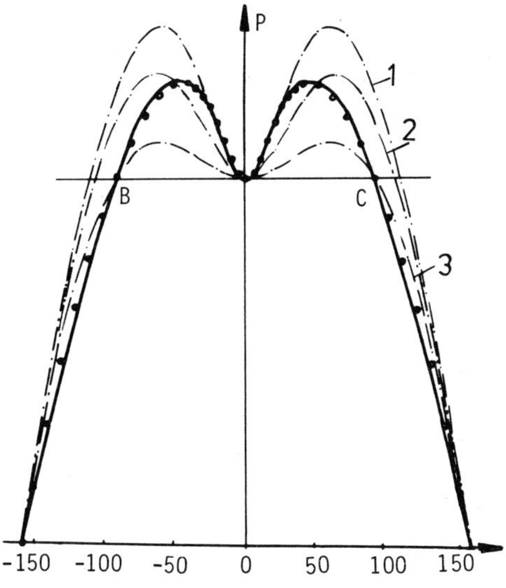

Fig.7.1 Dependence of Ne laser power at $\lambda = 1.15$ μm on frequency. The experimental data (points) were obtained with 0.7 torr Ne pressure and 6.4 mA discharge current. The calculated curves were obtained without allowing for radiation trapping (dotted curves: 1, $\Gamma/2\pi = 33$ MHz; 2, 44 MHz; 3, 70 MHz) and allowing for the trapping (continuous curve) with parameters $ku/2\pi = 435$ MHz, $\eta = 1.14$, $a = 0.5$, $\Gamma/2\pi = 23$ MHz

frequency dependence of the generated power and curves calculated from the Lamb formulae and (7.1). When the Lamb formula ((7.1) with $\tilde{\alpha} = 0$) is used, the calculated and experimental curves do not agree. The curve calculated from (7.1) is in good agreement with the experimental curve. The parameters $\tilde{\alpha}$ and Γ may be found by processing the experimental data with electronic computers.

A simple method to determine the parameters $\tilde{\alpha}$ and Γ from the Lamb dip without computers was used by MATYUGIN et al. [7.2]. For convenience, (7.1) can be reduced to

$$P(\Omega) \simeq A \frac{\eta - \exp(\Omega/ku)^2}{1 + \dfrac{a\Gamma^2}{\Omega^2 + \Gamma^2}} , \qquad (7.2)$$

where $a = 1/(1 + \tilde{\alpha})$. An expression of the form of (7.2) is frequently used to analyze the influence of collisions on the Lamb-dip shape [7.3]. Equation (7.2) comprises five parameters that must be determined from the experimental curves. Parameter A is a scale coefficient that is independent of frequency;

it is of no interest. The Doppler parameter ku can be determined from independent measurements. The threshold of generation is fixed by changing the discharge current in the laser. When the length of the discharge tube is increased and the current is not changed, the detuning frequency Ω_{th} at which generation is interrupted is measured. The ratio of appropriate discharge lengths is equal to that of gain to losses $\eta = \ell_2/\ell_1$ and

$$ku = \Omega_{th}/\sqrt{\ell n \frac{\ell_2}{\ell_1}} , \qquad ku >> \Gamma . \tag{7.3}$$

The frequency scale is usually found from the distance between orders of a laser interferometer. Because the effects of frequency pulling in a He-Ne laser are not essential as usual, the generated frequency can be considered to be equal to the frequency of an empty cavity. The parameters Ω and Γ are selected by the method of successive approximations. At $\Omega = 0$

$$P_o = A \frac{\eta - 1}{1 + a} . \tag{7.4}$$

For the frequency detuning Ω_a at which the generated power is equal to that in the line center P_o we have

$$\frac{\eta - \exp(\Omega_a/ku)^2}{1 + a \dfrac{\Gamma^2}{\Omega_a^2 + \Gamma^2}} = \frac{\eta - 1}{1 + a} . \tag{7.5}$$

Hence

$$a = \frac{\eta - 1}{\eta - \exp(\Omega_a/ku)^2} \left(1 + \frac{a\Gamma^2}{\Gamma^2 + \Omega_a^2}\right) - 1 . \tag{7.6}$$

Because $\Omega_a >> \Gamma$ and $a \approx 1$ the second term in parentheses of (7.6) may be neglected. Then, taking into consideration that $\eta = \exp(\Omega_{th}/ku)^2$ for the zeroth approximation of a, we have

$$a^{(o)} = \frac{\exp(\Omega_a/ku)^2 - 1}{\exp(\Omega_{th}/ku)^2 - \exp(\Omega_a/ku)^2} . \tag{7.7}$$

When $a^{(o)}$ is known, the power $P(\Gamma)$ that corresponds to the detuning $\Omega = \Gamma$ can be calculated. In the zeroth approximation, taking into account that $\Omega_{th} >> \Gamma$, we obtain

$$P(\Gamma) = P_o \frac{1 + a^{(o)}}{1 + a^{(o)}/2} . \tag{7.8}$$

When $P(\Gamma)$ is known it is easy to find $\Gamma^{(o)}$. The value of $\Gamma^{(o)}$ thus found is used to determine $a^{(1)}$ from (7.6) and then $\Gamma^{(1)}$. As the analysis has showed, the first approximation is usually sufficient. Higher approximations gave corrections within the limits of experimental accuracy.

When the Lamb-dip shape is investigated experimentally, it is necessary to give much consideration to a number of circumstances that significantly affect accuracy. It is very important to prevent instabilities of the discharge current, to minimize the influence of vibrations and to achieve a linear change of frequency with change of cavity length without modifying the high quality of the cavity. In the experiments carried out by MATYUGIN et al. [7.2] on the 1.15 μm line the installation consisted of a Ne laser with a dc discharge. The laser fittings were all steel and were placed on a massive plate. Owing to prevention of vibrations and to discharge stability they obtained records of generated power with very low noise level (less than 1%). The ratio of gain to threshold was within the range of 1.1 to 1.15. After having obtained the values of the dip width at different ratios, by extrapolation to zero field they found the value of homogeneous width at various Ne pressures. The natural width, which was equal to $\gamma_{rad}/\pi = 34 \div 2$ MHz was determined from the dependence of the homogeneous width on pressure (Fig. 7.2).

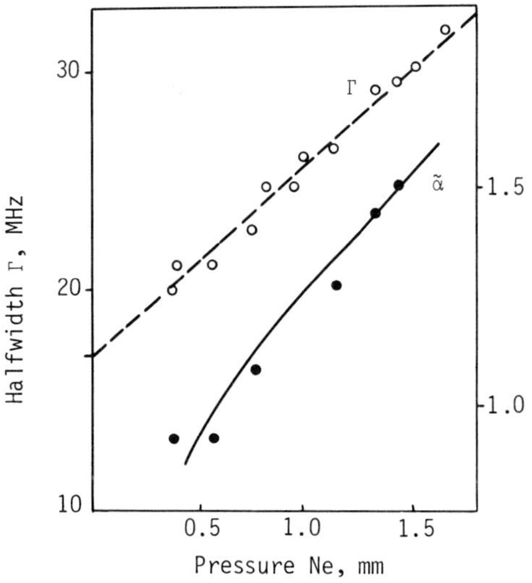

Fig.7.2 Dependence on Ne pressure of homogeneous halfwidth Γ of the gain Ne line at λ = 1.15 μm and the parameter $\tilde{\alpha}$, allowing for transition saturation homogeneity

The method of nonlinear absorption can be used both with an internal absorption cell and with an external one. For small absorptions, the cell inside the cavity is best. The peak amplitude depends on the loss-absorption ratio. With 1% losses in the cavity resonances can be recorded with absorption less than 0.1%. A width was first measured with a He-Ne laser with a nonlinear Ne absorber at 0.63 and 1.52 μm. The results obtained by different authors differ only slightly and are close to the values obtained by other methods. However, for the 0.63 μm line the differences of the values of γ_{rad} obtained by some authors are greater than the experimental errors. This may be connected with the difficulty of processing the data on the dependence of generated power on frequency. The shape of the peak of the generated power depends on many parameters that are difficult to determine. In analysis of the peak shape, taking saturation into account is complicated. Unlike an ordinary laser, in the general case the value of saturation in an absorption cell cannot be estimated from the excess of gain over threshold. Such an estimate requires additional experiments. The asymmetric position of the power peak presents some difficulties for processing the results.

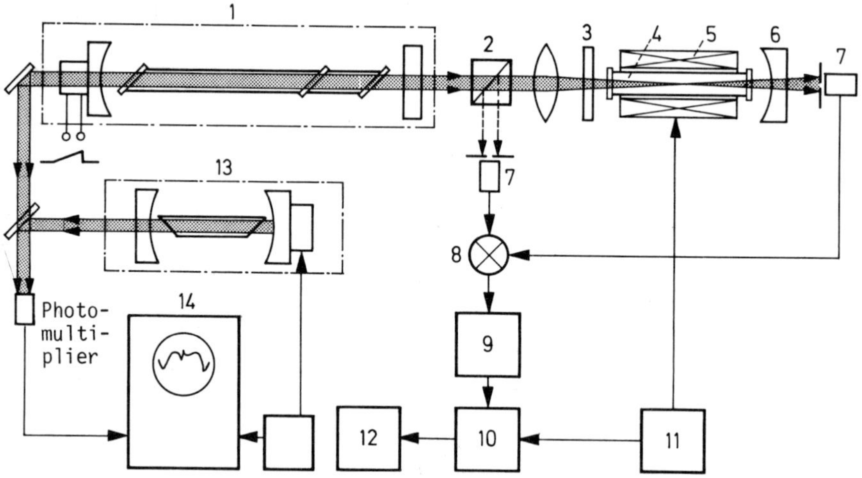

Fig.7.3 Experimental arrangement for investigating the absorption saturation in an external Ne cell located in the magnetic field by the oppositely travelling probe-wave method. 1 - high-intensity single-frequency He-Ne laser at $\lambda = 0.6328$ μm; 2 - polarization prism; 3 - $\lambda/4$ plate; 4 - external Ne absorbing cell; 5 - solenoid; 6 - weakly reflecting mirror; 7 - photodetector; 8 - summing element; 9 - selective amplifier; 10 - synchronous detector; 11 - audiofrequency generator; 12 - self-recorder; 13 - reference heterodyne laser; 14 - oscillograph

The above drawbacks can be easily removed by using the external saturated cell of absorption. Figure 7.3 shows the experimental design of MATYUGIN et al. [3.61,7.2] for measuring the homogeneous width of the 0.63 μm line by the method of a weak oppositely travelling wave and by use of the Zeeman effect. With different relaxation constants and rather small fields, the resonance shape of absorption of the weak oppositely travelling wave depends on the homogeneous line widths and on the value of saturation; it is described by (2.145). The width of the 0.63 μm line in a pure neon discharge was investigated with relatively weak optical fields, when saturation had the value $G \sim 1$. In order to achieve such saturation in tubes 50 cm long, it is necessary to have the high output power of the single-frequency He-Ne laser and the possibility of continuous frequency tuning, at least near the line center. These requirements were satisfied by a He-Ne laser with an amplifying tube about 100 cm long and mode selection produced by use of a nonlinear absorber. The radiation from this laser, which was not plane polarized, was passed through a polarizing prism, was circularly polarized by a quarter-wave plate and focused in the center of an external absorption cell. A weak reflector, placed behind the absorption cell, formed a beam that passed through the cell in the opposite direction and was deflected out to a photodetector by the polarizing prism. Simultaneously, the polarizing prism combined with the $\lambda/4$ plate provided optical isolation between the mirror and the laser cavity. Part of the radiation that passed through the cell in the forward direction was focused on a second photodetector that connected together with the first detector to a common load. By choosing signals at the outputs of the photodetectors in such a way that they are equal and opposite in phase, noise in a recording system can be decreased by more than an order of magnitude. In addition, with available absorption, the difference between the readings of the photodetectors corresponds to the absorption of an oppositely travelling wave. A signal was fed from the load to the recording system, which consisted of a selective amplifier, a synchronous detector and an automatic recorder. When the laser frequency was tuned, curves showing the line shape of absorption of a weak wave in the presence of a strong one and a dispersion curve of the absorption line were recorded. These curves are shown in Fig.7.4. A signal proportional to the absorption coefficient (Fig.7.4a) was obtained when the discharge current in the external cell was modulated. The dispersion curve of the absorption line (Fig.7.4b) was obtained by superimposing a longitudinal alternating magnetic field on the absorption cell. In the magnetic field, the absorption line splits into two Zeeman components with right- and left-hand circular polarizations. Scanning of the positions of these components by the varying magnetic field causes modulation of the absorption of the weak wave. With small scanning amplitudes, the dispersion curve is the first derivative of the absorption line with respect to frequency.

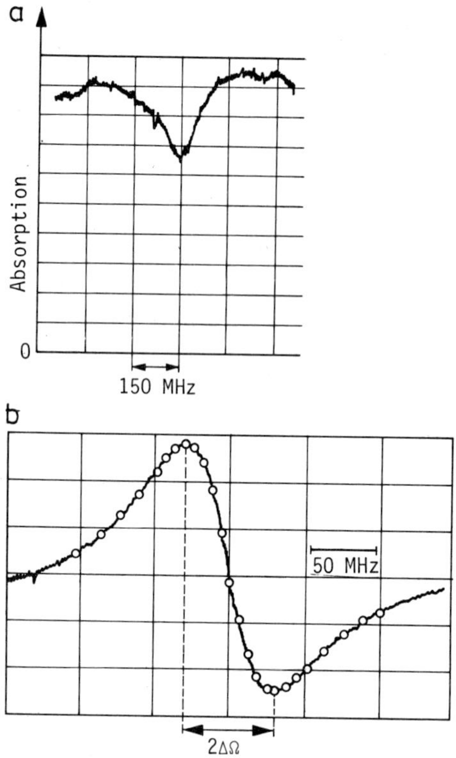

Fig.7.4a,b Resonance shape of nonlinear absorption of weak wave in the presence of the strong oppositely travelling one (a) and the shape of the first frequency derivative (b) of this resonance (Ne pressure is 1 torr, discharge current is 20 mA, $\lambda = 6328$ Å)

The distance between maxima $2\Delta\Omega$ on the curve (Fig.7.4b) is $\Delta\omega/\sqrt{3}$ where $\Delta\omega$ is the dip width of nonlinear absorption. Using this relation, we can find the value $\Delta\omega$ by the curves (Fig.7.4b).

The field broadening of the dip was taken into account in two ways. First, the dependence of $\Delta\omega$ on the intensity of the radiation that passed through the cell was plotted for each pressure, and the parameter 2Γ was determined by extrapolation to zero value of the field. Because this method is based on the saturation effect, the field in the cell cannot be reduced to a very small magnitude. Therefore, the form of the approximating function is important for obtaining the right value of the width extrapolated to zero field intensity. Equation (2.146) was used as the approximating function.

The second way consisted of measuring the absorption coefficient in the cell per one passage, in the line center for the strong field (saturated absorption coefficient $\kappa(P)$) and for the weak field (unsaturated absorption coefficient κ_0). When κ_0 was measured, the radiation intensity was reduced by a factor of less than 10^{-2}, because for the weak field G could be assumed to be zero. The saturation parameter for the strong field G was determined by using the known relation $\kappa = \kappa_0/\sqrt{1+G}$. When the resulting value of G is inserted in (2.146) Γ can be found.

The values of Γ found by these two different ways were approximately equal. A small increase of the line width with the discharge current, whose value was about $\partial\Gamma/\partial i = 0.15$ to 0.17 MHz/mA, was observed. The value of the natural line width of $\gamma_{rad} = 22 \pm 6$ MHz was obtained by extrapolating to zero pressure and current.

Measurement of Level Lifetimes

The methods of nonlinear laser spectroscopy provide comprehensive information on level relaxation constants. At present, the direct methods of measurement of level lifetimes by damping of the spontaneous radiation are widely used. The methods of photon counting by use of the technique of coincidence delay in many cases obtain reliable data with high accuracy. Nevertheless, spectroscopic methods of measurement of lifetimes may supplement the direct methods in cases where they, for some reasons, cannot be used. For instance, when levels are closely spaced, it is difficult to excite one level by a monokinetic electron beam. In essence, the direct methods determine values of the excitation lifetime on a level. Therefore when there is, for instance, resonant radiation trapping, the lifetime measured by this way is not equal to that of an isolated atom. Laser spectroscopic methods are free of these limitations.

Usually, in the optical spectral region, the decay constants of levels of an operating laser transition differ considerably. So, in many cases, the natural line width is determined by the lifetime of a short-lived level. If the lifetime of one level is measured in some way, the lifetime of another is determined from the natural width.

The measurement of level lifetimes by the laser spectroscopic methods is effective for short-lived levels that decay to the ground level. As an example, we can use a neon $2s_2$ level. The measured natural width of the $2s_2$-$2p_4$ transition is 34 ± 2 MHz. The lifetime of the $2p_4$ level has been measured by a number of researchers who used direct methods [7.4]; it is known to be equal to $\tau\,(2p_4) = 1.8 \times 10^{-8}$ s, to a high accuracy. The lifetime of an isolated atom in the $2s_2$ state, obtained from spectroscopic investigations [7.1, 2,5], is $\tau\,(2s_2) = 6.4 \times 10^{-9}$ s. The lifetime of the $2s_2$ level, measured by the damping of spontaneous radiation [7.6] at 10^{-1} torr pressure is 10^{-7} s. The disagreement is caused by resonant radiation trapping. Measurements of the lifetime by the delayed-coincidence method in vacuum ultraviolet, by LAWRENCE and LISZT [7.7] have recently yielded the value $\tau = 7.8 \times 10^{-9}$ s, which agrees with the values obtained spectroscopically [7.1,2,5] within the limits of the experimental errors.

If there is no additional information on level lifetimes simple experiments can be performed to measure the ratio of the lifetimes of the levels of a transition. They are based on the measurement of a relative change of level population under the action of a field [7.8-10]. The ratio of the changes of the populations of levels Δn_1 and Δn_2 under the action of the field is

$$\frac{\Delta n_1}{\Delta n_2} = \frac{\tau_1}{\tau_2} \; . \tag{7.9}$$

The population change is connected with the changes of intensity of the spontaneous radiation from the upper and lower levels of the transition by the ratio

$$\frac{\Delta I_1}{\Delta I_2} = \frac{\tau_1}{\tau_2} \; \frac{A_1\omega_{01}}{A_2\omega_{02}} \; , \tag{7.10}$$

where A_1 and A_2 are probabilities of the spontaneous transition of the corresponding lines from the levels, and ω_{01}/ω_{02} is the ratio of the transition frequencies. The ratio A_1/A_2 can be measured under conditions with which gain is zero, i.e., the level populations are identical, allowing for statistical scales. In this case the ratio of the absolute line intensities is equal to that of the transition probabilities A_1/A_2. By use of this value, the ratio τ_1/τ_2 can be found. The transition probabilities of neon on the lines 0.63 μm ($3s_2 - 2p_4$) and 0.60 μm ($2p_4 - 1s_4$) were measured by this technique. The ratio of the absolute probabilities of these transitions was 0.35 ± 0.035 [7.8-10]. The absolute probability of the transition $3s_2 - 2p_4$ was found, from the known probability of the transition $2p_4 - 1s_4$, to be $5.2 \times 10^6\ s^{-1}$. The resulting value of the lifetime of the level $3s_2$ is 3.2×10^{-8} s. The relative probability of the transition $3s_2 - 2p_4$, i.e., the ratio of the above transition probability to the total probability of the decay of the level $3s_2$ is 0.16.

The most comprehensive information on level relaxations can be gained when the line shape of absorption of the weak wave is observed in the presence of the strong wave. However, these methods are not widely used for measuring level lifetimes.

Three-Level Methods of Measurement

Unlike the methods of two-level spectroscopy, the methods of nonlinear spectroscopy with three levels are used to provide comprehensive information both on transition width, and on relaxation constants of individual levels. These data can be obtained when the line shapes of stimulated or spontaneous radiation of coupled transitions in an external field are studied. Comparison of resonance widths of probe waves that propagate along and opposite to an external wave provides data on the decay constant of a common level. The difference of halfwidths of the corresponding resonances is equal to the rate of decay of the common level,

$$\Gamma_+ - \Gamma_- = \gamma_0 \ . \tag{7.11}$$

This method is applicable to the case of "quenching" collisions that cause level damping and "strong" collisions that cause great change of direction of the velocity vector of a particle. Both types of collisions reduce the duration of the atom-field interaction. In the case of the "strong" collisions, an atom may stay on the level but after collision it ceases to interact with the field. From this point of view, the "strong" collisions are similar to the "quenching" ones. From (7.11) it is seen that the accuracy of measurement of γ_0 depends on the ratio of widths Γ_+ and Γ_-. Better accuracy is obtained with greatly different widths Γ_+ and Γ_- which is the case when $\gamma_0 \gg \gamma_1, \gamma_2$. When $\gamma_0 < \gamma_1, \gamma_2$ the difference between the widths Γ_+ and Γ_- is small; consequently, high accuracy of measurement of Γ_+ and Γ_- is required.

Because of rather large instrumental widths of Fabry-Perot interferometers, the accuracy of experiments for observing resonances of the spontaneous radiation on the coupled transition is usually insufficient to measure the rate of decay of the common level. Experiments that use induced transitions have better accuracy. The great possibilities for measuring lifetimes by this method were demonstrated in [7.5,11]. In the present work, we explore the gain line shape of the Ne coupled transition $2s_2 - 2p_4$ ($\lambda = 1.15$ μm) in the presence of the field (see Fig.5.14). The large difference between the resonance widths for the two opposite directions as well as the high accuracy of experiments enabled us to find the rate of decay of the level $2s_2$. The difference between the widths $\Gamma_+ - \Gamma_-$, which corresponds to the rate of decay of the level $2s_2$ was obtained by extrapolating the dependence of $\Gamma_+ - \Gamma_-$ to zero pressure. The measured value of the lifetime $(6.1 \pm 1.0) \times 10^{-9}$ s is in good agreement with results obtained from the other independent measurements.

7.2 Collision Broadening of Nonlinear Resonances

The methods of nonlinear laser spectroscopy of superhigh resolution enabled us to find a new approach to investigations of the influence of collisions on spectral line shapes. By use of narrow resonances inside a Doppler contour, collisions at low gas pressures can be studied when a collision broadening is considerably less than Doppler broadening. It permits measurement of the broadening and shift of a spectral line due to collisions at low pressures when a binary model of collisions is valid. These are the parameters that up to now have been the subject of numerous experiments with narrow resonances of absorption saturation of transitions of atoms and molecules. At a very low gas pressure, the collision frequency is comparable with the rate of decay of the levels. Under such conditions, elastic scattering of colliding particles can considerably affect the shape of narrow resonances. Therefore, study of resonance line shapes at low pressure provides information on characteristics of elastic scattering. The methods of laser spectroscopy enabled us to study perturbation of individual levels at collisions.

7.2.1 Principal Experimental Results

Thorough investigations on the collisional broadening of a Lamb dip by a foreign gas (helium) have been made on neon lines $\lambda = 0.63$ μm [7.12-15], $\lambda = 1.15$ μm [7.16] in a He-Ne laser. In [7.12,14] it was assumed that collisions only broadened the Lamb dip. A value of the dip width was selected such that an experimental dependence coincided with that calculated from the Lamb formula. The values of the Lamb dip width were found at different pressures. Then the effect of collisional broadening on the dip halfwidth was determined. A linear dependence of the collisional broadening on He pressure, with a slope of about 60 MHz/torr was obtained, within the limits of experimental errors. For 0.63 μm the results of various authors are in good agreement.

In some studies, the records of the Lamb dip were processed in accordance with (7.2). The Lamb-dip broadening was assumed to be principally associated with weak collisions and to depend linearly on pressure; saturation homogeneity and decrease of the dip depth were associated with "strong" collisions. The technique for recording the data was given in Section 7.1. The parameters Γ and a by which cross-sections of weak and strong collisions were determined were found for 1.15 μm lines [7.16] and for 0.63 μm lines [7.13,15].

The collisional shift of the 0.63 μm Ne lines, caused by collisions with He atoms has been measured by use of different methods. LISITSYN and CHEBOTAYEV [7.17] measured it by use of Lamb-dip-stabilized He-Ne lasers. When the pressure was changed, the frequency shift of the stabilized laser was associated with the collisional shift. In addition, they explored the shift of the center of the Lamb dip in a He-Ne laser relative to a peak in the spectrum of a He-Ne laser with a Ne absorption cell [7.19]. Experiments have indicated that there is practically no shift of the power peak at Ne pressure of 0.1 torr in the absorption cell. Therefore, the frequency shift of the He-Ne laser indicates that the shift is caused by collisions with He atoms. Linear shifts of the gain line towards the ultraviolet, of the order of 20 ± 2 and 16 ± 2 MHz per torr were obtained for ^{3}He and ^{4}He, respectively. Similar results were reported in [7.18].

Very simple experiments on the shift were made with the He-Ne/Ne laser. As seen in Fig.3.5, the power peak is shifted with respect to the gain-line maximum. This is caused, mainly, by the shift of the gain-line maximum in a He-Ne mixture. The shift measured in such a way in [7.18,20] was 21 ± 3 and 20 ± 3 MHz/torr, respectively. The results obtained on the shift and broadening of the 0.63 μm line caused by collisions with He atoms are rather reliable.

The large line shift indicates that interactions between colliding particles that result in phase randomization play a dominant role in the Lamb-dip broadening. The influence of weak collisions is slight. The linear dependence of the dip width on pressure also indicates the dominant role of collision interactions in Lamb-dip broadening. For an appreciable influence of "weak" collisions on dip shape, the dependence of dip width on gas density must be nonlinear. It is connected with the diffusion nature of widening of a nonequilibrium velocity distribution of particles in a great number of collisions [7.21]. "Strong" collisions are improbable. The observed saturation homogeneity and decrease of dip depth in the He-Ne laser, as indicated in [7.1], are caused, mainly, by resonant radiation trapping. Figure 7.5 shows the dependence of the parameters Γ and aΓ on pressure in the He-Ne mixture for the 0.63 μm line, according to the data of [7.13,15]. The solid curve, calculated from the model of resonant radiation trapping, is in good agreement

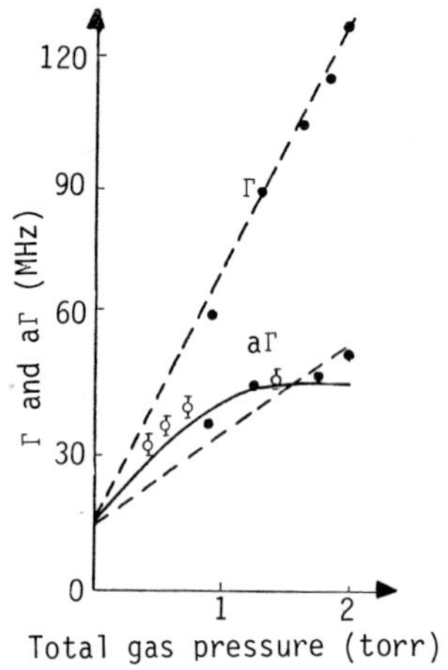

Fig.7.5 Dependence of parameters Γ and aΓ on pressure of a helium-neon mixture for $\lambda = 0.63\ \mu$. The continuous curve is calculated in [7.1], dotted curves in [7.13]. Points are experimental: • from [7.13], o from [7.15]

with the experimental results. The mechanism of collisions in the He-Ne mixture was analyzed theoretically by VDOVIN et al. [7.22], and by KOUTSOYANNIS and KARAMCHETI [7.23].

Narrow resonances may be used for investigating very-small line asymmetry when conventional spectroscopic methods cannot be used. The line asymmetry leads to the shift of the line maximum with respect to the center of the Lamb dip, which results in asymmetric dependence of generated power on frequency [7.16,24]. Analysis of this dependence can yield the parameter that describes the line asymmetry [7.16]. These experiments require rather careful tuning of laser systems, and maximum elimination of modification of the quality of the cavity when the frequency is changed.

The line asymmetry results in the dependence of the position of the dip center on the excitation level [7.19], which was used for investigating the 0.63 μm line asymmetry in the He-Ne laser.

The dependence of generated power on frequency, in an asymmetric line is

$$P(\Omega) \simeq \frac{\bar{\Omega}^2 - (\Omega - \Delta)^2}{1 + \Gamma^2/(\Gamma^2 + \Omega^2)} \ , \tag{7.12}$$

where $\bar{\Omega}$ is the frequency at which generation ceases, Δ is the shift of the maximum of the line, due to asymmetry.

The shift of the frequency that corresponds to the line maximum (a) and to the center of the Lamb dip agrees qualitatively with calculated curves (Fig.7.6). The line maximum is shifted about 2.5 MHz/torr towards the red region, relative to the center of the Lamb dip. The dip-center shift restricts the attainment of high values of frequency reproducibility. The

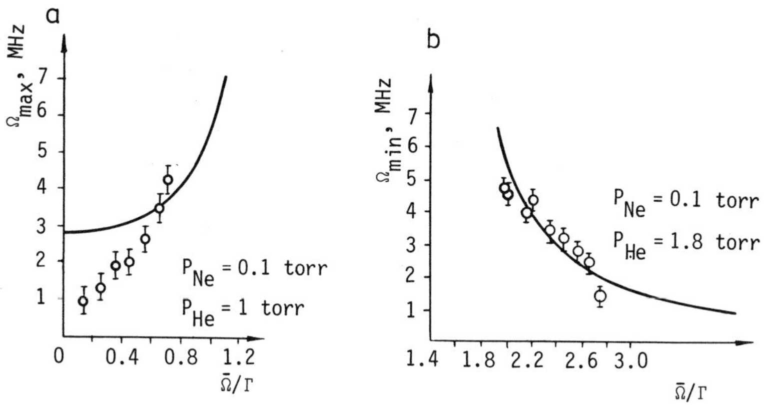

Fig.7.6a,b Dependence of asymmetry on the parameter $\bar{\Omega}/\Gamma$

results reported in [7.16] and [7.19] are in good agreement to each other but they are inconsistent with the data of [7.24]. The causes of this discrepancy are not yet clear.

A great number of studies have been devoted to investigations of collision broadening in neon lines, caused by collisions with neon atoms. The first results on resonance broadening in neon were obtained when the Lamb-dip shape was investigated in He-Ne lasers [7.16]. Acquisition of reliable data on broadening of neon lines in a He-Ne mixture is difficult because of masking broadening by collisions of Ne and He atoms and because of the small range of Ne operating pressures in the He-Ne mixture. So later investigations on broadening were made, mainly, in cells with Ne. For the 0.63 μm line, the results obtained by use of different methods agree well.

The data on collision broadening of lines from transitions 2s - 2p were obtained by use of different methods, also. The 1.5 μm ($2s_2 - 2p_1$) line broadening was measured, as reported in [7.25], by use of an internal absorption cell filled with Ne. The data on the 1.15 μm line broadening in a Ne laser were reported in [7.1,2,36]. The data reported in [7.1,2,5,25], obtained by different methods agree well (the collisional broadening of transitions 2s - 2p is equal to about 10 MHz/torr), but they are quite inconsistent with the data of [7.26]. The discrepancy appears to be attributable to the method of data processing, because [7.26] does not take into account the saturation homogeneity of the transition 2s - 2p due to radiation trapping.

A mercury laser at the wavelength of 1.53 μm ($7p\,^3P_1^0 - 7s\,^3S_1$ [7.27]) offers interesting possibilities for investigation of the collision broadening of narrow resonances by foreign gases (see Subsec.7.2.2). The laser can operate with admixture of all noble and many molecular gases. Figure 7.7 shows the dependences of the resonance broadening and shift in the 1.5 μm mercury line on He and Ne pressures [7.27].

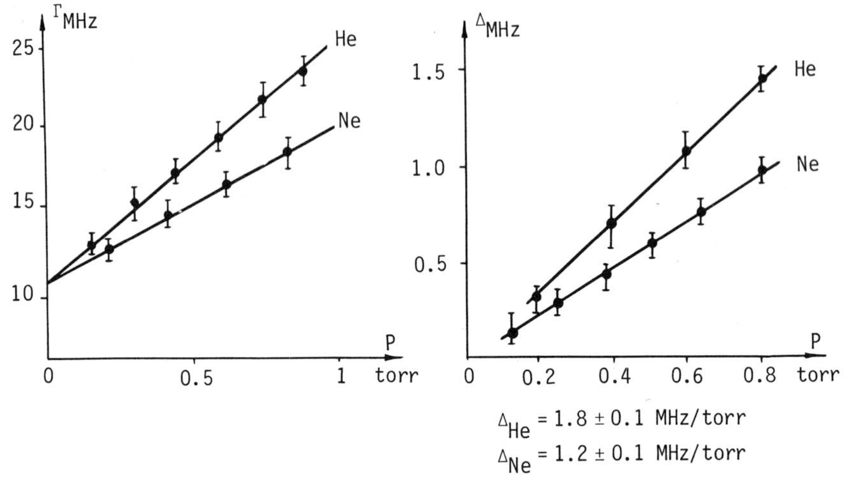

Fig.7.7 Dependence of broadening and shift of the Hg line at $\lambda = 1.5$ μm on He and Ne pressure

7.2.2 Mechanisms of Collisions

As the experimental results have indicated, data that characterize resonance broadening and shift to high accuracy are usually obtained with no principal difficulty. However the physical interpretation of the processes of collisions, and the selection of a model, proved to be difficult. Resonance shapes are affected by elastic angular scattering of atoms caused by collisions, by phase randomizing and quenching collisions, by dependence of line broadening and shift on atomic velocities, by resonant radiation trapping, and by field polarization. The influence of so many different factors makes it difficult to analyze experimental results; this presents difficulties for selection of a model of collisions and for determination of the constants that characterize the elementary processes.

It is evident that in order to determine the model of collisions and to get information on the constants of interaction it is necessary to carry out complex investigations such as

a) simultaneous measurements of the collisional broadening of a Doppler contour and of narrow resonances;

b) analysis of the narrow resonance shape and of the Doppler contour of a line;

c) investigation of "strong" collisions connected with resonance excitation exchange;

d) study of line shift and collisional broadening;

e) investigation of the temperature dependence of broadening and shift as well as the dependence of homogeneous width on detuning frequency;

f) investigation of the dependence of line broadening on the principal quantum number of one of the levels.

Conducting all of these investigations with one substance would be a complex and lengthy project so far not fully completed for any substance. The broadening of neon lines in the gas was explored in great detail [7.2]. Broadening of lines related to the levels $2p^5ns$ ($n = 3, 4, 5, 6, 7$) was explored in these experiments. Broadening of narrow resonances was investigated for the transitions $2p^54s - 2p^53p$ and $2p^55s - 2p^53p$. The Doppler contour shape of an absorption line was studied on visible lines by use of a Fabry-Perot interferometer. The line contours of spontaneous emission and absorption were analyzed.

Comparison of the collision broadening of the Doppler contour and of the narrow resonance on the 0.63 μm line led to a conclusion about the contribution of the elastic scattering. In collisions with no phase randomization, the Lamb-dip broadening is determined by the characteristics of angular scattering. At small collision frequencies and $ku\Theta \gg \gamma$ the collision broadening of the Lamb dip is determined by the total elastic scattering cross-section. When $ku\Theta < \gamma$ the collisions with no phase randomization, as noted previously, do not lead to dip broadening.

The behavior of the Doppler contour width in such collisions differs qualitatively from that of the Lamb dip. Collisions with no phase randomization result in reduction of the effective vector velocity of atoms and line narrowing, which is well known in the microwave range. Therefore the difference between the broadening of the two contours indicates the contribution of the elastic scattering. In collisions with phase randomization there is no difference between resonance broadening and broadening of the Doppler contour. An additional contribution to resonance broadening can be caused by angular scattering of atoms at collisions, which leads to smoothing of the dip in the velocity distribution of atoms [7.28]. The additional contribution depends nonlinearly on pressure and leads to a greater broadening than is exhibited by the Doppler contour. Comparison of experimental data for the 0.63 μm line has indicated that collision broadening of the Doppler contour even exceeds slightly the resonance broadening. This enabled us to arrive at the conclusion that the collision resonance broadening is mainly caused by collisions with phase randomization. The contribution of atom diffusion in a velocity space, which is due to the angular scattering by collisions, is negligible. The linear dependence of width on pressure also confirms this conclusion. The characteristic scattering angle of the model of solid spheres may be estimated from

$$\Theta = \hbar/Mua \ , \tag{7.13}$$

where a is the sphere radius, which may be taken equal to an optical one and can be found from the data on the line broadening.

When analyzing the broadening of different lines we can infer that the 0.63 μm line broadening is caused by perturbation of the upper level $2p^5ns$. Hence the characteristic scattering angle $\Theta \approx 10^{-2}$ rad. Because, when analyzing the resonance in the line center, we consider atoms with zero velocity projection onto the axis of observation, the angular scattering of such atoms leads to the appearance of v_z, a velocity component. The width of the velocity distribution of atoms for one event of scattering is $\Delta v = u\Theta$. An additional dip broadening has the value $ku\Theta$. For the 0.63 μm line, this broadening is less than the dip width. With an increase of pressure, owing to its diffusion nature, the collision angle increases in proportion to the square root of density, whereas the broadening caused by phase randomization increases in proportion to density. From this it follows that at pressures of several torr the contribution of atom diffusion is very small, compared to the broadening caused by phase randomization.

Saturation homogeneity and reduction of dip depth were explained by "strong" collisions in early investigations of the Lamb dip in He-Ne lasers. Investigations of excitation transfer by change of velocity in three-level schemes [7.5] have showed that the change of velocity of atom excitation is due to resonant radiation trapping, not to collisions. Radiation-trapping effects appear especially strongly on the 1.15 μm line.

Analysis of collision broadening of neon lines as a function of principal quantum number revealed the mechanism of interaction of colliding Ne atoms. Figure 7.8 shows the dependence of collision broadening as a function of the principal quantum number of one of the levels, for the lines 5852 Å ($n = 3$), 11522 Å ($n = 4$), 6328 Å ($n = 5$), 5349 Å ($n = 6$), 4892 Å ($n = 7$). The difference of the broadening of the lines can be attributed to perturbations of levels

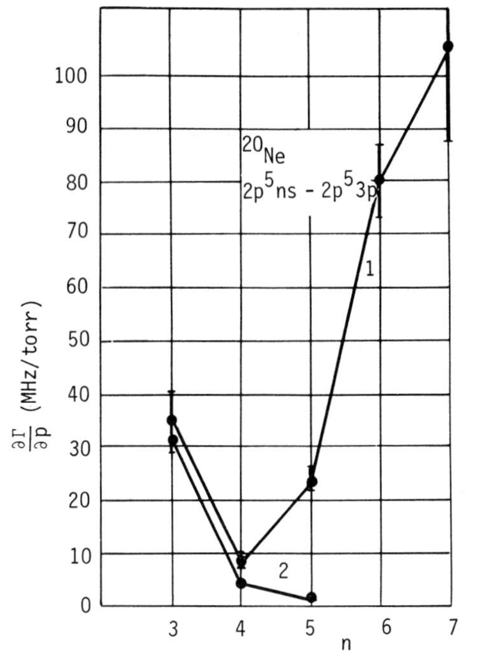

Fig.7.8 Dependence of collision broadening of Ne lines $2p^5ns$ - $2p^53p$ on the main quantum number of one of the levels. 1 - measured values; 2 - calculated values of the resonance broadening

that belong to a configuration ns. The distinctive feature of the dependence is a well-defined minimum that corresponds to $n = 4$. Because one of the levels of the transitions under study is resonant, it is reasonable to assume that one of the mechanisms of line broadening is resonant excitation exchange at collisions. For resonant collisions, the theory is well developed (see the latest works by KAZANTSEV [7.29], BERMAN and LAMB [7.30]); the experimental results can be compared with the calculated value of broadening. The line broadening caused by resonance exchange can be found from the formula

$$\frac{\partial\Gamma}{\partial P} = 5.7\bar{\lambda}^3 NA \ , \quad \bar{\lambda} = \lambda/2\pi \ , \tag{7.14}$$

where λ is the wavelength of the resonance radiation, and A is the resonance transition probability.

The collision broadening measured for the line 5852 Å ($n = 3$) is equal to the calculated value. With increase of the principal quantum number, the reduction of A leads to a decrease of the collision broadening due to resonance-excitation exchange. For the most highly excited levels, with quantum number $n = 7$, the resonant broadening becomes small. Exchange forces play a leading role here.

Nonresonant forces are characterized by the dependence of the broadening on the relative velocity of colliding atoms. This dependence may be used to find the type of interaction. Because the mean velocity of the colliding atoms depends on temperature, the collision broadenings and shifts are also functions of temperature. In some cases, the dependence of the broadening on the velocity may be obtained by measuring the homogeneous line width at different frequency detunings from the line center. The data on the neon line broadening obtained for the line center, on a half-height, and on the wings of the line, differ from each other and are equal to 24 ± 2, 30 ± 5, 44 ± 5 MHz, respectively. This difference can be explained if a dependence of the broadening on the atomic velocities is assumed. The nonlinear resonance in the line center is related to atoms whose velocity projections onto the axis of observation are zero ($v_z = 0$), and the Doppler contour of the spontaneous emission line is formed from the emission of atoms with different velocities $v_z \sim u$.

The dispersion part of the contour, measured by using the technique of [7.2], is sensitive to the change of the contour shape on the wings. On the line wings, the atoms with the velocity projections $v_z > u$ play a leading role in formation of the contour; the value of the lorentzian width found by this method is more than that found from the Lamb dip (for potential of the form $u \sim C_n/R^n$). The collision broadening measured on the line wings is also more than that measured on the half-height or in the center of the line by the method suggested in [7.32].

The dependence of the shift on velocity leads to line asymmetry. Reference [7.33] proposes a theory of the line shape that allows for the dependence of the shift and broadening of the line on the velocities of the colliding particles.

The data on the relative shift and broadening of the line may be also used for finding the atom-interaction potential. When the broadening and shift of the line by foreign gases is studied, it is especially interesting to explore the case in which the mass of the emitting atom is much more than that of atoms of the foreign gas. Illustrations of such studies are the measurements of the broadening and shift of the Lamb dip in a mercury laser by foreign gases, which were described previously. Because the atomic mass of Hg is usually large compared to the mass of the foreign atoms, when analyzing the narrow resonance broadening, we can neglect the rate of heavy-particle interaction and use the simplest theory of the collision broadening of the Lamb dip, which takes into consideration the phase randomization of only one oscillator. In this case, the broadening constant of the Lamb dip coincides unambiguously with that of the line contour of the emitting atoms in the conventional theory of collisional broadening, and rather precise measurements of collision broadenings and shifts of resonances can be used to calculate repulsive and attractive parts of the interaction potential. In order to calculate the interaction constants C_6 and C_{12} for the Lenard-Jones potential 6-12 by the ratio between the line shift and broadening the results of [7.34] can be used.

The broadening and shift for the potential 6-12 may be expressed in terms of the interaction constants,

$$\gamma = 4\pi\left(\frac{3\pi}{8}\right)^{2/5} Nu^{3/5} C_6^{2/5} B(\alpha) \tag{7.15}$$

$$\Delta = 2\pi\left(\frac{3\pi}{8}\right)^{2/5} Nu^{3/5} C_6^{2/5} S(\alpha) \ , \tag{7.16}$$

where

$$\alpha = 0.536 u^{6/5} C_{12}/|C_6|^{11/5} \ ,$$

N is concentration of surrounding particles, and $B(\alpha)$ and $S(\alpha)$ are integral functions that depend on the constants C_6 and C_{12}. The shift and broadening are thus connected through the functions $B(\alpha)$ and $S(\alpha)$. The functions $B(\alpha)$ and $S(\alpha)$ are plotted in Fig.7.9 [7.34]. In collisions with atoms of He and Ne, the shift-to-broadening ratio proved to be approximately 0.11 for the mercury line at 1.5 μm. From these experimental results, the functions $B(\alpha)$ and $S(\alpha)$ were found and the coefficients C_6 and C_{12} were determined. For collisions with the atoms of He, they are $C_6 = 5.2 \times 10^{-33}$ cm^6 s^{-1} and $C_{12} = 8.3 \times 10^{-77}$ cm^{12} s^{-1}; with the atoms of Ne, $C_6 = 10^{-33}$ cm^6 s^{-1} and $C_{12} = 2.1 \times 10^{-78}$ cm^{12} s^{-1}.

7.2.3 Investigation of Relaxation Processes on Individual Levels by TLS Methods

Up to now, we described the influence of collisions upon the shape of saturation resonances. The use of resonances in three-level systems offers quite new possibilities. They make it possible to study independently relaxation processes of individual levels, and to observe line broadenings on forbidden transitions. We illustrate it by investigations of the collision broadening of resonant SRS lines in neon.

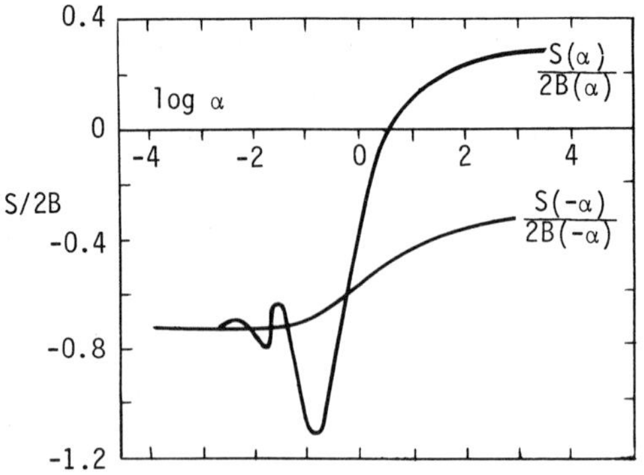

Fig.7.9 Function $S(\alpha)/2B(\alpha)$ plotted against $\log \alpha$

In the previous section, we indicated how important it was to find the physical cause of excitation diffusion in the velocity space, in order to interpret correctly the data on the influence of collisions upon the resonance shape. In many studies, it was assumed that the saturation homogeneity on neon transitions was due to "strong" collisions, in which the atomic velocities varied within the limits of thermal velocities.

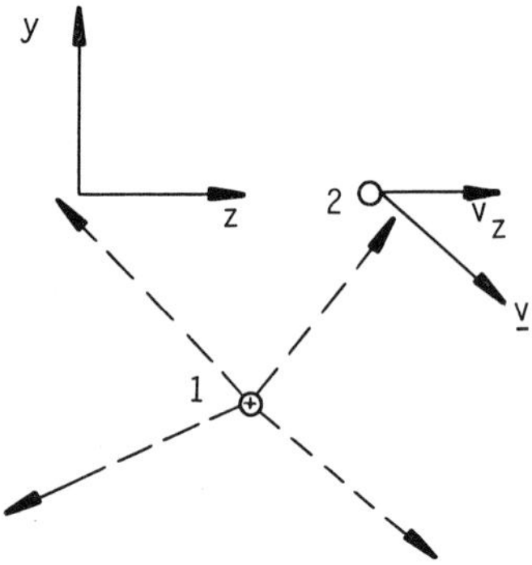

Fig.7.10 Transfer of excitation with change of atomic velocity due to resonant radiation trapping

Resonant radiation trapping results in the same effects as do "strong" collisions. Therefore, studies of resonant radiation trapping are important for investigations of collisions. At concentrations of neon atoms ($P \sim 10^{-1}$ torr) used under operating conditions of He-Ne lasers, a free-path length of an ultraviolet photon is far less than laser-beam dimensions. Therefore, the resonant radiation trapping takes place in a volume, and the lifetime of the level $2s_2$ is determined by the decay on lower-lying levels only 2p ($\tau \simeq 10^{-7}$ s). The influence of radiation trapping on the nature of the field-atoms interaction is explained with the aid of Fig.7.10. Let atom 1 move in the direction perpendicular to the plane of the figure. In this plane, the atom emits with the frequency of the line center. The different directions of photon emission are marked with dotted lines. The photons emitted by this atom may be absorbed by the other atoms whose directions of motion are perpendicular to that of photon propagation. Let us take atom 2 as an example. In the direction of the observer, it interacts with the field at the other frequency, and its velocity projection onto the axis z is not equal to zero. Therefore, the excitation exchange occurs between atoms that have different velocities. Though excitation of atoms occurs in a narrow velocity space Δv, atoms can be found with arbitrary velocities, due to overradiation and catching of photons. On the other hand, the excitation may be similarly transferred from the region of velocities of the atoms that do not interact with the field to the region of velocities that are resonant. This results in saturation homogeneity by interaction with the strong field. The same scheme of neon levels $2s_2 - 2p_4$, $2s_2 - 2p_1$ was chosen as the operating scheme. The laser field at $\lambda = 1.52$ μm, when absorbed, transfers atoms to the level $2s_2$ which is resonant. Because under experimental conditions and at gas pressure of about 1 torr the homogeneous width is considerably less than the Doppler width, the field interacts with the atoms whose velocities meet the condition $v_{res} = \pm\Omega/k$. As a result of this interaction, a peak arises in the velocity distribution of atoms on the level $2s_2$. Owing to the above processes of overradiation, atoms with arbitrary velocities appear in the distribution, as well.

A nonequilibrium velocity distribution of atoms and a wide band with a Maxwell distribution can be recorded by the line shape of radiation from the level $2s_2$. It is possible to use the line shape of spontaneous radiation on any coupled transition. The easiest way of investigation of the line shape of stimulated radiation is to use a single frequency tuned within the limits of a Doppler radiation line. The transition $2s_2 - 2p_4$ is the most convenient for these purposes. It is easy to get generation on this transition, in a wide range on the 1.15 μm line in a He-Ne laser. The nonequilibrium part of the velocity distribution of atoms is responsible for the narrow resonance in the gain (absorption) line in the transition that arises under the action of a pump field at $\lambda = 1.52$ μm. The wide band produces an ordinary Doppler contour of the gain (absorption) line. The amplitude relation of the sharp and wide parts of the contour of the gain line indicates the role of the resonant radiation trapping.

The first experiments were made on the line shape of stimulated radiation in a three-level laser (see Chap.9). The radiation of a high-power single-frequency He-Ne laser at $\lambda = 1.52$ μm was produced in a laser cavity at $\lambda = 1.15$ μm. Usually, the radiation absorption at $\lambda = 1.52$ μm is observed in a discharge of pure Ne. Depending on the discharge current and Ne pressure at

$\lambda = 1.15$ μm, both absorption and gain can be observed. It is more convenient to study resonant radiation trapping under conditions in which a small absorption is observed at $\lambda = 1.15$ μm. Then the gain and generation are fully due to the action of the field with $\lambda = 1.52$ μm. Figure 7.11a shows the velocity distribution of atoms that are on the level when the 1.52 μm laser is pumped. Figure 7.11b shows the gain line shape in the 1.15 μm laser. When the field frequency is detuned at $\lambda = 1.52$ μm and when the field intensity is increased, the generation has a complicated form (Fig.7.11c). There is a sharp generation peak associated with the resonance SRS and the dependence of generated power on frequency with the Lamb dip in the line center that is characteristic of usual gas lasers. That dependence is caused by resonant radiation trapping. Doppler "underlining" was observed within the range of neon pressures from 0.2 to 1.5 torr. When the neon pressure was varied by a factor of 8, the amplitude relation of peaks and the underlining changes slightly. Estimates have shown that, at a neon pressure of 0.1 torr, the frequency of "strong" collisions must be more than 10^8 s^{-1} in order to lead to an appreciable effect of excitation diffusion in the velocity space. On the other hand, at such collision frequencies, the sharp structure as well as the Lamb dip must vanish at pressures greater than 1 torr. These considerations and further investigations show that the observed excitation diffusion is due to resonant radiation trapping.

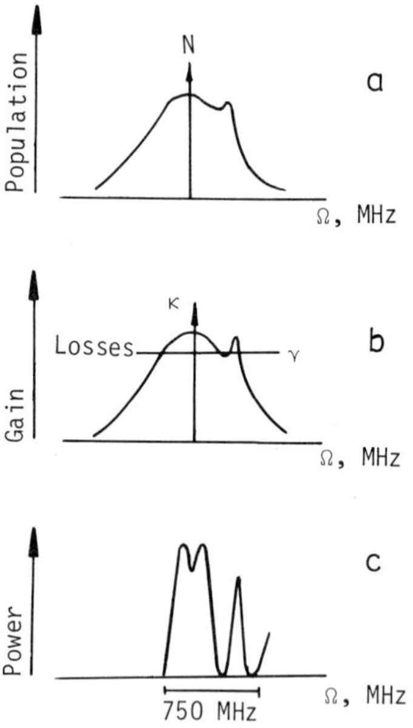

Fig.7.11a-c Velocity distribution of atoms on level $2s_2$ (a); gain line shape (b); dependence of generated power on frequency at $\lambda = 1.15$ μm when pumped by $\lambda = 1.5$ μm (c)

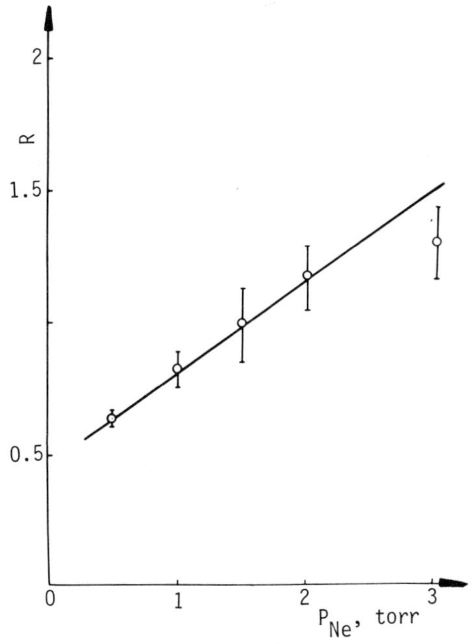

Fig.7.12 Dependence of ratio of resonance and underlining amplitudes on pressure

Quantitative investigations with the aid of a three-level laser are difficult, because of the complicated nature of the generation characteristics of the laser. The radiation trapping and the 1.15 μm line broadening, using three-level schemes, were thoroughly investigated by study of the line shape of the stimulated radiation. These experiments are described in Chapter 5. Figure 5.14 shows the records of a signal of the absorption line at $\lambda = 1.15$ μm under the action of the field at $\lambda = 1.5$ μm. Figure 5.14 shows records of signals with and without cancellation of Doppler underlining. The ratio of the underlining amplitude to the resonance amplitude is about 1. With an increase of pressure, the relative underlining amplitude increases. However, the analysis shows that its relative increase is associated mainly with the narrow resonance broadening and not with an increase of the frequency of the collisions that lead to formation of the underlining itself. Figure 7.12 shows the dependence of the ratio R of the amplitudes of resonance and underlining R in the gain line of a probe wave at $\lambda = 1.5$ μm, for oppositely travelling waves.

A convincing proof of the leading role of resonant radiation trapping in underlining at neon pressure of about 1 torr is that extrapolation of the dependence of the ratio R to zero pressure is a constant. If the underlining is caused by collisions only, the dependence on pressure should begin from zero. The experimental results shown in Fig.7.12 depend slightly on pressure; this once more confirms the slight role of collisions. From the dependence shown in Fig.7.12, we can see that, at about 1 torr, the contribution of collisions to excitation diffusion amounts to about 20%. Thus, at pressures greater than 5 torr, collisions play a role comparable with radiation trapping.

The method for observing a line of resonant stimulated Raman scattering (SRS) permits us to explore directly the processes of excitation relaxation with change of the velocity of the excited particles. Because the line shape of the coupled transition is sensitive to the velocity distribution of particles on the common level, from its change we can study the change of atomic velocities at collisions. The SRS line shape caused by two-quantum transitions is sensitive to line broadening by collisions. The difference of collision broadenings of the lines of forward and backward scattering clarifies the role of different levels in line broadening. This information could not be gained by use of traditional methods.

The experiments described for observing radiation trapping involve observations of excitation diffusion for velocities within the range of the mean velocity of particles v_0. It is obvious that the method can be used successfully in cases in which the Doppler shift at scattering by an angle equal to $ku\Theta$ is more than the emitted line width Γ. Then the corresponding underlining has a width of $\sim ku\Theta$. From the experimental data and theory of collisions, the angle Θ is 1° within an order of magnitude, which corresponds to an underlining width within a range of about 10 MHz. Such a magnitude of underlining is comparable with resonance widths; direct observation of it is therefore difficult. Analysis of the results requires very careful experiments and processing of their results. Such experiments have not been done yet. But even a simple analysis of collisional broadening of a SRS resonance yields much information on the mechanism of collisions. The nature of resonance broadening is rather critical to the type of interaction of colliding particles, and to the contribution of different levels to line broadening. Therefore, investigations of the effects of collisions on resonance SRS lines are much more informative than those on Lamb-dip broadening.

A simpler way is to use a phenomenological model of collisions for describing their influence upon a resonance SRS line. The system of (2.67,68) for a density matrix permits description of the influence of quenching and phase-randomizing collisions by the corresponding replacement of relaxation constants Γ_{ik} and γ_i.

As should be expected, the most interesting peculiarities arise in a forward-scattering line. In collisions with phase randomization, not only the parameters of resonance change but also its shape: phase-randomizing collisions lead to underlining with the line width of a step-by-step transition. The underlining intensity is proportional to the frequency of collisions. A similar phenomenon arises in a resonance fluorescence line, in collisions [7.35].

The line shape of RSRS at $k'/k > 1$ and $N_0 \simeq 0$ has the form

$$\kappa_- = \kappa_0' \left\{ \frac{\Gamma_{01} + \Gamma_{02} - \gamma_0 - \Gamma_{12}}{\Gamma_{01} + \Gamma_{02} - \Gamma_{12}} \cdot \frac{\Gamma_+}{(\Omega' - \frac{k'}{k}\Omega)^2 + \Gamma_+^2} + \right.$$

$$\left. + \frac{\gamma_0}{\Gamma_{01} + \Gamma_{02} - \Gamma_{12}} \cdot \frac{\Gamma_-}{(\Omega' - \frac{k'}{k}\Omega)^2 + \Gamma_-^2} \right\} , \qquad (7.17)$$

where

$$\Gamma_{ik} = \frac{\gamma_i + \gamma_k}{2} + \frac{\nu_i + \nu_k}{2} , \qquad \Gamma_- = \Gamma_{12} + (\frac{k'}{k} - 1)\Gamma_{10} , \qquad \Gamma_+ = \Gamma_{02} + \frac{k'}{k}\Gamma_{10} ,$$

γ is the rate of decay of the ith level, allowing for collisions, $\nu_i = n_0 \bar{v}\sigma_{sc}$, σ_{sc} is the total cross-section of elastic scattering of an atom on the ith excited level in collisions with surrounding particles, n_0 is the density of particles, and

$$\kappa_0' = \frac{8\pi^{3/2} k' p_{02}^2 V^2 N_1}{\hbar k u \gamma_0} .$$

From (7.17), we can see that only collisions on the upper level lead to the appearance of underlining, with a width Γ_+. Transform (7.17) to the form

$$\kappa_- = \kappa_0' \left[\frac{\gamma_0}{\gamma_0 + \nu_0} \cdot \frac{\Gamma_-}{(\Omega' - \frac{k'}{k}\Omega)^2 + \Gamma_-^2} + \frac{\nu_0}{\gamma_0 + \nu_0} \cdot \frac{\Gamma_+}{(\Omega' - \frac{k'}{k})^2 + \Gamma_+^2} \right] . \qquad (7.18a)$$

A backward-scattering line has a lorentzian form. The resonance width depends on line broadening of the transitions $0 \to 1$ and $0 \to 2$.

For forward scattering

$$\kappa_+ = \kappa_0' \left[\frac{\Gamma_+}{(\Omega' + \frac{k'}{k}\Omega)^2 + \Gamma_+^2} \right] . \tag{7.18b}$$

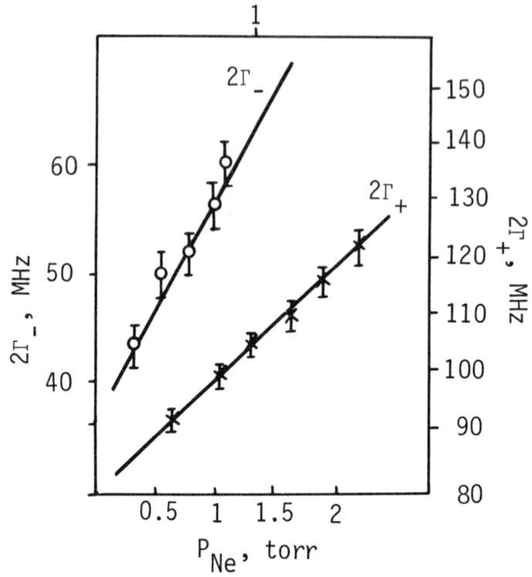

Fig.7.13 Dependence of line width of forward ($2\Gamma_-$) and backward ($2\Gamma_+$) scattering on Ne pressure

a) Broadening of Lines. Figure 7.13 shows the dependence of $2\Gamma_+$ and $2\Gamma_-$ on pressure. The backward line width is composed of the broadening of the transition $2s_2 - 2p_1$ (it determines the broadening of a Bennett hole in the velocity distribution of atoms), allowing for the differences of wavenumbers k and k', and of the broadening of the transition $2s_2 - 2p_4$. The two transitions belong to the same electron configuration and have a common resonant level, which must make a principal contribution to the line broadenings. Therefore, we can consider the line broadenings to be approximately equal. Hence the collision broadening of the neon transition 2s - 2p is equal to 10 MHz/torr. The accepted values of the broadening are in good agreement with those measured in [7.31]. If we use the earlier measurements of broadening of the transition $2s_2 - 2p_4$ by the Lamb dip (8.8 MHz/torr) we shall obtain 10 MHz/torr for the broadening of the transition $2s_2 - 2p_1$; this is in good agreement with the results of measurements of a power resonance in a laser with nonlinear absorption.

b) Broadening of a Forbidden Transition. The broadening of the forward line is composed of the broadening of a forbidden transition $2p_1 - 2p_4$ and of the broadening of a Doppler part of the SRS line. Collisions with phase randomization lead to underlining with a width $2\Gamma_+$. We estimate the influence of the underlinings on the effective width of the scattering line, assuming that $\gamma_0 \gg \nu_0$. The change of halfwidth caused by the underlining is

$$\Delta\Gamma \sim \Gamma_+ \frac{\Gamma_-^2}{\Gamma_+^2 + \Gamma_-^2} \cdot \frac{\nu_0}{\gamma_0} \cdot \frac{\Gamma_+}{\Gamma_-} \; . \tag{7.19}$$

This change may be taken for an additional resonance broadening. A tentative estimate of $\Delta\Gamma$ can be obtained from measurements of the collisional broadening of the 1.15 μm line. If it is assumed that the contribution of the phase-randomizing collisions is the highest possible value, 8 MHz/torr, then from (7.19) we obtain an estimate of the upper limit of the value of the additional resonance broadening, 2.5 MHz/torr. This value lies within the range of the experimental accuracy. Knowing the broadening of Γ_- and of the 1.15 μm line, we obtain 5 ± 2.5 MHz/torr for the broadening of the forbidden transition.

c) Relaxation of Individual Levels. As we have already indicated, for quenching and "strong" collisions the frequency of collisions on the level 0 can be found from the difference $\Gamma_+ - \Gamma_-$. The broadening of the common level is 6.5 ± 3 MHz/torr. The differences between the broadenings of the 1.15 and 1.5 μm lines and the broadening of the level $2s_2$ give the broadening of the levels $2p_1$ and $2p_4$, 1.5 and 3.6 MHz/torr, respectively. Their sum gives the broadening of the forbidden transition found previously. The values of the broadening of the levels $2p_1$ and $2p_4$ agree with the data gained from the measurements of relaxation of the magnetic moment (measurements were carried out for the level $2p_9$) at collisions [7.36].

7.3 Investigation of Level Structures

The most important aspect of the matter is the ability of the methods of nonlinear spectroscopy to find out the structure of absorption lines hidden by Doppler broadening. Until recently, a great part of the experiments have been with atoms and molecules whose absorption-line frequencies coincided accidently with frequencies generated by the most popular lasers. Large potentialities of nonlinear laser spectroscopy will be discovered when tunable single-frequency visible and infrared lasers are used.

7.3.1 Isotopic Structure

One of the first demonstrations of the potentialities of nonlinear spectroscopy was the work [7.37] in which narrow resonances in the line of spontaneous radiation from a cavity of the He-Ne 1.15 μm laser were used for precise measurements of isotopic shifts of two optical transitions in neon. Isotopic structure usually hidden by Doppler broadening was totally resolved. Measurements were made by use of a single-mode He-Ne laser oscillating at $\lambda = 1.15$ μm

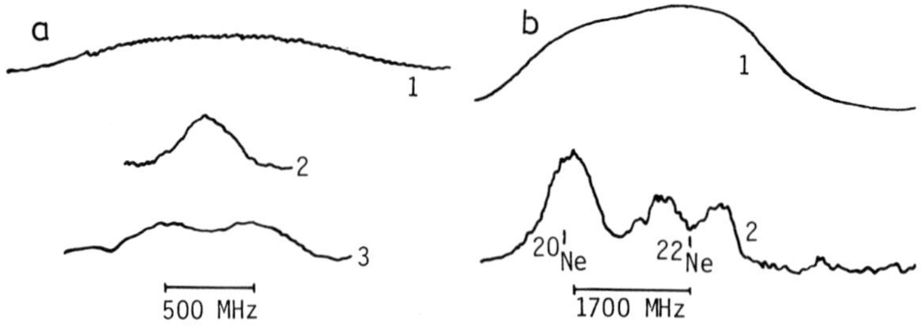

Fig.7.14a,b Shape of spectrum of spontaneous radiation of Ne at 0.6096 μm ($2p4 - 1s4$ transition) from cavity of a He-Ne laser operating at $\lambda = 1.15$ μm with ^{20}Ne isotope (a) and with mixture of ^{20}Ne and ^{22}Ne isotopes (b). 1 - Doppler-broadened contour of spontaneous radiation; 2 - structure observed when the generated frequency at 1.15 μm is tuned to the center of the gain line of ^{20}Ne; 3 - detuned from the center of the gain line of ^{20}Ne

($2p_4 - 2s_2$ transition). Spontaneous radiation ($\lambda = 6096$ Å) from a transition to a lower level was observed through one of the mirrors. The spontaneous-radiation spectrum was measured with a scanned Fabry-Perot interferometer. Figure 7.14 shows the spectrum of spontaneous radiation from the level $2p_4$ for the case where only one isotope ^{20}Ne was used in the laser (a), and for the case where a laser tube was filled with a mixture of isotopes ^{20}Ne and ^{22}Ne in the ratio 2:3 (b). In the first case, the spontaneous-radiation spectrum contains one peak when the laser frequency is tuned precisely to the center of the Doppler contour of the gain line of ^{20}Ne, or two symmetric peaks when the laser frequency is detuned with respect to the line center (the center of the Doppler contour is exactly midway between the two peaks). The emergence of the two peaks is, evidently, connected with the action of a standing-wave field of the coupled transition in causing the formation of two peaks in the velocity distribution of atoms of ^{20}Ne on the level $2p_4$. When the frequency of the laser that contained the mixture of the isotopes was tuned to the line center of the isotope ^{20}Ne, the spontaneous-radiation spectrum comprised three peaks: one due to ^{20}Ne and two due to ^{22}Ne. The center of the Doppler line of ^{20}Ne coincided with the peak of ^{20}Ne, and the center of the Doppler contour of ^{22}Ne lay midway between the two peaks. Thus, the isotopic shift of ^{20}Ne and ^{22}Ne for the 6096 Å line was directly measured. Its value was 1706 ± 30 MHz; for the 1.15 μm line it was 257 ± 8 MHz; the frequencies of the isotope ^{20}Ne were shifted towards the red.

The hyperfine structure of lines from the odd isotope ^{21}Ne was observed and the quadrupole moment of the ^{21}Ne nucleus was measured in [7.38]. In this work the output radiation from a sufficiently high-power single-frequency He-Ne laser ($\lambda = 1.15$ μm) was focused in an external gas-discharge cell that contained the odd isotope ^{21}Ne at a low pressure (about 0.1 torr). The spontaneous radiation at $\lambda = 6096$ Å from the cell was observed forward and backward and analyzed by use of a Fabry-Perot interferometer scanned over a spectral range of 4090 MHz. It is important to note that, unlike the case of a standing wave, the use of travelling waves makes the analysis of the observed line structure simpler, because in the latter case the number of narrow resonances is approximately half as many. The spectrum of the forward and backward spontaneous radiation may be observed separately.

Figure 7.15 shows the hyperfine splitting for the levels $1s_4$, $2p_4$ and $2s_2$ of ^{21}Ne, and Fig.7.16 shows the spontaneous-radiation spectra at 6096 Å observed in the direction of a strong field (a) on the coupled transition and in the opposite direction (b). The Doppler line shape of spontaneous radiation is also shown for each case. The calculation shows that the hyperfine interaction in ^{21}Ne yields 18 three-level systems. Eight of them were observed in the experiment, the intensities of the other ten resonances were negligible. However, the forward and backward spontaneous radiations yield two different spectra, which must be described with the same set of parameters. When the parameters are selected in a computer, shapes of the spectrum are calculated that are simultaneously in good agreement with the experimental data for the two cases. The results of the theoretical adjustment are shown in Fig.7.16. In the calculations, the differences of the resonance widths in a three-level system were taken into account, with due regard to the directions of propagation of the strong and probe waves. For the levels of ^{21}Ne, the width of the narrow resonance for observation in the direction opposite to that of the strong wave is $2\Gamma_- = 225$ MHz, and the width of the wider resonance in the opposite direction is $2\Gamma_+ = 265$ MHz. As a result of the calculation, the quadrupole moment of the nucleus ^{21}Ne was determined,

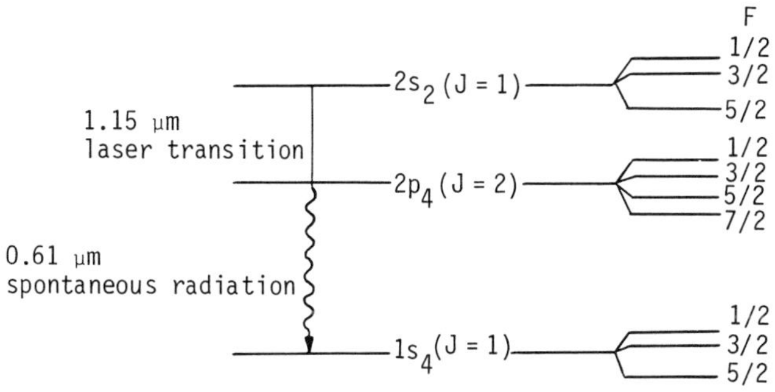

Fig.7.15 Hyperfine splitting of $1s_4$-, $2p_4$-, and $2s_2$-levels of ^{21}Ne

Q = (+0.1029 ± 0.0075) barn, which was in agreement with the value which was known previously with less accuracy [7.39].

The development of tunable lasers will permit use of these very effective methods for measuring fine, hyperfine, and isotopic structures of any atomic or molecular level.

7.3.2 Investigations of Zeeman and Stark Effects

In the optical spectral region, investigation of Stark and Zeeman effects has required strong magnetic and electric fields, because the splitting of the levels must exceed the Doppler width, at least. Use of nonlinear optical

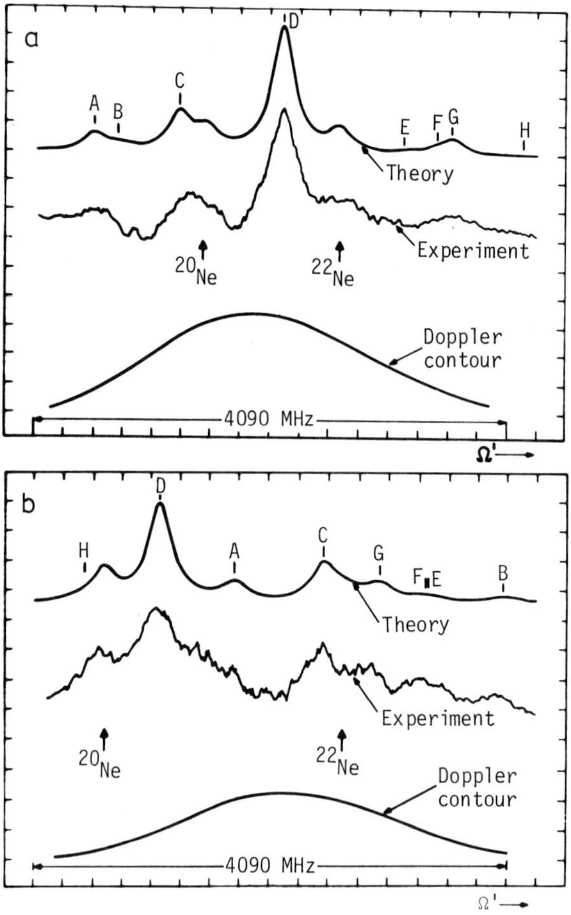

Fig.7.16a,b Experimental and theoretical shapes of the spontaneous-radiation spectrum of the 6096 Å line of ^{21}Ne in the strong travelling-wave field of the 1.15 μm transition. a - observation in the direction of the strong wave; b - observation in the opposite direction

resonances eliminates Doppler broadening; consequently, Zeeman and Stark effects can be studied under conditions where splitting of the components is far less than the Doppler width.

Outwardly, an electric or magnetic field can play a double role. For a nondegenerate, two-level transition a greater field changes only the transition frequency and can be used for tuning the central part of the Doppler-broadened line, i.e., of the frequency of a narrow nonlinear resonance to that of a laser field. Here the use of narrow nonlinear resonances permits more precise measurements in much weaker fields, because the magnitude of an external field needed to shift the line by the resonance width decreases by a factor of ku/Γ, compared with linear spectroscopy.

Real atomic and molecular transitions are degenerate and the external field, first of all, gives rise to level splitting. As a result, the Doppler-broadened line becomes a set of separate overlapping lines. By use of nonlinear resonances, this splitting can be perceived long before the splitting of Doppler-broadened lines, i.e., the splitting can be recorded inside the Doppler width. This possibility is well known in spectroscopy of atomic transitions in the magnetic field (Hanle effect, see Introduction), where an effect of splitting inside the Doppler width manifests itself by changing the polarization of the spontaneous radiation. In the case of nonlinear resonances, it manifests itself by changing absorption, i.e., not in a spontaneous but in an induced process. Therefore, the methods of nonlinear spectroscopy are especially important for transitions with rapid nonradiative relaxation when the spontaneous radiation is weak, for instance, for rotational-vibrational transitions of molecules.

The simplest effect that arises from action of the external field on the Doppler-broadened line is its splitting. For instance, in the external magnetic field an atomic transition splits into two components that correspond to left- and right-hand circular polarizations. If the magnitude of splitting exceeds the homogeneous width, nonlinear resonance occurs in the line center of each component. This was first observed in a He-Ne laser at λ = 1.5 μm with a nonlinear absorption cell placed in a longitudinal magnetic field [7.25]. When the laser frequency was scanned, two separate generated power peaks, situated in the centers of σ^+ and σ^- components with the corresponding polarization of generated radiation were observed. When the generated frequency was scanned, the nonlinear interaction of two waves with different polarizations (hysteresis polarization effect), caused by interaction of two transitions with a common level was observed.

When TLS methods are used, considerable improvement of resolution can be achieved. As shown in Chapter 5, a nonlinear resonance of a transition that has a level in common with the transition in the strong field can be considerably narrower than the homogeneous width of the transition. This was demonstrated in one of the first experiments of nonlinear spectroscopy [7.40] in which a narrow resonance was observed on an atomic transition in the external magnetic field. With small splitting of levels, a pair of coupled transitions with close frequencies is easily produced. In the experiment, an amplifying transition was acted upon by two travelling waves of a Xe laser, which had a difference of frequencies Δ (two axial laser modes); the steady magnetic field changed the difference of frequencies of the transitions $\Omega = \omega_1 - \omega_2$ between Zeeman sublevels of Xe. When the Zeeman splitting of pairs of transitions was tuned to Δ, a nonlinear resonance was observed in the transmittance of a Xe cell; it had a width of $\Gamma = 0.6$ MHz, which was two orders of magnitude smaller than the Doppler width. The shape of the resonance signal is shown in Fig.5.12. Its peculiarity is that the width $\Gamma_{res} = 1/2(\gamma_1 + \gamma_2)$ comprises only the constants of broadening of finite levels; the broadening of a common level γ_0, which is greater by an order of magnitude, is excluded.

Narrow lines of the resonance SRS can be used for exploring fine level structures. Here, we describe investigations of the RSRS line in neon in a magnetic field [7.41]. The longitudinal magnetic field removes degeneration of the levels 2s and 2p. By observation along the magnetic field H, three transitions with left-hand circular polarization (LCP $\Delta M = +1$) and three transitions with right-hand circular polarization (RCP $\Delta M = -1$) can be found in the 1.15 μm line. For light with LCP, the observed frequencies differ from an unshifted frequency ω_{02} by $-(g_0/\hbar)\mu_0 H$, $-(g_2/\hbar)\mu_0 H$, $-(2g_2 - g_0/\hbar)\mu_0 H$, and, for light with RCP, by $(g_0/\hbar)\mu_0 H$, $(g_2/\hbar)\mu_0 H$, $(2g_2 - g_0/\hbar)\mu_0 H$, where μ_0 is the Bohr magneton. Owing to thermal motion, the Doppler widths of these transitions overlap in the region where the splitting varies linearly with the field strength H; it is therefore difficult to distinguish between the g factors. Consequently, for simplicity, it is assumed for the 1.15 μm line that $g_0 = g_2 = g$. In this case, the radiated line observed along the field is the sum of two noninteracting Doppler-broadened lines, which shift in opposite directions when the magnetic field is increased.

The Zeeman effect manifests itself differently in a three-level system in the presence of resonant monochromatic light fields. In this case, the splitting of resonances can be observed with weak magnetic fields; this makes possible precise measurements of g factors of levels and intensities of individual Zeeman components. The possibility of selective excitation of magnetic

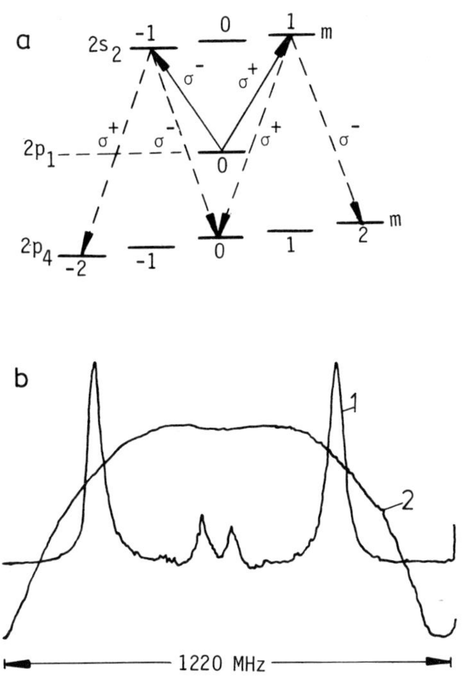

Fig.7.17a,b Zeeman effect on coupled transitions of Ne a - scheme of levels and transitions in the magnetic field; b - experimental line shape of stimulated emission at $\lambda = 1.15$ μm in the presence of the strong travelling wave at $\lambda = 1.52$ μm (curve 1); profile of probe signal (curve 2)

sublevels and of investigations of their exchange is attractive for investigations of the Zeeman effect.

Consider the Zeeman effect on the 1.15 μm line in the presence of an external monochromatic field at $\lambda = 1.5$ μm in the scheme given in Fig.7.17. Superposition of the longitudinal magnetic field leads to the splitting of an absorption line of the strong field into two σ components. A linearly polarized wave can be resolved into two components, with LCP and RCP, that have equal intensities. Absorption of the component with LCP leads to excitation of atoms to the sublevel with $M = -1$; with RCP, to the sublevel with $M = +1$; the field interacts with atoms whose velocities satisfy the resonance condition that their Doppler-shifted frequencies are equal to the detuning from the line center. The atoms that are in exact resonance have velocities $kv_{res} = \Omega + \Delta_0$ on the sublevel with $M = -1$ and $kv_{res} = \Omega - \Delta_0$ on the sublevel with $M = +1$. It is taken into account that a lower level of the transition $2s_2$ - $2p_1$ is nondegenerate ($J_1 = 0$) and $\Delta_0 = g_0\mu_0 H/\hbar$.

Two electric-dipole transitions of each magnetic sublevel of the transition $2s_2 - 2p_4$ in the longitudinal magnetic field are resolved, in accordance with selection rules for dipole radiation ($\Delta M = \pm 1$), with resonant frequencies $\omega_{02} + \Delta_0 - 2\Delta_2$ $(1 \to 2$, LCP), $\omega_{02} - \Delta_0$ $(1 \to 0$, RCP), $\omega_{02} - \Delta_0$ $(-1 \to 0$, LCP), $\omega_{02} - \Delta_0 + 2\Delta_2$ $(-1 \to 2$, RCP).

Experimental investigations were carried out according to the scheme described in Section 5.5, with the difference that an external absorption cell was placed in a longitudinal magnetic field. In order to eliminate polarization anisotropy, the external cell had windows perpendicular to the axis of the discharge tube and to the direction of the magnetic field. The polarizations of both the strong field at $\lambda = 1.52$ μm and the weak field at $\lambda = 1.15$ μm could be varied. Circularly polarized light was obtained by use of quarter-wave mica plates chosen for the wavelengths of 1.15 and 1.52 μm, respectively.

Figure 7.17 (with cancellation of a Doppler band by excitation diffusion and resonant radiation trapping) shows experimental records of the line shape of stimulated radiation of the 1.15 μm line in the presence of a plane-polarized travelling wave at $\lambda = 1.52$ μm and of a magnetic field $H = 80$ G, for forward scattering.

The splitting of the magnetic sublevels is determined from the formula $\Delta = 1.4\ gH$ (MHz), where H is the magnetic field in Gauss. For $H = 100$ G and $g = 1.3$, we obtain $\Delta \simeq 180$ MHz. With the normal Zeeman effect on the transition $2s_2 - 2p_4$, the distance between maxima of the σ components must be $2\Delta = 360$ MHz, which is appreciably less than the Doppler line width in a discharge, $\Delta\nu_D = 800$ MHz. Except for two strongly shifted components with different circular polarizations, there are only two components, which are little-shifted from the center of the transition. These are associated with the transition to the sublevel with $M = 0$, which does not shift in the magnetic field. It is obvious that at very closely spaced wavelengths, i.e., when $k'/k - 1 \ll 1$ these components will not shift, up to very strong fields, in spite of the fact that the common level shifts quite strongly in the magnetic field. This effect reflects the specific character of scattering in a three-level system. The frequency of the scattered light that has the same circular polarization as the external field, does not change in the magnetic field, whereas the change of polarization of scattered photons in the magnetic field is associated with the change of frequency of the scattered light.

The splitting of the stimulated-radiation line $\lambda = 1.15$ μm was used for precise determination of the ratio of the g factors of levels $2s_2$ and $2p_4$.

The measurements gave a ratio of g factors $g(2p_4)/g(2s_2) = 1.035 \pm 0.02$. Using the value obtained earlier for $g(2p_4) = 1.301$, we obtain for $g(2s_2) = 1.26 \pm 0.03$. This value agrees with that found by the double-resonance method [7.36]. From these measurements the absolute magnitudes of the g factors can be obtained by measuring precisely the magnitude of the magnetic field. The measurements yielded $g(2p_4) = 1.30 \pm 0.03$, which agrees with the earlier value [7.35]. These measurements were possible because of the narrowness of the peaks obtained from the stimulated forward scattering. The backward peaks are considerably wider. Consequently, the resolution of the method is less when the ordinary effect of population saturation is used.

7.3.3 Spectroscopy of Forbidden Transitions

Spectroscopy of forbidden transitions is a new, rapidly developing aspect of nonlinear laser spectroscopy. Two-photon absorption permits exploration for transitions and absorption lines that were previously out of the reach of investigations, because transitions allowed in a two-photon approximation are forbidden in a one-photon approximation. Resonances of two-photon absorption in a standing-wave field (see Chap.4) make it possible to exclude the influence of Doppler broadening and to obtain resolution determined by the natural width of a two-photon transition.

One of the first experiments, which explored the fine structure of a 4D level of Na by use of two-photon absorption with no Doppler broadening, was reported in [7.42]. The Doppler broadening was eliminated by use of an atomic beam. Investigating the 3S - 4D transition, the authors of [7.42] could record the two-photon absorption line, which was a superposition of the transitions $^2S\ (F=2) \to {}^2D_{5/2}, {}^2D_{3/2}$ and $^2S\ (F=1) \to {}^2D_{5/2}, D_{3/2}$. By use of a cw dye laser with an output power of 165 mW and 25 MHz width of the laser line the authors of [7.42] were the first to measure the difference between the energies of the hyperfine-structure components of the D level of Na, which they found to be $\Delta\nu_{hfs} = 1025 \pm 6$ MHz. At much the same time, a number of studies that explored resonances of two-photon absorption in a standing-wave field were reported. The authors of [7.43,44] explored a structure of the 5S level of Na. Using a pulsed dye laser with transverse pumping by radiation from a nitrogen laser the authors of [7.44] recorded the line shape of two-photon absorption of the 3S - 5S transition (Fig.7.18). The results gave a value for the hyperfine splitting of the 5S level of Na $\Delta\nu_{hfs} = 156 \pm 5$ MHz.

References [7.45,46] reported investigations of two-photon transitions with a cw dye laser. The 3S - 4D transition was reported in [7.45]. The width

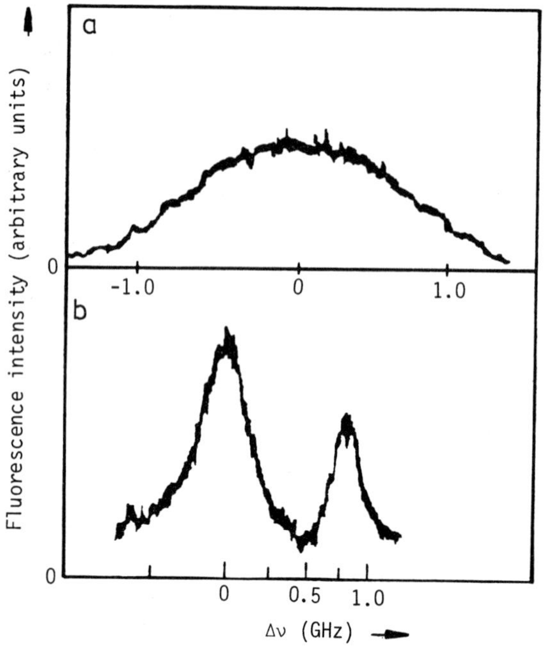

Fig.7.18a,b (a) Doppler profile of the two-photon absorption of the sodium 3s - 5s transition, from one beam with linear polarization. (b) Resolved hyperfine splitting of the same transition with two counter-propagating circularly polarized beams. The nonlinear horizontal frequency scale is characteristic of tilt tuning of the Fabry-Perot interferometer

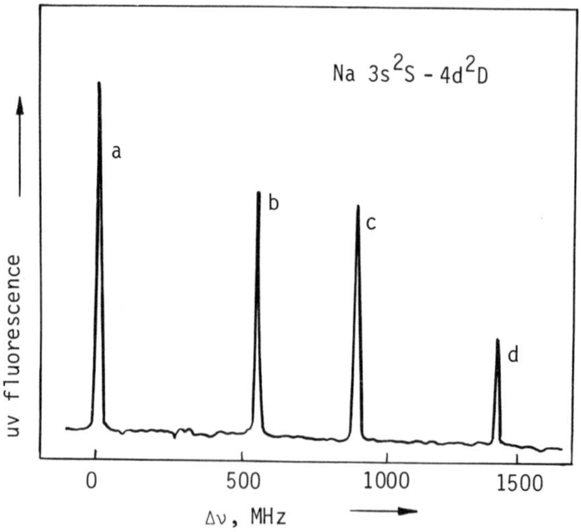

Fig.7.19 Record of the uv fluorescence light as the laser is scanned over the two-photon resonance. The frequency scale shows the actual laser-frequency shift; the atomic frequency differences are twice as large

of the laser radiation line was <30 MHz at power of about 30 mW. Figure 7.19 shows the line shape of the two-photon resonance. The measured value of splitting of the fine structure of the 4D level was 1035 ± 10 MHz, which coincides quite well with the results of [7.42]. Careful measurements of a two-photon resonance width are given in [7.46] for which a laser with a line width of about 10 MHz and power of 5 mW was used. The two-photon resonance width was about 24 MHz. The method of two-photon spectroscopy with no Doppler broadening found important application in investigations of the 1S - 2S transition in hydrogen; we shall return to this in Chapter 10.

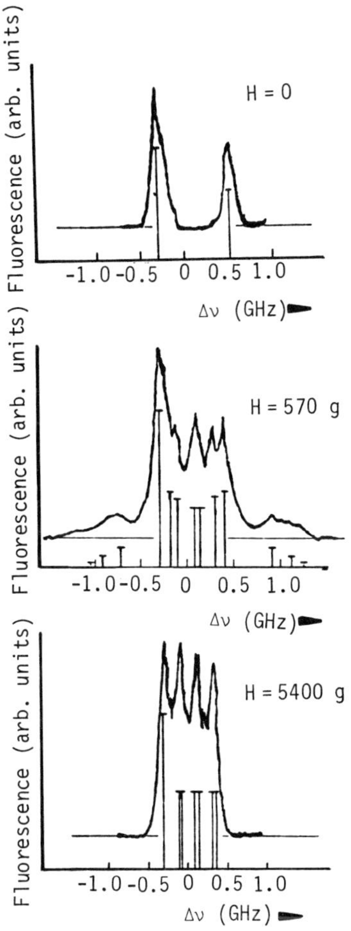

Fig.7.20 Hyperfine splitting of two-photon 3s - 5s transition in the ^{23}Na atom in zero magnetic field, and in magnetic fields of 570 and 5400 G, respectively. The vertical lines indicate the theoretically calculated line positions and their relative strengths. The experimental traces record the observed fluorescence intensity at 330 nm (4p - 3s transition) following the two-photon absorption process

The method of two-photon spectroscopy with no Doppler broadening made possible the observation of Zeeman splitting of the transition between two S states in Na [7.47]. The authors of that paper investigated the influence of the magnetic field on Zeeman splitting. They observed experimentally an approach of Zeeman components with an increase of the magnetic field, which agreed with theory. Figure 7.20 shows the results of measurements of hyperfine splitting of the two-photon transition 3S - 5S in Na with different values of the magnetic field. The resolution of the components was restricted by the laser line width (about 150 MHz), which was determined by the short duration of the laser pulse (about 4×10^{-9} s).

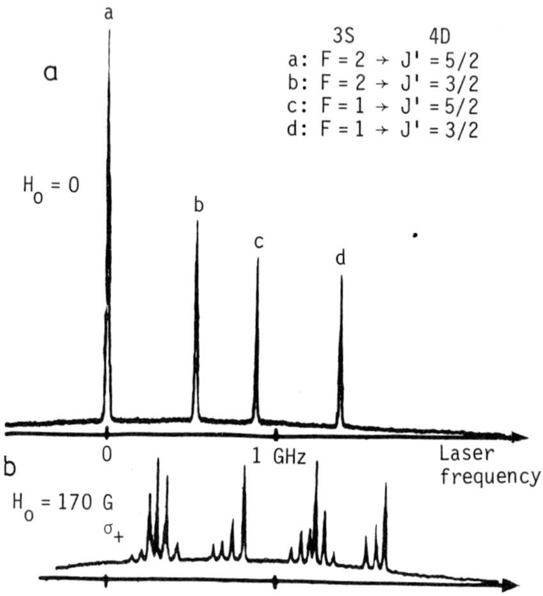

Fig.7.21 Two-photon absorption spectrum at 3s - 4d transition of Na at H = 0 and H = 170 G intensities of the magnetic field

Similar measurements, but with a cw dye laser were reported in [7.48]. The authors explored the influence of a strong magnetic field on the 3S - 4D transition in Na (so-called Paschen-Back effect). These measurements made it possible not only to follow the behavior of the Zeeman components when the magnetic field was varied but to calculate a fine structure of the 4D state, which was 1028.5 ± 3 MHz. These results are in good agreement with the results reported in [7.42,45]. Peak widths were <15 MHz. The results of the measurements are shown in Fig.7.21.

References

7.1 I.M. Beterov, Yu.A. Matyugin, S.G. Rautian, V.P. Chebotayev: Zh.Eksp.i Teor.Phys. 58, 1243 (1970)
7.2 Yu.A. Matyugin, A.S. Provorov, V.P. Chebotayev: Zh.Eksp.i Teor.Phys. 63, 2043 (1972)
7.3 A. Szoke, A. Javan: Phys.Rev.Lett. 10, 521 (1963)
7.4 W.R. Bennett, Jr., P.J. Kindlman: Phys.Rev. 149, 38 (1966)
7.5 I.M. Beterov, Yu.A. Matyugin, V.P. Chebotayev: Zh.Eksp.i Teor.Phys. 64, 1495 (1973)
7.6 W.R. Bennett, Jr.: Appl.Opt. Suppl.1, 24 (1962)
7.7 G.M. Lawrence, H.S. Liszt: Phys.Rev. 178, 122 (1969)
7.8 L.S. Vasilenko, V.P. Chebotayev: Zh.Prikladnoi Spektrosk. 6, 436 (1966)
7.9 A.S. Khaikin: Thesis, P.N. Lebedev Physical Institute of the USSR Academy of Sciences, 1969
7.10 Th. Hänsch, P. Toschek: Phys.Lett. 20, 273 (1966)
7.11 I.M. Beterov, Yu.A. Matyugin, V.P. Chebotayev: Zh.Eksp.i Teor.Phys.Pis. Red. 12, 174 (1970)
7.12 P.W. Smith: J.Appl.Phys. 37, 2089 (1966)
7.13 R.H. Cordover, P.A. Bonzyak, A. Javan: Bull.Am.Phys.Soc. 12, 89 (1967)
7.14 G.A. Mikhenko, E.D.Protzenko: Optica i Spectr. 30, 668 (1969)
7.15 W. Dietel: Phys.Lett. 29A, 269 (1969)
7.16 A. Szoke, A. Javan: Phys.Rev. 145, 137 (1966)
7.17 V.N. Lisitsyn, V.P. Chebotayev: Zh.Eksp.i Teor.Phys. 54, 419 (1968); S.N. Bagayev, Yu.D. Kolomnikov, V.N. Lisitsyn, V.P. Chebotayev: IEEE J. QE-4, 868 (1968)
7.18 T.R. Sosnovski, W.B. Johnson: IEEE J. QE-4, 56 (1968)
7.19 S.N. Bagayev, Yu.D. Kolomnikov, V.P. Chebotayev: Proceedings of SNIIM 9, Novosibirsk (1971), p.7
7.20 P.H. Lee, M.L. Skolnick: Appl.Phys.Lett. 10, 303 (1967)
7.21 S.G. Rautian: Proceedings of the Lebedev Physical Institute. Nonlinear Optics, V.43 (1968), p.3
7.22 Yu.A. Vdovin, V.M. Galitsky, V.M. Ermachenko: *Theory of Atomic Collisions*, Moscow (1970)
7.23 S.P. Koutsoyannis, K. Karamcheti: IEEE J. QE-4, 912 (1968)
7.24 A.L. Bloom, D.L. Wright: Appl.Opt. 5, 1528 (1966)
7.25 V.N. Lisitsyn, V.P. Chebotayev: Optica i Spectr. 26, 856 (1969)
7.26 H.K. Holt: Phys.Rev. A2, 233 (1970)
7.27 K.A. Bikmukhametov, V.M. Klement'ev, V.P. Chebotayev: Kvantovaja Electronika 3, 74 (1972)
7.28 Yu.V. Brzhazovsky, L.S. Vasilenko, V.P. Chebotayev: Zh.Eksp.i Teor. Phys. 55, 2096 (1968)
7.29 A.P. Kazantsev: Zh.Eksp.i Teor.Phys. 51, 1751 (1966)
7.30 P.R. Berman, W.E. Lamb, Jr.: Phys.Rev. 187, 221 (1969)
7.31 E.A. Ballik: Appl.Opt. 5, 170 (1966)
7.32 J.T. Davies, J.M. Vaughan: Astrophys.J. 137, 1302 (1963)
7.33 V.A. Alekseyev, T.L. Andreyeva, I.I. Sobel'man: Zh.Eksp.i Teor.Phys. 62, 614 (1972)
7.34 N.R. Hindmarsh, A.D. Petford, G. Smith: Proc.Royal Soc.London A297, 296 (1967)
7.35 Ye.V. Baklanov: Zh.Eksp.i Teor.Phys. 65, 2203 (1973)
7.36 C.G. Carnington, A. Gorney: Opt.Commun. 1, 115 (1970)
7.37 R.H. Cordover, P.A. Bonczyk, A. Javan: Phys.Rev.Lett. 18, 730 (1967)
7.38 T.W. Ducas, M.S. Feld, L.W. Ryan, Jr., N. Skribanowitz, A. Javan: Phys. Rev. A5, 1036 (1972)
7.39 K. Uehara, Jr.: Phys.Soc.Japan 34, 777 (1973)
7.40 H.R. Schlossberg, A. Javan: Phys.Rev.Lett. 17, 267 (1966)

7.41 I.M. Beterov, Yu.A. Matyugin, V.P. Chebotayev: Preprint of the Institute of Semiconductor Physics of the Siberian Branch of the USSR Ac.of Sci. No.21 (1971)
7.42 D. Pritchard, J. Apt, T.W. Ducas: Phys.Rev.Lett. 32, 611 (1974)
7.43 F. Biraben, B. Cagnac, G. Grynberg: Phys.Rev.Lett. 32, 643 (1974)
7.44 M.D. Levenson, N. Bloembergen: Phys.Rev.Lett. 32, 645 (1974)
7.45 T.W. Hänsch, K.C. Harvey, G. Meisel, A.L. Schawlow: Opt.Commun. 11, 50 (1974)
7.46 F. Biraben, B. Cagnac, G. Grynberg: Phys.Lett. 49A, 71 (1974)
7.47 N. Bloembergen, M.D. Levenson, M.M. Salour: Phys.Rev.Lett 32, 867 (1974)
7.48 F. Biraben, B. Cagnac, G. Grynberg: Phys.Lett. 48A, 469 (1974)

8. Nonlinear Molecular Laser Spectroscopy

Application of nonlinear optical resonances to molecular spectroscopy increased resolution by several orders of magnitude. But the importance of laser methods for molecular spectroscopy is not confined to this. Use of instruments of high resolution is difficult in many cases owing to overlapping of lines of electron vibrational-rotational transitions. Now the situation is sharply changed. Application of tunable lasers obtains high-resolution spectra of many complex molecules, explores hyperfine structure of transitions, and precisely measures isotopic structure. It made possible measurements of line shifts and broadenings of vibrational-rotational transitions of molecules. The potentialities of laser spectroscopy can be seen from Fig.8.1.

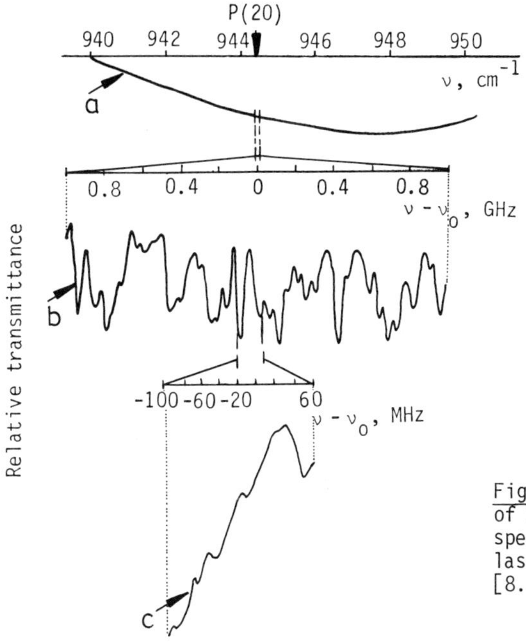

Fig.8.1a-c Absorption spectrum of SF_6, obtained in a - ordinary spectrograph; b - tunable diode laser [8.1]; c - tunable CO_2 laser [8.2]

Figure 8.1 shows the absorption spectrum of SF_6 obtained on a spectrograph [8.1]. Spectra with superhigh resolution were obtained in the same spectral region by use of a tunable semiconductor laser [8.1] (Fig.8.1b) and of a cw high-pressure CO_2 laser [8.2] (Fig.8.1c). Only in these cases it is possible to obtain undistorted spectra of absorption of molecules with resolution restricted by only the Doppler width.

8.1 Investigation of Level Structures

Nonlinear optical resonances permit exploration of line structures that are hidden by Doppler broadening. As in the optical spectral region, many experiments with nonlinear optical resonances have been performed with molecules whose frequencies coincided accidentally with those of more widely used lasers.

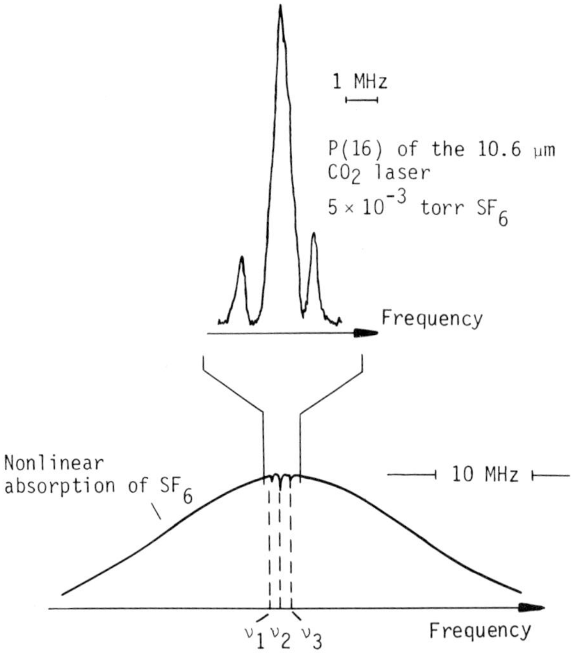

Fig.8.2 Complex structure of narrow resonances resolved when the absorption of SF_6 is saturated by the P(16) 10.6 μm line of the CO_2 laser

A typical illustration of such an observation is given in Fig.8.2 which shows an experimentally observed structure in one of the absorption lines of the molecule SF_6 in the field of a line P(16) radiated by the 10.6 μm band of a CO_2 laser [8.3]. The Doppler broadening for this transition is 30 MHz and the separation of the resonances in the triplet is about 300 kHz. The central resonance has a width of about 1.3 MHz, which appears to be caused by the internal structure of lines, which could not be resolved in this experiment. The spectral lines of atoms and molecules, on which narrow absorption resonances were obtained, are given in Tables 3.1 and 3.2.

8.1.1 Electron Vibrational-Rotational Transitions of Molecules

In their investigations, SCHAWLOW and coworkers explored the hyperfine structure of a number of absorption lines of electronic transitions of the molecules of I_2 [8.4-6]. They studied 11 transitions of the molecules of $^{127}I_2$ and $^{129}I_2$, which coincided with generated lines of ionic Ar (5145 Å, 5017 Å) and Kr (5682 Å, 5308 Å, 5208 Å) lasers. The authors of [8.7,8] have explored a hyperfine structure of the line P(33) of $^{127}I_2$, which coincides with the line radiated by the He-Ne 6328 Å laser.

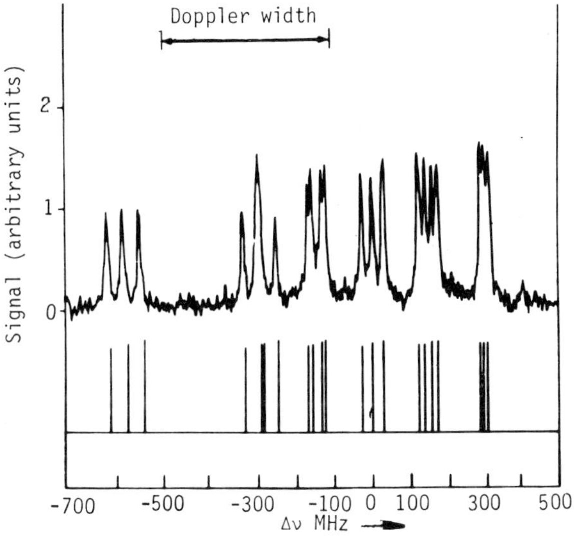

Fig.8.3 Hyperfine structure of P(117) line of the band 21-1 $^{127}I_2$, resolved by the method of absorption saturation using ionized-krypton laser emission

All of the investigated absorption lines corresponded to transitions between $\Sigma_g^+(X)$ and $^3\Pi_{ou}^+(B)$ electronic states, but the vibrational and rotational quantum numbers were quite different. A typical example of the observed nonlinear spectrum of a hyperfine structure for the line of the band P(117) 21-1 $^{127}I_2$ is shown in Fig.8.3. Investigation of the hyperfine structure made it possible to establish two mechanisms of its formation: 1) nuclear electric-quadrupole interaction, which changes slightly for different lines; 2) magnetic spin-rotation interaction, which depends critically on the vibrational energy in an excited electron state. The measured values when hyperfine splittings are observed are the differences between constants of the electric-quadrupole interaction in an initial and a final state,

$$\Delta eQg = eQq' - eQq^{0} \qquad (8.1)$$

and the difference between constants of the magnetic spin-rotation interaction,

$$\frac{\mu G}{I_1} = \frac{\mu}{I_1}\,(G' - G^{0}) \ , \qquad (8.2)$$

where I, Q and μ are spin, quadrupole and magnetic moments of a nucleus of I (for ^{127}I: $I_1 = 5/2$, $Q = -0.79$ barn, $\mu = 2.808\ \mu_{nuc}$; ^{129}I: $I_1 = 7/2$, $Q = 0.55$ barn, $\mu = 2.617\ \mu_{nuc}$).

8.1.2 Hyperfine Structure

A hyperfine structure can arise in a vibrational-rotational spectrum of molecules as a result of three effects: quadrupole, magnetic and isomeric structures. The quadrupole interaction causes splitting of rotational-vibrational lines by 10^5 to 10^7 Hz, dependent on the constant of the quadrupole interaction and on the angular momentum of a molecule. The magnetic interaction between the angular moment of the molecule and the nuclear spin causes a smaller splitting, ranging from 10^3 to 10^5 Hz. Resolution of the order of 10^9 to 10^{10} is required to find this value. Another effect that requires application of nonlinear spectroscopy is the change of the hyperfine structure of the vibrational-rotational transition when nuclei of a molecule are excited to an isomeric (metastable) state. In the isomeric state, the nucleus usually has a spin that differs significantly from that of the ground state, and the nuclear mass increases by $\Delta m = E_{exc}/c^2$.

High resolution (over 10^{10}) was achieved when the hyperfine magnetic structure of a component of the line P(7) of the band ν_3 of methane [8.9-11] was studied. Resonances with a width of about 6 kHz reported in [8.9] permitted resolution of the Lamb-dip structure in methane, which consists of a number of lines spaced about 10 kHz apart. An external cell, 13 m long, was used to obtain resonances with small widths. The diameter of the laser beam was 5 cm, and the methane pressure was 10^{-4} torr. The studies employed highly stable lasers, because weak components required 12 hours recording time. Figure 8.4 shows a record of the structure. The spectrum consists of three components, closely spaced but of different intensities, and two weaker lines.

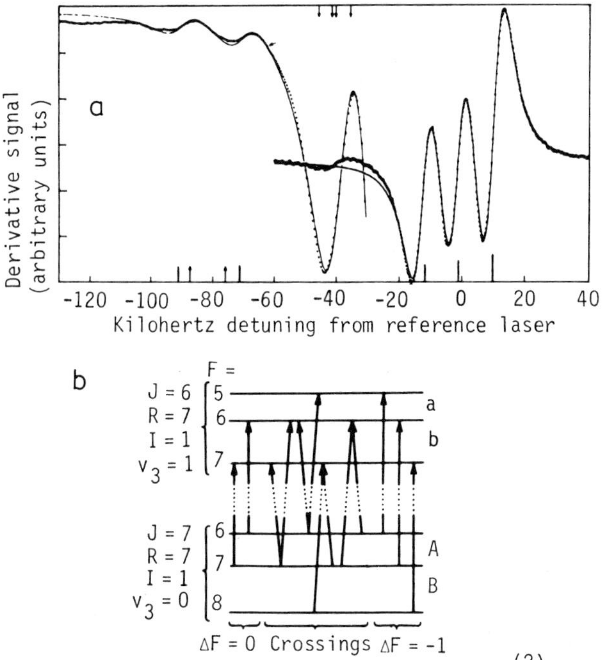

Fig.8.4a,b Magnetic hyperfine structure of $F_2^{(2)}$ component of the P(7) line of the ν_3 CH_4 band, obtained by the method of methane absorption saturation, using a He-Ne laser at $\lambda = 3.39$ μm (a); transitions responsible for formation of the structure (b)

The arrows in Fig.8.4a show the positions of four crossing resonances that arise at the expense of pairs of transitions to a common level, which are shown in Fig.8.4b. The crossing resonances are located between two usual transitions and have intensities proportional to the geometric mean of the intensities of the parent transitions. Resonances with a width of about 3 kHz were obtained later in [8.10] by use of a telescopic expander (see Chap.6). Use of the telescopic expander inside a cavity [8.11] considerably increased the resonance intensities so that the resonances could be recorded directly on an automatic recorder. The experimentally measured distances between the lines and the relative resonance intensities were close to the theoretical values. Three of the most intense lines correspond to the transitions with $\Delta F = -1$. The distance between them is about 11 kHz. The energy-level diagram of the transitions between the hyperfine structure components responsible for the observed resonances is given in Fig.8.4b.

The hyperfine quadrupole structure of a transition of $^{12}CH_3{}^{35}Cl$ was first explored in [8.12] by use of the line P(26) of a 9.4 μm CO_2 laser. The observed spectrum agrees with the calculated spectrum. Reference [8.29]

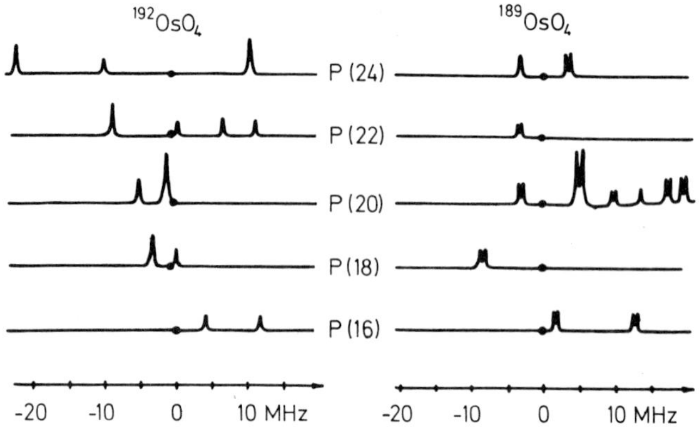

Fig.8.5 Saturated absorption spectra of molecules $^{192}OsO_4$ and $^{189}OsO_4$ measured with the aid of several lines of P branch of the CO_2 laser at 10.6 μm (circles mark the centers of gain lines of the CO_2 laser)

contains an analysis of all three effects, quadrupole, magnetic and isomeric structures, in the vibrational-rotational spectrum of monoisotopic molecules of OsO_4. For this purpose, the internal structure of Doppler-broadened absorption lines of multi-isotopic molecules of $^{187}OsO_4$, $^{189}OsO_4$, $^{190}OsO_4$ was experimentally explored by use of a number of lines of the P and R branches of a CO_2 laser. As a rule, narrow resonances that correspond to simultaneous transitions of several monoisotopic molecules are observed on a line contour of the radiation of an ordinary low pressure CO_2 laser. Figure 8.5 shows an example of the spectrum of saturated absorption of lines of the band ν_3 of molecules of $^{189}OsO_4$ and $^{192}OsO_4$ observed by use of radiation of several lines of the P branch of the CO_2 laser in the 10.6 μm region. The structure of the vibrational-rotational transitions of the four investigated monoisotopic molecules of OsO_4 has the peculiarity that the spectrum of $^{189}OsO_4$ consists of paired resonance doublets. A few single resonances in the spectrum of $^{189}OsO_4$ can be attributed to impurities, consisting of the other monoisotopic molecules. Unlike other isotopes of Os a nucleus of the isotope ^{189}Os has a spin $I = 3/2$ in the ground state and a rather large value of magnetic (μ_{nucl} = 0.65004 nucl. magneton) and quadrupole ($Q_{nucl} = 0.8 \times 10^{-24}$ cm^2) moments. In the cases of $^{192}OsO_4$ and $^{190}OsO_4$ owing to the fact that the nuclear spin $I = 0$, there is no hyperfine splitting and single resonances are observed in the spectrum. In the case of $^{187}OsO_4$ ($I = 1/2$) the absorption spectrum is complicated by magnetic hyperfine splitting, which could not be resolved under the experimental conditions [8.29]. The doublet structure observed in the spectrum of $^{189}OsO_4$ is due to a nuclear quadrupole structure. The estimated value of a constant of the quadrupole interaction is $eq_JQ \simeq 0.6$ MHz.

8.1.3 Stark and Zeeman Effects on Vibrational-Rotational Transitions

Nonlinear optical resonances permit exploration of Stark and Zeeman effects on individual lines when the Doppler lines overlap. The high resolution makes it possible to investigate the effects with rather slight electric and magnetic fields. Tuning of the laser-generated frequency over a resonance provides precise measurements of Stark and Zeeman line splittings.

In experiments of this kind, the transition frequency is tuned to the light-field frequency, and the line shift in an external field is precisely measured over a narrow nonlinear resonance. This technique was used, for instance, in [8.13] to measure precisely the Stark effect on the transition $4_{04}(a) \rightarrow 5_{14}(S)$ of the band $\nu_3 NH_2D$ with the help of the line P(20) of the 10.6 μm band of a CO_2 laser. Usually this line is not absorbed in the NH_2D but in a homogeneous electric field some rotational-vibrational lines can be tuned in resonance with the field of a CO_2 laser. When the center of an absorption line NH_2D is precisely tuned to the field frequency, a narrow nonlinear resonance, which permits precise measurement, arises on each line. The width of narrow resonances was 1.6 MHz; it was determined mainly by broadening due to collisions at the 30 mtorr gas pressure in the Stark cell. The

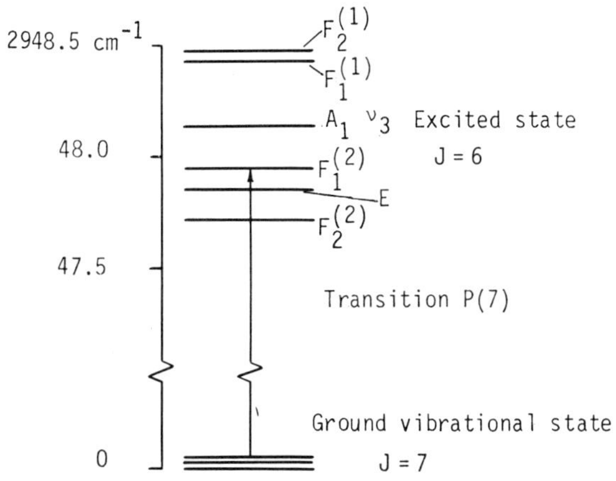

Fig.8.6 Diagram of the CH_4 energy levels associated with the transition P(7) ν_3 at $\lambda = 3.39$ μm

precision of such measurements is sufficient to determine the second-order Stark shift in rather weak fields (several kV cm^{-1}). Similar measurements were made with the molecules NH_2D [8.14] and NH_3 [8.15] by use of lines radiated by a CO_2 laser and with the molecule CH_3F by use of the 3.39 μm line of a He-Ne laser [8.16].

A series of interesting experiments on the Stark effect was performed with the CH_4 molecule and the 3.39 μm He-Ne laser. An excited state of the transition P(7) of the band ν_3 of CH_4 has six Coriolis sublevels [8.17,18]. The energy-level diagram of methane, for the 3.39 μm transition, is shown in Fig. 8.6. Two sublevels have first-order Stark effect, whereas four have only second-order Stark effects [8.19]. The transition P(7) $F_2^{(2)}$ is in exact resonance with the wavelength of the He-Ne laser; the transition P(7) E differs from it, toward red, by 0.096 cm^{-1}.

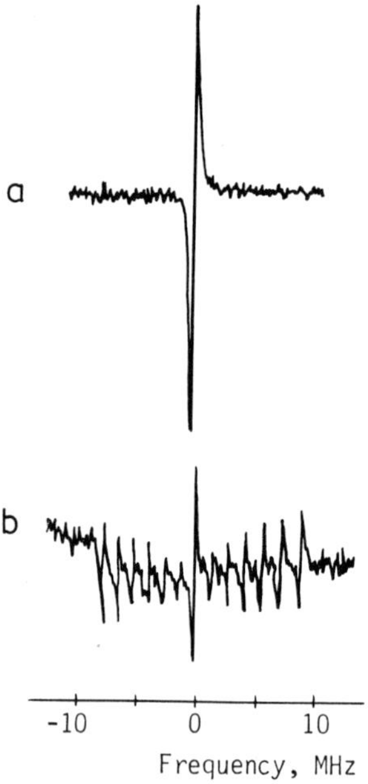

Fig.8.7a,b Stark structure of the E component of the line P(7) ν_3 CH_4 observed inside the Doppler line by the method of absorption saturation
a - in zero electric field;
b - in an electric field of 1660 V/cm

The linear Stark effect on the transition E was studied in [8.20,21]. The He-Ne laser frequency was tuned to 3 GHz by an external axial magnetic field and the Stark CH_4 cell was placed inside the laser cavity. The spectra of saturated absorption of the E component of P(7) of CH_4, in the absence of an electric field and in the presence of 1660 $V \cdot cm^{-1}$ are shown in Fig.8.7. Because of a slight electric dipole moment 0.0200 ± 0.0001 D in the excited vibrational state, each spectral line splits into $2J+1=13$ equidistant components which are hidden by Doppler broadening but become apparent in the nonlinear absorption spectrum. The resonances were recorded by the method of frequency modulation (the observed signal is the first derivative with respect to frequency of the nonlinear absorption line) and have a width of about 100 kHz. The number of components confirms unambiguously that the upper level has a rotational angular moment $J=6$, in accordance with the notation of the transition P(7). In addition, this experiment confirmed the theoretical prediction [8.22] that spherical-top molecules similar to CH_4 have a slight dipole moment in the excited vibrational state because of rotational-vibrational interaction.

The quadratic Stark effect on the transition $F_2^{(2)}$ of CH_4 was first reported in [8.19]. It was not possible to resolve the Stark splitting even in the 40 $kV \times cm^{-1}$ field, but the influence of the electric field was manifest in the emergence of asymmetric broadening of a narrow resonance. The smallness of the Stark shift for this line provided high reproducibility of a He-Ne/CH_4 laser that used this transition to stabilize the frequency.

The use of narrow resonances for investigating a Zeeman effect is more essential for particles whose magnetic moment is determined by nuclear magnetic moments only. Such a situation is realized for most molecules in the ground state. In particular, the ground state of the CH_4 molecule possesses a small magnetic moment, of the order of a fraction of a nuclear magneton. Magnetic fields of over several hundreds of kilogauss are required to study the Zeeman effect on a Doppler-broadened line by the methods of linear spectroscopy. Narrow nonlinear resonances allow investigations of the Zeeman effect in magnetic fields to two orders of magnitude weaker. Such experiments were performed as reported in [8.23] on the $F_2^{(2)}$ component of the line P(7) of CH_4. Nonlinear resonances on transitions with selection rules $\Delta M_J = \pm 1$ in the fields of light waves having left- and right-hand circular polarizations, as well as their interactions, were observed.

One narrow resonance that shifted by a value $\pm\mu_{nuc} g_J H$ as a function of the direction of polarization was observed in a circularly polarized light field. Figure 8.8a shows the nonlinear resonances for this case reported in [8.23]. The value of the g factor for a rotational magnetic moment was determined from the magnetic shift of the resonance frequencies. For CH_4, it was found that $g_J = +0.311 \pm 0.006$, which is in good agreement with the value $|g_J| = 0.3133 \pm 0.0002$ obtained in experiments with molecular beams. The peculiarity of the method of nonlinear spectroscopy is the possibility of measuring not only the absolute value but also the sign g_J.

The two transitions $\Delta M_J = \pm 1$ are resolved in the linearly polarized light field and in the axial magnetic one, so that two circularly polarized light waves interact with two coupled transitions. For this case, the nonlinear resonances are shown in Fig.8.8b. In this case, two narrow resonances split, owing to the Zeeman effect, are simultaneously observed; an additional crossing resonance also arises because of a common level (see Subsec.5.3.3).

In experiments with an external electric field acting upon molecules it is possible to measure dipole molecular moments in the ground and excited states to a sufficiently high accuracy. Consider such an experiment, which is frequently called "double optical resonance" performed by BREWER [8.24] with the CH_3F molecule and two CO_2 lasers that generated closely spaced frequencies Ω_1 and Ω_2 on the line P(20) of the 9.6 μm band. The experimental

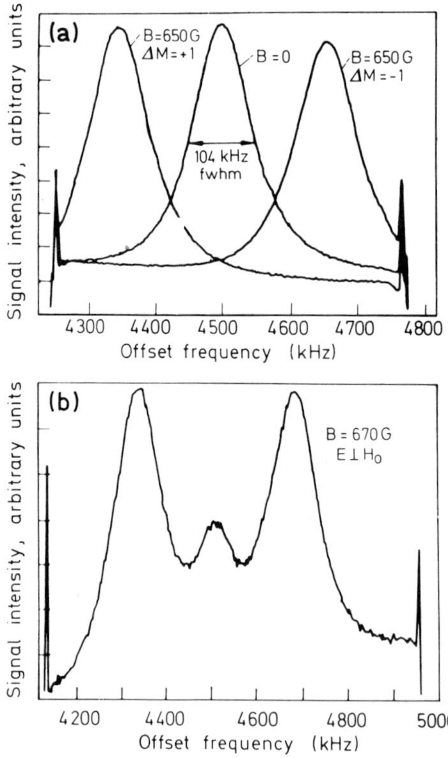

Fig.8.8a,b Zeeman structure of the $F_2^{(2)}$ component of the line P(7) CH_4 observed inside the Doppler line by the method of absorption saturation.
a - resonances in a circularly polarized light field for two different polarization directions; b - resonances in a linearly polarized light field

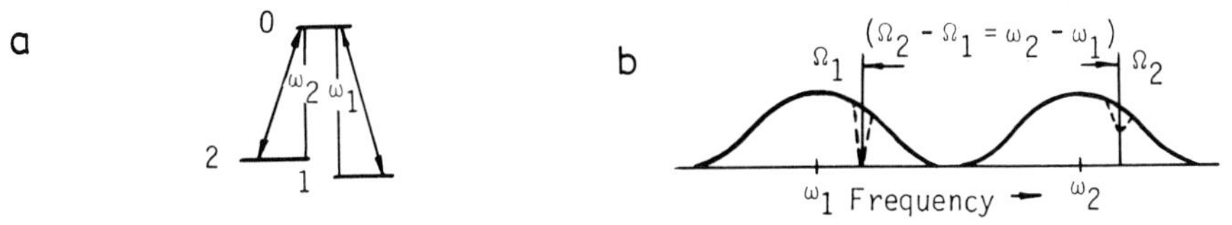

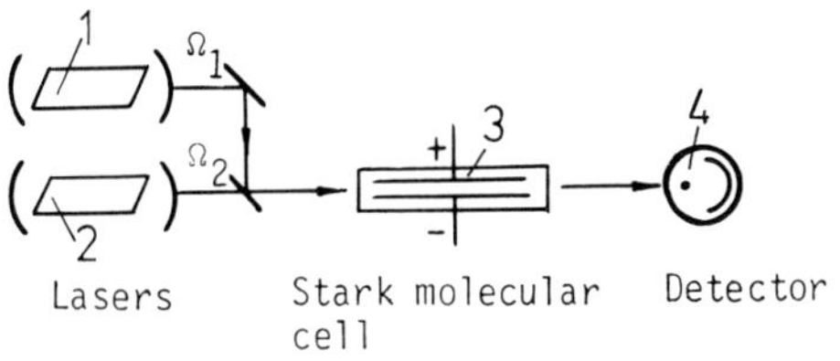

Fig.8.9a,b Experimental design for observing a double optical resonance by use of the Stark effect on the CH_3F molecule.
a - scheme of transitions; b - experimental design (1,2 - laser with closely spaced frequencies, 3 - Stark cell, 4 - photodetector);
b - formation of narrow resonance in the line of 0-2 transition when the coupled transition 0-1 is resonant

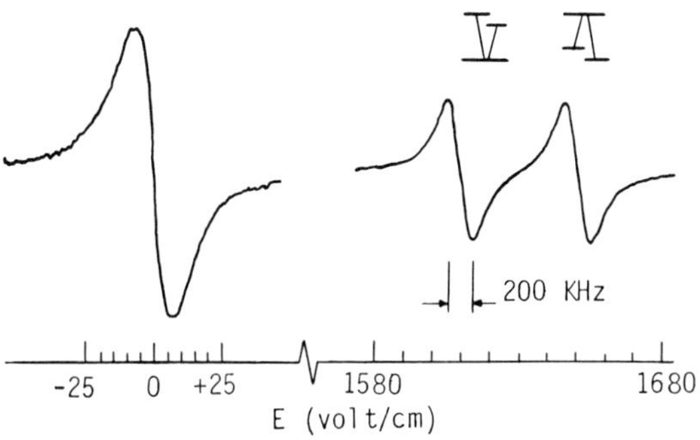

Fig.8.10 Signal of the double optical resonance on the $^{12}CH_3F$ molecule at 3.1 x 10^{-3} torr pressure

design is given in Fig.8.9. The CH_3F molecule is a symmetric top; in the external electric field it has a first-order Stark effect with the value of splitting,

$$\Delta\omega_{J,K} = \frac{2\mu}{\hbar} E \frac{K}{J(J+1)} , \tag{8.3}$$

where I is the rotational quantum number, K is a projection onto the main axis of the top symmetry. A double resonance signal at pressure of 3.1×10^{-3} torr is shown in Fig.8.10. The frequencies of CO_2 lasers coincided with the transition $(\nu_3, J, K) = (0, 12, 2) \rightarrow (1, 12, 2)$ of the band ν_3 CH_3F and the difference of their frequencies is $\Delta\Omega = 39.629$ MHz. The first resonance at zero field is a signal of level crossing, which arises when level degeneration is removed by the field. This corresponds to the fact that the condition of resonance

$$\Delta\omega_{J,K} = \Omega_2 - \Omega_1 \tag{8.4}$$

may be fulfilled at E = 0 with single-frequency laser radiation. Two signals arise, which correspond to resonance with finite level splitting. Because the dipole moments in the ground and excited states differ slightly, two resonances arise. This permits measurement of molecular dipole moments to high accuracy. Results of measurement of the electric dipole moment of CH_3F in such an experiment and of measurements made by radiospectroscopy [8.26] and by use of a molecular beam [8.27] are tabulated in Table 8.1. The accuracy of measurement by the method of nonlinear spectroscopy is 1/2000; it permits measurement of dipole moments in the excited state.

Table 8.1 Accuracy of measurements of electric dipole moment of molecule $^{12}CH_3F$ by different methods [8.28]

v	J	K	Dipole moment, D	Method
0	0, 1, 2	0, 1	1.8572	Radiospectroscopy [8.26]
0	1	1	1.85850	Molecular beam [8.27]
0	5	4	1.85852	Molecular beam [8.27]
0	12	2	1.8596	Double optical resonance [8.24]
$v_3=1$	12	2	1.9077	Double optical resonance [8.24]

8.1.4 Beat Measurements of Rotational Constants and Isotopic Shifts

The precision of measurements of rotational constants and of isotopic shifts in such molecular systems as CO_2, CO has been considerably improved recently by the development of quick-acting ir photoreceivers. In [8.31] a bulk GaAs mixer was used at room temperature in order to obtain beats between pairs of rotational-vibrational lines of CO_2 lasers. The frequencies of the beats ranged from 50 to 80 Hz, which were measured to an accuracy better than 1 MHz for thirty-seven pairs of transitions. The rotational constants for levels 00°1, 02°0, and 10°0 of a CO_2 molecule were determined from these measurements to an accuracy that is about two orders of magnitude better than that of measurements by classical spectroscopic methods. The measurement accuracy of line frequencies by conventional spectroscopic methods amounts to about 0.05 cm^{-1} (±1.5 GHz). The beat frequencies between seven pairs of oscillation lines of $^{12}C^{16}O_2$ and $^{12}C^{18}O_2$ lasers were reported in [8.32]; the measurement accuracy amounted to 5 to 10 MHz. A copper-doped germanium photoconductor operating at the temperature of liquid helium was used as a mixer.

The measurement accuracy of the rotational constants was improved when the beat technique with stabilized lasers was used. The beat frequencies between thirty pairs of $^{12}C^{16}O_2$ lines of the 10.4 μm branch and between twenty-six pairs of lines of the 9.4 μm branch were measured with Lamb-dip frequency-stabilized lasers [8.33]. This improved the measurement precision by a factor of 20-30 and measured the rotational constant of Hv for the first time. The technique of measurement consisted of shifting the oscillation frequencies of two CO_2 lasers with the nth harmonic of a klystron ($3 \leq n \leq 6$); the difference frequency was observed

$$f = \nu_{1,co_2} - \nu_{2,co_2} - n\nu_{kl} \ .$$

A Josephson junction was used to produce the shift. This method of measurement made possible precise measurement of frequencies ranging from 32 to 63 GHz. Because the level energy, as is known, is described by

$$T(v,J) = G(v) + B_v J(J+1) - D_v J^2(J+1)^2 + H_v J^3(J+1)^3 - L_v J^4(J+1)^4 + \cdots \ ,$$

the rotational constants are determined from the frequency difference.

Heterodyne measurements of the shift of frequencies in a point metal-metal contact diode, when using Q-switched CO and CO_2 lasers, permitted determination [8.34] of the frequency of the P(13) line of $^{12}C^{16}O$ with an accuracy of 5 MHz. Frequencies of other lines of $^{12}C^{16}O$ and $^{13}C^{16}O$ were obtained from the absolute frequencies of the lines of $^{12}C^{16}O_2$, which are known with an accuracy of 30 kHz. In this case, the radiation of CO lasers was beat with the second harmonic of CO_2, obtained in $CdGeAs_2$. Beat frequencies up to 11 GHz were measured by use of a high-resolution HgCdTe detector. This method requires lower laser power to obtain a high signal-noise ratio. The second harmonic of 30 μW was obtained with a CO_2 laser power of 1 w.

8.2 Collision Broadening of Resonances on Vibrational-Rotational Transitions

Until laser methods appeared, it was not practical to study infrared line shapes of vibrational-rotational molecular transitions. For a number of reasons, spectroscopic methods that use instruments with superhigh resolution have not been widely used in the ir.

Investigation of single lines was difficult because of overlap of lines of vibrational-rotational transitions. So it is not mere chance that few studies were devoted to investigations of the influence of collisions on vibrational-rotational transitions (see review [8.36]). The methods of laser spectroscopy have greatly changed the situation. A great many investigations of broadening of Doppler contours of lines and of nonlinear optical resonances have been reported quite recently. A qualitatively new phenomenon, a nonlinear dependence of the collision width of resonances on the density of particles, has been found in studies of collision broadening of nonlinear resonances.

8.2.1 Collision Broadening of Contours of Absorption Lines

Data on collision broadening of Doppler contours of absorption lines of molecules are of practical interest. They are also necessary for comparison with data on broadening of nonlinear resonances and for finding a model of collision broadening of lines. Therefore, in this section, we shall briefly consider the use of lasers for investigating contours of Doppler-broadened lines.

Single-frequency tunable lasers in the ir band allow investigation of contours of single absorption lines. Diode and spin-flip lasers were successfully used for this purpose [8.1]. A parametric generator was used in [8.37] to explore line contours of a HF molecule at pressures of about 10 atm. In some particular cases, the absorption-line contours can be investigated by use of gas lasers operating at electron transitions. The frequencies of these lasers are usually tuned by use of a magnetic field. In one of the first studies [8.38] the methane line shape and collision broadening were explored by use of a 3.39 μm He-Ne laser, whose discharge tube was located in a magnetic field.

The range of laser tuning may be increased by increasing the pressure of the amplifying medium. This method of broadening of a gain line and the large range of tuning of the frequency radiated by a single-frequency laser were used when the absorption-line contours in methane at $\lambda = 3.39$ μm were investigated with a He-Ne laser [8.39], and in CO_2 in the 10.6 μm range by use of a high-pressure CO_2 laser [8.40].

A simple method of measurement of collision broadening in molecular gases is based on the observation of the dependence of the absorption coefficient on the gas density. This method was used, mainly, when the collision broadening in CO_2 at $\lambda = 10.6$ μm [8.41-44] was investigated. The results of measurements of the collision broadening of CH_4 lines ($\lambda = 3.39$ μm) are summarized in Table 8.2. The collision broadening of lines of vibrational-rotational transitions is usually 10 MHz/torr. The line shift is small compared to the broadening; it is about 100 kHz/torr.

Table 8.2 Broadening of Doppler contour in CH_4 ($\lambda = 3.39$ μm)

Transition	Type of collisions	Broadening (MHz/torr)	Reference
Line P(7)	CH_4-He	3.6 ± 0.32	1
of the	CH_4-He	3.8-0.16	2
band ν_3	CH_4-Ne	3.2	1
	CH_4-Ar	4.3	1
	CH_4-N_2	5 ± 0.31	2
	CH_4-N_2	4.8	1
	CH_4-N_2	6.8	3
	CH_4-Kr	5.4 ± 0.4	4
	CH_4-Xe	5.8 ± 0.4	4
	CH_4-CH_4	7.4 ± 0.5	5,1
	CH_4-CH_4	6.3 ± 0.3	6
	CH_4-CH_4	5.92 ± 1.2	2
	CH_4-CH_4	7.8 ± 0.8	7

1 H.J. Gerritsen, M.E. Heller: Appl.Opt.Suppl. 2, 70 (1965)
2 P. Varanasi: J.Quant.Spectr.Radiat.Trans. 11, 1711 (1971)
3 S.C. Wait, P.J. Baldacchino, S.E. Wiberley: J.Molec.Spectr. 28, 490 (1968)
4 H. Goldring, A. Szöke, E. Zamir, A. Ben-Reuven: J.Chem.Phys. 49, 4253 (1968)
5 H.J. Gerritsen, S.A. Ahmed: Phys.Lett. 13, 41 (1964)
6 G. Hubbert, T.G. Kyle, C.J. Troop: JQSRT 9, 1469 (1969)
7 S.N. Bagayev, Ye.V. Baklanov, V.P. Chebotayev: Preprint No.22, Inst.of Semiconductor Physics, Novosibirsk (1972)

8.2.2 Broadening of Nonlinear Resonances

Collision broadening of the Doppler contour of absorption lines of molecules can be investigated for single lines of a vibrational-rotational spectrum. Nonlinear optical resonances permit investigations even in cases where lines of VRT overlap. The narrowness of resonances makes it possible to study the influence of collisions at low pressures ($<10^{-2}$ torr). Under these conditions, the influence of elastic scattering of colliding molecules on resonance shapes becomes appreciable. Thus, new interesting information can be obtained on the interaction of colliding particles, by use of spectroscopic methods.

Table 8.3 shows the principal results of investigations of collision broadening of resonances in various gases. Collision broadening of resonances was investigated by different methods. An internal absorption cell can be used to investigate the broadening in methane and in carbon dioxide. An oppositely

Table 8.3 Collision broadening and shift of rotational-vibrational lines of some molecules (at 0.1-0.01 torr pressure)

Molecule	Transition wavelength (μm)	Collision particle	Width broadening (MHz/torr)	Shift (MHz/torr)	Reference
CH_4	P(7), ν_3	CH_4	32.6 ± 1.2	0.075 ± 0.15	[3.27]
	3.39	Xe	27.8 ± 0.8	0.029 ± 0.15	[3.27]
SF_6	ν_3 10.6	SF_6	17.4 ± 4	0.3	[3.74,90]
CO_2	$\nu_2 \leftrightarrow \nu_3$, P(20)	CO_2	11 ± 1.8	0.13 ± 0.06	[9.58]
	9.6	He	-	0.07 ± 0.03	[9.58]
		N_2	7.6 ± 1.8	0.14 ± 0.06	[9.58]
NH_3	ν_2	NH_3	23	-	[3.86]
	10.8	Xe	8.4	-	[3.86]
OsO_4	ν_3 10.6	OsO_4	13 ± 3	-	[3.84]
SiF_4	ν_3	SiF_4	8.2 ± 3	-	[3.82]
	9.6		5.2 ± 4	-	
CO_2	$\nu_2 \leftrightarrow \nu_3$, P(20) 10.6	CO_2	7.5 ± 0.75	-	[3.6]

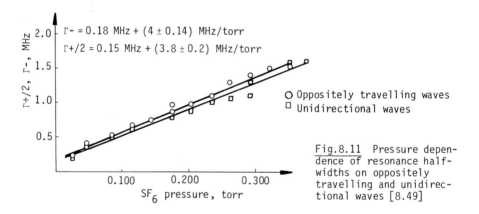

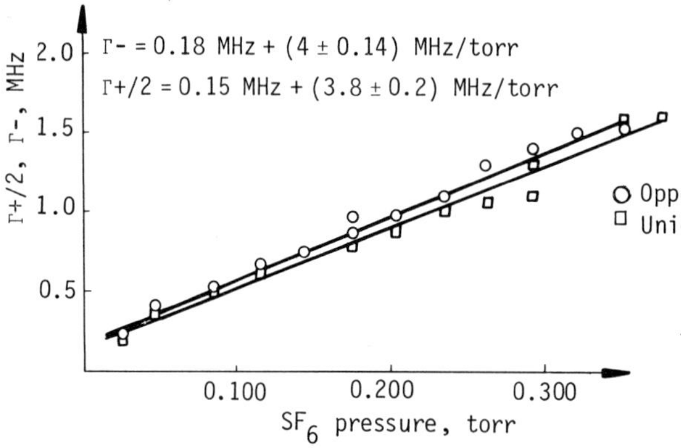

Fig.8.11 Pressure dependence of resonance halfwidths on oppositely travelling and unidirectional waves [8.49]

travelling weak probe wave in an external absorption cell is mainly employed when a CO_2 laser is used with other molecular gases. At fairly low pressures the shape of the Lamb dip in an active medium can be explored in CO_2 and CO lasers [8.45,46]. A fluorescing cell was used in an absorption cell [8.47].

The collision broadening in SF_6 was explored in [8.48] by the weak, oppositely travelling-wave method and a CO_2 laser at 10.6 μm (see Chap.3). Use of spaced light beams permitted an estimate of the scattering cross-section, and the free-path length, from the decrease of the resonance intensity. Resonances in SF_6 that resulted from interaction of unidirectional and oppositely travelling waves were explored in [8.49] in order to find mechanisms of collision broadening in molecular gases. Figure 8.11 shows the resonance widths in SF_6, obtained by use of a CO_2 laser at $\lambda = 10.6$ μm, for unidirectional and oppositely travelling waves. The difference of the resonance broadenings, 2 to 1, shows that the relaxation cross-sections of different levels of SF_6 molecules, during collisions, are approximately identical. The results of collision broadening measurements in SF_6 are similar to data gained from experiments on quantum echo in SF_6 [8.50].

The homogeneous width for various detunings of the frequency of the strong field from the line center was explored in [8.51] by the method of an oppositely travelling probe wave whose frequency was shifted relative to that of a strong wave by the value Δ. The dependence of the homogeneous broadening on the detuning frequency Ω (see Chap.7) and, hence, on colliding-particle velocities permits study, in more detail, of the nature of interaction of particles during collisions. Previously, this could be done only by varying the gas temperature. Figure 8.12 shows the dependence of the width on the gas density for two detunings $\Omega = 0$ and $\Omega = 75$ MHz in NH_3 as a function of Xe pressure. No frequency dependence of collision broadening in NH_3 gas was observed. This can be explained by the resonant nature of collisions, in which frequencies are independent of particle velocities.

The method is more effective if the emitting particle is lighter than the surrounding particles. It is well illustrated by the calculation made in [8.52]. Figure 8.13 shows the results of calculation of homogeneous width as a function of frequency detuning, for various ratios of the masses of emitting and surrounding particles.

A saturation homogeneity due to the collisions, which lead to smoothing of the Lamb dip in CO_2, was studied in detail in [8.53,54]. A similar phenomenon was explored in a CO laser in [8.46].

Collision broadening of nonlinear resonances exceeds that of the Doppler contour. The authors of [8.55] and [8.56] have observed nonlinear dependences of collisional broadening on pressure in CH_4 and CO_2, respectively, at low

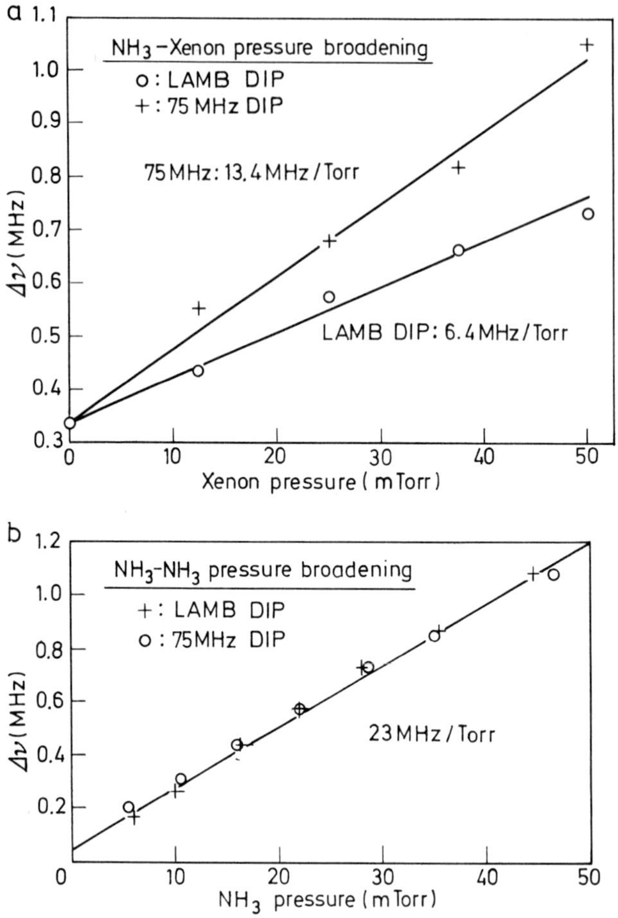

Fig.8.12a,b Plot of width of saturation resonances for NH_3 molecules Doppler shifted 75 MHz from line center and molecules with no z component of velocity (Lamb dip). (a) as a function of Xe pressure for broadening by Xe at fixed NH_3 pressure of 10 mtorr, and (b) as a function of NH_3 pressure for self-broadening [8.51]

pressures; it has been associated with the influence of the elastic scattering of particles by collisions.

These features of collision resonance broadening open up new opportunities for investigating collisions by spectroscopic methods.

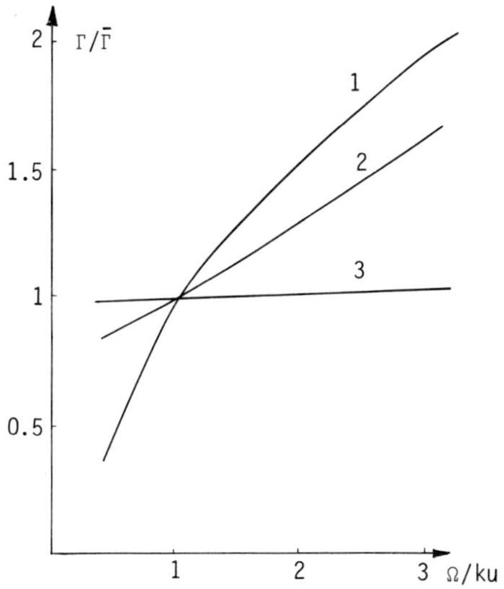

Fig.8.13 Calculated dependence of a homogeneous width on the frequency detuning from the line center for the interaction of C_6/R^6, for different ratios of masses of emitting and impinging particles. 1 - 50, 2 - 1, 3 - 0.02

8.2.3 Homogeneous Saturation of Vibrational-Rotational Transitions at Low Pressure

The lifetimes of vibrational states of molecules are usually long compared to the time intervals between collisions. Consequently, saturation of the transition is determined by diffusion in the velocity space. In this case the spectrum of an individual molecule is a sequence of wave packets with different frequencies within ku, i.e., the spectrum of an individual molecule is that of the whole ensemble. Therefore, during the lifetime of a vibrational state, any molecule can interact with the field, in spite of the fact that only molecules whose frequencies differ from that of the field by not more than Γ interact effectively with the field at each moment. Thus, homogeneous saturation takes place when the homogeneous width 2Γ is far less than the Doppler width.

In the first experiments [8.53] for investigating absorption resonances in a CO_2-He-N_2/CO_2 laser, the pressure in the cell varied from 0.4 to 2 torr. At such pressures, the homogeneous width of an absorption line in CO_2 is far less than the width of the Doppler gain line in the CO_2-He-N_2 mixture of the active medium in a CO_2 laser. Estimates have shown that, because of the inhomogeneous nature of saturation, a generated-power peak connected with the Lamb dip in the absorption line should be observed under the conditions of the experiments. Under cw conditions, the generated-power peak was not observed in [8.53]. This could be due to appreciable homogeneous saturation.

An absorption coefficient that takes into account a strong exchange between rotational sublevels due to collisions is given by

$$\kappa = \kappa_0 \left\{(1+gE^2)^{1/2} + \frac{\sqrt{\pi}P^2E^2}{\hbar^2 ku}\,[\tau_2 W(J) + \tau_1 W(J+1)]\right\}^{-1} , \qquad (8.5)$$

where κ_0 is nonsaturated gain in the line center, g is a parameter of the inhomogeneous saturation in the cell, and W(J) is the Boltzmann distribution of excited molecules over rotational sublevels. The value τ_2 is defined by

$$\tau_2 = \frac{\tilde{\nu}_2}{(\gamma_2 + \gamma_{2col})(\gamma_2 + \gamma_{2col} - \tilde{\nu}_2)} , \qquad (8.6)$$

where γ_2 is the probability of destruction of the vibrational state, owing to spontaneous transitions and diffusion towards walls, γ_{2col} is the probability of collision as a result of variations of velocity and of the rotational quantum number J, and $\tilde{\nu}_2$ is the transition probability by reason of collisions; τ_i, γ_i, γ_{icol}, $\tilde{\nu}_1$ are the corresponding values for a lower level. Provided that γ_{icol} and $\tilde{\nu}_i$ differ slightly (the difference between these magnitudes gives the destruction probability in volume) and that each of them is much greater than γ_2 (8.6) can be of the form

$$\tau_i = \frac{1}{\gamma_i + \gamma_{icol} - \tilde{\nu}_i} . \qquad (8.7)$$

In this case, τ_2 and τ_1 signify lifetimes of the corresponding vibrational levels.

The first term in (8.5) is connected with the inhomogeneous saturation. The second term describes the saturation homogeneity. A simple estimate shows that, at a pressure of several torr, the saturation is practically completely homogeneous. Figure 8.14 shows the dependence of the coefficient 1/g (recall that it connects the saturation parameter G with the field density, according to $G = gE^2$) on gas pressure. This dependence confirms the saturation homogeneity. The destruction rate of vibrational states in a laser beam calculated from (8.5) agrees with the data obtained by other methods. A decrease of the destruction rate with increase of pressure is due to a corresponding decrease of the rate of diffusion of vibrationally excited molecules. At

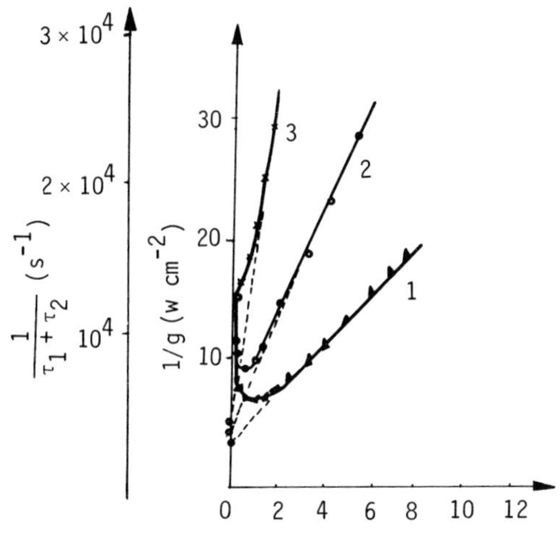

Fig.8.14 Dependence of saturation power 1/g and the value $(\tau_1 + \tau_2)^{-1}$ on the transition P(20) of the 001-100 band of CO_2 on mixture-gas pressure (1 - He, 2 - Ne, 3 - N_2)

high pressures, the rate of destruction of vibrational states increases, owing to collisions.

With short-time interaction with the field, the smoothing of a dip in a gain-, or absorption-line contour takes a time interval that is more than or, at least, equal to that between collisions. In the case of vibrational-rotational transitions, collisions result not only in diffusion in the velocity space, but in a strong exchange between rotational sublevels. This process increases the time for which the dip in the gain-, or absorption-line center is smoothed.

In the model of strong collisions, taking into account rotational relaxation, an expression for the condition of smoothing of the dip in the absorption line can be obtained,

$$W(J)\,\frac{\tilde{\nu}_i}{\gamma_i + \gamma_{icol} - \tilde{\nu}_i} \cdot \frac{\Gamma}{ku} > 1 \ . \tag{8.8}$$

In the case of a pulsed field, the lifetime in a vibrational state should be replaced with the time of action of a pulse τ_p. Then (8.8) can be of the form

$$\tau_p > \frac{ku}{W(J)\tilde{\nu}_i\Gamma} \ , \tag{8.9}$$

where Γ is mainly determined by collisions. For $J = 20$, $T = 800$ K, $P = 1$ torr, and for a cross-section of broadening of a vibrational-rotational transition $= 1.8 \times 10^{-14}$ cm^2 (8.8) is fulfilled at $\tau_p \simeq 10$ μs. This value for the time of smoothing of the dip in the absorption line is approximate, because a considerable contribution to its value can be provided by collisions with changing I and the processes associated with the resonant excitation exchange. Saturation with a pulsed field was studied in a Q-switched CO_2 laser. A mechanical interrupter was introduced into a cavity (modulation frequency ~2 kHz, generation time ~700 μs). A generated power pulse had a complicated shape, which consisted of two bursts, after which the generated power reached gradually the value corresponding to cw conditions. The first burst, of about 10 μs duration, is associated with the time of switching-on of the cavity Q. The second burst corresponds to the relaxation of a lower operating level, because the time of its appearance (about 20-30 μs after switching-on of the cavity Q) coincides with the lifetime of level 100. Full interpretation of this complex behavior of the output power of the cavity Q modulation requires additional investigation. The envelope of the first and of the second maxima, when the frequency is scanned, shows a power peak in the center of the absorption line. Figure 8.15 shows the envelope of the second burst when the frequency is scanned, with (curve 1) and without (curve 2) absorption. The level that corresponds to the cw condition is shown in curve 3. The power peak was observed 20-30 μs after switching the generation. This is in good agreement with that calculated above.

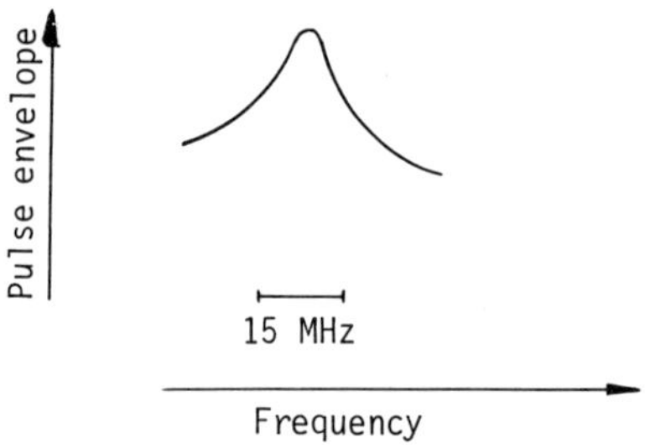

Fig.8.15 Envelope of power pulses against frequency

Introduction of sufficient absorption resulted in nonsteady-state conditions of laser action. In this case, the laser generated pulses of about 10 μs duration, whose amplitudes exceeded the level of continuous laser action by approximately an order of magnitude. When the laser frequency was scanned, the pulse amplitude was that of the power peak in the center of the absorption line. The halfwidth of the envelope depended on pressure. The collision broadening of the peak amounted to about 7.7 MHz/torr.

8.2.4 Nonlinear Dependence of Resonance Widths on Pressure

At low gas pressures (less than 10^{-2} torr) the influence of collisions that result in homogeneous saturation of vibrational-rotational transitions becomes insignificant. Inhomogeneous saturation turns out to be dominant. In this case, narrow resonances in an absorption line can be readily observed. Qualitatively new peculiarities, associated with elastic scattering of molecules by collisions, were found when the influence of collisions on the shape of narrow resonances in molecular gases was studied. These peculiarities are characteristic of long-lived levels and, hence, of vibrational-rotational transitions.

In Chapter 6 it was shown that the influence of elastic scattering of particles on the collisional broadening of resonances is manifest in the pressure range where resonance broadening is comparable with the Doppler frequency shift in scattering at the angle Θ, equal to $ku\Theta$. It is known from experimental data and results of the collision theory [8.57] that the characteristic angle Θ within which about 90% of all particles is scattered, ranges from 10^{-3} to 10^{-2}. So, for vibrational-rotational transitions, the resonance width must be, at least, by a factor of 10^{-3} to 10^{-2}, smaller than the Doppler width, i.e., it has a value of 10^4 to 10^6 Hz. Such resonance widths can be obtained at pressures of 10^{-2} to 10^{-4} torr. Note that optically allowed transitions have natural widths usually more than 10^{-3} to 10^{-2} of the Doppler width; the influence of elastic scattering is, therefore, difficult to observe.

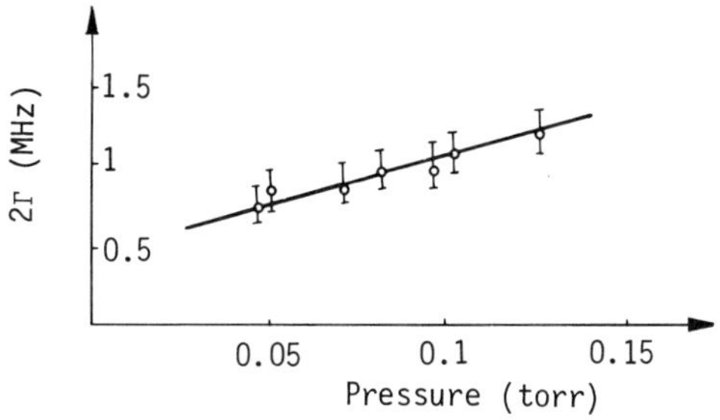

Fig.8.16 Dependence of P(20) line width of the 00°1-02°0 band of CO_2 on pressure

Some of the first evidence of the influence of elastic scattering was obtained when the Lamb dip in CO_2 at $\lambda = 10.6$ μm was studied [8.58]. The power peak in a CO_2 laser was used. Figure 8.16 shows the dependence of the Lamb-dip width in CO_2 on gas density. There are no obvious peculiarities of the observed dependence. The authors of [8.58] noticed that linear extrapolation of the dependence of the Lamb-dip width to zero pressure gives a width that exceeds by more than an order of magnitude the line width caused by flight effects. This means that there is an additional contribution of the process to the broadening which is not very dependent on particle density in the pressure range from 0.1 to 0.5 torr. Diffusion of excited particles by scattering at small angles might be such a process. After repeated scattering events the particle deviation has a diffusion nature. The angle Θ_d, at which a particle undergoing n collisions deviates, is equal to $\Theta_d \cong \bar{\Theta}\sqrt{n}$. When a Lamb dip is observed, of first importance is the number of elastic collisions during the lifetime of a vibrational level. In other words, n is the ratio of the cross-sections of elastic and nonelastic scattering. It can be readily seen that this value is independent of gas pressure. So the value, at which the Lamb dip is broadened supplementarily, is independent of pressure. To an order of magnitude $n = 10$, and $\Theta = 0.01$. So we have a constant contribution to broadening equal to ~1 MHz. A more detailed discussion of the experiment [8.58] is given in [8.59].

8.2.5 Dependence of Collision Broadening of Resonances on Gas Density

In the pressure range where $ku\Theta \sim \Gamma$, the angular scattering of molecules can influence the shapes and widths of resonances. In the pressure range where $ku\Theta \ll \Gamma$, collisional broadening is determined by the total elastic-scattering cross-section: a molecule does not interact with a field after collision. So, whether phase randomization takes place or not, collisions result in Lamb-dip broadening. This problem has been discussed in detail in Chapter 6. In the range $ku\Theta \sim \Gamma$ resonance broadening depends on the nature of the collisions. If phase-randomizing collisions play a small role, elastic scattering of particles at the angle Θ at which $ku\Theta < \Gamma$ does not result in resonance broadening. Collision broadening of resonance in this range depends nonlinearly on particle density; the broadening constant is determined by the contribution of nonelastic and phase-randomizing processes.

Finally, at sufficiently high pressures $ku\Theta \gg \Gamma$ the collision broadening constant is determined by the contributions of nonelastic and phase-randomizing collisions. Note that the collision broadening of a Doppler contour is defined by the same processes.

A nonlinear dependence of collision broadening of the Lamb dip at low gas pressures was first reported and explained in [8.55]. The Lamb dip in methane was investigated in the pressure range 10^{-3} to 10^{-2} torr, by use of a He-Ne/CH_4 laser. When the collision broadening in methane is explored the influence of field broadening should be taken into account. A value of the Lamb-dip width at each pressure was obtained when the field dependence of the Lamb-dip width was extrapolated to zero field. It is much more difficult to

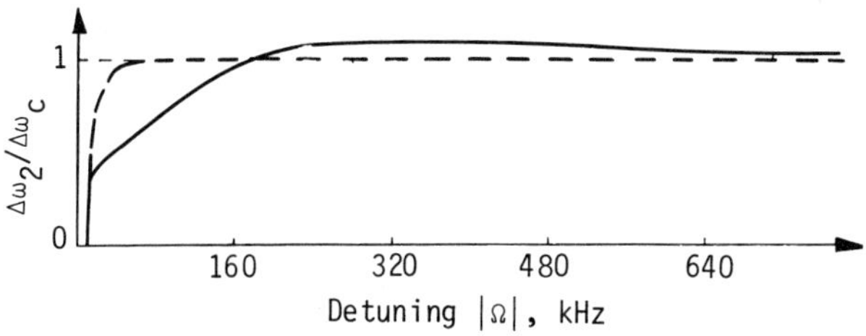

Fig.8.17 Dependence of the ratio of the frequency-modulation amplitude of the He-Ne/CH_4 laser oscillation to that of the cavity, on the frequency detuning Ω near the center of an absorption line of methane at $\lambda = 3.39$ μm (continuous curve). Dotted curve is obtained at $P_{CH_4} = 0$ torr

take into consideration the nonlinear pulling of the frequency generated by a He-Ne/CH_4 laser near the center of an absorption line of methane. The method used in [8.55] automatically took into account the nonlinear change of the generated frequency by the change of the cavity frequency. The essence of the experiment is similar to the observation of resonance in the spectrum radiated by a gas laser with nonlinear absorption, described in Chapter 3. When the cavity frequency is modulated by $\Delta\omega_c$ the generated frequency varies by $d\omega_2$; these values are connected by

$$d\omega_{os} = d\omega_c \left[1 - g \frac{1-\delta^2}{(1+\delta^2)^2}\right] , \qquad (8.10)$$

where $g = 2G\kappa_B c/\Gamma$, $\delta = \omega_{os} - \omega_0/\Gamma$ is the generation detuning of the generated frequency from the absorption-line center. It is seen from Section 8.10 that, when $\Omega = \Delta$, $d\omega_{os} = d\omega_c$. Figure 8.17 shows the dependence (curve 1) of the modulation of the laser-generated frequency on the generated-frequency detuning from the center of the absorption line of methane. Curve 2 shows the change of the generated frequency with the same amplitude of modulation of the cavity frequency, in the absence of an absorber. Detuning Ω, at which curves 1 and 2 cross, corresponds to the halfwidth Γ of the line.

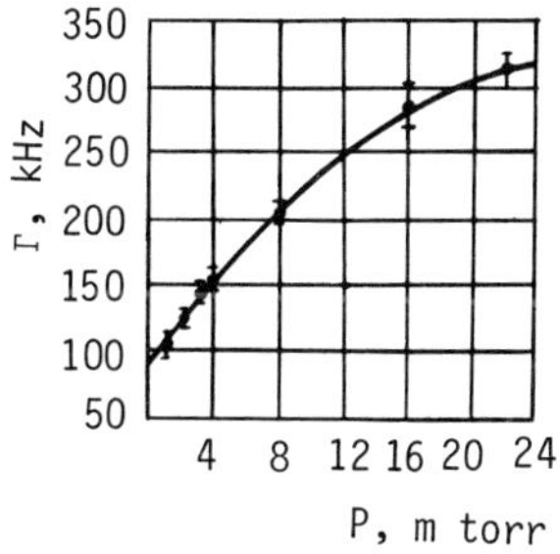

Fig.8.18 Dependence of Lamb dip width in CH_4 at $\lambda = 3.39$ μm on pressure

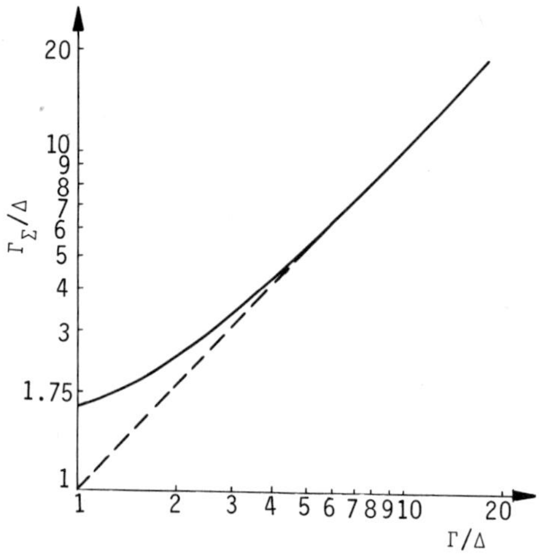

Fig.8.19 Calculated dependence of the contour width Γ_Σ/Δ of the line $F_2^{(2)}$ of methane on the width of an individual component

Figure 8.18 shows the dependence of the collision broadening of the Lamb dip in methane. Extrapolation to zero pressure of the Lamb part of curve $\Gamma(P)$ yields a value of width that is in good agreement with the calculated flight width (see Chap.6). Estimates show that the influence of slow atoms in the pressure range under investigation can be neglected. The hyperfine structure of the methane line under study can also influence the dependence of the width of the resulting contour on pressure. Figure 8.19 shows the dependence of the contour formed by the three strongest components on the width of each individual component (the widths of the components are assumed to be equal). It can be seen that at a contour width of about 100 kHz the collision broadening of the resulting contour is practically equal to the broadening of each component. The dependence corresponds to the analysis shown in more detail in Chapter 6. According to [8.57] 90% of CH_4 molecules are scattered at angles $\Theta \lesssim 10^{-2}$; therefore, in the range $\Gamma \sim 10^4$ to 10^5 Hz Γ depends linearly on pressure. The decrease of the slope of the curve at large P shows that some of the atoms begin scattering at angles $\Theta \sim \Gamma/kv$ with no loss of coherence. The difference between the slopes of the curve at low and high pressures indicates a great role of elastic scattering with no phase randomization. This means that the scattering probabilities for both levels are approximately equal. Additional evidence of the slight role of phase-randomizing collisions is the difference between broadenings of the Lamb dip and of the Doppler contour. According to the data of [8.38,39] the broadening of the Doppler contour is 7.4 ± 0.4 MHz/torr and 7.8 ± 0.8 MHz/torr, respectively. This broadening can be due to nonelastic scattering as well as to some difference of the scattering probabilities. Phase-randomizing collisions result in identical broadening of the Lamb dip and of the Doppler contour. Elastic scattering with no phase randomization causes narrowing of the Doppler contour, which is determined by the transport cross-section σ_{tr}. A slight line shift Δ confirms to some extent the conclusion that phase randomization plays a small role.

The results of investigations have been used by the authors of [8.55] to find elastic-scattering cross-sections of molecules in excited states. At very low pressure the broadening is determined by the sum of the elastic scattering cross-sections of molecules on the two levels, which is $(10 \pm 0.3) \times 10^{-14}$ cm^2. It is necessary to find the mechanism responsible for the broadening of the dispersion part of the Doppler contour, in order to determine the scattering cross-section of molecules on each level with the same accuracy. Because this mechanism cannot be considered known, the accuracy of measurements of the scattering cross-section on each level, according to the preceding considerations, is determined by the ratio of the collisional broadening of the Lamb dip and of the Doppler contour. The sum of the scattering cross-section and the elastic-scattering cross-section on each level is $\sigma = (5 \pm 1.2) \times 10^{-14}$ cm^2. A bend occurs in the curve at $\Gamma(P) = 500$ kHz; it corresponds to a characteristic scattering angle $\bar{\Theta} = 0.5 \times 10^{-2}$ rad. If it is assumed that the interaction is proportional to c/R^6, the total cross-section is associated with the transport relation $\sigma_{tr} \sim \sigma\sqrt[3]{\bar{\Theta}}$. The calculation gives $\sigma_{tr} = 8 \times 10^{-15}$ cm^2. This estimate agrees with the accepted gas-kinetic cross-section of methane [8.60].

In accordance with the above analysis, peculiarities should be expected in the behavior of the Lamb-dip shift at low pressures, which are of great importance for production of optical standards with high reproducibility of frequency. In the region of low pressures, when $ku\bar{\Theta} \gg \Gamma$, in spite of whether phase randomization occurs or not, after collision an atom goes out of the interaction region and, hence, does not make a contribution to the Lamb-dip shift Δ_L. Only a small portion of the atoms scattered into the range of angles γ/ku give the shift Δ_L. Taking into account that $\gamma \sim Nu\sigma$, we should expect that $\Delta_L \sim (Nu\sigma/ku\Theta)$. Because Δ is linearly dependent on pressure, Δ_L is quadratically dependent on density.

Collision resonance shifts can be measured by use of highly stable lasers. Figure 8.20 shows the dependence of the frequency of the power generated by a peak-stabilized He-Ne/CH_4 laser, on He and Xe pressures. In the pressure range from 1 to 5 mtorr, the shift is negligible; it begins increasing rapidly at pressures of more than 10 mtorr. The shift in methane was 10 Hz/mtorr. Note that the influence of the hyperfine structure in methane on the

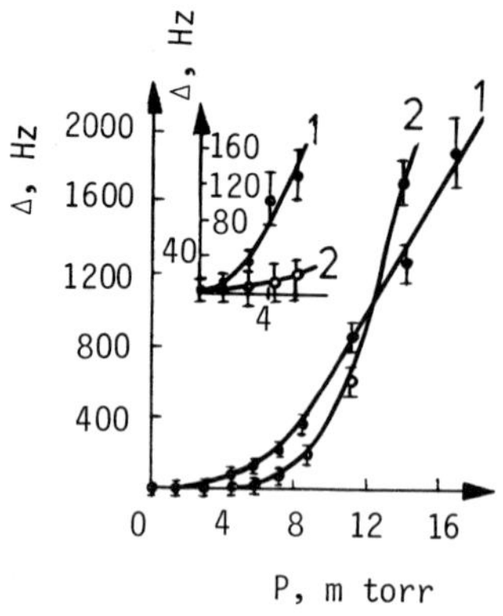

Fig.8.20 Generated-frequency shift of the He-Ne/CH_4 laser at $\lambda = 3.39$ µm stabilized at the power peak, as a function of the He pressure (curve 1) and of Xe (curve 2) in the absorbing cell. CH_4 pressure in the cell was 10^{-3} torr

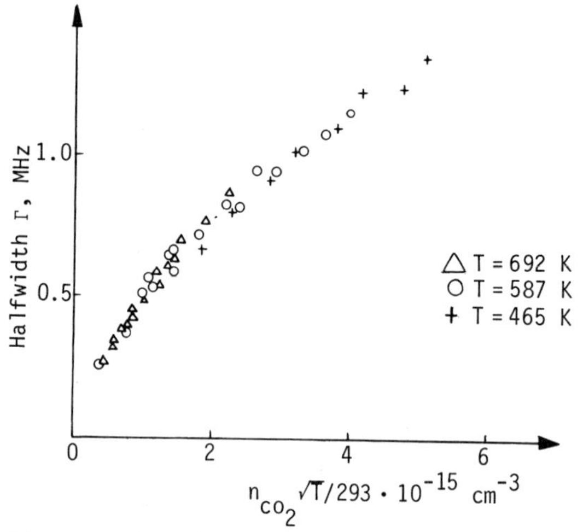

Fig.8.21 Dependence of the resonance halfwidth in CO_2 on gas density

resonance is greater than on the broadening. This circumstance complicates the interpretation of data on the resonance shift.

In conclusion, we note that collision broadening of resonance in CO_2 at $\lambda = 10.6$ µm in a wide pressure range was investigated in [8.56]. Nonlinear dependence of collision broadening of resonance on the particle density is observed at $\lambda = 3.39$ µm, as in the case of methane (Fig.8.21). In the range of low pressures, the collision-broadening constant is three times as great as that at high pressures. Thus, the nonlinear dependence of resonance on gas pressure should be rather the rule than an exception.

8.3 Use of Resonances in Investigations of Nonelastic Collisions

Resonances that arise at the interaction of unidirectional waves can be used in investigations of nonelastic processes. Resonances in an absorption line of a weak wave in the presence of a strong one have widths corresponding to lifetimes of the upper and lower levels and the width of the absorption line (see Chap.3). The relaxation of molecular states is determined, as a rule, by collisions. So the study of the interaction of unidirectional waves with oppositely travelling waves in molecular gases provides information on collisions that result in level relaxation and broadening.

The interaction of unidirectional waves with an inhomogeneously broadened line for nondegenerate systems has been studied in detail in [8.61] (see Chap.3). Reference [8.62] offers a treatment of polarization phenomena at the interaction of unidirectional fields when collisions are taken into account. It has been shown that at rather great degeneracy $I \gg 10$ the differences of line shapes of absorption that result from polarization effects are negligible. The absorption coefficient in a weak field is given by

$$\kappa \sim 1 - X \frac{V^2}{2\Gamma - \gamma_1} \cdot \frac{\gamma_1}{\gamma_1^2 + \Delta^2} - X \frac{V^2}{2\Gamma - \gamma_2} \cdot \frac{\gamma_2}{\gamma_2^2 + \Delta^2} -$$

$$- (2\Gamma - \gamma_1 - \gamma_2) \frac{2\Gamma}{(2\Gamma)^2 + \Delta^2} \cdot XV^2 \left[\frac{1}{\gamma_2(2\Gamma - \gamma_1)} + \frac{1}{\gamma_1(2\Gamma - \gamma_2)} \right] , \qquad (8.11)$$

where X is a coefficient dependent on the degeneracy of the interacting levels and the polarizations of the fields, $V = Ed/2\hbar$, Δ is the detuning between frequencies of the weak and strong fields, γ_1 and γ_2 are the rates of decay of the levels, 2Γ is the radiated line width. The second and third terms describe interference dips that have widths that correspond to the lifetimes of the upper and lower states, the fourth term describes the population dip, with width 4Γ.

This method was used in [8.49] for investigating the broadening of resonances in SF_6. Experiments have showed that a resonance shape in unidirectional waves is described by a dispersion curve with accuracy comparable to the experimental errors (see Fig.8.22). The collision broadening of resonance in unidirectional waves was half as much as that in oppositely travelling waves. This shows that lifetimes on vibrational levels of the upper and lower states are equal to the reciprocal of line width $1/\Gamma$, to an accuracy comparable with measurement errors. Thus, $\tau = \tau_2 \approx \tau_1 \approx 1/\Gamma$. This shows that scattering probabilities of the upper and lower vibrational-rotational states

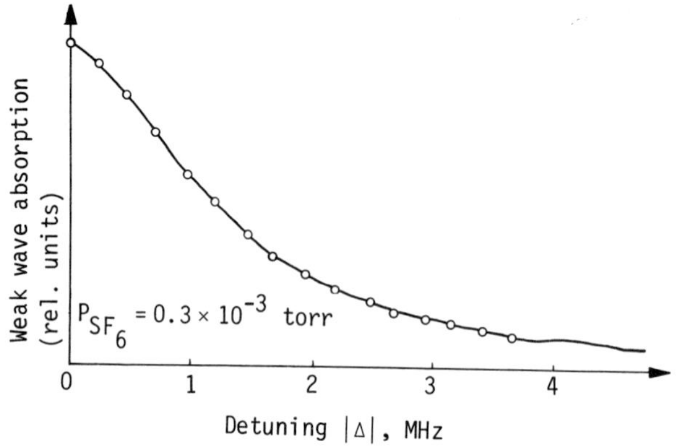

Fig.8.22 Dependence of absorption of a weak wave in the presence of a strong wave propagating in the same direction in SF_6, on the frequency detuning

are identical, and that phase randomization at collisions is negligible. The value $1/\Gamma = T$ measured in [8.49] agrees with the value T measured for SF_6 in [8.50,63] by the method of photon echo. Values of $\tau = 22$ ns · torr and 24 ns · torr were obtained in [8.50,63], whereas a value of about 40 ns · torr was obtained from spectroscopic measurements in [8.49].

If the strong and weak waves are formed from the radiation of two independent lasers, the unidirectional-wave method has a limiting resolution equal to the sum of the line widths of the two lasers. Reference [8.49] suggests a modification of the unidirectional-wave method, in which the amplitude of the strong wave is slightly modulated. Two weak components, which are at the modulation frequency Ω away from the strong component, can be used as weak probe waves. Using this technique, the authors of [8.49] succeeded in obtaining a resonance in SF_6 with a width of the order of 1 kHz. The width of this resonance is practically independent of pressure and is explained by the thermal conductivity of the medium. At a pressure of the order of several torr and with a cell whose dimensions were of the order of several centimeters, the temperature relaxation of the medium is the slowest in the process of relaxation of $V \to T$.

Some nonelastic processes occur with no essential changes of the velocities of the colliding particles. In this case, the nonequilibrium velocity distribution, which arises at one pair of levels connected by a strong field, can be detected at the second pair of levels by use of the second monochromatic field when certain conditions are fulfilled.

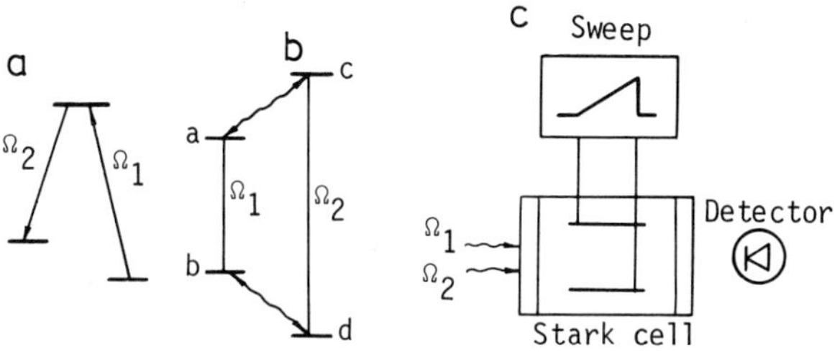

Fig.8.23a-c Traditional double-resonance level configuration (a). Collision-induced double-resonance level configuration; wavy lines indicate collision-induced transition (b). Experimental arrangement for monitoring optical double-resonance signals (c)

Figure 8.23 shows the system of four levels connected by resonant fields that have frequencies Ω_1 and Ω_2.

Consider the case when the field at the frequency Ω_1, resonant with the transition ab, causes saturation of this transition. Particles with velocities near $v_{21} = (\omega_{ab} - \Omega_1)/k_{ab}$ interact effectively with the field at the frequency Ω_1. Particles with velocities near $v_{21} = (\omega_{cd} - \Omega_2)/k_{cd}$ interact with the field on the transition cd. The transitions ab and cd will be coupled if the condition $\omega_{ab} - \Omega_1 = \omega_{cd} - \Omega_2$ is fulfilled (the factor k_{cd}/k_{ab} is omitted here).

These double resonances, induced by collisions, were first reported in [8.64-66]. References [8.64,65] report study of the change of velocity ΔV_z at collisions that vary the rotational quantum number I on the transitions of the band 001-100 of CO_2. Two pairs of stable single-mode CO_2 lasers were used. Two of them were stabilized on the center of transitions P(J) and P(J'). The two other lasers were offset locked to the stable lasers by use of frequency-offset lock. The frequency difference between each pair of lasers can vary. The laser radiation was directed into an absorption cell with CO_2 to which buffer gases H_2, He, CO_2, and Kr could be added. The resonances were recorded over the spontaneous radiation at 4.3 μm. The resonances induced by collisions with H_2 were observed for $|J - I'| = 2$ and 4. The resonances with $|J - J'| > 4$ were not observed. For collisions $J = 20 \rightarrow J' = 18$. From experimental results it was found that $\Delta v_z = 3 \pm 2 \times 10^3$ cm/s. The main contribution to the angular dependence of the interaction is given by anisotropic dispersion forces for collisions of CO_2 - He and CO_2 - Kr. The quadrupole-quadrupole interaction is added to these anisotropic dispersion forces for collisions of CO_2 - He and CO_2 - CO_2.

The double resonance induced by collisions was observed on transitions of the band ν_3 of $^{13}CH_3F$ molecule $(J,K) = (4,3) \rightarrow (5,3)$ in [8.67]. One laser was offset locked to the frequency of another with a detuning of 30.008 MHz. The radiation of the two lasers was directed into an absorption Stark cell. The collision-induced resonances were observed when the condition of double resonance was satisfied; the frequencies of the levels were changed by varying the Stark field. According to the experimental results the cross-section of the collision with $\Delta M = \pm 1$ for $(J,K) = (4,3)$ or (5.3) is about 100 $Å^2$. For this case, about 15% of the collisions vary K, while K remains unchangeable. The remaining 85% of the collisions change both J and K.

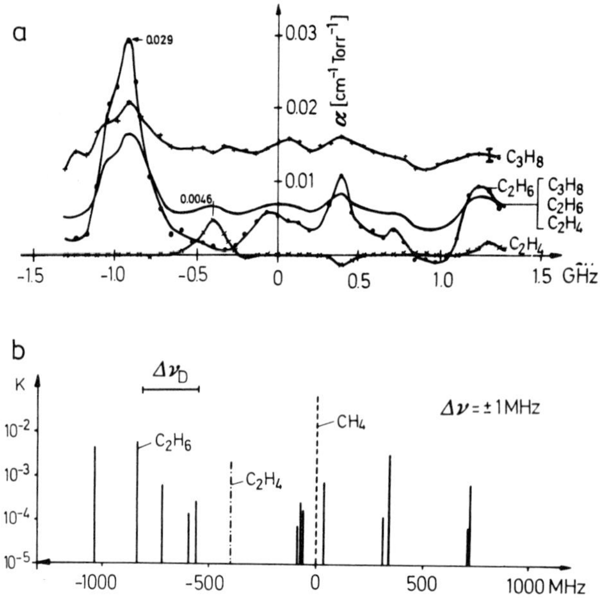

Fig.8.24a,b Nonlinear spectrum with no Doppler broadening obtained by the method of absorption saturation of molecules of methane, C_2H_4, C_2H_6 inside the He-Ne laser cavity at $\lambda = 3.39$ µm whose frequency is tuned in the vicinity of ±1000 MHz of the broadening line by variation of the magnetic field [8.68]

8.4 Applied Molecular Spectroscopy

At present, the technique of absorption saturation spectroscopy is the most exact method for study of details of molecular spectra for scientific purposes. However, this technique has such great advantages as, first, the extremely high information capacity of a unit frequency interval; second, the capability fo differentiate between details of a spectrum inside continuous, wide, vibrational bands of linear absorption of complicated molecules. These advantages encourage the prediction of practical use of methods of nonlinear spectroscopy methods for applied molecular spectral analysis, following the development of simple lasers with tunable frequencies in the ir band, similar to dye lasers in the visible band (it is possible that lasers in the near-ir band at F centers could be the basis for such lasers) [8.67].

The first successful experiments on application of a nonlinear spectrometer for these purposes are described in [8.68]. The ir bands lie in the range of 3.4 μm, utilizing the C-H hydrocarbon bond. When the frequency of a He-Ne 3.39 μm laser in the range of only 0.2 cm^{-1} is scanned by a magnetic field, narrow resonances are found, with widths of 0.5 MHz, arising from rotational-vibrational lines of CH_4, C_2H_4, C_2H_6, and other molecules. Figure 8.24 shows the experimentally observed absorption-saturation spectrum of a gaseous mixture of a number of hydrocarbons. The sensitivity of certain detection of molecular impurities by this method ranges from 10^{-2} to 10^{-3}. Development of simple ir lasers will probably make it possible to produce commercial nonlinear laser spectrometers for more complete and effective molecular spectral analysis.

References

8.1 E.D. Hinkley: Appl.Phys.Lett. 16, 351 (1970)
8.2 I.M. Beterov, V.P. Chebotayev, A.S. Provorov: Opt.Commun. 7, 410 (1973)
8.3 O.N. Kompanets, V.S. Letokhov: Zh.Eksp.i Teor.Phys. 62, 1302 (1972)
8.4 T.W. Hänsch, M.D. Levenson, A.L. Schawlow: Phys.Rev.Lett. 26, 946 (1971)
8.5 M.D. Levenson, A.L. Schawlow: Phys.Rev. A6, 10 (1972)
8.6 M.D. Levenson: Thesis, Stanford University (1972)
8.7 G.R. Hanes, J. Lapierre, P.R. Bunker, K.C. Shotton: J.Molec.Spectros. 39, 506 (1971)
8.8 G.R. Hanes, C.E. Dahlstrom: Appl.Phys.Lett. 14, 362 (1969)
8.9 C. Borde, J.L. Hall: Phys.Rev.Lett. 30, 1101 (1973)
8.10 C. Borde, J.L. Hall: Report on 2nd Symposium on Gas Laser Physics, Novosibirsk, June 1975
8.11 S.N. Bagayev, et al.: Report on 2nd Symposium on Gas Laser Physics, Novosibirsk, June 1975
8.12 T.W. Meyer, J.F. Brilando, C.K. Rhodes: Chem.Phys.Lett. 18, 382 (1971)
8.13 R.G. Brewer, M.J. Kelly, A. Javan: Phys.Rev.Lett. 23, 559 (1969)
8.14 M.J. Kelly, R.E. Francke, M.S. Feld: J.Chem.Phys. 53, 2979 (1970)
8.15 R.G. Brewer, J.D. Swalen: J.Chem.Phys. 52, 2774 (1970)
8.16 A.C. Luntz, J.D. Swalen, R.G. Brewer: Chem.Phys.Lett. 14, 512 (1972)
8.17 K.T. Hecht: J.Molec.Spectros. 5, 355 (1972)
8.18 L. Henry, N. Husson, R. Andia, A. Valentin: J.Molec.Spectros. 36, 511 (1970)
8.19 K. Uehara: J. Phys.Soc.Japan 34, 777 (1973)

8.20 A.C. Luntz, R.G. Brewer, K.L. Foster, J.D. Swalen: Phys.Rev.Lett. 23, 951 (1969)
8.21 A.C. Luntz, R.G. Brewer: J.Chem.Phys. 54, 3641 (1971)
8.22 M. Mizushima, P. Venkateswarlu: J.Chem.Phys. 21, 705 (1953)
8.23 E.E. Uzgiris, J.L. Hall, R.L. Barger: Phys.Rev.Lett. 26, 289 (1971)
8.24 R.G. Brewer: Phys.Rev.Lett. 25, 1639 (1970)
8.25 R.G. Brewer: Science 178, 247 (1972)
8.26 P.A. Steiner, W. Gordy: J.Molec.Spectros. 21, 291 (1966)
8.27 S.C. Wofsy, J.S. Muenter, W. Klemperer: J.Chem.Phys. 55, 2014 (1971)
8.28 R.G. Brewer: In *Fundamental and Applied Laser Physics*, Proceedings of the Esfahan Symposium, ed. by M.S. Feld, A. Javan, N.A. Kurnit (John Wiley & Sons, New York 1973), p.421
8.29 O.N. Kompanets, A.R. Kukudzhanov, V.S. Letokhov, V.G. Minogin, Ye.L. Mikhailov: Zh.Eksp.i Teor.Phys. 69, 32 (1975)
8.30 K. Shimoda: Japan.J.Appl.Phys. 12, 1393 (1973)
8.31 T.J. Bridges, T.Y. Chang: Phys.Rev.Lett. 22, 811 (1969)
8.32 R.T. Menzies, M.S. Shumate: IEEE J. QE-9, 862 (1973)
8.33 F.R. Peterson, D.G. McDonald, J.D. Cupp, B.L. Danielson: Phys.Rev.Lett. 31, 573 (1973)
8.34 D.R. Sokoloff, A. Sanchez, R.M. Osgood, A. Javan: Appl.Phys.Lett. 17, 257 (1970)
8.35 R.S. Eng, H. Kildal, J.C. Mikkelsen, D.L. Spears: Appl.Phys.Lett. 24, 231 (1974)
8.36 S.Y. Chen, M. Takeo: Rev.Mod.Phys. 29, 20 (1957)
8.37 J. Wormhodt, L. Marabella, J.I. Steinfeld: Report on 2nd Symposium on Gas Laser Physics, Novosibirsk, June 1975
8.38 E.J. Gerritsen, M.E. Heller: Appl.Opt.Suppl. 2, 73 (1965)
8.39 S.N. Bagayev, Ye.V. Baklanov, V.P. Chebotayev: Preprint of Institute of Semiconductor Physics, Novosibirsk, No.22 (1972)
8.40 L.S. Vasilenko, A.A. Kovalev, A.S. Provorov, V.P. Chebotayev: Sov.Quant. Electr. 2, 12 (1975)
8.41 T.K. McCubbin, Jr., R. Darone, J. Sorrell: Appl.Phys.Lett. 8, 118 (1966)
8.42 C. Rossetti, P. Barchewitz: C.R. Acad.Sci. 262, 1199 (1966)
8.43 E.T. Gerry, D.A. Leonard: Appl.Phys.Lett. 8, 227 (1966)
8.44 V.V. Danilov, E.P. Kruglyakov, Ye.V. Shun'ko: Preprint of Institute of Nuclear Physics, No.36-72, Novosibirsk
8.45 C. Borde, L. Henry: IEEE J. QE-4, 874 (1968)
8.46 C. Freed, H.A. Haus: IEEE J. QE-9, 219 (1973)
8.47 C. Freed, A. Javan: Appl.Phys.Lett. 17, 53 (1970)
8.48 O.N. Kompanets, A.P. Kukudzhanov, V.S. Letokhov, E.L. Mikhailov: Kvantovaja Electronika 16, 28 (1973)
8.49 L.S. Vasilenko: Report on 2nd Symposium on Gas Laser Physics, Novosibirsk, June 1975
8.50 C.K.N. Patel, R.E. Slusher: Phys.Rev.Lett. 20, 1087 (1972)
8.51 A.T. Mattick, A. Sanchez, N.A. Kurnit, A. Javan: Appl.Phys.Lett. 23, 675 (1973)
8.52 A.S. Provorov: Diploma work, Inst. of Semiconductor Physics, Novosibirsk (1970)
8.53 Yu.V. Brzhazovsky, V.P. Chebotayev, L.S. Vasilenko: IEEE J. QE-5, 146 (1969); Zh.Eksp.i Teor.Phys. 55, 2096 (1968)
8.54 T. Kan, G.J. Volga: IEEE J. QE-7, 141 (1971)
8.55 S.N. Bagayev, Ye.V. Baklanov, V.P. Chebotayev: Zh.Eksp.i Teor.Phys. Pis.Red. 16, 15 (1972)
8.56 L.S. Vasilenko, L. Kachanov, V.P. Chebotayev: Opt.Commun., in press
8.57 H.S.W. Massey, E.H.S. Burhop: *Electronic and Ionic Impact Phenomena* (Clarendon Press, Oxford 1952)
8.58 S.N. Bagayev, L.S. Vasilenko, V.P. Chebotayev: Preprint No.15, Inst. of Semiconductor Physics, Novosibirsk (1970)
8.59 A.P. Kolchenko, A.A. Puhov, S.G. Rautian, A.M. Shalagin: Zh.Teor.i Eksp.Phys. 63, 1173 (1972)

8.60 E.W. McDaniel: *Collision Phenomena in Ionized Gases* (John Wiley & Sons, Inc., New York, London, Sydney 1964)
8.61 Ye.V. Baklanov, V.P. Chebotayev: Zh.Eksp.i Teor.Phys. 61, 922 (1971)
8.62 S.G. Rautian, L.I. Smirnov, A.A. Shalagin: Zh.Eksp.i Teor.Phys. 62, 2097 (1972)
8.63 S.S. Alimpiyev, N.V. Karlov: Zh.Eksp.i Teor.Phys. 63, 483 (1972)
8.64 T.W. Meyer, C.K. Rhodes: Phys.Rev.Lett. 32, 637 (1974)
8.65 T.W. Meyer, W.K. Bischel, C.K. Rhodes: Phys.Rev. A10, 1433 (1974)
8.66 R.G. Brewer, R.L. Shoemaker, S. Stenholm: Phys.Rev.Lett. 32, 63 (1974)
8.67 L.F. Mollenauer, D.H. Olson: Appl.Phys.Lett. 24, 386 (1974)
8.68 W. Radloff, E. Below: Opt.Commun. 13, 160 (1975)

9. Nonlinear Narrow Resonances in Quantum Electronics

The narrow resonances of nonlinear absorption have a number of promising and effective applications in quantum electronics. The most impressive application is for stabilization of the laser frequencies. Essentially, the methods for obtaining narrow nonlinear resonances have been found in attempts to stabilize frequency. The first experiments showed the effectiveness of inhomogeneous saturation of a Doppler-broadened transition for selection of longitudinal laser modes. Finally, under certain conditions, instead of an absorption line with a narrow resonance dip, inversion of populations in a narrow spectral interval can be obtained, i.e., a narrow gain line. Narrow-gain-line lasers have very valuable properties, in particular, high short-term frequency stability. Thus, there is a great number of applications of resonances for problems of quantum electronics. We will consider some of them.

9.1 Selection of Modes by Nonlinear Absorption

9.1.1 Longitudinal Modes

Absorption inside the cavity influences, in an essential manner, the mode interaction and the stability of single-frequency generation. Even the first studies of nonlinear-absorption lasers showed a sharp increase of power under single-frequency conditions with absorption and attention has been paid to the selective properties of nonlinear absorption [9.1]. Subsequently, experimentalists have obtained convincing proofs of the effectiveness of selection of modes by use of nonlinear absorption in a laser with a large excess of gain over threshold [9.2,3].[8]

In spite of the variety of regimes in a laser with nonlinear absorption, its radiation spectrum displays certain regularities when the absorption is introduced. Owing to some homogeneity of saturation caused in a He-Ne laser by trapping of the resonant radiation in the absence of absorption, the width

[8] Mode selection by use of the splitting of an unsaturated absorption line in a magnetic field has been considered by BORISEVICH et al. [9.4].

of the generated spectrum is 1000 MHz. The regime is a free generation of nonsynchronous modes. Absorption of about 0.1% leads to self-synchronization of modes.[9] The frequency interval between modes depends on their positions relative to the line center. With an increase of absorption, the spectrum becomes spaced out. Unlike usual lasers, in which generation of some types of modes is suppressed, the generation of modes symmetrically situated relative to the center of a gain line appears to be stable. With some absorption, the single-frequency regime occurs, the laser frequency being smoothly scanned over a sufficiently wide interval. Owing to the strong saturation with absorption, the peak contrast of the generated power is very small. The selection effectiveness is very high. For example, in [9.6], 30 mW of power was generated in a single-frequency regime. It was 80% of the total power generated in all modes, in the absence of absorption. When the field density in an absorption cell was 4-6 times as great as that in the amplifying medium (it was obtained by suitable design of the cavity geometry), the increased absorption transformed the operation from the single-frequency regime to generation of more than 40 synchronized modes with a spectrum width of about 1500 MHz. Further increase of absorption was followed not by further widening but by narrowing of the spectrum, finally to a single mode. It is the regime where a practically interesting phenomenon of self-stabilization of the generated frequency was observed: when the length of the cavity was changed, a jump from one mode to another occurred while the generated frequency remained near the center of the absorption line, within the accuracy of the intermode interval (45 MHz).

The stability of single-frequency generation and the conditions for the selection of modes can be simply analyzed for large saturations on the basis of the model of dips in absorption and gain lines for a weak signal caused by the variation of the level population under the action of a strong field [9.6-8]. The advantage of the approach consists of the opportunity to make simple calculations of the high excesses of gain over threshold and to clarify some important regularities observed in the experiment.

A nonlinear absorber is the optimum selector of modes, because it does not cause losses at the generated frequency but introduces them at other frequencies. If the saturation parameters of the active medium and absorber differ considerably, the saturated absorption at the generation frequency can be

---

[9] Irrespective of selection of modes, the regime of mode self-synchronization by nonlinear absorption has been reported in [9.5] and subsequently has been used in a number of studies for generation of supershort-time pulses in a He-Ne laser.

far less than the saturated gain and, hence, the cavity losses. At the same time, owing to saturation inhomogeneity, the absorption at other frequencies can be sufficient to compensate the gain. The curves in Fig.9.1 show the way in which the resulting gain decreases with the increase of the absorption. And, finally, beginning from the curves with parameter $\beta = \kappa_{bo}g_b/\kappa_{ao}g_a = 0.7$ the gain at all other frequencies is less than the losses, and the single-frequency-generation regime is stable. The model described shows the triviality of condition of selection of far-type modes. The difference between saturated gain and absorption must be less than or equal to the losses; this was confirmed by the experiment. The selection condition does not contain the maximum relaxation constants which influence the effectiveness of mode selection.

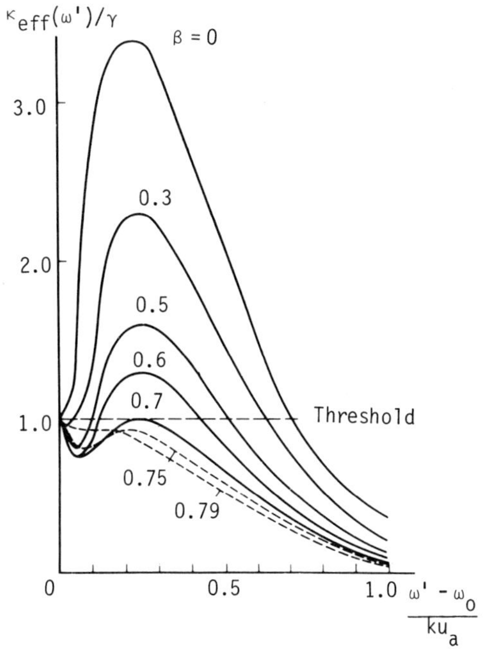

Fig.9.1 Dependence of effective amplification $\kappa_{eff}(\omega') = \kappa(P,\omega') - \kappa(P,\omega')$ on frequency ω', when generation occurs at the center of the amplification line ($\omega = \omega_a = \omega_b$). The effective amplification is expressed in units of linear losses γ [9.7]

To give the complete solution of the problem of single-frequency stability it is not sufficient to consider only the effects of population in the gain and absorption lines, in particular, when considering the regime of symmetrically situated modes. Coherence effects at the interaction of two waves cause the qualitative variations of the weak-signal line shape that are essential for solution of the problems of generated-frequency stability [9.8].

Considerations of population effects describe well the regions where the interaction of two waves appears to be weak, and explains the selection of far-type modes. However this consideration appears to be qualitatively insufficient when the modes are situated symmetrically relative to the line center. The difference between gain and absorption both for intense and for weak fields is equal to losses in the cavity, and the problem of single-mode-generation stability remains open to question. The Lamb theory, which is restricted to weak fields, also does not decide the problem.

Let $\Omega \gg \Gamma_0$. Then the standing-wave field at the generated frequency is the sum of two travelling waves that interact with different atoms. In this case, the velocity distribution of atoms in amplifying and absorbing media displays two dips symmetrically situated relative to the velocity $v = 0$. The wave travelling in the positive direction of the z axis produces a dip in the vicinity of Ω/k and the wave travelling in the opposite direction produces a dip in the vicinity of $-\Omega/k$. The weak-wave field at the mirror frequency can be represented in the form of two travelling waves, as well. Note that the waves with different frequencies, moving in opposite directions, interact with the same atoms. Thus, the intense and weak running waves are the oppositely travelling ones; the results of Chapter 2 can be used. It is not difficult to see that the additional term in (2.157) connected with splitting effects is equal to the difference between the gain (absorption) of the weak signal and the saturated gain (absorption) of the intense one. Therefore, the ratio of these additional terms in the gain and the absorption of the weak signal gives the solution of the problem of single-mode stability. If the additional contribution of the splitting effects to the absorption of weak signal is more than that in the gain, the single-mode generation regime is stable.

Complete analysis of these processes for several modes also appears to be very complicated. Even in the case of weak saturation and two modes it can be carried through only by computational methods. The interaction of two modes and generation stability of a nonlinear absorption laser have been analyzed in the approximation of weak saturation, the method developed for the theory of usual laser in two-mode regime being used [9.9,10].

The magnetic field applied to an internal nonlinear absorbing cell was discovered to influence the laser-radiation spectrum; the frequency composition depended on the position of the mode and on the magnitude of the magnetic field [9.10]. In the general case, the magnetic field makes it difficult to obtain single-frequency generation when the cavity frequency is scanned. Single-frequency generation in a narrow frequency range (350 kHz) has been reported in [9.10]. Outside this range, generation occurred at two frequencies.

This method of selection of axial modes is not used so universally as, for example, methods based on special resonant systems that suppress undesired modes. Wide application of this method is restricted by the limited availability of suitable absorbers. However, in cases where it is possible, this method is simpler and, what is very important, permits smooth scanning of frequency in a wide range. It is no mere chance that in the first commercial single-frequency He-Ne laser GL-159 mode selection is accomplished by nonlinear absorption [9.11].

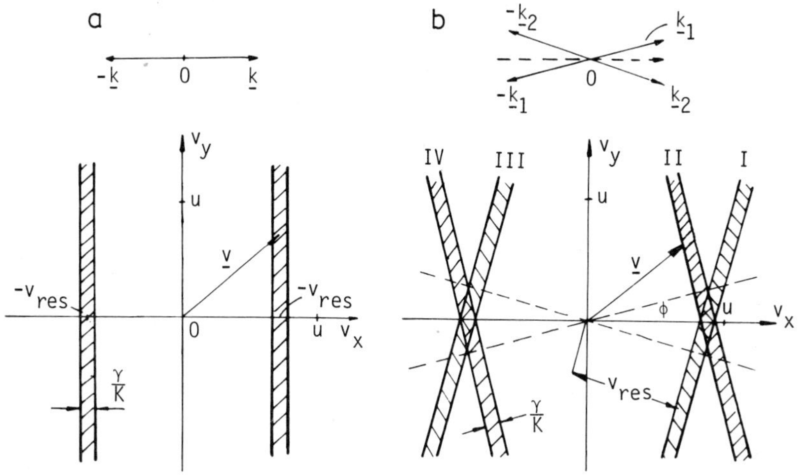

Fig.9.2a,b Projections of the particle velocities $\underline{V}$ onto the (X,Y) plane. In the shaded areas lie the velocities of the particles that interact resonantly with (a) a plane standing wave with wave vector $\underline{k}$ and frequency ω, and (b) with two plane standing waves with wave vectors $\underline{k}_1$ and $\underline{k}_2$ and frequency ω

9.1.2 Transverse Modes

A nonlinear absorbing cell can also favor self-selection of the lowest transverse mode, because the saturation of absorption in a low-pressure gas is sensitive to the configuration of the light field [9.12]. Let, for example, the light field be the superposition of two plane waves of the same frequency, whose wave vectors $\underline{k}_1$ and $\underline{k}_2$ are in the plane (X,Y) and form the angles $\pm\phi$ with the x axis (Fig.9.2). This configuration of the field arises, for example, when a uniaxial mode in the direction OY is excited in a plane-parallel cavity whose axis coincides with OX, i.e., the mode TEM_{on}. The transverse component of the wave vector is determined by the transverse size of a mirror and the transverse index of the mode n,

$$q = \frac{\pi}{a}(n+1) . \tag{9.1}$$

As usual, $a \gg \lambda$; therefore, the angle between the wave vectors,

$$2\phi = 2\frac{q}{k} \approx (n+1)\frac{\lambda}{a} , \tag{9.2}$$

is very small. Molecules whose velocity projections onto the plane (X,Y) are in the shaded parts of Fig.9.2b interact effectively with this light field. When $|\omega - \omega_0| \gg \Gamma_B$ and $\phi \gg \Gamma_B/ku$ in the velocity space there are four scarcely overlapping groups of particles that interact with the field. When the angle ϕ between the standing waves decreases, the degree of overlap of regions I and II (III and IV, respectively) increases, and when $\phi \ll \Gamma_B/ku$ the regions I and II (III and IV) overlap completely (Fig.9.2). In this case, particles are affected by the travelling waves in both standing waves. As a result, the degree of saturation of absorption is doubled. When $\omega = \omega_0$ the degree of saturation is again doubled, owing to the confluence of regions I-II and III-IV. Then the particles are affected by all four travelling waves. The condition that the particle interacts simultaneously with two standing waves on account of (9.2) can be represented by

$$\frac{u}{\Gamma_B} > \frac{a}{\pi(n+1)} . \tag{9.3}$$

The left-hand side of the inequality (9.3) is the mean distance of coherent particle-field interaction. The right-hand side is the characteristic size of the transverse inhomogeneity of the field. Thus, the degree of saturation is sensitive to the field configuration when the particle can be affected by the inhomogeneity of the transverse distribution of the field during the coherent interaction with the field.

The saturation of absorption in the field of two standing waves with arbitrarily directed wave vectors $\underline{k}_1$ and $\underline{k}_2$ can be considered to find the dependence of the nonlinear absorption coefficient on the angle between the wave vectors. The corresponding calculation is simple for the field that is represented as the superposition of the two standing waves,

$$E(r,t) = \mathcal{E}(\sin \underline{k}_1\underline{r} \pm \sin \underline{k}_2\underline{r}) \cos\omega t \tag{9.4}$$

where $\underline{k}_{1,2} = k\underline{e}_x \pm q\underline{e}_y$, $\underline{e}_x$ and $\underline{e}_y$ are unit vectors along axes X and Y; the positive sign in (9.4) corresponds to modes that are even in Y; the negative sign corresponds to odd modes. In the weak-saturation approximation, the absorption coefficient is [9.12]

$$\kappa(\omega) = \kappa_0(\omega)[1 - \frac{G}{2} F(\frac{\Omega}{\Gamma}, \frac{qu}{\Gamma})] , \tag{9.5}$$

where G is the saturation parameter, the function F describes the dependence of the degree of saturation on the detuning $\Omega = \omega - \omega_0$ and the transverse component q of the wave vectors $\underline{k}_1, \underline{k}_2$. The function F can be approximated when the homogeneous width is less than the Doppler width

$$\Gamma \ll ku \tag{9.6}$$

and when the angle between $\underline{k}_1$ and $\underline{k}_2$ is also small

$$q \ll k . \tag{9.7}$$

Condition (9.6) is necessary for the effect to appear; (9.5) restricts only the range of large angles, which are of no interest for the problem considered. The final expression for $F(\Omega,q)$ is

$$F(\frac{\Omega}{F}, \frac{qu}{\Gamma}) = \left[1 + \frac{1}{1 + (\frac{\Omega}{\Gamma})^2}\right] + \sqrt{\pi}\,\frac{\Gamma}{ku}\,[1 - \Phi(\frac{\Gamma}{qu})]\, e^{(\frac{\Gamma}{qu})^2} + $$

$$+ \frac{1}{\pi}\int_{-\infty}^{\infty} \frac{e^{-\xi^2} d\xi}{1 + (\frac{qu}{\Gamma}\,\xi + \frac{\Omega}{\Gamma})^2} , \tag{9.8}$$

where

$$\Phi(x) = \frac{2}{\sqrt{\pi}}\int_0^x e^{-\xi^2} d\xi$$

is the integral of the error function.

The first term in (9.8) describes the frequency dependence of the degree of saturation, the second term describes the angle dependence, and the third the frequency-angle dependence. In the particular case of standing waves that have coincident directions ($q = 0$) the second term is equal to unity, and the third term is equal to the first term minus unity. As a result, (9.8) transforms to (2.102), $F(\Omega/\Gamma, 0) = 2[1 + L(\Omega/\Gamma)]$. For exact resonance ($\Omega = 0$) the third term is equal to the second and (9.8) takes the form,

$$F(0,\frac{qu}{\Gamma}) = 2\left\{1 + \sqrt{\pi}\,\frac{\Gamma}{qu}\,[1 - \Phi(\frac{\Gamma}{qu})]\, e^{(\frac{\Gamma}{qu})^2}\right\} . \qquad (9.9)$$

Far from the center of the line ($|\Omega| \gg \Gamma$) the third term is negligible and the function $F(0,qu/\Gamma) = 2F(\infty,qu/\Gamma)$. Figure 9.3 shows the dependence of the degree of saturation on the transverse component of the wave vector for two limiting values of the field-frequency detuning ($\Omega = 0$ and $|\Omega| \gg \Gamma$). The degree of saturation of absorption in the field of two standing waves is seen to be sensitive not only to the field-frequency detuning relative to the line center but also to the angle between the waves. It is important that the angle dependence is the same for any value of Ω, i.e., for any part of a Doppler line.

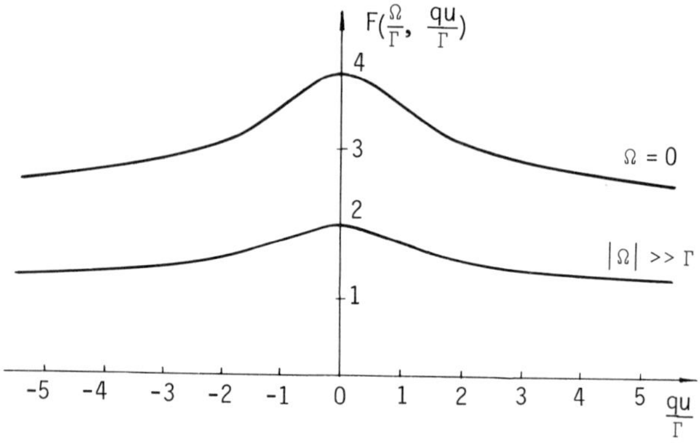

Fig.9.3 Relation between absorption saturation and transverse components q of the wave vectors of two standing waves (inside ($\Omega = 0$) and outside the Lamb dip ($|\Omega| \gg \Gamma$))

The dependence of the degree of saturation of absorption on the angle between two standing waves can be used for self-selection of the lowest transverse laser mode. For this purpose, a nonlinear-absorption gas of low pressure is introduced in a cell in the cavity. The self-selection of the lowest transverse mode requires that the molecular free path in the cell, with regard to variation of state due to collisions and radiation relaxation, is more than the transverse dimension of the beam.

9.2 Production of Narrow Gain Lines

When a coherent light wave interacts with an absorption line not only a narrow "hole" can be obtained but the population in a narrow frequency range of absorption lines can be inverted. The inversion can take place either at the pumping transition (two-level inversion) or at the transition connected with one of the levels of the saturated transition (inversion in three-level scheme). It corresponds to creation of narrow gain resonances. This possibility was theoretically investigated in 1965 in [9.13], before narrow Lamb dips in a saturated absorption line were studied.

Narrow gain lines with the width $\Delta\omega_a$, which is far less than the bandwidth of the cavity $\Delta\omega_c$, are necessary to produce lasers with high frequency stability. A narrow gain line provides good frequency stabilization even without servo control of the generated frequency. This line is connected with the fundamental cavity frequency ω and with the frequency of the gain-line center ω_0 by the relation

$$\frac{\omega - \omega_0}{\omega_c - \omega} = \frac{\Delta\omega_a}{\Delta\omega_c} . \tag{9.10}$$

As is generally known, for lasers $\Delta\omega_a \approx 10^{-2}$ to 10^{-4} $\Delta\omega_c$ as usual. Hence, the generated frequency coincides to a good accuracy with the peak of a spectral gain line without automatic frequency control. It is important for obtaining high short-time stability, i.e., for times less than the typical servo-control time constant. In the optical range, the bandwidth of the Fabry-Perot cavity is usually $\Delta\nu_c = 10^6$ to 10^8 Hz; the width of the narrowest spectral lines is determined by the Doppler effect and is $\Delta\nu_a = 10^8$ to 10^9 Hz. Conventional ways of excitation of atoms and molecules, in particular, by electron impact, resonance optical pumping, incoherent light, photodissociation of molecules, etc., give the excitation probability that is on the average independent of the direction of motion of the atom or molecule. Hence, the gain-line widths as well as the widths of absorption lines are determined by the Doppler effect. All these ways of pumping can be called "incoherent". Using atomic or molecular beams with "incoherent" pumping (for example, by photodissociation of molecules [9.14] or heating [9.15]) we can, in principle, diminish the Doppler-gain-line width by one or two orders of magnitude. But this is quite insufficient for essential change of the ratio between the gain line width $\Delta\nu_a$ and the cavity bandwidth $\Delta\nu_c$. The methods of obtaining narrow gain lines considered in the next section can be called "coherent" methods of optical pumping. In principle, they should obtain very narrow gain resonances that satisfy the condition $\Delta\nu_a \ll \Delta\nu_c$. However, the methods of obtaining narrow gain resonances in two-level systems differ essentially from those in three-level systems; they will be considered separately.

9.2.1 Narrow Gain Resonances in Two-Level Scheme

The idea of obtaining narrow gain resonances in a two-level scheme is as follows. Particles (atoms or molecules) cross a coherent light beam whose frequency lies within a Doppler absorption line. If the free path of a particle is much more than the beam diameter a and the absorption line corresponds to the transition of a particle to the long-lived excited state, then the particles interact with the field coherently during all of the time necessary for them to cross the beam $\tau \simeq a/u$. If the light-wave field satisfies the condition

$$\frac{pE}{\hbar}\tau \approx \pi \qquad (9.11)$$

then the level population is inverted. This inversion is similar to that of spin population obtained by a 180°-pulse method, which is well known in NMR-spectroscopy. In our case, the inversion has the impulse character due to the finite time necessary for a particle to cross a light beam, rather than that due to pulse switching of the field. The inhomogeneous character of absorption-line broadening leads to the highly peculiar deformation of the absorption line by a strong field shown in Fig.1.11.

When a pumping beam inverts the populations of levels of molecules in the beam, the gain appears to arise when the beam of molecules going out of the pumping beam satisfies the conditions of resonance and inversion (9.11). When the molecules in an equilibrium gas are saturated the possibility of gain is not so obvious, because together with inverted molecules that cross the pumping beam, noninverted molecules also enter the light beam. However, in this case the gain can be obtained by use of a hollow light beam [9.16]. With the pumping geometry shown in Fig.9.4 the molecules, while getting into the inner region of beam, first must cross the light field that causes the inversion of level population of the molecules, if (9.11) is fulfilled. It should be noted that, because of frequency detuning, thermal distribution of molecular velocities, and the dependence of time in the field on the impact parameter ρ, the "180°-inversion" condition cannot be satisfied for all of

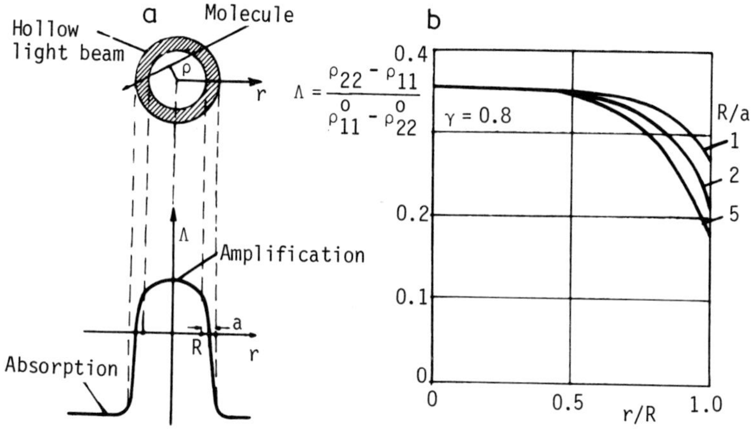

Fig.9.4a,b Formation of narrow amplification peak in the Doppler-broadened absorption line of a low-pressure gas at the interaction of molecules and the hollow tubular light beam (a - geometry of the region of amplification, b - radial distribution of the inversion degree Λ)

the molecules that enter the beam. Figure 9.4b shows the spatial distribution of inversion over the ray cross-section. Maximum inversion, $\Lambda = (\rho_{22} - \rho_{11})/(\rho_{11}^0 - \rho_{22}^0)$, ($\rho_{ii}$ and ρ_{ii}^0 are the population probabilities of the ith level, in the presence of the pumping field, and without it, respectively) occurs in the center of the beam when the inversion time $\tau_{inv} = \pi\hbar/pE$ is approximately equal to that of transverse-crossing time of a hollow-beam wall $\tau_{tr} = a/u$, where $u = \sqrt{2kT/M}$ is the most probable velocity of a molecule, i.e., when

$$\gamma = \frac{\tau_{tr}}{\tau_{inv}} \approx 1 \ . \tag{9.12}$$

For the particles that are in exact resonance with the field $(\omega_0 - \omega + kv = 0)$, $\Lambda = 0.36$. Near the beam wall, the inversion decreases, owing to an increase of the number of molecules that enter the beam at large impact parameter ρ. Naturally, the decrease depends on the ratio of the inner beam radius R to the beam-wall thickness a. Outside of the hollow beam, inversion is not reached at all, because the particles have not intersected the beam.

The degree of inversion Λ depends essentially on the ratio of the inversion time τ_{inv} to the flight time τ_{tr}, i.e., on the parameter γ. The maximum possible inversion $\Lambda = 1$ is not achieved, owing to thermal distribution of the molecular velocities. Figure 9.5 (dotted line) shows the dependence of the degree of inversion of an equilibrium gas illuminated by a hollow beam on the parameter for the two-level nondegenerate system. In fact, rotational-

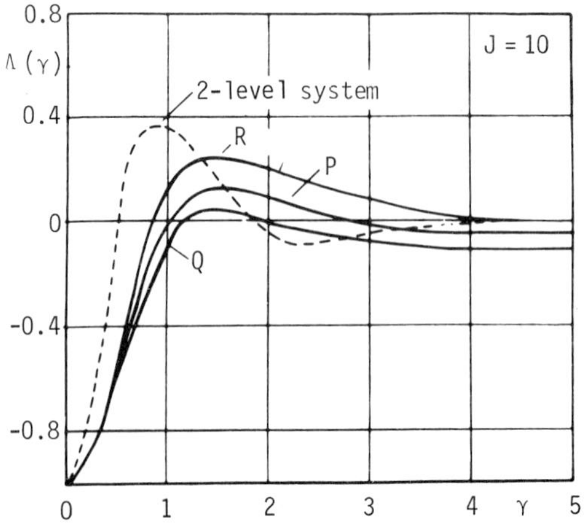

Fig.9.5 Degree of inversion of populations of a molecular gas as a function of the saturation parameter $\gamma = \tau_{tr}/\tau_{inv}$ for the nondegenerate two-level system (dotted line) and that for the transitions $(J = 10) \to (J = 9,10,11)$ of P, Q, and R branches

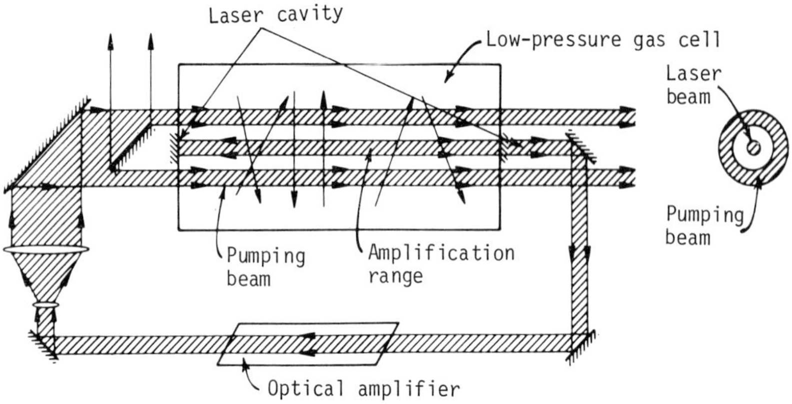

Fig.9.6 Scheme of the two-level cw gas laser pumped by its own amplified radiation

vibrational molecular transitions are usually highly degenerate. Therefore, there are several ensembles of two-level particles, with different dipole moments of transition P_m, that interact with the field, owing to level degeneration. It is also difficult to invert simultaneously the populations of all of the molecules that satisfy the resonance condition. This problem has been considered in detail in [9.17]. The most favorable situation is that for transitions from the R branch, for practically any J.

The narrow gain peaks that result from the interaction of a coherent beam with a gas absorption line make possible an unusual two-level cw laser. Figure 9.6 shows the principle of this laser. This is a two-level gas laser in which coherent pumping is produced by its own amplified radiation. In an absorbing low-pressure gas cell, the hollow pumping beam inverts the molecules. In the inner region of the pumping beam, the inverted molecules are used as the amplifying medium of the laser. The beam generated by the laser has considerable gain; from the generated beam, a hollow pumping beam is formed by use of a telescope and an aperture. A quantum amplifier is the energy source in this laser. This laser has rather unusual characteristics, calculated in [9.19,20]; the laser-generated frequency is automatically, and with high accuracy, tuned to the center of a gas absorption line; an initial field with nonzero amplitude is necessary for self-excitation of the laser. From a practical point of view, this scheme is of interest for production of coherent radiation with unusually high short-time frequency stability. Because of considerable experimental difficulties, this scheme has not yet been made to work.

Attention should be paid to the effect of anisotropy of gain by a narrow peak. Gain is possible only for a wave with just the same direction as that of the pumping wave; for the oppositely travelling wave, absorption alone is possible. An exception is the case in which the pumping wave is precisely tuned to the center of the Doppler line. The effect in that case is responsible for the frequency stabilization of a laser pumped by its own radiation at the center of the transition [9.20].

The unidirectional nature of bleaching of a Doppler-broadened absorption line with the field-frequency detuning relative to the line center can be, even without gain, used for optical isolation of a laser from its own reflected or scattered radiation [9.21]. The idea of this is quite obvious. The strong laser wave saturates the transition without significant weakening. The oppositely directed weak wave at $|\omega - \omega_0| >> \Gamma$ interacts with unsaturated molecules and is absorbed in the cell. Experimentally, this effect has been demonstrated with SF_6 as an absorber for the line P(18) of the CO_2 laser at 10.57 μm [9.21].

9.2.2 Narrow Gain Resonances in Three-Level Scheme

In the presence of a third level connected with either an upper level or a lower one, gain can be obtained at transitions connected with the pumped transition. This is easy to understand by considering either a peak in the

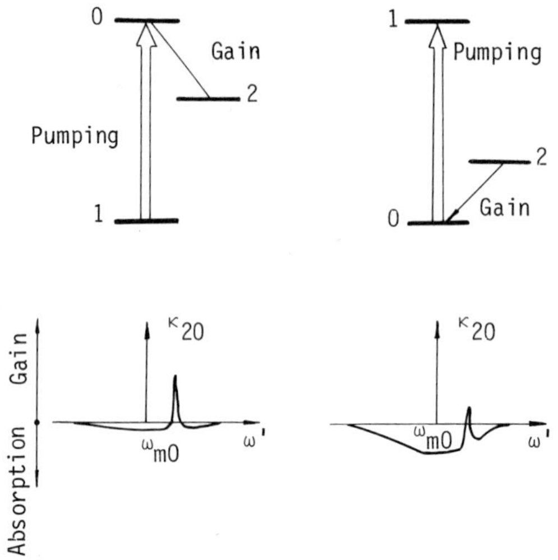

Fig.9.7 Formation of the narrow gain line by influence of a strong travelling light wave on a coupled transition

velocity distribution of molecules at the upper level of the pumped transition, or a hole in the velocity distribution at the lower level (Fig. 9.7). The gain occurs because of the presence of particles at the excited level or because of the absence of particles at the lower level; the velocity projections of these particles onto the direction of propagation of the pumping field on the transition $m-n$ lie within a narrow interval $v \pm \Delta v = (\Omega \pm \Gamma)/k$, where Ω is the frequency detuning of the pumping field from the center of the absorption line, k is the wavenumber, and Γ is the homogeneous halfwidth of the line. The population can be inverted only for particles whose velocities satisfy the resonance condition; full inversion between the levels can be absent.

The qualitative picture based on population effects appears to be true only for cases in which the lifetime of atoms at the common level is sufficiently long, that is, when the condition for inversion is fulfilled. Then a three-level gas laser can be considered simply as a laser with a particle beam in effect created by an external field.

As has been shown in Chapter 5, in the other limiting case of the ratio of relaxation constants ($\gamma_0 >> \gamma_1$) the radiation at the coupled transition arises mainly from two-quantum transitions, such as Raman scattering in the scheme under consideration. In a sense, this case can be considered as a gas laser based on an induced resonant Raman scattering, because two-quantum transitions change to a great extent all of the characteristics of a three-level gas laser and those of an amplifier.

Anisotropic properties of an induced radiation (absorption) line of a Doppler-broadened transition with the field at the coupled transition offer the possibility of production of unidirectional light amplifiers [9.21-23]. A unidirectional amplifier has gain only for waves travelling in given direction, whereas an oppositely travelling wave is absorbed in the whole frequency range. A unidirectional amplifier can operate even in weak fields. For the case $k' \gtrsim k$, the absorption coefficient of the wave that propagates in the direction that coincides with that of pumping field at the coupled transition is given by (5.50). The absorption of an oppositely travelling wave is described by (5.50). For fixed widths Γ_0 and Γ_-, we obtain the condition for unidirectional gain,

$$\Gamma_- < \Gamma_+ \,, \quad \frac{N_2 - N_0}{N_1 - N_0} < 2\,\frac{k'}{k}\,\frac{|V|^2}{\gamma_0 \Gamma_+} \,. \tag{9.13}$$

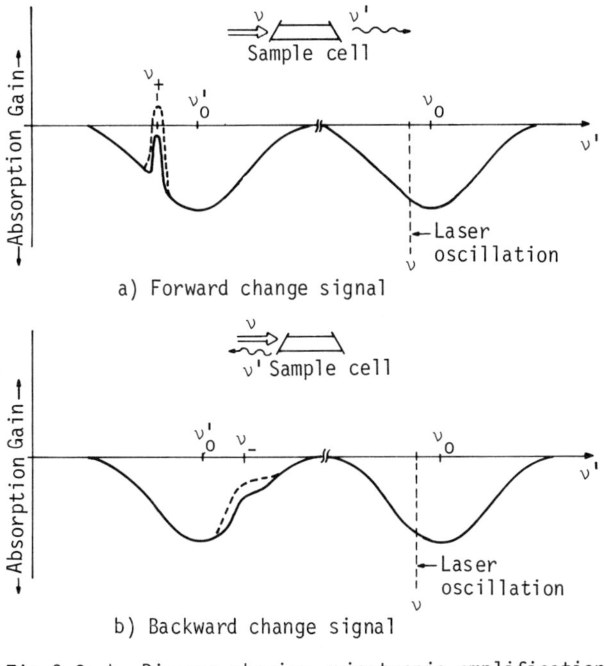

Fig.9.8a,b Diagram showing anisotropic amplification on a coupled transition. a - amplification in the direction of the pumping wave; b - amplification in the opposite direction (ν_0 is the frequency of the center of the line of the pumping transition, ν_0' is the frequency of the center of the line of the coupled transition, ν is the laser frequency)

Figure 9.8 shows the gain line shape of the coupled transition for unidirectional and oppositely travelling waves, which illustrates the effect of anisotropic gain. Figure 9.9 shows the experimental record of the gain shape of the unidirectional amplifier in Ne at the $2s_2 - 2p_4$ transition ($\lambda = 1.15$ μm) obtained by optical pumping with the 1.52 μm field ($2s_2 - 2p_1$) [9.23]. Note that the effects of saturation of population do not produce the gain under these conditions; the gain is achieved without inversion of the populations of levels 0 and 2. The gain is caused by two-quantum transitions of the Raman-scattering type, in the scheme considered. The greatest degree of unidirectivity of the amplifier is achieved in the cases in which two-quantum processes give the main contribution to the gain.

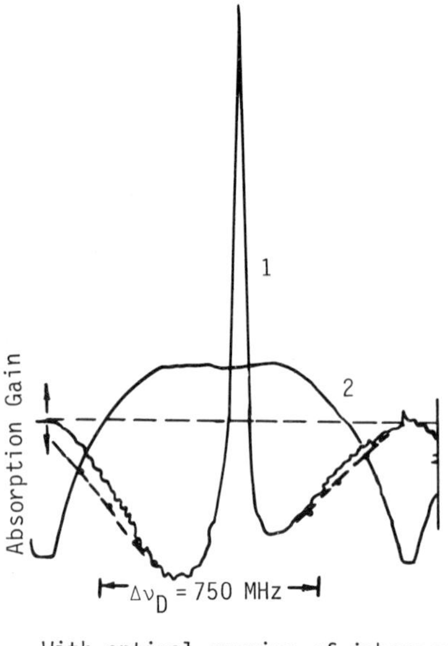

Fig.9.9 Experimentally observed line shape of gain of a unidirectional amplifier (probe field at $\lambda = 1.15$ μm, strong field at $\lambda = 1.52$ μm, unidirectional waves, gas discharge in ^{20}Ne) - curve 1; probe signal - 2

With optical pumping of inhomogeneously broadened transitions by a monochromatic field, the unidirectivity of the gain can arise only from population effects and is caused by the Doppler shift. In fact, with sufficiently large frequency detuning of a pumping field from resonance, gain of the probe field of a definite frequency occurs only for a wave that propagates in one direction. Nevertheless, when the effects of coherence are taken into account, unidirectivity of gain can be obtained over a wide frequency range. This can be done by optically pumping vibrational transitions of molecules at low pressure with lines of the P branch $P_1(J)$ of a pulsed 2.7 μm HF laser [9.24]. The gain was observed at the coupled rotational transitions $(1, J-1) \rightarrow (1, J-2)$ in the excited vibrational state. The gain anisotropy was investigated in a ring cavity. The intensities of forward and backward generation were studied separately. In this scheme, the maximum power generated in the wave that propagated in the same direction as the pumping field was 40-400 times as great as the power generated in the wave that propagated in the opposite direction. When the pumping-field intensity was decreased, the generation regime of a moving wave only was observed.

The properties of gain unidirectivity and anisotropy must significantly influence the generation of a standing wave in a laser that operates with this amplifying medium. There should be a sufficiently sharp dependence of the gain coefficient on the frequency detuning of the pumping field Ω. This dependence arises because, at $\Omega \approx 0$, both components of the standing wave experience gain, and at sufficiently great Ω one of the components of the standing wave is absorbed. Consequently, at $\Omega = 0$ and $\Omega \neq 0$, with the same amplitudes of an external field and the same absorption coefficient, the threshold conditions for generation of a standing-wave field differ essentially. These effects were first explored experimentally [9.23,25] in a three-level gas laser at the transition $(2s_2 - 2p_4)$ $\lambda = 1.15$ μm when scanned frequency from a He-Ne laser at $\lambda = 0.63$ μm $(3s_2 - 2p_4)$ or $\lambda = 1.52$ μm $(2s_2 - 2p_1)$ passed through a Ne gas-discharge cell (Fig.9.10). Generation at $\lambda = 1.15$ μm occurred at the fundamental TEM oscillations, only in the presence of the external field. The space between the mirrors of the three-level laser allowed observation of the generated line shape within 750 MHz. The generated line shape differs essentially from the usual shape. At strong saturation, it has the form of

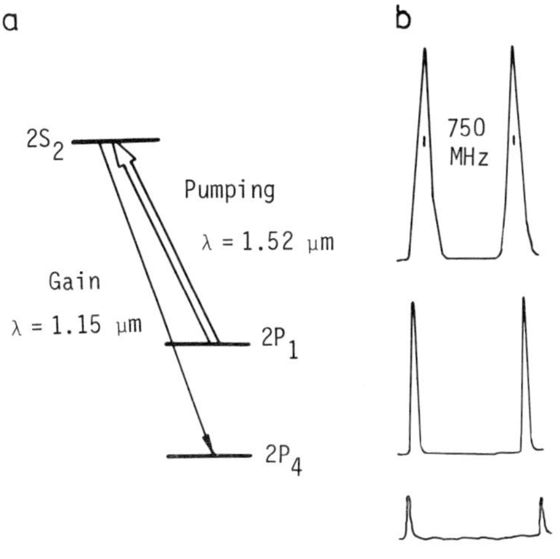

Fig.9.10a,b Three-level gas Ne laser ($\lambda = 1.15$ μm) pumped by an external monochromatic field ($\lambda = 1.52$ μm). a - scheme of transitions; b - line shape of generation at various excesses of gain over generation threshold

a Doppler contour with a Lamb dip in the center. When the frequency of the external field coincides with the center of the absorption line a narrow generation peak is observed. A width of the generation region depends on the excess of gain over threshold. Under certain conditions, very narrow generation regions may be obtained. The properties of three-level lasers have been described in more detail in Chapters 5 and 7.

9.3 Laser-Frequency Stabilization by Means of Narrow Resonances

One of the outstanding achievements of the quantum electronics of the microwave range was the development of the cesium passive frequency standard (the transition between the hyperfine structure levels $F = 4$, $m_F = 0 \to F = 3$, $m_F = 0$ of the ground state of ^{133}Cs in the absence of external fields) [9.26,27]. This is accepted as the international standard with frequency 9 192 631 770.0000 Hz. Another outstanding achievement was the development of a hydrogen laser (transition $F = 1$, $m_F = 0 \to F = 0$, $m_F = 0$ between hyperfine structure levels of the ground state of H, the nominal frequency of which is 1 420 405 751.7864 ± 0.0017 Hz) [9.27], with extremely high frequency stability (2×10^{-14} per day).

The development of quantum generators of coherent optical radiation opened up the opportunity, in principle, to use quantum transitions of atoms and molecules to develop lasers with highly stable frequencies in the optical range (10^{13} to 10^{15} Hz). Mastering of the optical range by use of generators with stable-frequency radiation is important in many respects. First, a quantum generator in the optical range, unlike a microwave generator, can serve as a standard of both length and time, simultaneously. Because the length of a light wave is much less than the characteristic dimensions of devices, length can be measured precisely, by interference. For this reason, a wavelength of an optical spectral line is adopted as the international standard of length (the international meter contains 1650 763.73 wavelengths of the radiation from the $2p_{10} - 5d_5$ transition of ^{86}Kr atoms $\lambda = 6056.9$ Å [9.28]). In the optical range, however, as distinct from the microwave range, it was impossible up to now to measure the frequency, i.e., time. A quantum generator of optical radiation with a stable frequency can serve as a quantum standard of time only together with an optical-frequency measuring instrument. Development of a quantum standard of frequency in the optical range would make possible a single quantum standard of time and length. Thus, the difficulty that the international units of time and length are defined in terms of spectral lines of different atoms (^{133}Cs and ^{86}Kr) could be eliminated. Furthermore, by use of a standard in the optical range, the time required to measure time or frequency to a given accuracy could be substantially decreased, because the accuracy of frequency measurements for a given time interval is inversely proportional to the frequency.

In order to develop a laser with high frequency stability it is necessary to have, in the optical range, an atom or molecule that satisfies two conditions:

1) The reference frequency must be stable and reproducible; at least as much stability as that of a laser is required.

2) The width of the resonance curve of the reference frequency should not exceed the required stability by a factor of more than 10^3 to 10^4. This condition is not strict, because it depends upon the state of the art of automatic setting on the center of the resonance curve by use of servosystems.

From this it follows that resonance with the same stability and reproducibility as the line center and with a relative width of about 10^{-9} to 10^{-10} is required, to obtain frequency stability and reproducibility of the order of 10^{-13}. This corresponds to the stability and reproducibility already achieved in the microwave range. Such stability can be achieved, in principle, on many quantum transitions of atoms and molecules whose initial and finite levels are only slightly disturbed by external fields and collisions. However, it is notorious that Doppler broadening of spectral lines in a gas, which usually amounts to a relative value of 10^{-5} to 10^{-6} does not permit fulfillment of condition 2. Therefore, it is necessary to develop methods of narrowing optical spectral lines. The narrowing of a spectral line by the molecular-beam method is very effective in quantum electronics devices in the radio-frequency range. But in the optical range it does not solve the problem, because it requires extremely intense beams with angular divergences of 10^{-3} to 10^{-4} rad, production of which is rather difficult.

Attainment of narrow optical resonances induced by a coherent light wave acting upon a quantum transition with long-lived levels was proposed in [9.13], which was devoted to the problem of production of highly stable-frequency lasers. An essential step was the proposal to obtain narrow resonances by saturating the absorption of a low-pressure gas cell. The first independent proposals and experiments on the use of nonlinear-absorption resonances were made in the Lebedev Physical Institute of the USSR Academy of Sciences [9.18], in the Institute of Semiconductor Physics of the Siberian Branch of the USSR Academy of Sciences [9.1] and in the USA [9.29]. Subsequent rapid development of this method resulted in considerable progress in generation of light oscillations with high frequency stability. Suffice it to say that in the period 1967-1972 the best values of long-term relative stability of laser frequency increased from 10^{-8} to 10^{-14}. At the same time, methods of absolute measurement of optical oscillation frequencies were developed. This development was initiated by the work of JAVAN [9.30] and led to the measurement of the 3.39 μm He-Ne laser oscillation frequencies [9.31]. Thus, the principal part of the problem of production of the united quantum time and length standard was solved, and within the next few years technical elaborations can be expected in this subject.

The physical principles of production of highly precise-frequency lasers and the state-of-the-art till 1968 are described in the survey by BASOV and LETOKHOV [9.32]. In the next section, we consider only the methods that are widely used in laboratories and have favorable prospects.

Before beginning the description, we review definitions of terms frequently used [9.33]: frequency accuracy is the degree with which a generator frequency corresponds to the adopted definition of the frequency unit (Hertz); frequency reproducibility is the degree with which a generator of the given type reproduces the same frequency from one period of operation to another; frequency stability is the degree with which a generator emits the same frequency during continuous working. The definition of stability requires statement of the time interval during which the stability is measured.

9.3.1 Method of Internal Nonlinear-Absorption Cell

The block scheme of laser-frequency stabilization of a power peak with an internal nonlinear-absorption cell is given in Fig.9.11. The resonance peak is tuned by use of laser-frequency modulation $\bar{\nu} = \nu + \delta \cos(2\pi ft)$ in the neighborhood of the resonance peak. When the laser frequency is detuned with respect to the resonance peak, an amplitude-modulation signal of the laser output, at the frequency f, arises. This signal is used to control the cavity

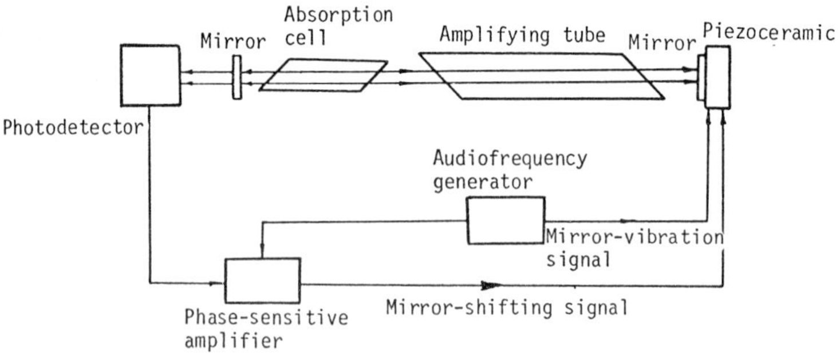

Fig.9.11 Scheme of laser frequency stabilization over a power peak with an internal nonlinear-absorption cell

frequency, i.e., to tune the generated frequency to the resonance peak. This is done by a servosystem that controls the position of one of the cavity mirrors, which is mounted on a piezoceramic.

He-Ne Laser with a Ne Cell

The first experiments on frequency stabilization of a gas laser with nonlinear absorption over an output power peak were carried out with a He-Ne laser at 6328 Å [9.34]. A discharge tube with pure Ne at a pressure of 0.1 torr was used as the absorption cell. An electric discharge populated an upper-lying level of Ne with an energy of 19 eV. Thus, the same quantum transition as in a He-Ne laser was used. The width of the output peak was 20-40 MHz; this width is determined, mainly, by natural (radiative) broadening (20 MHz) as well as by a small contribution of collisional broadening due to collisions of atoms of Ne (24 MHz/torr). The decrease of pressure to 0.1 torr does not therefore lead to peak narrowing but increases considerably its amplitude. Optimal pressures range from 0.1 to 0.3 torr. Under these conditions, the contrast of the output peak (the ratio of the peak amplitude P to the level of power P_0 from which it rises) is close to 100%. In order to decrease discharge noises, it is better to use a hf discharge. The noise level of the output signal and of the recording scheme permits the generated frequency to be tuned to the center of the power peak by regulation of the cavity length to an accuracy of about 0.1 MHz, i.e., 1/300 of the resonance width.

The frequency reproducibility of He-Ne/Ne laser (here and subsequently, the laser is designated ahead of the solidus (/) and the nonlinear absorber after it) was 10^{-9} [9.34,35]. The frequency reproducibility is limited by a collisional line shift that depends on the discharge current and by the asymmetric power-peak position. Under usual laboratory conditions (mass-stabilized optical table) the short-term frequency stability amounted to 10^{-9} in 10^{-3} s. The value of short-term stabilization in this and other types of stabilized lasers is determined mainly by environmental conditions: mechanical disturbance and acoustic noise levels. The long-term stability depends on the degree of discharge permanence in the amplifying and absorbing cells.

The limitations of frequency stability and reproducibility caused by a large natural width and collisional shifts are inherent in an absorbing cell of this type. A number of investigators tried to select a molecular absorber to obtain a narrow resonance on the 6328 Å line of a He-Ne laser. The best results were obtained with an I_2 absorbing cell.

He-Ne Laser with an I_2 Cell

The authors of [9.36] discovered the coincidence of an absorption line of I_2 vapor line R(127) of the band 11-5 and a radiation line of the He-Ne laser at 6328 Å and have obtained a narrow output peak of the He-Ne/I_2 laser with a width of 4.5 MHz. The contrast of the power peak amounts to 0.1% at a pressure of 0.1 torr; there are some difficulties of frequency stabilization over resonances of such small amplitude, mainly due to the large contribution of the slope of the Doppler contour of the gain line to the output-peak position. It is possible to overcome this difficulty by stabilizing the frequency on the third-harmonic signal that arises from laser frequency modulation in the neighborhood of the top of the power peak, which has an amplitude inversely related to the width of the peak (see Fig.3.11). By this method the frequency stability of 10^{-11} with an averaging time of 10^3 s was reported in [9.37]. The frequency reproducibility that resulted from stabilization on the i component of the spectrum of saturated absorption of I_2 was 2×10^{-10} in [9.37] and 4×10^{-11} in [9.38].

The properties of a He-Ne laser stabilized on narrow peaks in $^{129}I_2$ were reported in detail in [9.39]. Broadening due to iodine pressure was about 13 MHz/torr and the resonance width, extrapolated to zero pressure was 2.6 MHz. The frequency shift due to pressure was about 2×10^{-9} ν_0/torr. The broadening and shift due to the field intensity were small. The peak width varied only 10% and the peak frequency only 2×10^{-11} ν_0 when the intensity in the I_2 cell was varied by a factor of 5-6. The frequency stability of such lasers amounted to 2×10^{-12} with an averaging time of 10 s; the frequency reproducibility was about 10^{-10}. Wavelengths of the laser stabilized on narrow resonances in $^{127}I_2$ and $^{129}I_2$ were measured. The values obtained are tabulated in Table 9.1. The ^{86}Kr length standard, which has, as is well known, a wide and asymmetric emission line was used in the measurements of the wavelengths. Such measurements have some uncertainty caused by the method used to set on the line of the Kr standard. Wavelengths measured by setting on the center of gravity of the ^{86}Kr line are given in the first column. Wavelengths measured by setting on some position intermediate between the maximum and the center of gravity of the ^{86}Kr line are given in the second column.

Some laboratories in the USA, France, Canada, and England have measured the wavelength of a ^{3}He-^{20}Ne laser stabilized on the i component and have obtained similar values by use of the same method of setting on the krypton standard. Taking this into account as well as the great difficulties of using the krypton standard for precise measurements of wavelengths, compared to the

Table 9.1 Wavelengths of He-Ne laser stabilized on narrow resonances of saturated absorption of an iodine molecule

Laser	Absorber	Component	Wavelengths (pm)	
			of the center of gravity	of the intermediate point
^3He-^{20}Ne	$^{129}I_2$	k	632 991.2670 ± 0.0009	632 991.2714 ± 0.0009
^3He-^{22}Ne	$^{129}I_2$	B	632 990.0742 ± 0.0009	632 990.0786 ± 0.0009
^3He-^{20}Ne	$^{127}I_2$	i	632 991.3954 ± 0.0009	632 991.3998 ± 0.0009

stabilized laser, at its meeting in 1973, the Consultative Committee on the Definition of the Meter recommended use of 632 991.399 ± 0.0025 pm as the wavelength in vacuum of the ^3He-^{20}Ne laser stabilized on the i component of $^{127}I_2$ [9.40]. This value is based on the assumption that an intermediate point on the profile of the radiation line of the krypton standard between the maximum and the center of gravity of the line represents the wavelength of the krypton standard. Thus, narrow resonances of saturated absorption provide a length standard that is much more convenient in use than the official standard.

Table 9.2 Frequency stability of He-Ne laser with CH_4 cell

Averaging time (s)	10^{-3}	10^{-2}	10^{-1}	10^{0}	10^{1}	10^{2}
Length of run (min)	30	30	30	120	120	120
Average difference frequency between the two lasers (Hz)				3	2	
Fractional standard deviation per laser $\frac{\sigma}{\sqrt{2}\nu}$	6×10^{-12}	3×10^{-12}	5×10^{-13}	1×10^{-13}	3×10^{-14}	5×10^{-15}

He-Ne Laser with a CH_4 *Cell*

The best results on frequency stability have been achieved with a He-Ne laser with a methane absorption cell at $\lambda = 3.39$ μm (Table 9.2). The power peak in the He-Ne/CH_4 laser was first reported in [9.41]. A component $F_2^{(2)}$ of the rotational-vibrational line P(7) of the methane band ν_3 is about 50-80 MHz away from the center of the gain line of the ordinary 3.39 μm He-Ne laser. Precise coincidence of the gain and absorption lines is achieved by increasing the He pressure in an amplifying medium or by use of ^{22}Ne [9.42].

Frequency stability and reproducibility of 10^{-11} was obtained in the first work. These results were obtained in the He-Ne/CH_4 laser with a cavity 100 cm long, with an absorption cell about 50 cm long, and about 10^{-2} torr of pressure of CH_4. The width of a typical power peak was 200-300 kHz (at 300 K in the CH_4 cell), the peak contrast was several percent. When the time of observation was increased, the long-term stability of the He-Ne/CH_4 laser was 10^{-13}. If the frequency difference is determined by use of two statistically identical generators, the frequency stability, when the duration of measurement is τ, is characterized by the root-mean-squares spread of the frequency of one laser $\delta\nu/\nu$, which is associated with fluctuations of the frequency difference $\Omega = \nu_2 - \nu_1$ of two lasers by

$$\left(\frac{\delta\nu}{\nu}\right)_\tau = \sqrt{\frac{1}{2}\left(\frac{\Omega}{\nu_0}\right)^2_\tau} \quad . \tag{9.14}$$

Frequency stability is often described by

$$\left(\frac{\delta\nu}{\nu}\right)_\tau = \frac{1}{\nu}\sqrt{\langle\sigma^2(2,\tau)\rangle} \ , \quad \sigma(2,\tau) = \frac{1}{2}\,[\Omega_{n\tau} - \Omega_{(n+1),\tau}] \ , \tag{9.15}$$

where $\sigma(2,\tau)$ is the so-called Allan parameter [9.43], $\Omega_{n\tau}$ is the frequency difference of two lasers at the moment $n\tau$ measured during the time interval τ. The dependence of the Allan parameter for the He-Ne/CH_4 system, with the averaging time τ is shown in Fig.9.12. The long-term stability increases proportionally to $\sqrt{\tau}$, where τ is the averaging time, up to the time of the order of 100 s. This dependence may be explained by an advantageous contribution of fluctuations of the type of white noise. At $\tau > 100$ s the increase of stability stops, which can be explained as the influence of a flicker noise. The influence of fluctuations of different spectra on the frequency stability of a narrow-resonance-stabilized laser, involving an effect of frequency autostabilization is theoretically considered in [9.45].

Frequency reproducibility of the He-Ne/CH_4 system has been explored in some laboratories. Experiments performed in the US National Bureau of Standards (NBS) in Boulder, Colorado, gave a reproducibility of 0.5×10^{-11} [9.46]. The main contribution to fluctuations arises from frequency-selective changes of intensity of the background against which the power peak is observed. These changes are caused by return of a very small portion of the radiation into the cavity. Careful optical isolation of the laser from the measurement device is required. A collision shift (about 10^{-12} Hz mtorr) also contributes to the He-Ne/CH_4 frequency reproducibility. A large shift ~1 kHz of the

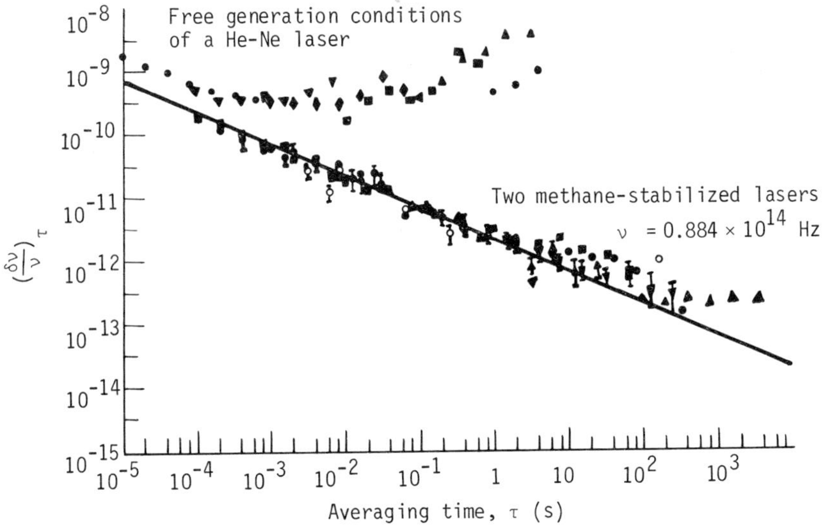

Fig.9.12 Relation between frequency stability of the He-Ne laser at $\lambda = 3.39$ μm and the averaging time, for the free generation (1) and with power-peak stabilization (2) with an internal absorption cell. The continuous curve shows the calculated dependence based on the assumption of white noise

stabilized laser frequency, dependent on radiation intensity, was reported in [9.47]. The cause of this dependence remained unclear till recently. The authors of [9.48] found a fine structure of the CH_4 line $F_2^{(2)}$, caused by the nuclear spin of the CH_4 molecule, on which the laser frequency is stabilized (see Chap.8). The line consists of three usually unresolved components of different intensities spaced about 10 kHz apart. Each component has different saturation of absorption; hence, this structure may be one of the causes of the peak-frequency shift depending on the field intensity. This effect can limit frequency reproducibility of the He-Ne/CH_4 laser to from 10^{-11} to 10^{-12} [9.48]. Significant progress in obtaining high stability and reproducibility of the He-Ne/CH_4 system was made in the Institute of Semiconductor Physics (Novosibirsk, USSR) where frequency reproducibility of about 3×10^{-14} was achieved [9.49-51]. The dependence of long-term stability of the He-Ne/CH_4 system on observation time is shown in Fig.9.13. In order to obtain these values of frequency stability and reproducibility it is necessary to operate with an intense peak at a methane pressure as low as 10^{-3} torr. This is possible by use of a long absorption cell (about 200 cm) and matching the parameters of saturation and gain by expanding the diameter of the beam in the methane cell. In this case, the power peak may have a contrast of about 100% and a width of 40-50 kHz at a power of some milliwatts. In order to measure the frequency difference of two He-Ne/CH_4 lasers, the frequency of one of

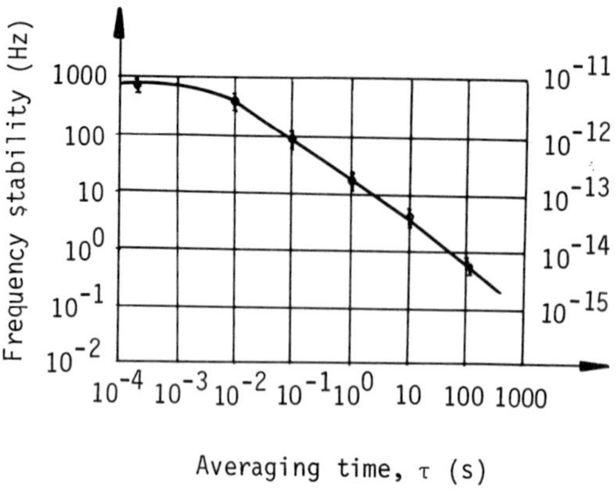

Fig.9.13 Frequency stability $(\delta\nu/\nu)_\tau$ of He-Ne/CH_4 laser at $\lambda = 3.39$ μm as a function of averaging time

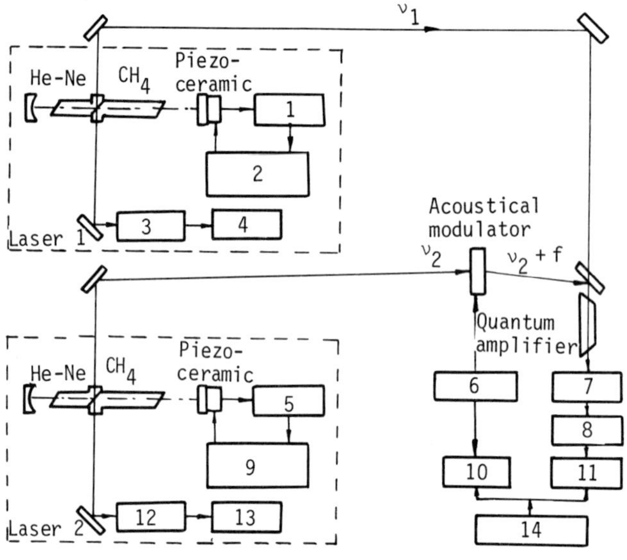

Fig.9.14 Experimental arrangement for measuring frequency stability and reproducibility of two He-Ne/CH_4 lasers. 1,3,5,7,12 - photodetectors; 2,9 - automatic-frequency-control units; 4,13 - power meters; 6 - generator; 8 - selective amplifier; 10,11 - frequency meters; 14 - synchronizer

Table 9.3 Frequency reproducibility of He-Ne laser with CH_4 cell

Averaging time (s)	Number of independent alignments	Frequency stability per laser	Standard deviation of difference frequency (per laser) (Hz)
30	30	1×10^{-14}	3.2

them was shifted by 29 MHz by use of an optico-acoustical modulator. The scheme of measurement of the frequency difference of the two lasers, which provides a high degree of the optical isolation of the lasers from each other and from the measurement system is shown in Fig.9.14. Results of recent studies of reproducibility are given in Table 9.3. The relative root-mean-square deviation of the frequency difference from the mean of one laser was 3.5×10^{-14} for 30 independent laser tunings. Each independent tuning involved switching off and switching on the AFC system, retuning the cavity mirrors, and renewal of the gas in the amplifying and absorbing tubes. Owing to collisions, the shift of the center of the Lamb dip decreases in a low-pressure nonlinear-absorption cell (see Chap.8). If, in the pressure range greater than 10^{-2} the shift is of the order of 100 Hz/mtorr [9.41], in the pressure range of about 10^{-3} the shift was 7 ± 5 Hz/mtorr in the experiments [9.49]. Thus, at pressures of 10^{-3}, which were controlled to an accuracy of 10%, the relative shift of the frequency of the dip center ranges up to 10^{-14}. Change of He pressure in the amplifying tube also influenced the position of the dip center. The experimentally observed shift of the stable laser frequency amounted to about 10 Hz/torr when the He pressure was changed in the amplifying tube.

Relativistic and quantum effects become significant when a frequency reproducibility of 10^{-14} with the He-Ne/CH_4 system is sought. The relativistic (second-order) Doppler effect is a red shift of the frequency of moving oscillators and causes the temperature shift of the Lamb-dip center (see Chap. 10) which for CH_4 amounts to 150 Hz at 300 K. Experimental investigations have indicated that in methane the temperature shift is of the order of 0.5 Hz/degree [9.52]. The mean effective velocity of molecules v_{eff} responsible for the Lamb-dip shift, due to the second-order Doppler effect ($\Delta_L = -\omega_0 u/2c^2$) ω_0 is the transition frequency) may be less than the value of an average thermal velocity $u = \sqrt{2kT/M}$; it is caused by slow atoms. The calculation made in [9.53] showed that the effective velocity depends on the mode of generation of a laser with nonlinear absorption and on the method used for frequency stabilization. Under conditions of experiment [9.52] v_{eff} is close to u.

In photon emission and absorption by the CH_4 molecule the quantum-recoil effect splits the power peak, owing to the difference between the absorption and emission frequencies. The splitting amounts to about 2 kHz, i.e., is within the power peak. Because the amplitude of each unresolved component depends on the change of level population in the field, the position of the resulting peak in turn depends on the level relaxation constants [9.54] and on the saturation parameter [9.55]. This effect is considered in detail in

Chapter 10. The influence of the effect on the frequency reproducibility disappears at very low pressures when the free-path length exceeds the beam diameter. Estimates show that in the range of low methane pressures (10^{-3} torr) and low saturations ($G \simeq 0.1$) the relative power-peak shift due to the recoil effect ranges from 10^{-14} to 10^{-15}.

The magnetic hyperfine structure (MHFS) of vibrational-rotational levels of methane molecules can badly influence the resonance position and the shape of nonlinear absorption in a gas and restrict the frequency reproducibility to from 10^{-11} to 10^{-12} [9.48]. The theoretical analysis of the influence of MHFS on frequency reproducibility of gas lasers made in [9.50,56] indicated that, under certain conditions of He-Ne/CH_4 laser operation, hyperfine splitting does not prevent achievement of frequency reproducibility of the order of 10^{-13}. In the experiments reported in [9.49,51] lasers were tuned to the resonance peak by reducing to zero the error signal in the AFC system by frequency modulation (FM) of the laser radiation. The modulation frequency was close to the resonance halfwidth, and the modulation was far less than unity. In this case, the radiated spectrum contains two side frequencies. The position at which the absorption of weak side components of an FM signal is equal and there is no amplitude modulation of an output signal corresponds to the stabilized frequency of the laser radiation. With the contour asymmetry associated with MHFS, this frequency does not coincide with the resonance maximum.

In [9.50,56] the problem of absorption of the frequency-modulated signal in the standing-wave field was solved by allowing for three strong MHFS components to an accuracy of the fourth order of the field. The detuning of the stabilized frequency Ω relative to the central MHFS component is

$$\Omega = t \cdot \Delta[F_1(\Gamma,\Delta,\eta) + GF_2(\Gamma,\Delta,\eta)] \ , \qquad (9.16)$$

where Δ denotes the frequency spacing between the MHFS components, Γ is the collision-determined halfwidth of a separate component, $\eta = f/2\Delta$, $t = t_1 - t_{-1}$, $t_k = (|P_k|^2 - |P_0|^2)/|P_0|^2$ $(K = 0,\pm 1)$, f is the modulation frequency, $G = 2|P_0|^2 E^2(\gamma_2^{-1} + \gamma_1^{-1})/\hbar^2\Gamma$ is the saturation parameter at the transition $K = 0$, P_k is the dipole matrix element of the K component of the transition, γ_2^{-1}, γ_1^{-1} are the lifetimes of the upper and lower levels, 2E is the amplitude of the standing-wave field.

The calculation results for the function $\Omega_2/Gt\Delta = F_2(\Gamma,\Delta,\eta)$ at $\eta = 0$ and $\eta = 0.55$ are given in Fig.9.15. When $\eta = 0$, the curves describe the position of the resonance peak, which largely depends on the pressure (i.e., Γ) and

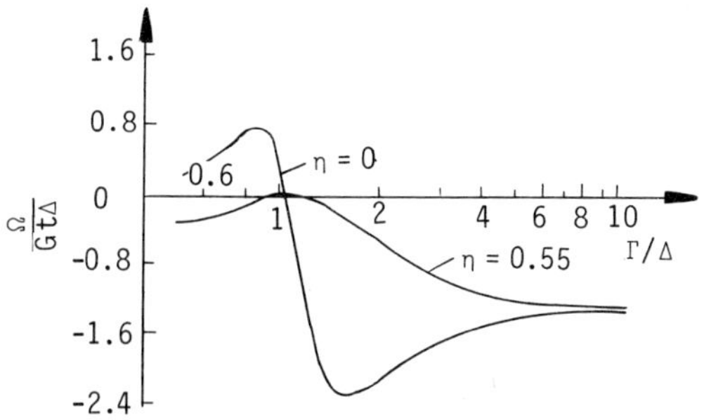

Fig.9.15 Shift of stabilized-laser frequency caused by methane MHFS when the field in the cavity is varied (Γ is halfwidth of an individual component of the MHFS, dependent on pressure, Δ is frequency detuning between the components, f is modulation frequency $\nu\eta = f/2\Delta$)

the laser field. On the contrary, when $\eta = 0.55$, in the range $\Gamma/\Delta \sim 1.5$ the dependence of the stabilized frequency shift on both the pressure and the laser field becomes slight. The relative intensities of three strong components of a methane line $F_2^{(2)}$ are related as 0.85 : 1 : 1.15. Then the relative intensities of resonances must be approximately $0.85^2 : 1 : 1.15^2$. However, allowing for level degeneracy, as has been recently shown by BAGAYEV et al., the resonance intensities are 0.87 : 1 : 1.17, which is in good agreement with the experiment. The numerical values of function F_2 along the ordinate axis corresponds to the effective values $t = 0.15$ and $t = 0.1$, allowing for level degeneracy. Note that flight effects and gaussian distribution of the field in the beam also decrease the values of the functions F_1 and F_2.

Estimates have indicated that for the total resonance width in methane of 30 to 50 kHz ($\Gamma/\Delta \sim 1.5$), frequency of the laser modulation about 15 kHz, and value of the field in the cavity $G = 0.2$ to 0.3, the stabilized-frequency shift is about 5 Hz when the methane pressure and the field in the cavity are changed 10%. This is in good agreement with experimental results and explains the frequency reproducibility of about 3×10^{-14} reported in [9.50]. When $\Gamma \sim 100$ kHz and $G \sim 1$, the laser field change of 10% leads to shifts of about 100 Hz; it restricts the reproducibility to the range from 10^{-11} to 10^{-12}.

CO_2-N_2-He Laser with a CO_2 Cell

The first attempts to obtain a narrow resonance of nonlinear absorption in a molecular gas by use of a CO_2-N_2-He laser at 10.6 μm and of a CO_2 absorbing cell were made in [9.57]. The main difficulty consisted in the automodulation of radiation; it is overcome only by reducing CO_2 pressure to 0.1 torr.

In order to increase absorption in CO_2 on the transition between excited vibrational levels of CO_2 molecule it is necessary to heat a CO_2 cell up to 400°C. With an absorbing cell 100 cm long, the power-peak contrast is about 2%; its width is 1 MHz. Under such conditions, the frequency reproducibility reaches 10^{-10}, the short-term stability 10^{-10}, and the stability with averaging time $\tau = 100$ s approaches 4×10^{-11} [9.58].

9.3.2 Method Using an External Nonlinear-Absorption Cell

Achievement of a narrow resonance in a nonlinear absorber placed in a laser cavity requires a special choice of saturation parameters and absorption and gain coefficients. In many cases it is impracticable, owing to great difference between the saturation parameters of the amplifying medium and of the absorber. In such cases, the saturation of absorption by a laser light field outside the cavity can be obtained outside the cavity by producing either the Lamb dip in a standing-wave field or a dip in the absorption of a weak, oppositely travelling wave. When an external absorption cell is used, the influence of the nonlinear absorber on the laser oscillation frequency and amplitude is eliminated and the spatial configuration of the light field can be controlled and varied. Laser-independent control of the field characteristics permits optimization of a narrow resonance width and amplitude. The possibility of external laser-frequency modulation to search for a peak maximum permits exclusion of frequency pulling to the top of a gain line. Finally, the external-cell dimension can be made large, in order to give the amount of absorption required. In spite of these obvious advantages of the method that uses an external nonlinear-absorption cell, unlike nonlinear spectroscopy, it has not yet been widely used for frequency stabilization.

The scheme of stabilization with an external cell is in general similar to the scheme shown in Fig.9.11 for the case of an internal cell. The difference is that the frequency can be modulated outside a laser by use of a special phase modulator.

He-Ne Laser with a Ne Cell

An external Ne cell with a dc discharge was used in [9.59] to stabilize a relatively high-power (15 mW) single-frequency He-Ne laser at 6328 Å. Optical isolation (a polarizer and a $\lambda/4$ plate) was used to eliminate the influence of the backward wave on the laser and to produce a wave with circular polarization in the absorbing cell. The position of the center of the absorption line was scanned by a longitudinal, variable magnetic field. It allowed automatic laser frequency control without modulation. Frequency reproducibility of about 10^{-9} was achieved; it was limited mainly by the frequency shift of the center of the absorption line at varying discharge current.

CO_2-N_2-He Laser with a SF_6 and OsO_4 Cell

Stabilization of a CO_2 laser on the line P(18) of the 10.6 μm band of SF_6, by a narrow absorption resonance of a weak oppositely travelling wave in an external SF_6 cell was reported in [9.60]. High absorption of SF_6 for several lines of the CO_2 laser (~0.5 to 1.5 cm^{-1}/torr) permits operation at SF_6 pressures of the order of 10^{-3} to 10^{-2} torr and resonance widths from 0.3 to 1.0 MHz. Figure 9.16 shows the shape of a typical transmission resonance of a SF_6 cell and its position relative to the gain line P(18) of the CO_2 laser. The transmission peak is shifted with respect to the gain-line center by 10 MHz. Figure 9.17 depicts the data on the stability of this system in the Allan variable. The long-term stability amounted to 10^{-12} for 100 s averaging time [9.61]; it was limited mainly by pulling of the transmission-peak top to the maximum of the laser output (Fig.9.16). This may be eliminated by use of external frequency modulation of the CO_2 laser to tune to the

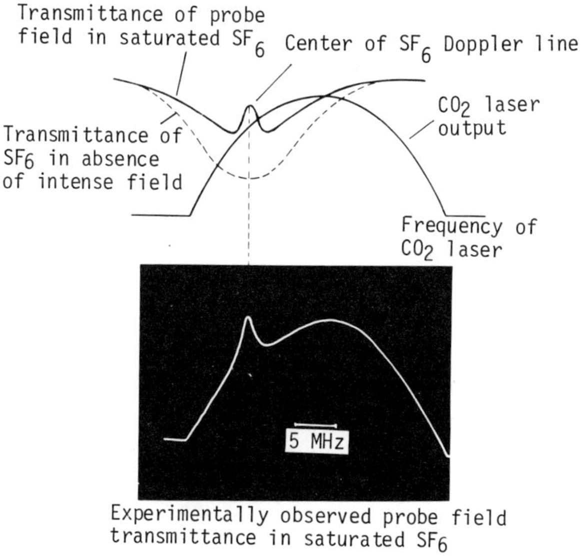

Fig.9.16 Experimentally observed narrow resonance of transmission of a weak, oppositely travelling wave of the CO_2 laser (line P(18) of the 10.6 μm band) in an external SF_6 cell, with SF_6 pressure of 4×10^{-2} torr, and explanation of its position.

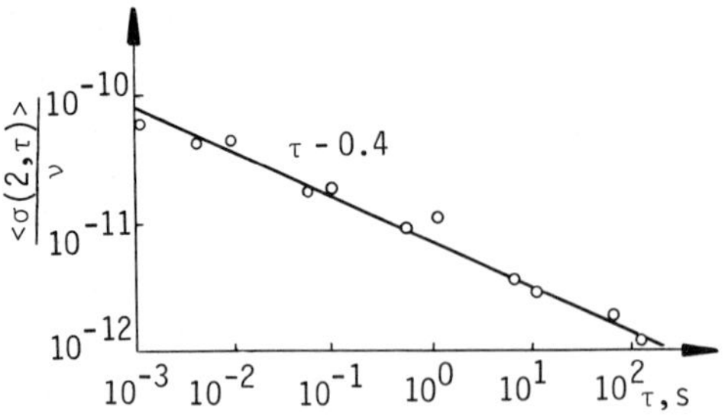

Fig.9.17 Dependence of frequency stability of a CO_2 laser, stabilized on the narrow resonance in an external nonlinear-absorbing SF_6 cell, on averaging time τ

transmission-peak top. The frequency reproducibility of the CO_2 laser has not yet been explored in detail. The upper limit of the collision shift of the SF_6 absorption line (less than 300 Hz/mtorr) estimated in [9.60] indicates a frequency reproducibility of 10^{-11} when the SF_6 pressure (10^{-2} torr) is controlled with an accuracy of 10%.

Detection of narrow resonances in the methane-like molecule OsO_4 in the field of CO_2 laser radiation [9.62] permits use of this molecule to stabilize the CO_2 laser frequency. The first experiments of this kind were reported in [9.61]: frequency reproducibility is about 10^{-11} and long-term frequency stability is about 10^{-12}.

9.3.3 Modification of the Nonlinear-Absorption-Cell Method

Fluorescence Resonances

The nonlinear-absorption-cell method (especially when using the cell outside a cavity) is not applicable for weakly absorbing transitions for which $\kappa_{ob} l_b \ll 10^{-2}$ (κ_{ob} is initial absorption per unit length, l_b is the absorbing-cell length). BASOV and one of the authors of [9.63] have proposed to observe intensity resonances of the total fluorescence at saturation of absorption in a standing light wave. When the strong-field frequency passes through the center of a Doppler-broadened transition, the fluorescence intensity shows a narrow resonance dip caused by that in the saturated-absorption coefficient in a standing wave. The peculiarity of the nonlinear-fluorescence method is the absence of strong-wave radiation in the photodetector. This fact gives a high sensitivity of the method of weakly absorbing transitions.

Narrow fluorescence resonances have been discovered, as reported in [9.64], when the absorption of CO_2 at low pressure was saturated by CO_2 laser radiation. Fluorescence was observed on an upper level of the laser transition in the range of 4.3 μm. The sensitivity of this method is such that it allows the use of the transition between excited levels of the CO_2 molecule with an absorption coefficient of 1.5×10^{-6} cm^{-1} at 300 K. Frequency stability of about 10^{-11} was achieved [9.64] when the CO_2 laser frequency was stabilized by this method. In the case of CO_2 this method has an important possible application--frequency stabilization on any rotational-vibrational lines of the P or R branches of the 9.6 or 10.6 μm transitions. This was the basis of measurements of the absolute frequency of the 3.39 μm He-Ne laser and for precise measurement of the speed of light: a CO_2 laser stabilized on the 9.3 μm line R(10) was used in the sequence of frequency multiplications; its frequency coincided with that of the third harmonic of the 28 μm H_2O laser [9.65].

Competitive Resonances

BASOV and collaborators [9.66,67] proposed to use an effect of competition of oppositely travelling waves in a ring laser with a nonlinear absorbing cell, for further narrowing of a power resonance. Competitive resonances were considered in Chapter 3.

The authors of [9.67,68] have found competitive resonances with a width of 30 kHz within an absorption line of CH_4 that had a homogeneous width of about 300 kHz, which is a 10-fold narrowing of the power peak. The competitive resonances have considerable absolute amplitude and high contrast, which makes them attractive for laser-frequency stabilization. The frequencies of two independent He-Ne/CH_4 lasers (CH_4 pressure of 10 mtorr) were stabilized on competitive resonances with a width of 60 kHz and a contrast of about 50%, as reported in [9.68,69]. Frequency stability was about 5×10^{-14} within 30 min., with averaging times of 10 s. The He-Ne/CH_4 frequency reproducibility in a ring cavity has not yet been determined.

Competitive resonance occurs also in a laser with a linear cavity when it operates in two axial modes that are symmetrically located relative to a gain (absorption) line. Owing to strong nonlinear mode coupling, the competitive resonances can have widths less than the homogeneous width. A 4- to 5-fold narrowing of the competitive power peak in the He-Ne/CH_4 laser, as compared to the usual power peak was reported in [9.70]. Experiments on frequency stabilization of a two-mode He-Ne/CH_4 laser on the competitive resonance were first reported in [9.71]; in those experiments, the long-term frequency stability was 3×10^{-13}. Nonlinear effects that are characteristic of mode competition were reported to have an additional influence on frequency reproducibility, to about 10^{-12}. All of the conditions that reduce frequency reproducibility of the He-Ne/CH_4 system with an ordinary cavity have similar effects in all of these cases.

Laser-Frequency Autostabilization

Nonlinear absorption inside a laser cavity can significantly affect the frequency characteristics of the laser, even when no system of automatic frequency control is used. Nonlinear frequency pulling to the center of an absorption line occurs because of formation of a narrow peak of effective gain [9.18]. This phenomenon is called frequency autostabilization and is considered in detail in Chapter 3.

Frequency autostabilization is of some interest for eliminating rapid fluctuations of laser frequency that are difficult to eliminate by automatic frequency control on a power peak. Comparison of frequency fluctuations of an ordinary laser with those of a laser with an internal nonlinear-absorbing cell under conditions of frequency autostabilization shows that fluctuations in the latter are far less than in the former [9.45]. The mean deviation of frequency fluctuations is reduced by about a factor $1/(1+S)$ where S is the factor of autostabilization introduced in Subsection 3.1.3.

If frequency is stabilized over a power peak, frequency autostabilization gives no gain of long-term stability. This is because the increase of nonlinear frequency pulling leads to just the same decrease of sensitivity of the frequency control on a power peak. In this case, the change of cavity frequency $\delta\omega_c$ leads to a change of the generation frequency that is $\delta\omega = \delta\omega_c/(1+S)$, which stretches the power peak along the scale of the cavity-frequency drift. So-called tandem systems of frequency stabilization were proposed in [9.72] in order to achieve simultaneously high short- and long-term frequency stability by use of nonlinear absorption. In such systems, large, slow drifts of generated frequency that result from cavity-frequency instability are removed by frequency stabilization on an external reference frequency (for example, by use of a nonlinear-absorbing cell), and slight but very rapid frequency drifts are removed automatically by autostabilization by use of a nonlinear-absorbing cell inside the cavity. Together with this, the system of two-frequency tandem stabilization of frequency was proposed in [9.13], also. In this scheme, two frequencies ν_1 and ν_2 are generated in the cavity. Frequency ν_1 is generated under conditions of frequency autostabilization on the center of a nonlinear-absorption line at the frequency ν_1 of an internal cell; frequency ν_2 is used for automatic frequency control of a common cavity or for stabilization of the optical length of the cavity. This scheme is optimal for simultaneous high short- and long-term frequency stability of a gas laser.

As was noted in Chapter 3, the factor of autostabilization on atomic transitions can reach very large values. For an absorbing cell with an initial optical density $\kappa'_{ob} l_b = \kappa_{ob} L \approx 1$ (κ'_{bo} is the absorption coefficient per unit length in the cell, κ_{ob} is the absorption coefficient distributed over the cavity length L), with the optimal saturation parameter $G_b \approx 2.4$, the factor of autostabilization according to (3.32) is

$$S = 0.17 \frac{\kappa_{bo} c}{\Gamma_b} \approx 0.1 \frac{(c/2L)}{(\Gamma_b/\pi)} , \tag{9.17}$$

i.e., one-tenth of the ratio of the intermode distance to the homogeneous width of the absorption line. For forbidden atomic transitions (for example, for lines of MgI 4571.1 Å, CaI 6572.78 Å, SrI 6892.59 Å and others) this ratio can amount to 10^3 to 10^5. This method is therefore rather favorable for obtaining high short-term stability of cw dye lasers whose frequencies are kept in the vicinity of the center of an absorption line on such autostabilizing transitions.

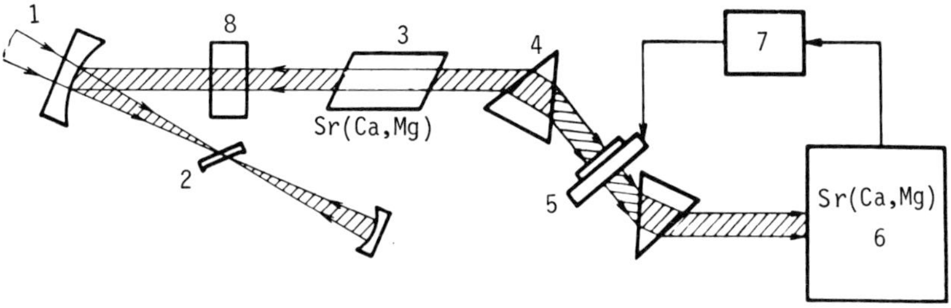

Fig.9.18 Possible simplified scheme of quantum frequency standard in visible range, based on cw dye lasers: 1 - pumping beam, 2 - dye-solution jet, 3 - intracavity nonlinear-absorption cell with vapors of Sr (or Ca, Mg), 4 - dispersive element, 5 - piezoceramic-controlled output mirror, 6 - external nonlinear-resonance reference frequency of atoms of Sr (or Ca, Mg), 7 - servosystem, 8 - frequency-selective element for producing single-mode oscillation of dye laser

Recently, the authors of [9.73] conceived the idea of combining the method of frequency stabilization in cw dye lasers based on the effect of automatic frequency stabilization of a laser with an internal atomic nonlinear-absorption cell with subsequent frequency locking by resonance with an external reference frequency from the same atoms used in the external nonlinear-absorption cell. The general scheme of such a combined quantum-frequency standard is shown in Fig.9.18. The regime of frequency autostabilization makes it possible to keep the oscillation frequency within the homogeneous line width of the absorbing medium, whereas the cavity frequency may vary over a wide range. In particular, when forbidden (intercombination) transitions in alkaline-earth metals such as Mg (4571.1 Å), Ca (6572.78 Å), Sr (6892.59 Å) are used, as shown in [9.73], very high factors of autostabilization S, from 10^2 to 10^5, can be obtained. This means that if the natural cavity frequency is maintained within several MHz, the oscillation frequency of an autostabilized laser will be stable with an accuracy to 0.1 to 10 kHz. The width of the inverted Lamb dip in an atomic beam may vary over this range. Indeed, there are practically no collisions in the beam, and the homogeneous line width of beam absorption is determined by the radiative width and the finite time of particle flight through the light beam. This time can be reduced by cooling the atomic beam, selecting slow atoms and widening the light beam. After the laser frequency falls within the inverted Lamb dip in the atomic beam, it may be locked to the center of dip with a high accuracy (10^{-3} to 10^{-4} of dipwidth) by means of a servosystem. So, the frequency of such dye lasers will be determined accurately to 0.1 to 10 Hz by the center frequency of the atomic-beam absorption line.

In this way, the effect of autostabilization makes it possible to increase the short-term frequency stability of cw dye lasers from two to four orders; the absence of hyperfine structure and other characteristics of optimal transitions (the same in the atomic beam and in the nonlinear absorber) permits us to hope for frequency reproducibility of 10^{-14} to 10^{-13} in the visible region of the spectrum.

Such an approach to development of a quantum frequency standard in the optical range is, in essence, a further development of the method suggested in [9.13]. It consists in producing a narrow nonlinear resonance inside the Doppler absorption line at intercombination transitions $n^1S_0 - n^3P_1$ of an atomic beam of alkaline-earth elements; up till now it has not been made to work for lack of appropriate amplifying media. Progress made since it was proposed has permitted definition of a specific way to make this method work.

9.3.4 Frequency Stabilization in Two-Mode Operation

For various practical purposes, for instance in interferometry of large distances, frequency stabilization in two-mode operation is of interest. If simultaneous oscillation at two frequencies is possible in a single laser cavity, then by stabilizing one of the radiated frequencies good stability of the frequency difference is achieved. Two-frequency oscillation can be obtained on a single Doppler-broadened transition. In this case, the difference of the oscillation frequencies is within the width of a Doppler-broadened transition. This is the simplest way to obtain oscillation of two frequencies of two uncoupled transitions, when a Doppler line width is less than the distance between the centers of the two transition lines. The most important factor that influences the stability of the oscillation-frequency difference is anomalous dispersion near the transition. The influence can be eliminated or considerably reduced by suitably choosing the laser operating conditions. If the oscillation-frequency difference considerably exceeds the gain line width on each transition, the cavity length L may be selected so that the oscillation frequencies are located in the line centers of each transition [9.74].

We describe in brief the results of investigations of the frequency stability of two-frequency lasers.

He-Ne Laser with a CH_4 Cell

A stable mode of generation of two symmetric axial modes in a He-Ne laser with a methane absorber was first reported in [9.75]. The frequency difference between generated axial modes obtained in this work was about 450 MHz, and the region of continuous tuning of the generated frequencies was about 100 MHz. When the cavity length was scanned near the symmetric position of the modes, a peak with a contrast of 3% and a width of 300 kHz was found in the generated power. When the cavity length was scanned over the maximum of the power peak the position of the generated modes was stabilized relative to the line center and, consequently, the radiated frequency of each mode as well as their differences was stabilized. In this case, the half-sum of the oscillation frequencies was equal to the frequency of the center of the absorption line.

The position and absolute frequency of each of two generated modes are determined by the distance between the mirrors and by the difference between their longitudinal indices. In the case of two lasers (two stabilized lasers were used in experiments on stabilization) at the symmetric position of the modes in each laser, the frequency detuning between two closely spaced laser modes is

$$\delta\Omega = (\Omega_1 - \Omega_2) \simeq \frac{1}{2}\left(\frac{c}{2L_1}\Delta n_1 - \frac{c}{2L_2}\Delta n_2\right) , \tag{9.18}$$

where c is the speed of light, L_1,L_2 are optical lengths, and Δn_1,Δn_2 are the differences between the longitudinal indices of the generated modes of the two lasers. When the optical characteristics of the beat signal that results from mixture of the radiation of two lasers at the frequency $\delta\Omega$ are measured, the stability of a gas laser at each generated mode can be determined.

The authors of [9.75] achieved frequency stability on each oscillation mode of a symmetric regime of the order of 3×10^{-12} with an averaging time of 1 s.

References

9.1 V.N. Lisitsyn, V.P. Chebotayev: Zh.Eksp.i Teor.Fiz. 54, 419 (1968)
9.2 V.P. Chebotayev, I.M. Beterov, V.N. Lisitsyn: IEEE J. QE-4, 788 (1968)
9.3 P.H. Lee, P.B. Schafer, W.B. Barker: Appl.Phys.Lett. 13, 373 (1968)
9.4 N.A. Borisevich, A.P. Voitovich, A.Ya.Smirnov, A.N. Krasovsky: Zh. Prikladnoi Spektroskopii 8, 588 (1969)
9.5 A.G. Fox, S.E. Schwarz, P.W. Smith: Appl.Phys.Lett. 12, 371 (1968)
9.6 I.M. Beterov, V.N. Lisitsyn, V.P. Chebotayev: Optika i Spektroskopia 30, 932,1108 (1971)
9.7 W.R. Bennett, Jr.: Comments Atomic Molec.Phys. 2, 10 (1970)
9.8 Ye.V. Baklanov, V.P. Chebotayev: Zh.Eksp.i Teor.Fiz. 61, 922 (1971)
9.9 M.S. Feld, A. Javan, P.H. Lee: Appl.Phys.Lett. 13, 424 (1968)
9.10 A.P. Voitovich, N.I. Kabayev, A.Ya. Smirnov, A.P. Shkadarevich: Optika i Spektroskopia 30, 940 (1971)
9.11 S.A. Alyakishev, S.P. Borisovsky, Ye.P. Ostapchenko: Abstracts of the All-Union Symposium on Gas Laser Physics, Novosibirsk, 1969, p.9
9.12 V.S. Letokhov: Zh.Eksp.i Teor.Fiz. 56, 1748 (1969)
9.13 N.G. Basov, V.S. Letokhov: Zh.Eksp.i Teor.Fis.Pis.Red. 2, 6 (1965)
9.14 J.P. Singer, I. Gorog: Bull.Amer.Phys.Soc. 7, 14 (1962)
9.15 N.G. Basov, A.N. Orayevsky, V.A. Shcheglov: Zh.Eksp.i Teor.Fiz.Pis.Red. 4, 61 (1966)
9.16 N.G. Basov, V.S. Letokhov: Zh.Eksp.i Teor.Fiz.Pis.Red. 9, 660 (1969)
9.17 V.S. Letokhov, B.D. Pavlik, S.P. Fedoseyev: Prepring of the Institute of Spectroscopy of the USSR Acad.of Sci., No. 86 (1971)
9.18 V.S. Letokhov: Zh.Eksp.i Teor.Fiz.Pis.Red. 6, 597 (1967)
9.19 V.S. Letokhov: B.D. Pavlik: Zh.Eksp.i Teor.Fiz. 53, 1107 (1967); Preprint of the Institute of Spectroscopy of the USSR Acad.of Sci., No. 108 (1966)
9.20 V.S. Letokhov, B.D. Pavlik: Zh.Tekhnicheskoi Fiziki 40, 1638 (1970)
9.21 F. Keilmann, R.L. Scheffield, M.S. Feld, A. Javan: Appl.Phys.Lett. 23, 618 (1973)
9.22 N. Skribanowitz, M.S. Feld, R.E. Francke, M.J. Kelly, A. Javan: Appl. Phys.Lett. 19, 161 (1971)
9.23 I.M. Beterov: Thesis, Institute of Semiconductor Physics, Siberian Branch of the USSR Acad. of Sci. (1970)
9.24 N. Skribanowitz, I.P. Herman, R.M. Osgood, Jr., M.S. Feld, A. Javan: Appl.Phys.Lett. 20, 428 (1972)
9.25 I.M. Beterov, V.P. Chebotayev: Zh.Eksp.i Teor.Fiz.Pis.Red. 9, 216 (1969)
9.26 R.E. Beehler, R.C. Mockler, J.M. Richardson: Metrologia 1, 114 (1965)
9.27 R. Vessot, H. Peters, J. Vanier, R. Beehler, D. Halford, R. Harrach, D. Allan, D. Glaze, C. Snider, J. Barnes, L. Cutler, L. Bodily: IEEE Trans. IM-15, 165 (1966)
9.28 K.M. Baird, L.E. Howlett: Appl.Opt. 2, 455 (1963)
9.29 P.H. Lee, M.L. Skolnick: Appl.Phys.Lett. 10, 303 (1967)

9.30 A. Javan: Ann.N.Y.Acad.Sci. 168, 715 (1970)
9.31 K.M. Evenson, J.S. Wells, F.R. Petersen, B.L. Danielson, G.W. Day: Appl.Phys.Lett. 22, 192 (1973)
9.32 N.G. Basov, V.S. Letokhov: Uspekhi Fiz.Nauk 96, 585 (1968)
9.33 A.O. McCoubrey: Proc. IEEE 54, 116 (1966)
9.34 S.N. Bagayev, Yu.D. Kolomnikov, V.N. Lisitsyn, V.P. Chebotayev: IEEE J. QE-4, 868 (1968)
9.35 G.M. Strakhovsky, V.M. Tatarenkov, A.N. Titov: Izmeritel'naya Tekhnika 12, 25 (1970)
9.36 G.R. Hanes, C.E. Dahlstrom: Appl.Phys.Lett. 14, 362 (1969)
9.37 G.R. Hanes, K.M. Baird, J. DeRemigis: Appl.Opt. 12, 1600 (1973)
9.38 A.J. Wallard: J.Phys. E5, 926 (1972)
9.39 W.G. Schweitzer, Jr., E.G. Kessler, Jr., R.D. Deslattes, H.P. Layer, J.R. Whetstone: Appl.Opt. 12, 2927 (1973)
9.40 CCDM Recommendation, M1 (1973)
9.41 R.L. Barger, J.L. Hall: Phys.Rev.Lett. 22, 4 (1969)
9.42 N.G. Basov, M.V. Danileiko, V.V. Nikitin: Zh.Prikladnoi Spektr. 54, 2217 (1969)
9.43 D. Allan: Proc.IEEE 54, 221 (1966)
9.44 R.L. Barger, J.L. Hall: Proc.23rd Ann.Symp.Frequency Control, Ft. Monmouth, New Jersey, 6-8 May 1969, p.30
9.45 V.S. Letokhov, B.D. Pavlik: Kvantovaya Elektronika N4(10), 32 (1972)
9.46 J.L. Hall, G. Kramer, R.L. Barger: Rept.Conf.Precise Electromagnetic Measurement, June 1972, Boulder, Colorado
9.47 N.B. Koshelyayevsky, V.M. Tatarenkov, A.N. Titov: Zh.Eksp.i Teor.Fiz. Pis.Red. 13, 592 (1971)
9.48 C. Borde, J.L. Hall: Phys.Rev.Lett. 30, 1101 (1973)
9.49 S.N. Bagayev, Ye.V. Baklanov, V.P. Chebotayev: Zh.Eksp.i Teor.Fiz.Pis. Red. 16, 344 (1972)
9.50 S.N. Bagayev, Ye.V. Baklanov, Ye.A. Titov, V.P. Chebotayev: Zh.Eksp.i Teor.Fiz.Pis.Red. 20, 292 (1974)
9.51 S.N. Bagayev, V.P. Chebotayev: Appl.Phys. 7, 71 (1975)
9.52 S.N. Bagayev, V.P. Chebotayev: Zh.Eksp.i Teor.Fiz.Pis.Red. 16, 614 (1972)
9.53 Ye.V. Baklanov, B.Ya. Dubetsky, Ye.A. Titov, V.M. Semibalamut: Kvantovaya Elektronica 2, 2518 (1975)
9.54 A.P. Kol'chenko, S.G. Rautian, R.I. Sokolovsky: Zh.Eksp.i Teor.Fiz. 55, 1864 (1968)
9.55 Ye.V. Baklanov: Proc.3rd Vavilov Conf.Nonlinear Optics, Novosibirsk, June 1973, p.117
9.56 Ye.V. Baklanov, Ye.A. Titov: Kvantovaya Elektronika 2, 1781 (1975); 2, 1893 (1975)
9.57 Yu.V. Brzhazovsky, L.S. Vasilenko, V.P. Chebotayev: IEEE J.QE-4, 23 (1968); QE-5, 146 (1969); Zh.Eksp.i Teor.Fiz. 55, 2096 (1968)
9.58 L.S. Vasilenko, M.I. Skvortsov, V.P. Chebotayev, G.I. Shershnyova, A.V. Shishayev: Optika i Spektroskopia 32, 1123 (1972)
9.59 S.N. Bagayev, L.S. Vasilenko, V.M. Klement'yev, Yu.A. Matyugin, B.I. Troshin, V.P. Chebotayev: Optika i Spektroskopia 32, 802 (1972)
I.M. Beterov, Yu.A. Matyugin, B.I. Troshin, V.P. Chebotayev: Avtometriya 5, 59 (1972); 5, 71 (1972); 6, 55 (1972); 6, 64 (1972)
9.60 O.N. Kompanets, A.R. Kukudzhanov, V.S. Letokhov, Ye.L. Mikhailov: Kvantovaya Elektronika 4, 28 (1973)
9.61 V.M. Gusev, O.N. Kompanets, A.R. Kukudzhanov, V.S. Letokhov, Ye.L. Mikhailov: Kvantovaya Elektronika 1, 2465 (1974)
9.62 Yu.A. Gorokhov, O.N. Kompanets, V.S. Letokhov, G.A. Gerasimov, Yu.I. Posudin: Opt.Commun. 7, 320 (1973)
9.63 N.G. Basov, V.S. Letokhov: Report on URSI Conference "Laser Measurement", Sept. 1968, Warsaw, Poland; Elec.Tech. 2, 15 (1969)
9.64 C. Freed, A. Javan: Appl.Phys.Lett. 17, 53 (1970)

9.65 K.M. Evenson, J.S. Wells, F.R. Petersen, B.L. Danielson, G.M. Day: Appl. Phys.Lett. 22, 192 (1973)
9.66 N.G. Basov, E.M. Belenov, M.V. Danileiko, V.V. Nikitin: Zh.Eksp.i Teor. Fiz. 57, 1991 (1969)
9.67 N.G. Basov, E.M. Belenov, M.V. Danileiko, V.V. Nikitin, A.N. Orayevsky: Zh.Eksp.i Teor.Fiz.Pis.Red. 12, 145 (1970)
9.68 N.G. Basov, E.M. Belenov, M.I. Vol'nov, M.V. Danileiko, V.V. Nikitin: Zh.Eksp.i Teor.Fiz.Pis.Red. 15, 659 (1972)
9.69 N.G. Basov, E.M. Belenov, M.V. Danileiko, V.V. Nikitin: Kvantovaya Elektronika 15, 42 (1971)
9.70 M.A. Gubin, A.I. Popov, Ye.D. Protsenko: Kvantovaya Elektronika 3, 99 (1971)
9.71 N.G. Basov, M.A. Gubin, V.V. Nikitin, Ye.D. Protsenko, V.A. Stepanov: Zh.Eksp.i Teor.Fiz.Pis.Red. 15, 525 (1972)
9.72 S.N. Bagayev, L.S. Vasilenko, V.P. Chebotayev: Optika i Spektroskopia 24, 156 (1970)
9.73 V.S. Letokhov, B.D. Pavlik: Kvantovaya Elektronika 3, 60 (1976)
9.74 V.P. Chebotayev: Radiotekhnika i Elektronika 11, 1712 (1966)
9.75 S.N. Bagayev, A.K. Dmitriyev, V.P. Chebotayev: Zh.Eksp.i Teor.Fiz.Pis. Red. 15, 91 (1972)

10. Narrow Nonlinear Resonances in Experimental Physics

The discovery of narrow optical resonances induced by laser radiation has formed the basis for superhigh-resolution nonlinear spectroscopy and precision spectroscopy. This is now one of the most promising trends in laser spectroscopy. Some laboratories are running and have already carried out experiments on measurement of fundamental constants by use of narrow nonlinear resonances. Narrow molecular resonances have made it possible to create electromagnetic oscillators with very high frequency stability; this and the elaboration of methods for direct measurement of optical frequency has opened the way for introducing a single joint quantum standard of time and length.

The effects and methods studied in most of the preceding chapters are based on the ability of an intense monochromatic beam to change the level population of quantum transition of particles whose velocity lies in a narrow range. The possibility to change velocity distributions is not a new idea in physics. As far back as a hundred years ago J.C. Maxwell imagined a demon who could select molecules that have velocity higher than the mean thermal velocity. An intense laser field is a realization of the demon imagined by Maxwell. A laser "demon" can be applied to other areas of physics when it is necessary to change the velocity distribution of atoms or molecules in a gas. Specifically, laser radiation can induce narrow resonances of absorption and emission of γ radiation in a gas. Laser methods for production of such resonances can add much to the method of narrow-resonance production based on the Mössbauer effect. This will enable us, in principle, to bridge the gap between the gamma-ray range and the optical range of electromagnetic waves just as the gap between the radio-frequency and optical ranges has been bridged by lasers.

10.1 Measurement of Fundamental Physical Constants

Narrow nonlinear resonances permit, first, more precise measurements of the central frequencies of Doppler-broadened optical spectral lines, thus increasing substantially the accuracy of determinations of fundamental constants based on such measurements. Second, narrow and frequency-stable molecular resonances are very effective for stabilization of laser-oscillation frequencies. Again, this leads to substantial increase of the accuracy of measurement of fundamental constants which needs highly frequency-stable sources of optical radiation. Both of these applications of narrow nonlinear resonances are being successfully developed by dozens of laboratories in several countries. In this section, three experiments of the kind will be considered: precision measurement of the speed of light by lasers stabilized on narrow resonances, precision measurement of the Rydberg constant and the 1s - 2s transition frequency of hydrogen and positronium atoms. Also some prospects of application of narrow resonance to measurement of stability of fundamental constants will be discussed.

10.1.1 Speed of Light

Lasers with frequencies stabilized on narrow molecular resonances are excellent sources of coherent light. They have provided new measurements of the speed of light by simultaneous measurements of the wavelength and frequency of light oscillations (TOWNES [10.1]). For such an experiment, it is necessary to measure the absolute frequency of light oscillation. The basic idea of direct measurement of electromagnetic oscillation frequency in the optical range consists of multiplying the radiation frequency of microwave oscillators up to light frequencies. A set of frequency-stabilized cw lasers that radiate in the submillimeter and ir ranges and nonlinear elements practically inertialess up to the ir range is needed to carry out this idea.

The first experiments on mixing laser radiation with harmonics of klystron were performed by use of common silicon point contacts, which had been widely used to generate harmonics of the submillimeter range [10.2]. During these experiments, the laser output and klystron radiation were simultaneously fed to the diode. The intense laser radiation could be considered as a reference oscillator in heterodyne detection of a weak signal of high harmonics of the klystron radiation. By use of this method, the frequencies of HCN (337 and 311 μm) [10.2] and DCN (194 and 190 μm) [10.3] lasers were measured. In the latter case, the laser frequency was heterodyned with harmonics 22 and 23 of a klystron operating at a frequency of 70 GHz. By mixing the frequencies of a HCN laser (337 μm) and a D_2O laser (84 μm) the region of frequency measurements could be extended up to $3 \cdot 10^{12}$ Hz [10.4] but further extension was not possible with a silicon mixer, because of the high-frequency limit of mixing determined by the lifetime of its current carriers.

In [10.5] an idea was proposed to use the strong optical nonlinearity at the point contact of the p-n junction of a tunnel diode, which should be inertialess up to optical frequencies. Subsequent development of nonlinear mixers employed the tunnel effect, not at the p-n junction, but at the contact between two metals.

Radical improvements of point-contact diodes, which allow their application up to the optical range, have been made by JAVAN at MIT in Cambridge, Massachusetts, and by EVENSON at the NBS in Boulder, Colorado, USA. Unlike previous constructions, silicon was replaced by a metal; nonlinearity occurred at the point of mechanical contact between a thin metal (tungsten) filament about 1-2 μm in diameter and several mm long and a metal (nickel) surface. The contact diode was placed off the waveguide, and the mixed laser radiations were focused directly on to the filament near the contact. The harmonic frequency difference of mixed signals lying in the microwave range was detected by heterodyning it with a microwave signal from a klystron whose frequency was close. The metal-contact diode is sometimes called a "cat's whisker" diode. High-frequency mixing in such a diode was proved up to frequencies in the range of 5 μm [10.6,7].

After EVENSON et al. had optimized the operation of the "cat's whisker" as an antenna for infrared waves [10.8], he measured the absolute frequencies of cw lasers on H_2O (78 and 28 μm) [10.9], CO_2 (P(18) and P(20) lines of the 10.6 μm band) [10.10], and a He-Ne laser at 3.39 μm [10.11]. This work has been discussed in a recent article by EVENSON and PETERSEN [10.12]. Here we present only brief information. Figure 10.1 shows schematically the frequency-synthesis chain of lasers and klystron used to measure the frequency of the He-Ne laser. The experiment was performed in several steps. First, the frequency of the CO_2 laser at the line R(10) of the 10.6 μm band was measured. This measurement included measurement of the frequencies of HCN (337 μm) and H_2O (28 μm) lasers. The third harmonic of a CO_2 laser at the R(30) line of the 10.6 μm band lies 45 GHz from the frequency of the He-Ne/CH_4 stabilized laser. Determination of that difference was the last step of the experiment. The frequency difference between the two lines of the CO_2 laser in use was measured (second step of experiment) by comparing that difference with the frequency of an HCN laser at 337 μm. Of course, the actual procedure of measurements was much more complicated. The three saturated-absorption-stabilized lasers (upper right part of Fig.10.1) served as reference oscillators for the He-Ne and the two CO_2 lasers, which were locked at a frequency of a few MHz apart from the frequencies of the stabilized lasers. The frequency scale was calibrated by reference to the cesium frequency

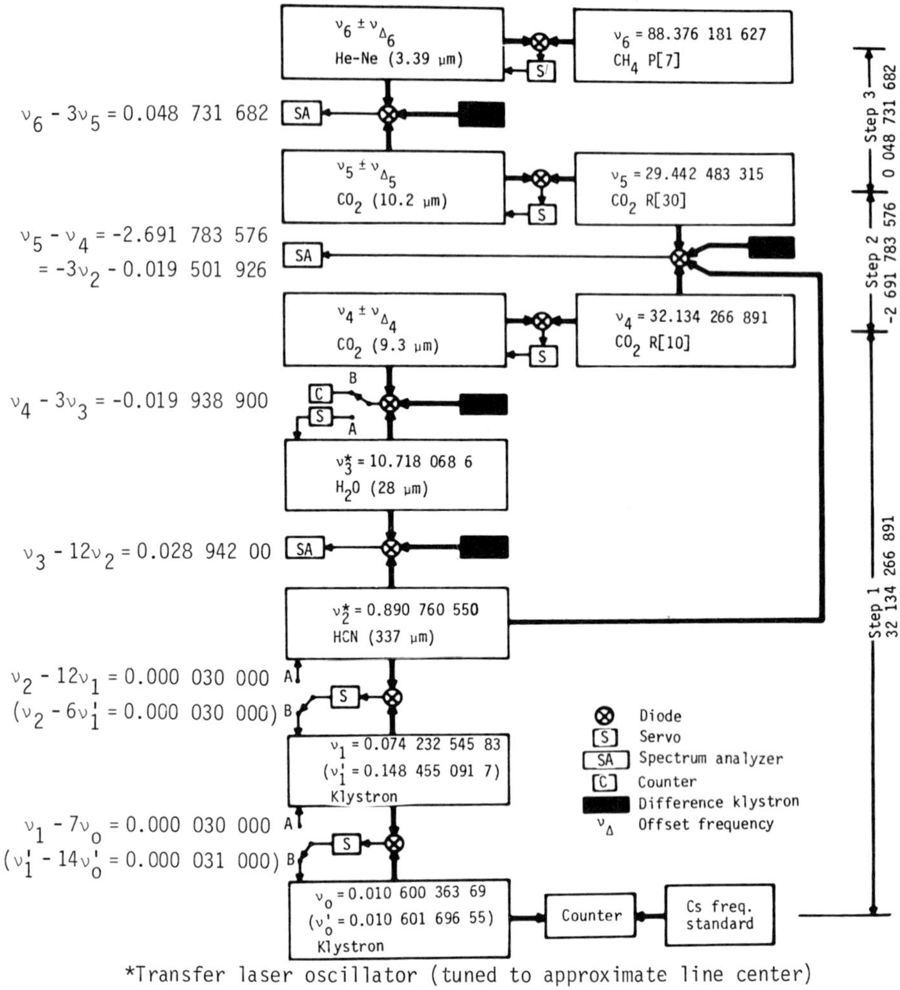

Fig.10.1 Stabilized-laser-frequency chain for absolute measurement of He-Ne/CH_4 (3.39 μm) laser frequency. All frequencies are given in THz (EVENSON and PETERSEN [10.12])

Table 10.1 Absolute frequencies of He-Ne and CO_2 laser lines [10.12,13]

Molecule	Line	λ (μm)	Frequency (THz)[a]
$^{12}C^{16}O_2$	R(30)	10.18	29.442 483 315 (25)
$^{12}C^{16}O_2$	R(10)	9.3	32.134 266 891 (24)
$^{12}CH_4$	R(7)	3.39	88.376 181 627 (50)

[a] The numbers in parentheses in the right-hand column are 1-standard-deviation errors, which indicate uncertainties in the last two digits.

standard. The final results of this frequency measurement of He-Ne and CO_2 lasers are given in Table 10.1 [10.12,13].

Independent measurement of the He-Ne/CH_4 laser wavelength, by interferometric comparison with the wavelength of the international length standard ^{86}Kr gave $\lambda\ (CH_4) = 3.392\ 231\ 376\ (13)\ \mu m$ [10.14]. This made possible the determination of a new value of the velocity of light [10.13]: $c = 299\ 792\ 456.2\ (1.1)$ m/s to an accuracy of $\delta c/c = \pm 3.6 \cdot 10^{-9}$.

The main error in this speed-of-light measurement is attributable to the error of wavelength measurement. At the 5th session of the Consultative Committee on the Definition of the Meter (CCDM) [10.15], results of BARGER and HALL [10.14] as well as measurements made at the International Bureau of Weights and Measures in Paris [10.16], the National Research Council of Canada [10.17] and the National Bureau of Standards at Gaithersburg [10.18] were compared to give a recommended value for the wavelength of the transition of methane:

$$\lambda_{CH_4}[P(7),\ \text{band}\ \nu_3] = 3\ 392\ 231.40 \times 10^{-2}\ \text{m}. \tag{10.1}$$

Multiplying this recommended wavelength of methane by the measured frequency gives, for the value for the speed of light:

$$C = 299\ 792\ 458\ \text{m/s}\ \left(\frac{\delta c}{c} = \pm 4 \times 10^{-9}\right)\ . \tag{10.2}$$

This result is in agreement with the recommended value of $C = 299\ 792\ 500\ (\pm 100)$ m/s and is about *100 times more accurate*. This has confirmed a widespread opinion that advances in the field of quantum frequency standards in the optical range must lead to substantial increase of the measurement accuracy of some fundamental constants. What is more, this measurement of the frequency of a more-stable source of optical radiation, which a He-Ne/CH_4 laser is, has demonstrated the possibility, in principle, of developing a single joint standard for length and time.

10.1.2 Rydberg Constant and Transition Frequencies in Hydrogen Atom

Accurate measurement of the structures and positions of the center frequencies of spectral lines is necessary for determination of the Rydberg constant, which is related to fundamental constants by $R_\infty = me^4/2\hbar^2$. R_∞ has been determined in many cases by the use of spectral lines of hydrogen, its isotopes, and helium. Hydrogen is more suitable for such measurements because it permits accurate comparison of experimental data with the conclusions of quantum theory. The accuracy of measurement of the Rydberg constant, when narrow nonlinear resonances are not used, is about ±0.01 cm^{-1}: R_∞ = 109 737.312 ± 0.007 cm^{-1} [10.19]; that is, the relative accuracy is $\pm 10^{-7}$. In the visible range, the Balmer red line H_α is best suited for measurements. Its Doppler broadening, however, is almost 6000 MHz, which masks the most essential details of the fine structure of the line. Although most details of the fine structure of the H_α line have been revealed by the methods of radio spectroscopy and level-crossing spectroscopy, the inadequate accuracy of measurement limits the accuracy of determination of the Rydberg constant. At Stanford University (USA) an experiment on measurement of the Rydberg constant has been done, which is based on accurate determination of the line center of a particular component in the H_α line (6563 Å) when absorption is saturated by a pulsed, tunable dye laser [10.20].

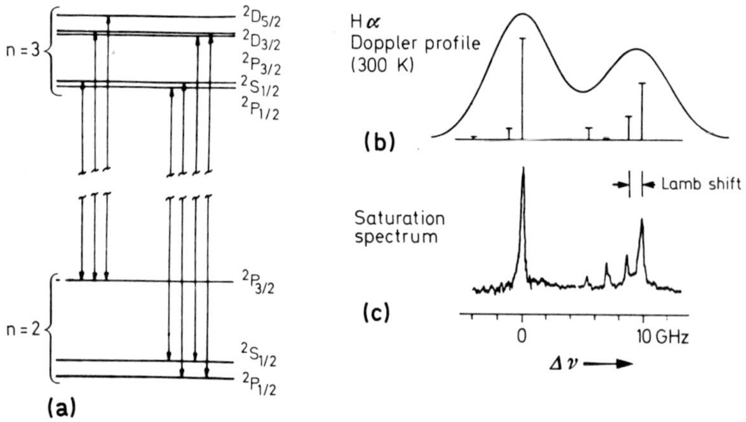

Fig.10.2a-c Saturation spectroscopy of the hydrogen Balmer line H_α: a) energy level with theoretical fine structure; b) linear spectrum with positions of fine-structure components; c) saturation spectrum (HANSCH et al. [10.22])

Part of this measurement is, of course, a measurement of Lamb shifts in the fine structure of the H_α line, based on the same technique of saturation of the H_α line absorption by a pulsed, tunable dye laser. Figure 10.2a shows the theoretical level diagram for the fine structure of the H_α line. The state 2 $^2S_{1/2}$ is 1058 MHz higher than the 2 $^2P_{1/2}$ state. This shift is not given by the Dirac theory but is explained by quantum-electrodynamic corrections. It was first discovered, in experiments by LAMB and RETHERFORD [10.21], by use of the methods of radio spectroscopy and is now called the Lamb shift.

By use of a simple gas-discharge Wood tube at room temperature, the authors of [10.22] managed for the first time to resolve the fine- and hyperfine-structure components of the H_α line and to observe the Lamb shift, directly, in the optical spectrum of nonlinear absorption. With a Doppler line width of about 6000 MHz (Fig.10.2b) the resonances in a weak counter-running wave absorption attained by the method discussed in Subsection 3.2.1 were only 25 MHz wide. The spectrum of the observed saturated absorption of H_α is shown in Fig.10.2c.

By use of a pulsed laser in this experiment, it was possible to observe the saturated absorption in the afterglow of the pulsed discharge and thus to minimize Stark shifts and broadening of the observed transition. Under conditions that provide maximum resolution (small width of laser-pulse spectrum, and small angle between the intense and probe beams) HÄNSCH et al. managed also to measure the hyperfine splitting ($\Delta\nu = 177$ MHz) of the $2^2S_{1/2}$ state in the component $3^2P_{3/2} - 2^2S_{1/2}$ of H_α by the optical method. Measurement of wavelength of the resolved components of the H_α line permits more-precise determination of the Rydberg constant. For precise measurement of the transition wavelength, a He-Ne laser was used, stabilized by the inverted Lamb dip in I_2, with its wavelength calibrated against the krypton length standard. These experiments have enabled us to measure the frequency difference between the fine-structure components of the H_α line, accurate to 1 MHz. The new value of the Rydberg constant is

$$R_\infty = 109737.3143\ (10)\ \text{cm}^{-1} \tag{10.3}$$

that is, its relative accuracy is $1 \cdot 10^{-8}$, which is one order of magnitude better than previous measurements.

Observation of narrow resonances of two-photon transitions at forbidden transitions of hydrogen atoms is another promising opportunity for precision measurement of transition frequencies [10.23]. The forbidden transition 1S - 2S of hydrogen at $\lambda = 1215.68$ Å is best suited to this purpose. Because the 2S state is metastable, this ensures extremely small width of resonance. In the standing wave of a laser at $\lambda = 2431.36$ Å, owing to a narrow resonance at the center of the Doppler line of the two-quantum transition, considered in Chapter 4, it is possible to produce a density resonance of hydrogen atoms excited to the 2S level, which is effected by only the second-order Doppler effect (Chap.4). A narrow resonance with width of about 130 kHz is quite attainable, which corresponds to the resolution $\nu/\Delta\nu \simeq 10^{10}$ (see Subsec.4.1.2 and Fig.4.4b).

The two-photon absorption probability at the transition 1S - 2S of hydrogen can be estimated from the equations given in Chapter 4 (the coefficient B in formula (1) from [10.23] for the two-photon transition rate for hydrogen must, however, be 243, rather than 6).

The first experiments on production of narrow resonances of two-photon absorption at the transition 1S - 2S of hydrogen and deuterium atoms were reported in [10.24,25]. They gave, immediately, valuable spectroscopic information on the Lamb shift in the 1S level of H and D. A pulsed dye laser at 4860 Å with a spectral width of 120 MHz and peak power from 30 to 50 kW was used as a source of coherent radiation. In a lithium formate crystal, the second harmonic, at 2430 Å with a peak power of about 600 W, was generated. The hydrogen cell was placed in the area of the standing light wave of 2430 Å, which produced two-photon excitation of the 2S state. During atomic collisions, the 2P state was populated and fluorescence of 2P - 1S at the line L_α was observed, with wavelength 1215 Å. In experiments, resonances were produced whose widths were only 2% of the Doppler width, which for H at 300 K is 30 GHz.

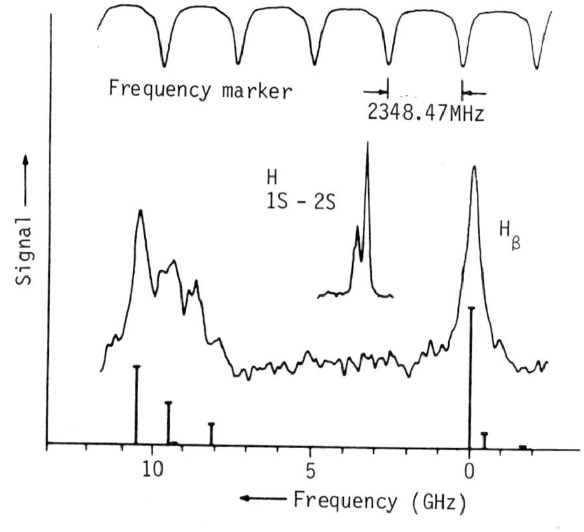

Fig.10.3 Saturation spectrum of the hydrogen Balmer β line with theoretical fine-structure spectrum, and simultaneously recorded 1S - 2S two-photon spectrum. The top curve gives the resonances of the calibration interferometer (LEE et al. [10.25])

Simultaneous measurement of the spectra of linear absorption [10.24] and saturated absorption [10.25] of the H_β line, by use of the radiation at the first harmonic wavelength 4860 Å is a feature of the brilliant experiments by HÄNSCH et al. [10.24]. Figure 10.3 shows the spectra of saturated absorption at the H_β line and two-photon absorption at the transition 1S - 2S of deuterium reported in [10.25]. The dashed line in Fig.10.3a shows the estimated positions of the components of the H_β line. Comparison of the spacings of the frequencies in the 1S - 2S and $2P_{3/2}$ - $4D_{5/2}$ lines enabled HÄNSCH et al. to evaluate the Lamb shift of the ground states 1S of hydrogen (8.20 ± 0.10 GHz) and deuterium (8.25 ± 0.11 GHz) atoms [10.25]. Before that experiment, only the Lamb shift of the 1S level of deuterium had been known (7.9 ± 1.1 GHz), which was estimated by HERZBERG after difficult measurements of the absolute wavelength of L_α [10.26].

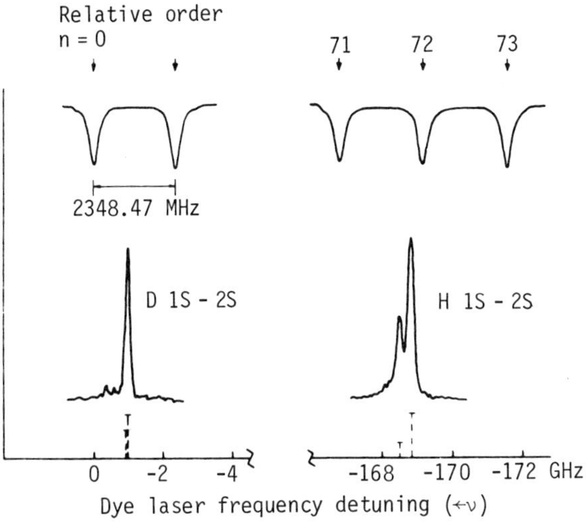

Fig.10.4 Doppler-free two-photon spectra of the 1S - 2S transition in H and D with transmission maxima of the calibration interferometer (LEE et al. [10.25])

Spectra of two-photon absorption at the transition 1S - 2S of hydrogen and deuterium were measured simultaneously [10.25]; consequently, the isotopic shift was measured precisely. Such spectra, with corresponding calibrated frequency scales, are shown in Fig.10.4. The spectrum of hydrogen has, as expected, a doublet due to hyperfine structure. The hyperfine splitting of deuterium is only one-fourth as wide and is not resolved. The measured isotope shift for the transition 1S - 2S is 670.933 ± 0.056 GHz, which is several orders of magnitude better than that measured previously, by use of methods of linear spectroscopy.

The increased accuracy of measurement of Doppler-free spectra of the L_α and H_β lines of hydrogen and deuterium atoms opens the way for further increase of accuracy of measurement of the Rydberg constant and of the electron-photon mass ratio (from isotope-shift measurements).

10.1.3 Optical Transitions in Positronium Atoms

Studies of the structure of the energy levels of electron and positron bound state, i.e., the positronium atom (Ps), and precise measurements of transition frequencies of the Ps atom enable us also to test the relativistic quantum theory. At present, there are two experimental methods to study the energy levels of the Ps atom: the direct microwave method of measuring the intervals of the fine structure in the first excited state [10.27], which allows direct comparison of experimental results with relativistic calculations, and the indirect method of measuring the fine-structure interval $1^1S_0 - 1^3S_1$ in the ground state, based on comparison between the experimental quenching of the 3γ annihilation radiation in the magnetic field and that theoretically predicted [10.28].

A lot of information on energy-level structure and more-precise comparison between experiment and relativistic theory can be obtained by studying the Ps atom by the methods of laser Doppler-free spectroscopy considered in this book. Spectroscopic measurements by use of standard methods of detecting either laser-radiation absorption or optical fluorescence of excited states are not possible with the extremely small available densities of Ps atoms. With Ps, the methods of excited-state detection must be based, as in microwave experiments, either on annihilation γ quantum counting or on counting coincidences (anticoincidences) of annihilation γ quanta and optical quanta. A number of schemes, and corresponding calculations, for precise measurement of the optical transitions of the Ps atom by use of the methods of laser Doppler-free spectroscopy have been suggested in [10.29].

Figure 10.5a shows schematically the fine-structure levels of the ground and first excited states of the Ps atom; the figure gives the lifetimes of the levels with respect to 2γ and 3γ annihilation decay [10.30,31]. The lifetimes of the fine-structure levels of the first excited state with respect to optical radiative decays can be determined easily from the corresponding results for hydrogen atoms. The lifetime of the 2P levels depends on L_α decay and is $\tau_{L_\alpha} = 3.2$ ns; that of the 2S levels is determined by two-photon decay and is $\tau_{2S} = 0.24$ s.

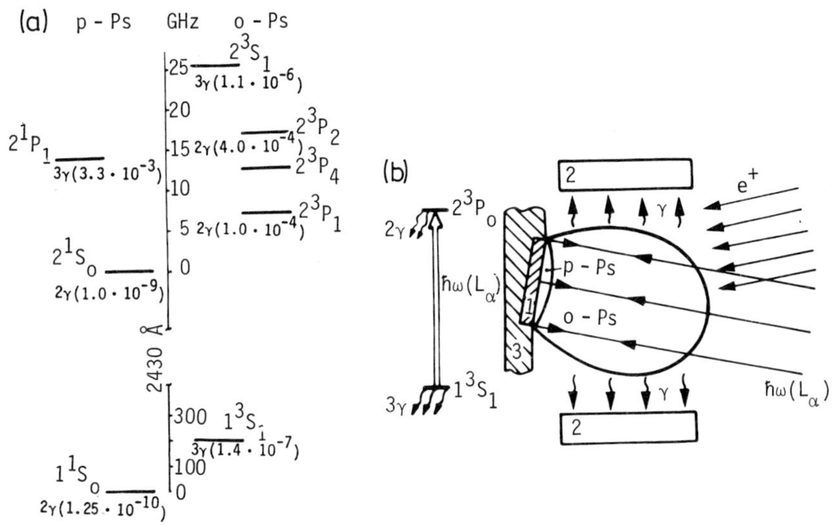

Fig.10.5a,b Diagram of fine-structure levels of ground and first excited states and annihilation lifetimes for positronium atom (a) and a scheme of a possible experiment for observation of the narrow saturation resonance at $1^3S_1 - 2^3P_0$ transition of ortho-positronium (b). 1 - target and mirror, 2 - counter of γ quanta, 3 - protective screen (LETOKHOV and MINOGIN [10.29])

Saturation Spectroscopy of 1S - 2P Transitions

In a standing light wave of laser radiation ($\lambda = 2430$ Å) the saturated-absorption spectrum of the Ps atom must contain four narrow L_α resonances, which correspond to one transition $1\,^1S_0 - 2\,^1P_1$ of para-positronium (p - Ps) and three transitions $1\,^3S_1 - 2\,^3P_J$ ($J = 0,1,2$) of ortho-positronium (o - Ps) [10.31]. The width of all the resonances is determined by spontaneous L_α decay; it is $2\Gamma = (2\pi\tau_{L_\alpha})^{-1} = 50$ MHz, with Doppler width $\Delta\nu_D = 700$ GHz for the average thermal velocity of Ps atoms $u = 10^7$ cm/s.

It is known that three times more o - Ps atoms are formed than p - Ps atoms; also, the dimension of the cloud formed by vacuum-diffusing o - Ps atoms, when the target is irradiated by a beam of positrons [10.31,32], is $\tau_T^{(0)}/\tau_S^{(0)} = 1.1 \cdot 10^3$ times larger than the corresponding dimension of the cloud of p - Ps, where $\tau_S^{(0)}$ and $\tau_T^{(0)}$ are the lifetimes of p - Ps and o - Ps atoms in the ground state. For example, when the average thermal velocity of Ps atoms $u = 10^7$ cm/s, the dimension of the cloud of o - Ps is $L_T^{(0)} = u\tau_T^{(0)} = 1.4$ cm, whereas that of p - Ps is $L_S^{(0)} = u\tau_S^{(0)} = 1.2 \cdot 10^{-3}$ cm, only. Therefore of main interest for laser Doppler-free spectroscopy are o - Ps atoms.

Because the annihilation lifetimes of the upper levels of the 1S - 2P transitions considered are all different, the most appropriate method can be chosen for each level, to detect narrow resonances of saturated absorption. Let us consider detection of the narrow resonance at the transition $1\ ^3S_1$ - $2\ ^3P_0$. Because the upper level of the transition annihilates into two γ quanta (three-photon annihilation of o-Ps in the P state is strictly forbidden [10.30]) and the lower level annihilates into three γ quanta, the most convenient scheme for detection of narrow resonance must consist of decreasing the intensity of 2γ annihilation radiation of the $2\ ^3P_0$ state as the laser frequency passes through the center of the Doppler line, or the scheme must employ the narrow density resonance of excited particles in a standing light wave (see Sec.3.3).

A potential scheme for such an experiment is shown in Fig.10.5b. The o-Ps atoms formed by irradiation of target 1 by a beam of slow positrons (as in [10.32]) are excited by the standing wave (target 1 serves at the same time as a mirror). The excited atoms are detected from the signal of 2γ annihilation radiation by γ counters 2. The counters of γ quanta are used in a coincidence scheme that registers only coincidences from two γ quanta that have energy $E_\gamma = 0.5$ MeV. The main background sources in this case are 2γ annihilation radiation of o-Ps atoms formed in the states $2\ ^3P_0$ and $2\ ^3P_2$ (they are only 10^{-4} of the total of Ps atoms [10.33]) and the part of the 3γ annihilation radiation that contains two γ quanta whose energy is $\simeq 0.5$ MeV. According to estimates from [10.29] appreciable saturation ($G \simeq 1$) of the transition $1\ ^2S_1 - 2\ ^3P_0$ requires $I_s = 2.7\ \mathrm{W/cm^2}$; with a positron source intensity of 10^{-3} curie (this corresponds to a flow of Ps atoms of $3 \cdot 10^5$ atom/s) about an hour will be needed to detect a narrow resonance (about 70 MHz wide) of density of o-Ps atoms with $S/N = 5$. Because of recoil, every L_α resonance must be split into two resonances that have a frequency difference of $\Delta = 6.2$ GHz.

Two-Photon Spectroscopy of 1S - 2S Transition

Two-photon spectroscopy of the transitions $1\ ^1S_0 - 2\ ^1S_0$ p-Ps and $1\ ^3S_1$ - $2\ ^3S_1$ o-Ps is quite possible by use of laser radiation at 4860 Å. Let us consider, for instance, the case of two-photon absorption in o-Ps. Because the level $2\ ^3S_1$ is metastable for optical radiative decay and the type of annihilation decay is the same for the upper and lower levels, it is difficult to detect two-photon absorption resonances from variation of 3γ annihilation-radiation intensity. But, when Ps atoms are in a microwave field that induces transitions from the $2\ ^3S_1$ state to any of $2\ ^3P_J$ ($J = 0,1,2$), it becomes possible to detect narrow resonances of two-photon absorption from the coincidence signal of L_α quantum of the transition $2\ ^3P_J - 1\ ^3S_1$ and from the

γ quanta of 3γ annihilation radiation of the $1\ ^3S_1$ state. The external magnetic field that mixes the sublevels $2\ ^1S_0$ (m = o) and $2\ ^3S_1$ (m = o), p - Ps and o - Ps can also be used. In this case, two-photon absorption resonances can be detected from 2γ annihilation of the mixed state $2\ ^3S_1$.

Estimates of two-photon absorption probability for positronium in a standing light wave, the line shape and detection schemes are given in [10.29]. The rate of two-photon absorption of resonance is

$$W = 4.3 \cdot 10^{-3} I F(2\omega - \omega_{1S,2S})\ s^{-1}\ , \qquad (10.4)$$

where the intensity I is expressed in W/cm^2, F is the two-photon resonance-line shape normalized to unity at a maximum, which was calculated ($\varepsilon = \Delta\omega_q/\Gamma = 200 \gg 1$) in Chapter 4 (see (4.25)). The two-photon resonance shape is determined by the second-order Doppler effect and has the form given in Fig.4.5b with full half-height width 245 MHz and resonance shift -68 MHz. In order to conduct such an experiment with a source of positrons that provide 10^5 atom/s flow of Ps atoms with the thermal velocity $u = 10^7$ cm/s a laser with a peak power of from 10^4 to 10^5 W and repetition rate from 10^3 to 10^4 Hz, at 4860 Å should be used. Then it will require about ten minutes to detect two-photon absorption by the L_α-photon-γ-quantum coincidence signal at microwave-field-induced transitions $2\ ^3S_1 - 2\ ^3P_2$.

10.1.4 Experimental Check of Constancy of Fundamental Constants

The progress achieved and expected in measuring frequencies and wavelengths of spectral lines of different transitions of atoms and molecules opens the way for experimental, laboratory checking of time constancy of fundamental constants. When considering dimensionless combinations of fundamental physical constants, the radius and age of the Universe included, DIRAC [10.34] suggested that physical constants may vary in time because of expansion of the Universe. The Hubble reverse constant is $1.4 \cdot 10^{10}$ years, that is, during 1.4 years all distances are increased by a relative value of 10^{-10}. Within this time, some dimensionless physical constants may be also subjected to changes that differ not too greatly from 10^{-10} in order of relative magnitude. For instance, two standards of length ($L_1 = n_1 \cdot a_0$ is the length standard, in the form of a ruler; $L_2 = n_2(\hbar c/R_\infty)$ is the optical length standard, where $a_0 = \hbar^2/me^2$ is the Bohr radius, $R_\infty = me^4/2\hbar^2$ is the Rydberg constant, n_1 and n_2 denote integers) that differ by the dimensionless factor $\hbar c/e^2 = 137.03\,602\,(21)$ will no longer coincide. To have permanence of the world constant checked experimentally, these two length standards should be compared to an accuracy much better than 10^{-10}.

Another possibility for experimental check of the permanence of fundamental constants suggested by DICKE [10.35,36] consists of comparison of the frequencies of two highly frequency-stable quantum oscillators that operate at quantum transitions of different types, that is, which depend on fundamental constants in different ways. Lasers that use narrow resonances as references for frequency stabilization are the most suitable sources for experiments on such a scheme. Several versions of experiments of this type have been suggested by BASOV and LETOKHOV [10.5]. Assume that there is a quantum oscillator that operates at the transition between the levels of a hyperfine structure, say, a hydrogen maser at the line $\lambda = 21$ cm, and a submillimeter laser that is stabilized by a narrow resonance at the transition between the levels of a rotational structure. What will the comparison between the frequencies of these quantum oscillators result in? The hydrogen-maser frequency is given by

$$\omega_{hfs} = \frac{4}{3}\alpha^3 g_I \frac{m}{m_p}\frac{c}{a_0} , \qquad (10.5)$$

where a_0 is the Bohr radius, g_I is the gyromagnetic ratio of a photon. The frequency of rotational transition, for instance, for the singlet term of a diatomic molecule is

$$\omega_{rot} = \frac{\hbar}{I}(K+1) , \qquad (10.6)$$

where I is the molecular moment of inertia, and K is the rotational quantum number. The molecular moment of inertia $I \sim Ma_0^2$; hence the ratio between the frequencies of hyperfine and rotational transitions is determined by

$$\frac{\omega_{hfs}}{\omega_{rot}} \sim \alpha^2 g_I . \qquad (10.7)$$

This suggests that precise comparison between the frequencies of the hydrogen maser and the laser at the rotational transition during a long period of time can give information on variation with time of $\alpha^2 g_I$, that is the hyperfine-structure constant and the gyromagnetic ratio of a photon.

An analogous experiment might be done with two lasers stabilized with narrow resonances at transitions between rotational and vibrational molecular levels. The frequency of a vibrational molecular transition is determined by

$$\omega_{vib} = (K_0/M)^{1/2} , \tag{10.8}$$

where K_0 is the force constant and M is the molecular mass. The force constant can be expressed in terms of the known constants $K_0X^2 \sim E_{el}$, where X is the normal vibration amplitude ($X \sim a_0$), and E_{el} is the electronic energy ($E_{el} \sim R_\infty$). As a result, the ratio between the frequencies of rotational and vibrational transitions requires

$$\frac{\omega_{rot}}{\omega_{vib}} \sim \left(\frac{m}{M}\right)^{1/2} . \tag{10.9}$$

Precise comparison between the frequencies of two lasers stabilized with narrow rotational and vibrational resonances during a long period of time can give information on permanence of another dimensionless value, that is, the ratio between the masses of the electron and the nucleon. The same information can be obtained from experiments on comparison between the frequencies of vibrational transitions and electron atomic transitions.

In order to run such experiments for checking the permanence of fundamental constants, lasers stabilized with narrow resonances in the submillimeter, infrared and visible ranges to an accuracy of better than 10^{-11} during a month are needed, and schemes for comparison of their absolute frequencies with the same accuracy. We can hope that this problem will be solved within the next decade.

10.2 Quantum and Relativistic Effects

The fact that nonlinear molecular resonances are extremely narrow and stable makes it possible to observe relativistic and quantum effects during emission or absorption of optical photons, in particular, the red shift of frequency due to the second-order Doppler effect and line splitting due to recoil. Also, narrow resonances make possible, in principle, detection of isomeric shifts of vibrations of excited-nucleus molecules.

10.2.1 Second-Order Doppler Effect

As a consequence of relativistic time reduction, the frequency of radiation or absorption of a moving particle is subjected to an additional red shift that depends on its absolute velocity. In the laboratory-frame system, the radiation frequency is given by the known expression,

$$\omega = \omega_0[1 - (\frac{V}{c})^2]^{1/2} \left(1 - \frac{v_z}{c}\right)^{-1} , \quad (10.10)$$

where ω_0 is the radiation frequency of a stationary atom or molecule, V is the absolute velocity of the particle, and v_z is the projection of that velocity onto the direction of observation. Particles with $|v_z| \ll V$ contribute to the narrow resonance at the center of a Doppler line. Therefore V in (10.10) means the transverse (radial) molecular velocity v_r; its distribution function due to thermal motion is given by

$$W(v_r) = \frac{2v_r}{u^2} \exp\left(-\frac{v_r^2}{u^2}\right) . \quad (10.11)$$

Therefore, the average value of the red shift in a molecular resonance is

$$\Delta\omega = -\frac{1}{2}\overline{\left(\frac{v_r}{c}\right)^2}\omega_0 = -\frac{kT}{Mc^2}\omega_0 , \quad (10.12)$$

where k is the Boltzmann constant, M is the molecular mass, and v_r is averaged with the distribution of $W(v_r)$ determined by (10.11). In some studies (see, for example, [10.37]) no account is taken of the decrease of the velocity V of molecules that participate in the formation of the Lamb dip; due to their preferential motion at right angles to the light beam, the expression for $\Delta\omega$ shift differs from (10.12) by the factor 3/2.

For CH_4, the red shift of the Lamb-dip frequency should equal 0.52 Hz/grad. To observe such a frequency shift, the relative accuracy of tuning to the center of the dip should be of the order of 10^{-13}. Such an experiment has been done by BAGAYEV and CHEBOTAYEV [10.38]. They observed a temperature shift of the Lamb-dip center in CH_4 when they heated the absorbing cell of one of two stabilized lasers. They measured the frequency shift by observing the beat frequency. The average value of the beat frequency was measured accurate to several Hertz. Figure 10.6 shows the results of the experiment. Within the limits of experimental error, the observed dependence of frequency shift on temperature is linear; its slope is 0.5 ± 0.05 Hz/grad, which is in good agreement with the estimated value.

Recently SNYDER and HALL [10.39] have performed a more precise experiment to check (10.10) for the relativistic Doppler effect. Their experiment was

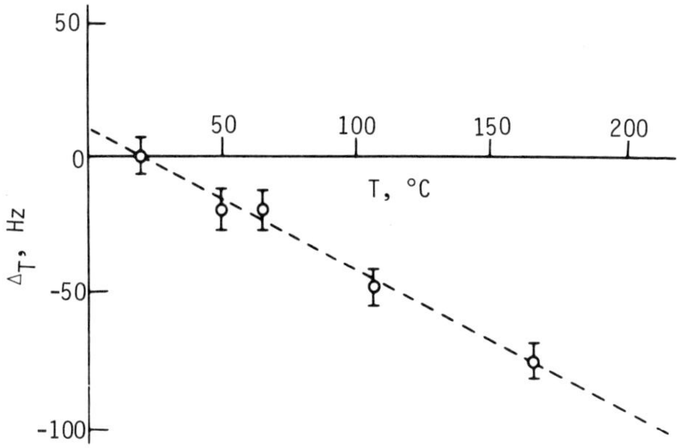

Fig.10.6 Temperature shift of frequency of the Lamb dip in CH_4 obtained by comparing the frequencies of two stabilized He-Ne/CH_4 lasers (BAGAYEV and CHEBOTAYEV [10.38])

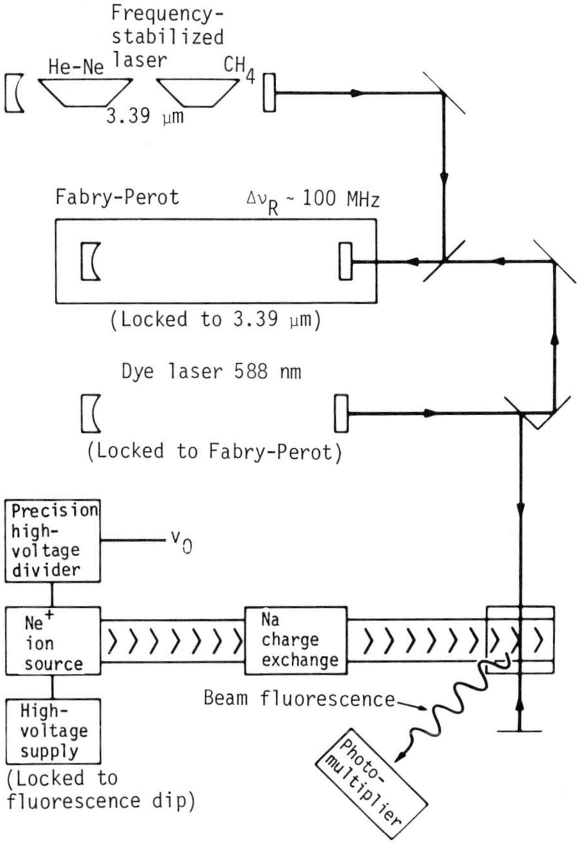

Fig.10.7 Schematic diagram of the experimental apparatus for measurement of the relativistic Doppler effect (SNYDER and HALL [10.39])

based on measuring the dependence of the quantum-transition frequency between two levels of fast Ne atoms on their velocity, with strictly transverse observation. The Lamb dip in the standing wave observed in fluorescence of excited atoms was used for precise measurement of the center of the transition line (see Sec.3.3). The scheme of this experiment is shown in Fig.10.7. The beam of metastable ($1s_5$) Ne atoms with velocity $10^{-3} \cdot c$ was formed by transfer of the charge of Ne^+ ions in a beam with an energy $\gtrsim 50$ keV in a chamber filled with sodium vapor. The transition $1s_5 \rightarrow 2p_2$ of Ne at 588 nm was excited by a standing wave 2 cm in diameter with limiting (diffraction) divergence of a frequency-stabilized dye laser with 0.1 W power. Nonlinear-absorption resonances at the transition $1s_5 - 2p_2$ were observed by measurement of fluorescence intensity at the transition $2p_2 \rightarrow 1s_2$ at 660 nm. The standing light wave was directed strictly perpendicular to the atomic beam of Ne, which eliminated the linear Doppler effect. By this method, resonances were produced whose natural widths were about 10 MHz. At the same time, a red shift in the nonlinear-resonance frequency at 1368 MHz occurred, owing to relativistic decrease of time for fast Ne atoms that had 50 keV kinetic energy. To achieve a sufficiently high accuracy of measurement of frequency shifts, the frequency of the dye laser was stabilized by a Fabry-Perot interferometer, the frequency of which, in its turn, was controlled by a He-Ne/CH_4 stabilized-frequency laser.

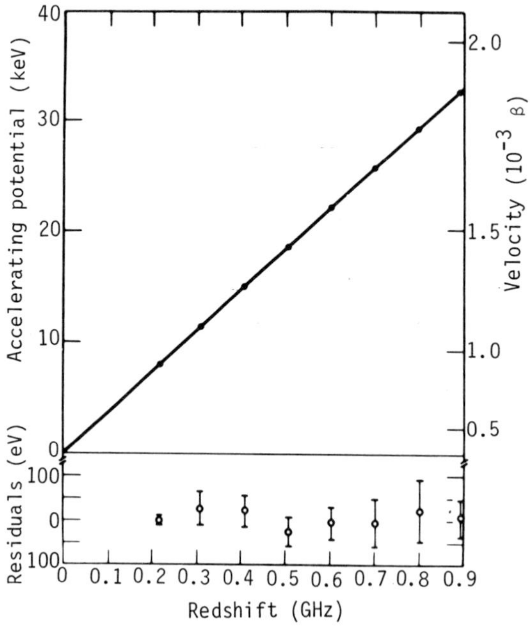

Fig.10.8 Experimental results of measurement of the relativistic Doppler effect. The solid line is a least-squares fit to the eight data points. The residuals, and one-sigma error bars are shown on an expanded scale in the lower part of the figure (SNYDER and HALL [10.39])

Figure 10.8 shows experimental data for the frequency shift, as a function of the energy or velocity of the Ne atoms. These data were compared with

$$\omega = \omega_0(1 - \beta^2)^\gamma + \omega_0\sqrt{\beta}\alpha \ , \tag{10.13}$$

where $\beta = V/c = (2eU/Mc^2)^{1/2}$, u is the accelerating potential, ω_0 is the transition frequency for a stationary atom, and α is the divergence of the running waves that form the standing wave, which produces a residual linear Doppler shift. The best agreement between (10.13) and the experimental results was obtained with $\gamma = 0.502 \pm 0.003$ and $\alpha = (2.5 \pm 10) \cdot 10^{-6}$ rad. So, the measurement accuracy for the parameter γ is 0.5%, which compares favorably with that of the best experiment reported previously, in [10.40]. The authors of [10.39] hope to increase the accuracy of the experiment by 30 times, which will make it possible to check the relativistic Doppler effect with unprecedented accuracy.

10.2.2 Recoil Effect

Because of the momentum of the optical photon $\hbar\omega/c$, the frequency of absorption ω_{ab} differs from that of emission ω_{em} by

$$\frac{\Delta\omega_{rec}}{\omega_0} = \frac{2\delta}{\omega_0} = \frac{\hbar\omega_0}{Mc^2} \ . \tag{10.14}$$

For example, for methane $\delta = 1.22 \cdot 10^{-11}\ \omega_0$; therefore, the recoil effect should occur, in principle, in experiments on narrow nonlinear resonances. As a consequence of the discrepancy between absorption and emission frequencies, the peak in the velocity distribution of particles at the upper level of the transition $n_2(v)$ does not agree with the hole in the velocity distribution at the lower level $n_1(v)$. Figure 10.9 shows distributions for this case, which differ from those (Fig.1.9b) for the case of a running wave. In the standing-wave field, there are two noncoincident peaks and holes in velocity distribution, owing to two running waves (Fig.10.9b). The structure of the nonlinear absorption spectrum has been studied in [10.41]; the recoil effect during photon emission or absorption was allowed for. The dependence of nonlinear absorption on the frequency of a standing-wave field at weak saturation has the form

$$\frac{\kappa(\omega)}{\kappa_0} = 1 - \frac{G}{2}\frac{\gamma_1}{\gamma_1 + \gamma_2}\left[1 + \frac{\Gamma^2}{\Gamma^2 + (\omega - \omega_0 - \delta)^2}\right] -$$

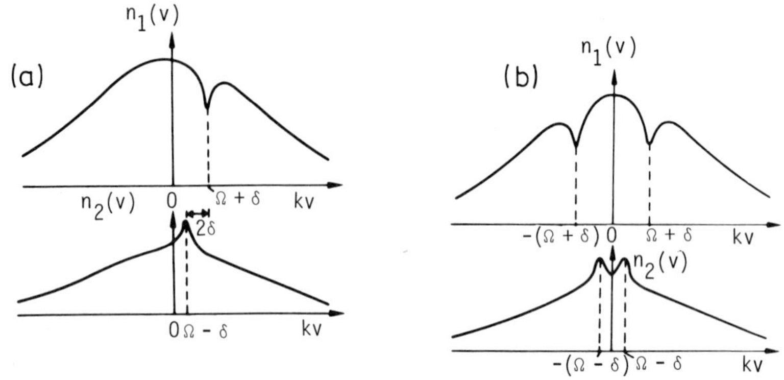

Fig.10.9a,b Velocity distributions of particles in the lower and upper levels of the transition, taking recoil into account: a) in a travelling-wave field, b) in a standing-wave field (KOL'CHENKO et al. [10.41])

$$- \frac{G}{2} \frac{\gamma_2}{\gamma_1 + \gamma_2} \left[1 + \frac{\Gamma^2}{\Gamma^2 + (\omega - \omega_0 + \delta)^2} \right] , \tag{10.15}$$

where $1/\gamma_i$ is the lifetime of a particle at the ith level. As seen, instead of one narrow resonance, two resonances with different amplitudes arise. The difference of the amplitudes of the narrow resonances is caused by the difference between the peaks and minima in the velocity distribution shown in Fig.10.9, that is, by the difference of the lifetimes of particles at the levels. Two narrow resonances are shifted about each other by the value 2δ, but, since this value is small as against the homogeneous width 2Γ, they almost always become coincident. Therefore, only one unsymmetrical resonance is observed; the shift of its frequency $\tilde{\omega}$ with respect to the center of the Doppler line depends on the relation between the relaxation constants,

$$\tilde{\omega} - \omega_0 = \frac{\gamma_2 - \gamma_1}{\gamma_2 + \gamma_1} \delta . \tag{10.16}$$

The level relaxation constants being equal, the recoil effect by no means causes any shift of the narrow resonance frequency.

The recoil effect is worthy of investigation, in view of its influence on the frequency reproducibility of lasers stabilized with molecular resonances. In these applications, we often have fields that are so strong that a weak-saturation approach breaks down. It is then of interest to know how the results of KOL'CHENKO et al. [10.41] are modified. The next term in the power series has been calculated by BAKLANOV [10.42], who displayed the shift of the Lamb dip as a function of the ratio between the relaxation rates of the two levels (γ_1/γ_2). Beyond this result, no analytical expressions are available; numerical calculations are necessary. STENHOLM [10.43] derives a set of equations for the density matrix, which includes exactly the effect of recoil. In general, the equations cannot be solved exactly. For one travelling wave of arbitrary amplitude, the solution can be found; it is then straightforward to develop a power series for an oppositely travelling weak field.

With recoil effect, the field no longer couples all levels of atoms that have the velocity v, but the lower level at v is coupled to $v + v_{rec}$ by the one running wave E_+ and to $v - v_{rec}$ by the other E_-. The recoil velocity is

$$v_{rec} = \frac{\hbar\omega}{cM} = c\,\frac{\Delta\omega_{rec}}{\omega_0} \, . \tag{10.17}$$

AMINOFF and STENHOLM [10.44] calculated the change of the shape of the inverted Lamb dip when one field is increased. The general result is that the recoil splitting is harder to resolve, owing to power broadening, but the asymmetry of the peak value caused by a difference of the relaxation rates ($\gamma_1 \neq \gamma_2$) is not much affected by the increased intensity.

No exact solutions have been obtained for the case with two strong fields. However, AMINOFF and STENHOLM [10.45] generalized the rate-equation approximation (REA) for the strong-signal case with the recoil effect. This approach (GREA) allowed them to consider cases in which both fields are too strong to allow a perturbation treatment. The calculations show that the general trends of the perturbation calculation are preserved. It has been found that REA overestimates the situation in the recoil-free case; here, it is also expected to exaggerate spectral features. When the recoil energy δ is of the order of the decay rates, $\delta \simeq \gamma$, the recoil-induced features are poorly resolved. It is then of interest to know when the splitting can be considered to be resolved and the line maxima to be shifted. AMINOFF's and STENHOLM's considerations [10.45] involve arbitrary criteria, but can be defined well enough to provide some idea of the resolution of the system.

Direct resolution of the recoil doublets, using saturated absorption techniques have been reported [10.46]. In these experiments, for the methane line at 3.39 μm the recoil splitting (10.14) is $1.4 \cdot 10^{-11}$ (2,163 kHz). The experimental installation with a telescopic beam expander inside a cavity was described in Chapter 6. At methane pressure of less than 20 μtorr we observed splitting of resonances on components of a hyperfine structure. Fig.

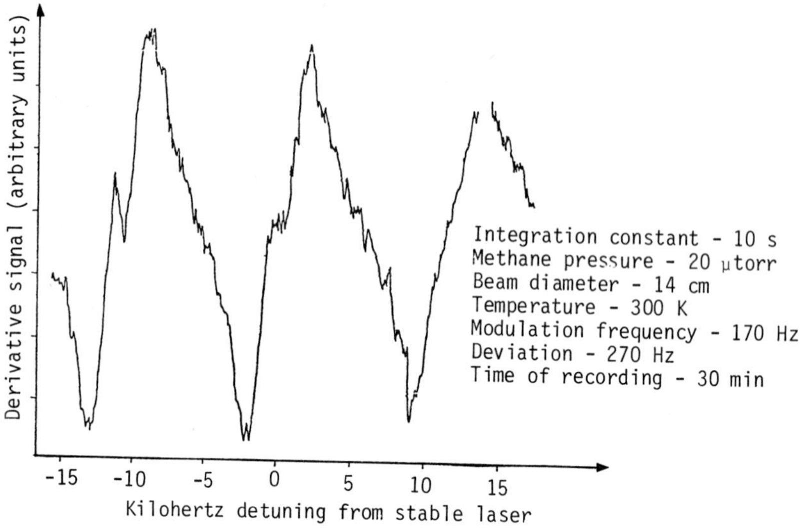

Fig.10.10 Record of the saturation resonance in methane on the $F_2^{(2)}$ line at 20 μtorr in the laser with a telescopic cavity

10.10 shows the record of resonance in a hyperfine structure of the $F_2^{(2)}$ line of methane. As can be seen in this figure, components of the hyperfine magnetic structure are of different widths. The width of the $6 \rightarrow 5$ component is the narrowest, and the resonance splitting on this width is more well-defined due to the recoil effect. Differences in the component widths are due to a non-compensated magnetic field of 0.5 Oe. An anomalous Zeeman effect occurs in weak magnetic fields and the splitting of components is therefore different. The $6 \rightarrow 5$ component suffers less Zeeman splitting and, hence, less broadening. Recently, direct observation of splitting of lines due to the recoil effect was reported in [10.70].

10.2.3 Mass-Energy Relation

The sensitivity of narrow resonances to negligible shifts of quantum-transition frequencies may form the basis for quite a new method for measuring the energy of excited metastable nuclei (isomeric nuclei) suggested by LETOKHOV [10.47]. The excitation of an atomic nucleus in a molecule must cause changes of the vibrational frequencies of the molecule, because, as the nuclear energy is increased by ΔE, the nuclear mass increases by

$$\Delta m = \frac{1}{c^2} \Delta E . \tag{10.18}$$

In a diatomic molecule AB, the excitation of the nucleus of atom A shifts the vibrational frequency by

$$\frac{\Delta\nu_{is}}{\nu_o} = - \frac{\Delta E}{2c^2} \frac{M_B}{M_A(M_A + M_B)} , \tag{10.19}$$

where M_A and M_B denote the masses of the unexcited nuclei, $M_A^* = M_A + \Delta m$ is the mass of the atom A with the excited nucleus. When the lighter atom ($M_A << M_B$) is excited, the isomeric shift reaches its maximum,

$$\frac{\Delta\nu_{is}^{max}}{\nu_o} = - \frac{\Delta E}{2M_A c^2} = 5.56 \cdot 10^{-4} \frac{\Delta E}{A} , \tag{10.20}$$

where A denotes the atomic weight of atom A, ΔE is the energy of nuclear excitation (MeV). With $\Delta E \simeq 0.3$ MeV for an atom with $A = 10^2$, the isomeric shift of the vibrational spectrum $\Delta\nu_{is}^{max}/\nu_o \simeq 1.67 \cdot 10^{-6}$, that is, of the same order as the Doppler width of the rotational-vibrational molecular line in a gas at temperature T, $\Delta\nu_D/\nu_o = 0.72 \cdot 10^{-6}(T/A)^{1/2}$. If the radioactive atom A is heavier than B, the shift decreases by approximately the factor $M_B/(M_A + M_B)$.

Nonlinear spectroscopy is the most effective method to detect isomeric shifts in the vibrational spectrum of a molecule. In this case, the resolution depends on the finite time of molecular transit through the light beam (Chap.6), that is, on

$$\frac{\delta\nu}{\nu} \simeq \frac{\lambda_o}{a} \frac{u}{c} , \tag{10.21}$$

where λ_o is the wavelength of vibrational transition, a is the light beam diameter, u is the mean molecular velocity. The minimum-detectable isomeric shift $\Delta\nu_{isom} = \delta\nu$, and thus the minimum-possible value of nuclear excitation energy ΔE_{min}, according to (10.19) and (10.21), is

$$\Delta E_{min} = 2 \frac{\lambda_o}{a} \frac{M_A}{M_B} (3kTM_{AB}c^2)^{1/2} . \tag{10.22}$$

For $M_A \simeq M_B \simeq 10^2$ at.un. (atomic units of mass), $T = 300$ K, $\lambda_o = 20\text{-}30$ µm, $a \simeq 1$ cm, $\Delta E_{min} = 0.5\text{-}0.7$ keV, that is, much smaller than characteristic energies of nuclear excitation. The accuracy of measurement of energy by this method is much better than the value of ΔE_{min}, because the distance between narrow resonances can be measured more accurately than their widths, at least up to 1% of resonance width. Thus, it is hoped that the energy of nuclear excitation may be measured accurate to $\delta E \cong 10^{-2}\ \Delta E_{min}$. For the above numerical example $\delta E \simeq 5$-7 eV, which, with the typical energy of nuclear excitation $\Delta E = 0.5$-0.7 MeV, corresponds to the relative accuracy $\delta E/\Delta E = 10^{-5}$. This exceeds the accuracy of measurement of energy of nuclei that emit γ rays, obtained by the use of the best precision γ spectrometers [10.48]. The method described can be applied also to nonradiatively decayed levels whose energy is evaluated by conventional techniques to no better than 10^{-2} [10.48].

In fact, these simplified estimates are too optimistic. Owing to nuclear excitation, the hyperfine structure of the spectrum changes, because the isomeric nuclear states differ from the ground state in nuclear spin. The potentialities of such an approach are estimated more realistically in [10.49] for the particular case of ^{189}Os nuclei in the OsO_4 molecule. For the ^{189m}Os isomer with energy $\Delta E = 30.8$ keV, the nuclear spin $I_f = 9/2$; that in the ground state is $I_g = 3/2$. The infrared absorption spectrum of $^{189m}OsO_4$ molecule consists of hyperfine components of vibrational transitions that are affected by nuclear quadrupolar interaction; the changes of the hyperfine structure are caused by variations of the same order of magnitude as quadrupolar splitting in the spectrum of $^{189}OsO_4$ with an unexcited nucleus. It is evident that the contribution made by nuclear mass variation can be determined only after measurement and identification of the whole hyperfine structure of the rotational-vibrational transition with given quantum numbers (v, J, K), for molecules with excited and unexcited nuclei.

There is one more difficulty in the case of polyatomic molecules, because the isomeric mass shift (like the isotope shift) is distributed over several normal vibrations; therefore, there is no simple relation (10.19) between the excitation energy ΔE and the isomeric frequency shift. In addition, it is also necessary to know the distribution of isotope shift over normal vibrations. For example, on the basis of Teller-Redlich product rule for isotopic frequencies, and known force constants of the OsO_4 molecule, it has been found that the isotope shift per 1 at. un. of mass for the vibration ν_3 is equal to 0.26 cm^{-1}. The corresponding isomeric frequency shift of the transition $v(\nu_3) = 0 \rightarrow v(\nu_3) = 1$ $(J, \Delta J = 0)$ in the $^{189m}OsO_4$ molecule is $\Delta\nu_3^{isom} = \Delta m \Delta\nu_3^{isot} = 264$ kHz (Δm is expressed here in atomic units of mass).

Thus, it is quite possible to detect the isomeric shift and isomeric structure of vibrational-rotational transitions of $^{189m}OsO_4$ molecules by the methods of nonlinear laser spectroscopy but it seems impossible to relate them to the excitation energy of ^{189m}Os nucleus with an accuracy better than 0.1%. This is also true for other polyatomic molecules, which are characterized by intervibrational coupling and by radiation absorption at transitions with large angular momentum J. To measure accurately the energy of excited metastable nuclear states, it is advisable to use tunable lasers for nonlinear spectroscopy of diatomic molecules whose isomeric nuclei have high excitation energy ($\Delta E \simeq 1$ MeV).

It should be noted that experiments on measuring the saturated-absorption spectrum of molecules with radioactive nuclei will be, perhaps, difficult because of high radioactivity of a cell with an absorbing gas. For example, for $^{189m}OsO_4$ molecules (decay period $T_{1/2} = 5.7$ hrs) at a pressure of 10^{-4} torr, the cell being 10 cm in diameter and 250 cm long, the gas radioactivity is about 10^2 curie. To do such experiments, it is necessary that new methods should be developed to detect narrow nonlinear resonances inside the Doppler profile when the number of molecules in the absorbing cell is very small. Detection of narrow density resonance of excited molecules, by selective photoionization of excited molecules, seems to be the most promising method.

10.2.4 Splitting of Energy Levels of Left- and Right-Hand Molecules

There is another very fine effect in molecular spectra, which lies beyond the scope of present experiments but in the future may be detected by methods of nonlinear laser spectroscopy. Two molecules that are mirror images of each other have slightly different energy levels because of parity violation due to weak interactions between the electrons and nucleons in a molecule, as predicted by LETOKHOV [10.50]. In other words, due to a very small admixture of the P-odd potential of the ep interaction, the energies of electronic, vibrational and rotational states for two molecules, which are mirror images of each other (L and D molecules), differ from one another by a negligible value. This energy difference has been estimated in [10.50].

The odd to inversion (P-odd) part of the potential of the ep interaction makes a small odd addition to the energy of coulomb interaction between the electrons and the nucleus. In order of magnitude, this addition is

$$\langle V_{ep} \rangle \simeq K_{ep} G \left(\frac{\lambda_p}{a_0}\right)^2 V_{coul} \sim 10^{-16} K_{ep} V_{coul} \quad , \qquad (10.23)$$

where G is the Fermi constant, K_{ep} characterizes the relative force of interaction between neutral and charged currents, λ_p is the Compton wavelength of the proton, a_0 is the Bohr radius, V_{coul} is the energy of coulomb interaction; the simplest case of interaction of one electron with one proton will be considered here for a basis of evaluation. This addition splits the electronic level of the E_{el} energy of left and right molecules by a negligible value [10.50],

$$\Delta E_{el} \simeq \langle V_{ep} \rangle \simeq 10^{-16} K_{ep} E_{el} , \tag{10.24}$$

where $V_{coul} \simeq E_{el}$ is the molecular electronic energy. In other words, the P-odd disturbance caused by the ep interaction removes the degeneracy of the energy levels of left- and right-hand molecules, so that the energies of the levels become different by the value shown in (10.24). The vibrational and rotational levels of energies E_{vib} and E_{rot} differ, respectively, by

$$\Delta E_{vib} \simeq \left(\frac{m}{M}\right)^{1/2} \Delta E_{el} \simeq 10^{-16} K_{ep} E_{vib} , \tag{10.25}$$

$$\Delta E_{rot} \simeq \left(\frac{m}{M}\right) \Delta E_{el} \simeq 10^{-16} K_{ep} E_{rot} , \tag{10.26}$$

where M is the molecular mass. The relative value of level splitting is about 10^{-16}.

Because the atomic nuclei usually contain neutrons, the P-odd en interaction, i.e., weak charged currents, may also be important for the effect under consideration. In this case, the total effect of weak currents will depend on the total action of neutral and charged currents. In order to estimate the total effect, the relative values of the constants of the ep and en interactions need to be known.

Although the splitting is 10^5 times the smallest that can be resolved by the most advanced methods of nonlinear laser spectroscopy, the progress in laser-spectroscopy methods which are discussed in this book enables us to measure directly the difference between the frequencies of left- and right-hand molecules. The frequencies of two lasers with extremely monochromatic radiation can be stabilized on the centers of the absorption lines of L- and D-type molecules, respectively. The frequency difference of the absorption-line centers of the L and D forms results in a difference between the frequencies of the lasers, which can be detected by measurement of their frequency difference. By use of low-pressure cells that have narrow nonlinear resonances inside the Doppler absorption line (of the inverted-Lamb-dip type) it is now possible to stabilize laser frequencies with absorption-line centers of rotational-vibrational molecular transitions with an accuracy not worse than 10^{-13}. Studies are in progress to improve the accuracy of stabilization of laser frequency to from 10^{-15} to 10^{-16}, which is of the order of magnitude of the estimate in (10.25). For these experiments, a methane molecule is used in which the hydrogen atoms are replaced by various haolgen atoms. These molecules can exist in two different, mirror-symmetric forms (i.e., two optical isomeric states) with a very long time of mutual transformation. The first step of this experiment (saturation spectroscopy of a CHFClBr racematic gas mixture by use of CO_2 cw lasers at 10.6 μm) has been reported in [10.51]. The next steps will include measurement of nuclear hyperfine structure of some rotational-vibrational bands with a resolution of about 10^{11}, separation of left- and right-hand forms of CHFClBr monoisotopic molecules and frequency stabilization of two CO_2 lasers with two independent absorption cells filled with left- and right-hand molecules of CHFClBr.

10.3 Laser γ Spectroscopy Without Doppler Broadening

The effect of a coherent light wave to excite the atoms or molecules in a gas that have a definite velocity has extensive applications, not only in optical spectroscopy. The ideas of nonlinear laser spectroscopy, in particular, may be applied to the region of nuclear γ radiation. The resolution of nuclear spectroscopy in a gas phase, which is usually limited by the Doppler effect can be increased substantially in a similar way. Some effects of laser nuclear spectroscopy, considered first in papers by LETOKHOV [10.52-55] are based on the relation between nuclear transitions and atomic electron transitions, if the nucleus is surrounded by electrons, and molecular vibrational transitions if the nucleus is bound in a molecule.

10.3.1 Connection Between Optical and Nuclear Transitions

At least two effects enable us to combine atomic and molecular transitions with nuclear transitions [10.55]. They are the recoil effect and the Doppler effect.

For a bare nucleus, the lines of nuclear emission and absorption are shifted with respect to each other by the recoil energy

$$R = E_0^2/2Mc^2 , \qquad (10.27)$$

where M is the nuclear mass and $E_0 \ll Mc^2$. The emission- or absorption-frequency shift is caused by changes of nuclear translational state when a γ quantum is emitted or absorbed, owing to the recoil effect. If the nucleus is located in an atom or in a molecule, the law of conservation of momentum and angular momentum governs not only the change of the translational state, for instance, of the molecule but also any change of its internal (electronic, vibrational, or rotational) state.

Laws of conservation of momentum and energy for the system--nucleus in atom of molecule + γ quantum--have the nonrelativistic form

$$M\underline{v}_0 \pm \hbar k_\gamma = M\underline{v} , \quad \pm\hbar\omega_\gamma + E_i + \frac{1}{2} Mv_0^2 = E_f \pm E_0 + \frac{1}{2} Mv^2 , \tag{10.28}$$

where $\underline{v}_0$ $\underline{v}$ denote the initial and final velocities of translational motion of a particle (an atom or a molecule), E_0 is the energy of the nuclear transition under consideration, and E_i and E_f are the initial and final internal energies of the particle; the signs <<+>> and <<->> correspond to absorption or emission of a γ quantum, respectively. It follows from (10.28) that the energy of the emitted or absorbed γ radiation is determined by

$$\hbar\omega_\gamma^\pm = E_0 \pm R + \hbar \underline{k}_\gamma \underline{v}_0 \pm (E_f - E_i) , \tag{10.29}$$

where the first term corresponds to the nonshifted transition; the second term gives recoil shifts due to change of particle translational state; the third gives frequency shift of emission and absorption lines due to the Doppler effect, and the last term gives line shifts caused by changes of atomic or molecular internal states.

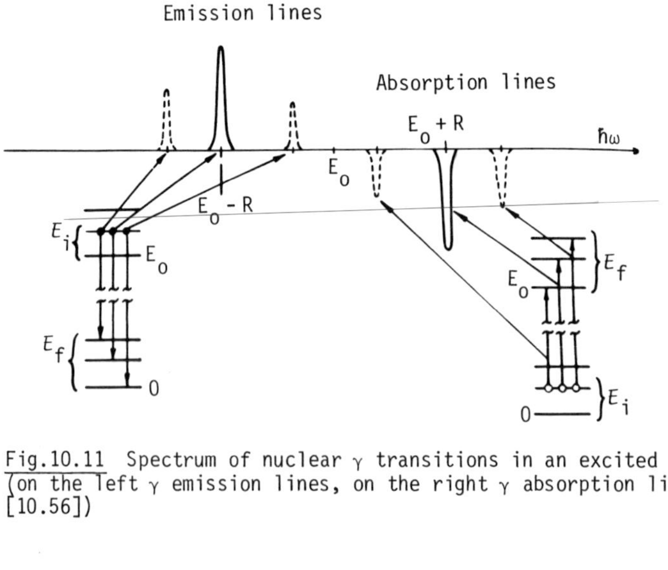

Fig.10.11 Spectrum of nuclear γ transitions in an excited atom or molecule (on the left γ emission lines, on the right γ absorption lines) (LETOKHOV [10.56])

Figure 10.11 shows the spectrum of γ transitions in absorption or emission for the nucleus in an initially excited atom or molecule ($E_i > 0$). During γ quantum emission a small part of the nuclear excitation energy may be transferred to the internal state of particle ($E_f > E_i$), in which case a γ satellite appears that is shifted towards the red from the γ-emission line (energy E_0 - R) from a particle in which the internal state remains unchanged. In a like manner, some of the particle-excitation energy, together with the nuclear excitation energy may be transferred to the γ quantum; in such case, a satellite appears that is shifted towards the blue from the line $E_0 - R$. Analogous shifts may also occur during γ quantum absorption.

For the nucleus in an atom, satellites of the γ line arise from electronic-nuclear transitions. In the case of a nucleus in a molecule, changes may occur in the electronic, vibrational or rotational energies of the molecule; for this reason, electronic-vibrational-rotational-nuclear transitions may occur. Of course, the intensities of such satellites depend on the probability of such composite transitions for the system--nucleus in an atom or nucleus in a molecule.

By changing the populations of atomic or molecular excited states by laser radiation we can, first, control the intensities of composite γ transitions and, second, induce new γ transitions that are shifted to longer wavelengths than the normal γ-absorption line $E_0 + R$ or shifted to shorter wavelengths than the γ-emission line $E_0 - R$ (Fig.10.11).

The frequency of nuclear γ transition is shifted by the value $\underline{k}_\gamma \underline{v}_0$ by the Doppler effect. If the distribution of nuclear velocities, that is of atomic and molecular velocities, is thermal (equilibrium), the term $\underline{k}_\gamma \underline{v}_0$ in (10.29) gives the Doppler broadening of the γ lines. By laser radiation we can excite atoms or molecules that have a specific projected velocity on the chosen direction (the direction of propagation of the laser wave); that is, we can change the velocity distributions of particles at the levels connected by the laser field (Fig.10.12a). For example, it is possible to have excited atoms (or molecules) with the velocity $\underline{v}_{res}$ determined by the optical-resonance condition,

$$\underline{k}_0 \underline{v}_{res} = \omega - \omega_0 , \qquad (10.30)$$

where $\underline{k}_0$ is the laser-wave vector, ω is the laser-field frequency, $E_i = \hbar\omega_0$ is the atomic (or molecular) transition energy. The spectral line of the composite γ transition, in which atoms (or molecules) with a nonequilibrium

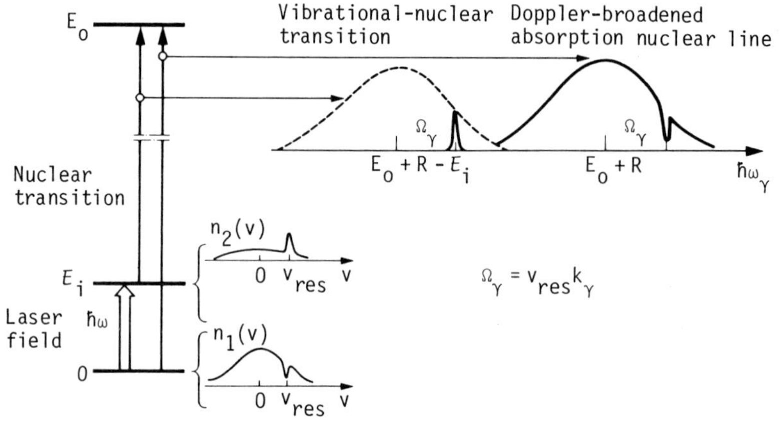

Fig.10.12 Formation of narrow resonances of γ absorption when an atom or a molecule is excited by a coherent light wave in a low-pressure gas (LETOKHOV [10.56])

velocity distribution participate, will consequently have a narrow resonance peak (Fig.10.12b) rather than an ordinary Doppler profile. The frequency of this peak is shifted from the center of the line $(E_0 + R - E_i)$ by

$$\Omega_\gamma = \underline{k}_\gamma \underline{v}_{res} = \frac{\omega - \omega_0}{\omega_0} \omega_\gamma \quad . \tag{10.31}$$

The peak can be tuned within the Doppler contour of γ line by tuning the laser-field frequency along the Doppler line of the optical-transition absorption. The absolute value of the γ-resonance detuning range of the ratio ω_γ/ω_0 is larger than the frequency-detuning range of the optical absorption. Therefore, by fine scanning the light-wave frequency in the range $\Delta\omega/\omega_0 \simeq 10^{-5}$ the frequency of γ resonances may be detuned by values of the order of 0.1 + 1 eV.

The idea to obtain narrow tunable γ resonances of absorption and emission was proposed in 1972 in [10.52,53]. The occurrence of vibrational satellites of nuclear γ transitions in a molecule was also considered in the simplest classical model in those papers. It is evident that γ lines free of Doppler broadening can be obtained, not only at the frequencies of composite γ transitions, but also in any case in which the nuclear-velocity distribution is changed in some way by laser radiation [10.54]. Therefore, both methods of changing the γ-transition spectrum by laser radiation (induction of satellites and narrow resonances in the Doppler profile of γ lines) may be used both together and separately.

Let us consider now specific quantum systems (an atom, a diatomic molecule, a polyatomic molecule, a positronium) where these ideas can be realized, and calculate for them the probabilities of such composite quantum transitions.

10.3.2 Electronic-Nuclear Transitions in Atoms

The possibility of electronic-nuclear γ transitions for the nucleus in an atom and their intensities are considered in a simple model in [10.57]; a rigorous calculation, with the same results, is given in [10.58]. The cause of electronic-nuclear transitions is that the center of inertia of the nucleus does not coincide with that of the whole atom; therefore, the nuclear recoil affects the electron motion, and vice versa.

The optical-electron coordinate $\underline{r}$ is related to the coordinate of the center of mass of the nucleus $\underline{R}$ by

$$\underline{R} + \frac{m}{M}\underline{r} = 0 \ , \tag{10.32}$$

where the origins of the coordinates are at the center of mass of the nucleus, and m and M are the masses of the electron and atom, respectively. The probability of the γ transition $a \to b$ with the change in optical electron quantum state $i \to f$ is given by

$$W_{fi}^{ba} = A_{ba}\left|\langle\psi_f^*(\underline{r})\right| e^{-i\underline{k}_\gamma \underline{R}} \left|\psi_i(\underline{r})\rangle\right|^2 = A_{ba}P_{fi} \ , \tag{10.33}$$

where A_{ba} is the probability of the γ transition between two levels of the bare nucleus, $\underline{k}_\gamma$ is the wave vector of γ quantum, $\psi_{i,f}(\underline{r})$ denotes the wave functions of the electron state, and the coordinates $\underline{r}$ and $\underline{R}$ are connected by (10.32). The vibration amplitude of the center of mass of the nucleus in an atom is much smaller than $\lambda\!\!\!^{-}_\gamma$, that is $\underline{k}_\gamma\underline{R} \ll 1$, and the expression for the transition probability P_{fi} reduces to [10.57],

$$P_{fi} = \left(k_\gamma \frac{m}{M}\right)^2 \left|\langle\psi_f^*(\underline{r})|\underline{n}_\gamma \underline{r}|\psi_i(\underline{r})\rangle\right|^2 = \left(k_\gamma \frac{m}{M}\right)^2 (\underline{r}_{if}\underline{n}_\gamma)^2 \ , \tag{10.34}$$

where $i \neq f$, $\underline{n}_\gamma$ is the unit vector in the $\underline{k}_\gamma$ direction, and $e\underline{r}_{if}$ is the matrix element of dipole moment of the transition $i \to f$. The probability of conservation of the initial atomic state is $P_{ii} \simeq 1$. Owing to the condition $\underline{k}_\gamma\underline{R} \ll 1$, the probabilities for $i \neq f$ will be $P_{fi} \ll 1$.

The γ-transition frequency with a change of electron state is determined by (10.29). If, before emission of a γ quantum the atom is in the ground state ($E_i = 0$), all electronic satellites are located on the long-wave side of the basic emission line $E_0 - R$; they are spaced at intervals equal to the excitation energy of the corresponding states of the electron shell. The intensities of different satellites, according to (10.34), are proportional to the squares of the electron-transition dipole moment and, hence, decrease in proportion to n_f, where n_f denotes the principal quantum number of the final state. The ratio of the intensities of the satellite and the basic line can be estimated from

$$K_{on} = E_{on}(\text{eV})[f_{on}E_\gamma(\text{MeV})/A]^2 , \qquad (10.35)$$

where E_{on} is the energy (eV) of the electronic transition $o \to n$, f_{on} is the oscillator strength of the transition $o \to n$, A is the atomic mass (in atomic units). For instance, for a nuclear transition in the ^{21}Ne isotope, with energy $E_\gamma = 6$ MeV, the intensity of the satellite that corresponds to the excitation of the resonance level Ne I with $E_{01} = 16.7$ eV, will be $K_{01} \simeq 5 \cdot 10^{-3}$. Atomic excitation can be detected in γ emission by the subsequent fluorescence. The relative intensities of electronic satellites are low because of the weak connection between electron motion and nuclear motion during recoil.

10.3.3 Molecular-Nuclear Transitions

It is clear from the simplest considerations that, when the nucleus in a molecule absorbs or emits a γ quantum, it sustains a shock that excites molecular vibrations, in which the nucleus takes part. If the emitting or absorbing nucleus n is off the center of mass of the molecule, such a shock inevitably gives rise to molecular vibrations. The probability of changes of the vibrational or rotational molecular states, unlike changes of the atomic electronic state, is not small and should be taken into account even in a zero-order approximation. Vibrational-nuclear transitions in a polyatomic molecule, involving a nucleus that is at the center of mass of the molecule (at the nuclear coordinate $\underline{r} = 0$ about the center of mass of the molecule), in which case the molecular rotational state remains constant, are studied in detail in [10.59,60]. Calculations for symmetrical polyatomic molecules of various symmetry groups are presented in [10.61]. Here only some of the results are given.

Polyatomic Symmetrical Molecule

The probability of vibrational-nuclear transition is determined by an expression like (10.33), where ψ_{if} designates wave functions of molecular vibrational states that depend on the coordinate $\underline{R}$ of the nuclear center of inertia. The nuclear center of inertia $\underline{R}$ in a molecule is positioned so that

$$\underline{R} = \underline{R}_o + \underline{r} + \underline{u} , \tag{10.36}$$

where $\underline{R}_o$ is the position of the center of mass of the molecule, $\underline{r}$ is the vector that connects the center of mass of the molecule with the equilibrium position of the nucleus, and $\underline{u}$ is the vibrational displacement of the center of mass of the nucleus about the equilibrium position. With $\underline{r} = 0$, the matrix element that gives the change of molecular state reduces to

$$<\psi_f^*(\underline{R})|e^{-i\underline{k}_\gamma\underline{R}}|\psi_i(\underline{R})> = <\psi_f^*(\underline{u})|e^{-i\underline{k}_\gamma\underline{u}}|\psi_i(\underline{u})> , \tag{10.37}$$

where the wave function $\psi(\underline{u})$ gives the vibrational displacement of the nucleus n. The nuclear vibrational displacement can be given as a combination of shifts due to different normal molecular vibrations,

$$\underline{u}_n = \sum_l Q_l \underline{\rho}_{ln} , \tag{10.38}$$

where Q_l is the normal <<l>>-th molecular coordinate, and $\underline{\rho}_{ln}$ is the shift of the nth nucleus due to the lth normal vibration.

In a harmonic approximation, it is easy to deduce the expression for the probability of the γ transition that is followed by the vibrational transition $0 \to v_l$ of an initially unexcited molecule [10.59,60],

$$P_{v_l,o} = \frac{1}{v_l!} Z_l^{v_l} e^{-Z_l} , \tag{10.39}$$

as well as for the vibrational transition $1 \to v_l$ of a molecule initially excited to the first vibrational level,

$$P_{v_l,1} = \frac{1}{v_l!} Z_l^{v_l-1} (Z_l - v_l)^2 e^{-Z_l} . \tag{10.40}$$

Parameter Z_l is the characteristic parameter that defines the average number of vibrational quanta accepted by the molecule,

$$Z_1 = \frac{R}{\hbar\omega_1}\,\frac{M - M_x}{M_x}\ , \tag{10.41}$$

where M_x is the mass of the γ-emitting (absorbing) nucleus, and M is the total molecular mass; the x nucleus is at the center of mass of the molecule.

When $Z_1 \gg 1$, instead of single emission and absorption lines of bare-nucleus γ radiation, a great number of vibrational-nuclear satellites arise at the frequencies $E_0 \pm R$, which can consume all of the energy of the base line. For nuclei initially in the vibrationally excited state ($E_i = \hbar\omega_1$) vibrational satellites arise with $(E_0 + R - \hbar\omega_1)$ and $(E_0 - R + \hbar\omega_1)$ energies and correspond to both emission and absorption lines. Thus, vibrational excitation of molecules with γ-emitting (absorbing) nuclei may become a new method to compensate for the shift due to recoil of emission and absorption lines of nuclear γ radiation.

In a low-pressure gas, coherent light waves excite molecules of velocities $\underline{v}$ that satisfy condition (10.30). As a result, the molecular velocity distributions of both levels of the vibrational-rotational transition under consideration can depart significantly from equilibrium. The absorption line shape of one vibrational satellite of the gamma transition at frequency $\omega_\gamma^{(v1)}$, caused by nuclei in molecules in the mth level (m = 1,2), is given by

$$\phi_m(\omega_\gamma) = \int \sigma\left[\omega_\gamma - \omega_\gamma^{(v_1)} - \underline{k}_\gamma\underline{v}\right] n_m(\underline{v})d(\underline{v}\underline{n}_\gamma)\ . \tag{10.42}$$

Here $\underline{n}_\gamma$ is a unit vector in the direction of propagation of the gamma quantum, n_m are the velocity distributions of the population of the m levels that are distorted by the laser wave, shown on Fig.10.12, and $\sigma(x)$ is the cross-section for resonant absorption of a gamma quantum,

$$\sigma(x) = \sigma_0\,\frac{\Gamma_\gamma^2}{x^2 + \Gamma_\gamma^2} \tag{10.43}$$

with σ_0 the peak value of the absorption cross-section and Γ_γ the natural line width of the nuclear gamma transition.

The typical shape of the spectra $\phi_1(\omega_\gamma)$ and $\phi_2(\omega_\gamma)$ is shown in Fig.10.13 for $\underline{n}_\gamma = \underline{n}$; the distributions of molecular velocities $n_1(v_z)$ and $n_2(v_z)$ in the ground and excited levels are also shown. The halfwidth Γ_{nucl} of narrow dips and peaks is

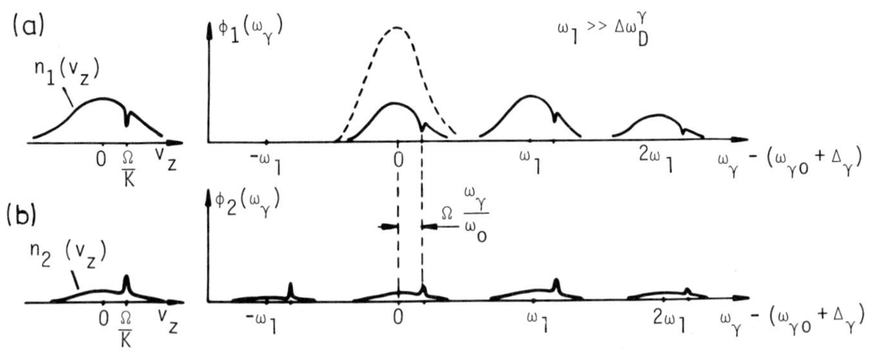

Fig.10.13a,b Spectrum of nuclear γ transitions for molecules in (a) unexcited and (b) excited molecules. The distribution of v_z, the component of molecular velocity in the propagation direction of the light wave (which coincides with the direction of γ-radiation observation), is shown at the left (LETOKHOV [10.60])

$$\Gamma_{nucl} \simeq \Gamma \frac{\omega_\gamma}{\omega_0} >> \Gamma_\gamma \ , \tag{10.44}$$

where Γ is the homogeneous half width of the vibrational transition.

Reference [10.61] contains the calculations of the Z_1 parameters for polyatomic symmetrical molecules of various symmetry groups, which are convenient for estimates.

Diatomic Molecule

When a nucleus in a diatomic molecule emits or absorbs a γ quantum simultaneous changes of the electronic, vibrational and rotational states may occur. Unlike polyatomic molecules, each of which has a center of symmetry, a change of rotational state is inevitable in this case. The full calculation for electronic-vibrational-rotational-nuclear transitions in a diatomic molecule has been published in [10.62]. As it shows, the probability of electron-state change is determined by an expression similar to (10.34) for the atom; that probability is very small.

The probability of change of vibrational state is given by the same expression as for one normal vibration of a polyatomic molecule. The expression for probability of the vibrational transition $v_i \to v_f$ takes the form [10.62]

$$P_{v_f,v_i} = \frac{v_i!}{v_f!} e^{-Z} Z^{v_f - v_i} \left[L_{v_i}^{v_f - v_i}(Z) \right]^2 \ , \tag{10.45}$$

where $Z = Z_0 \cos^2 v$, $Z_0 = (R/\hbar\omega_0)(m_2/m_1)$, v is the angle between $\underline{k}_\gamma$ and the molecular axis, $\hbar\omega_0$ is the vibration frequency, m_1 and m_2 denote the atomic masses, and L is the Laguerre polynomial. With $v_i = 0,1$ (10.45) reduces to (10.39) and (10.40), respectively.

The probability of change of rotational state in a rigid-rotator approximation can also be calculated exactly. In the case of the transition $J_i = 0 \to J_f = J$, the expression for the transition probability takes the form

$$\overline{P_{J,0}} = \frac{\hbar}{2a}(2J+1)J_{J+1/2}(a) \ , \qquad (10.46)$$

where $a = k_\gamma r_0 m_2/(m_1 + m_2)$, r_0 is the internuclear spacing, m_1 is the mass of the nucleus that interacts with the γ radiation, J is the Bessel function, the overbar indicates orientation averaging. The parameter a is equal to the average value of the angular momentum (in units of $\hbar$) transferred to the molecule.

The vibrational and rotational degrees of freedom of a diatomic molecule give up the same energy, which may be either more (with $m_1 \ll m_2$) or less (with $m_1 \gg m_2$) than the recoil energy R. It is apparent that, when the nuclear masses differ greatly and γ radiation interacts with the lighter nucleus, vibrational-rotational satellites will shift absorption (emission) lines distances larger than correspond to the energy R of recoil to change of molecular translational motion.

10.3.4 Double γ and Optical Resonance

Narrow resonances inside the Doppler-broadened line of transition absorption can be also produced by selective removal from a gas of molecules that have a given velocity, that is by a change of the velocity distribution of the total number of particles per unit volume

$$n(\underline{v}) = \sum_i n_i(\underline{v}) \ , \qquad (10.47)$$

where $n_i(\underline{v})$ is the velocity distribution of particles at the ith level. This technique has been suggested to produce narrow resonances at nuclear transitions by use of a coherent light field that acts upon the optical transition of the atom [10.54]; actually, it can be applied to any Doppler-broadened transitions (optical or nuclear). The technique is based on selective excitation of particles that have a particular value of v_z and removal of the excited particles out of the region of the interaction field.

Let a coherent light wave be in resonance with the electron transition of an atom or molecule. At strong saturation, a light wave excites about half of the particles at the low level, whose velocities comply with the resonance condition $kv_z = \omega_1 - \omega_0$, where ω_1 is the running-wave frequency, ω_0 is the center frequency of the transition. During such excitation, only the velocity distributions of particles at each level $n_1(v)$ and $n_2(v)$ vary so that the total distribution $n_1(v) + n_2(v)$ remains constant. This corresponds to the fact that in Fig.10.13 the depth of the hole in the distribution $n_1(v)$ equals the height of the peak in $n_2(v)$.

For a resonant hole to be formed in the total velocity distribution, the excited particles should be removed from the region of their interaction with the field.

Photoionization of Excited Atoms or Molecules

The region of low-pressure gas under excitation may be irradiated by an additional laser beam, the frequency of which ω_2 is only sufficient to photoionize excited atoms or molecules, but not unexcited ones (Fig.10.14a).

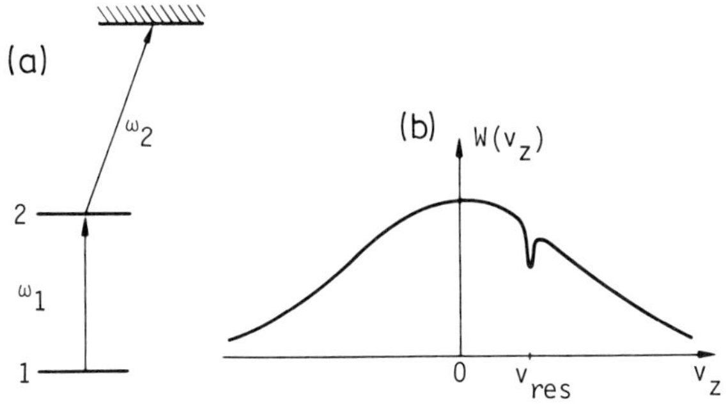

Fig.10.14 (a) Selective excitation of atoms with a specific velocity component v_{res}, followed by photoionization of the excited atoms, giving rise to (b) a hole in the velocity distribution of atoms on all levels

When the additional beam intensity $P_2 \simeq (\sigma_{exc}/\sigma_{ph})P_1$ (σ_{exc} is the particle-excitation cross-section, σ_{ph} is the excited-particle photoionization cross-section, and P_1 is the power of saturation of the electron transition), the photoionization probability of an excited particle throughout its lifetime at the excited level will be of the order of unity. The ions formed in a low-pressure gas ($\lesssim 10^{-2}$ torr) can be easily removed from the laser-beam area by use of a weak electric field. This will result in a resonant dip, for the velocities $kv_{res} = (\omega_1 - \omega_0)$, in the distribution of projected velocities of nuclei, nuclei in atoms or molecules (Fig.10.14b).

Photodissociation of Excited Molecules

An analogous effect can be attained by photodissociation of excited molecules with additional laser radiation, the frequency of which is only sufficient for photodissociation of excited molecules, but not for unexcited ones. During photodissociation, the molecular fragments are scattered at velocities much greater than the average thermal velocity. Because of this, the dissociated molecules leave the region of interaction between the low-pressure gas and the field.

Let us consider now an experiment in which this method of hole production in the velocity distribution $W(\underline{v})$ of, say, atoms may be used for spectroscopy inside the Doppler line of nuclear absorption. A collimated beam of γ quanta from a Mössbauer source passes through an atomic gas irradiated by two lasers that excite atoms and photoionize the excited atoms. A hole in the velocity distribution of atoms, and hence of nuclei, results in a hole in the Doppler line of absorption of any nuclear transition at the frequency

$$\omega_{nucl} = (\omega_{\gamma 0} + \Delta_\gamma) + k_\gamma v_{res} , \qquad (10.48)$$

where $k_\gamma v_{res} = (\omega_1 - \omega_0)(\omega_\gamma/\omega_0)$ determined by the velocity of excited atoms, Δ_γ is the recoil frequency, $\omega_{\gamma 0}$ is the center of the nuclear transition line when the recoil effect is ignored (Fig.10.15a). When the detuning of the laser frequency ω_1 with respect to the center of the Doppler line of the optical transition ω_0 is

$$\omega_1 - \omega_0 = \Omega = -\omega_0 \frac{\Delta_\gamma}{\omega_\gamma} , \qquad (10.49)$$

the dip in the nuclear Doppler line coincides with the Mössbauer-radiation line. The resonant minimum that appears in the absorption contour of γ quanta can be observed by detection of the intensity of resonant scattering of γ quanta (Fig.10.15b).

In such experiments, the nuclear-recoil energy may be evaluated accurate to 10^{-4} to 10^{-5}. If a nuclear transition consists of several lines masked by Doppler broadening, several resonant dips occur, simultaneously. Therefore, the method may be regarded as nuclear spectroscopy inside the Doppler

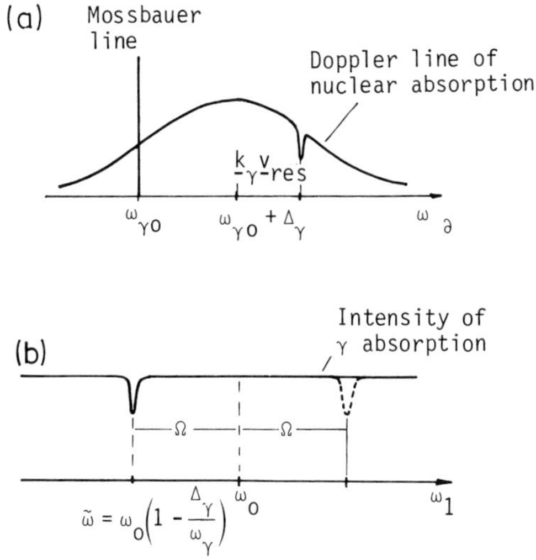

Fig.10.15 (a) Doppler-broadened γ-absorption line with hole, induced by selective remoavl of atoms with a specific velocity v_Z, and (b) the absorption of Mössbauer radiation by gas as a function of laser frequency (LETOKHOV [10.54])

line, with resolution determined by the relative width of narrow optical resonances, that is by $\Gamma_{nucl}/\omega_\gamma = \Gamma_o/\omega_o \simeq 10^{-8}$ to 10^{-9}.

10.3.5 Narrow Laser-Induced Resonances of 2γ Annihilation Radiation

Let us consider now methods of producing narrow and tunable lines of positronium-annihilation radiation at $\hbar\omega_\gamma = 0.511$ MeV [10.63].

As known [10.31], a considerable part of slow positrons annihilate by formation of positronium atoms, 25% by formation of para-positronium atoms (p-Ps) and 75% by formation of ortho-positronium atoms (o-Ps). The p-Ps atoms quickly reach the ground singlet state $^1S_0$, where they annihilate in a time $\tau_S^o = 1.25 \cdot 10^{-10}$ s, emitting two 0.511 MeV γ quanta (Fig.10.5a). The o-Ps atoms quickly reach the ground triplet state $^3S_1$, where they live 10^3 times longer than the p-Ps atoms and annihilate in a time $\tau_T^o = 1.4 \cdot 10^{-7}$ s, emitting three γ quanta with total energy of 1.022 MeV. Thus, the positronium atoms emit a line of 2γ annihilation in a short time τ_S^o and continuous spectrum of 3γ annihilation in a longer time τ_T^o. The line of 2γ annihilation is inhomogeneously broadened by the Doppler effect. The relative Doppler width is $\Delta\omega_D^\gamma/\omega_\gamma = (W_o/m_oc^2)^{1/2} = 10^{-3}$ to 10^{-4}, where W_o is the positronium kinetic energy. The natural width of the annihilation line is only $\Gamma = 1/\tau_S^o \simeq 10^{10}$ s^{-1}.

In the ground triplet state, the o-Ps atoms are distributed over three magnetic sublevels ($m = 0, \pm 1$). When a stationary magnetic field of several KG is switched on, the sublevels $m = 0$ of the states $^1S_0$ p-Ps and $^3S_1$ o-Ps get mixed. Owing to this mixing, the o-Ps atoms on the sublevel $m = 0$ undergo 2γ annihilation [10.31]. As a result, only o-Ps atoms on states $m = \pm 1$ stay in a long-lived triplet state, which must be velocity-selectively converted to the sublevel $m = 0$.

Narrowing and tuning of the positronium 2γ annihilation line by laser radiation consists of converting the o-Ps atoms, for which 2γ annihilation is forbidden, to o-Ps atoms; the conversion mechanism should make possible a selective conversion of the o-Ps atoms that have a definite projected velocity in the chosen direction. Under these conditions, for observation in the direction of (or opposite to) of propagation of the light wave, it is possible to observe a narrow resonance of 2γ-annihilation radiation of selectively excited Ps atoms against the background of the Doppler-broadened line of the annihilation radiation of nonselected Ps atoms. The width of the resonance will depend on the width 2Γ of the optical resonance at the transition ω_0,

$$2\Gamma_\gamma = 2\Gamma_B \frac{\omega_{\gamma 0}}{\omega_0} , \quad \Gamma_B = \Gamma\sqrt{1 + G} , \tag{10.50}$$

and the frequency

$$\omega_\gamma = \omega_{\gamma 0} \pm k_\gamma v_{res} \tag{10.51}$$

will depend on the laser-radiation frequency. As the laser frequency is tuned within the Doppler profile of the optical transition of positronium, the narrow resonance of 2γ annihilation radiation is tuned within the Doppler profile of 2γ line.

There are two ways for producing narrow resonances of 2γ annihilation radiation in the ground and first excited states of the Ps atom. The first method is direct excitation of o-Ps atoms by the running light wave to the state $2\,^3P_0$ or $2\,^3P_2$, which decays into 2γ quanta. The second method is selective excitation of o-Ps atoms in a magnetic field, for example, at the transition $1\,^3S_1$ ($m = \pm 1$) - $2\,^3P_1$ ($m = 0$), with their subsequent conversion to the state $2\,^3P_1$ ($m = \pm 1$) by the microwave field. As a result of spontaneous L_α decay, the selectively excited and converted atoms fall within the mixed state $1\,^3S_1$ ($m = 0$), which decays into two γ quanta in a sufficiently intense magnetic field. A microwave field is necessary in this case, because the quantum number m_s of the state $1\,^3S_1$ (m = 0) differs from that of $1\,^3S_1$ (m = ±1) by unity. The selection rule for optical transitions requires that Δm_s should be equal to zero; therefore radiative relaxation of atoms excited to $2\,^3P_1$ (m = 0) to the mixed state $1\,^3S_1$ (m = 0) is impossible. However, if the microwave field induces magnetic transitions ($\Delta m_s = \pm 1$) between the states $2\,^3P_1$ (m = 0) and $2\,^3P_1$ (m = ±1), L_α decay to the state $1\,^3S_1$ (m = 0) is possible from the latter.

Let us consider the first method [10.29] and take, for definiteness, the level $2\,^3P_0$. This method is more efficient than the second, as far as the maximum signal-noise ratio in detecting narrow γ resonances is concerned. A potential experimental scheme to observe narrow resonances of 2γ annihilation radiation of the $2\,^3P_0$ state is shown in Fig.10.16a. The coincidence scheme operates with γ-quantum counters, to eliminate most of the wide Doppler profile of annihilation radiation.

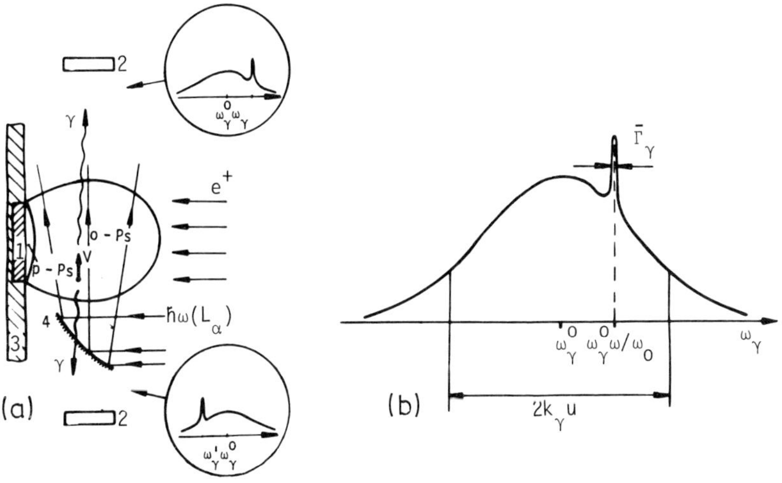

Fig.10.16 (a) Possible scheme for production and detection of narrow resonance of 2γ annihilation radiation of ortho-positronium atom (1 - target; 2 - counter of γ quanta; 3 - protective screen; 4 - defocused mirror); (b) narrow resonance of 2γ annihilation radiation on background of the Doppler-broadened 2γ annihilation line (LETOKHOV and MINOGIN [10.29])

The narrow-resonance line shape of 2γ-annihilation radiation will depend on the number of Ps atoms that have the velocity $\underline{v}$ in the range d^3v and that are excited to the state $2\,^3P_0$, which is given by

$$dn_T^{(1)}(\underline{v}) = N_T^{(o)} \frac{\tau_T^{(1)}}{\tau_T^{(o)} + \tau_T^{(1)}} \frac{G}{1+G} \frac{\Gamma_B^2}{(\omega' - \omega_o)^2 + \Gamma_B^2} W(\underline{v}) d^3v \ , \qquad (10.52)$$

where $N_T^{(o)} = (3/4)\tau_T^{(o)}N$ is the density of stationary o - Ps atoms in the state $1\ ^3S_1$, N is the total density of Ps atoms; $\tau_T^{(o)}$ and $\tau_T^{(1)}$ are the lifetimes of the $1\ ^3S_1$ and $2\ ^3P_0$ states, respectively, G is the saturation parameter of the optical transition of o - Ps, Γ_B is the halfwidth of the Bennett hole in the velocity distribution of the o - Ps atoms, ω' is the resonance frequency of the optical transition for a moving atom of Ps with allowance for the linear and second-order Doppler effect,

$$\omega' \left(1 - \frac{V^2}{c^2}\right)^{1/2} = \omega(1 - \underline{n}\ \frac{\underline{v}}{c}) \ . \qquad (10.53)$$

By introducing the frequency of the γ quantum detected in the direction $\underline{n}_\gamma = \underline{k}_\gamma/k_\gamma$,

$$\omega_\gamma = \omega_{\gamma o} \left(1 - \frac{V^2}{c^2}\right)^{1/2} / (1 - \underline{n}_\gamma \frac{\underline{v}}{c}) \ , \qquad (10.54)$$

which is related to the distribution (10.52) by (10.53), and integrating over the angular variables, we can derive an expression for the intensity of 2γ annihilation radiation in the direction $\underline{n}_\gamma$.

When the narrow resonance of 2γ annihilation radiation is observed in the direction $\underline{n}_\gamma = \underline{n}$, the resonance shape is lorentzian, accurate within the second-order Doppler effect; the central resonance frequency is ω_γ^o (Fig.10.16b). The line shape in the opposite direction ($\underline{n}_\gamma = -\underline{n}$) is more complex, but to a first (V/c) approximation it may be considered lorentzian.

Thus, the width of the annihilation-radiation narrow line is determined by the natural width of the L_α line,

$$\Gamma_\gamma/\omega_\gamma = (\omega_o \tau_{L_\alpha})^{-1} = 4 \cdot 10^{-8} \qquad (10.55)$$

and the second-order Doppler broadening,

$$\frac{\Delta\omega_{sec.D}}{\omega_\gamma} = u^2/c^2 \simeq 10^{-7} \ , \qquad (10.56)$$

where u is mean positronium velocity. The interval of tuning is determined by the relative width of the Doppler profile (the same for the L_α line as for the 2γ annihilation line), i.e.,

$$\frac{\Delta\omega_{tun}}{\omega_\gamma} = \frac{\Delta\omega_D^{opt}}{\omega_0} \simeq \frac{u}{c} \simeq 3 \cdot 10^{-4} , \qquad (10.57)$$

with $u = 10^7$ cm/s.

The basic difficulty of the method of 2γ annihilation line-shape narrowing considered here is to develop a cw laser in the region of $\lambda_0 = 2430$ Å with power from 10^{-2} to 10^{-1} W. Yet we may hope that rapid progress in tunable dye lasers and in the technique for frequency doubling in nonlinear crystals will remove this temporary difficulty.

10.3.6 Some Prospects of Laser Nuclear Spectroscopy

Let us consider the large gap in the resolution of methods for γ spectroscopy. In the close vicinity of the γ-radiation lines of Mössbauer nuclei γ spectroscopy may have a very high resolution, which exceeds that of the methods of nonlinear laser spectroscopy without Doppler broadening. Mössbauer spectroscopy enables us to study the spectral structure of nuclear transitions close to the center of the γ line $\omega_{\gamma 0}$, in approximate accordance with the condition

$$\frac{\Delta\omega_\gamma}{\omega_{\gamma 0}} = \left|\frac{\omega_\gamma - \omega_{\gamma 0}}{\omega_{\gamma 0}}\right| \lesssim 10^{-9} . \qquad (10.58)$$

Experiments in a wider range of $\Delta\omega_\gamma$ present a problem, because it is difficult to produce fast motion of the γ source relative to the target. Resonance γ spectroscopy without recoil is limited also by the value of maximum quantum energy 10^5 eV.

On the other hand, crystal γ spectrometers permit studying the γ-radiation spectrum over a wide range but with low resolution,

$$\frac{\Delta\omega_\gamma}{\omega_{\gamma 0}} \gtrsim \frac{1}{R_{max}} \simeq 10^{-5} . \qquad (10.59)$$

The value $\Delta\omega_\gamma \simeq 10^{-5}\ \omega_{\gamma 0}$ is determined by the maximum resolution of a γ spectrometer, $R_{max} \simeq 10^5$ [10.48].

Therefore, there are no experimental methods to measure the γ-radiation spectrum and the structure of nuclear levels in the range between the values given by (10.58) and (10.59). For $\hbar\omega_{\gamma 0} = 10^6$ eV, this unexplored region of nuclear-transition widths corresponds to energies from 10^{-3} to 10 eV, which are comparable with the typical *energies of atomic and molecular transitions*. From this point of view, extension of the scope of γ spectroscopy by using atomic-nuclear and molecular-nuclear transitions and laser radiation seems to be quite desirable. Potential laser methods of γ spectroscopy without Doppler broadening correspond to a resolution of the order of 10^9, and the Doppler width is of the same order (10.59). So the whole difficult range within each nuclear line can be, basically, studied by methods of laser-nuclear spectroscopy.

Apart from nuclear γ spectroscopy, laser radiation can be used successfully also for other γ-emitting particles, for exotic atoms in particular. We have considered previously the possibility to control the spectrum of positronium 2γ annihilation by use of laser radiation on optical transitions of positronium. Quite recently, the use of tunable lasers radiation has been shown, at CERN [10.64] to induce transitions between the levels $2P_{3/2}$ and $2S_{1/2}$ of the muonium hyperfine structure. That method has also been used to measure precisely the Lamb splitting of muonium atoms.

The ideas considered have many applications to various specific problems, such as composite transitions for "dressed" nuclei, stimulation of transitions between very close levels of nuclei or exotic atoms by laser radiation, and velocity-selective control over the populations of certain sublevels of "dressed" nuclei and exotic atoms. The possibility of velocity-selective optical orientation of nuclear spins in low-pressure gases is an example of this. In this case, we can control to a certain extent not only the angular distribution of γ radiation but also its spectrum. Among other things, by use of optical nuclear orientation in atoms that have specific projected velocities we can eliminate Doppler broadening and tune narrow resonances of γ radiation in certain directions. There are apparently many other interesting possible combinations of laser (optical) and nuclear spectroscopies [10.65].

Unfortunately, most of the effects considered are difficult to detect in a purely laser laboratory, because work with intense radioactive sources presents a problem, especially in the case of short decay times which necessitate rapid transportation of γ sources. It is more probable that this experiment can be conducted at nuclear laboratories where appropriate laser engineering can be used. At present, tunable lasers are not sufficiently reliable and available. Therefore, such laser-nuclear experiments can be carried out only by skilled laser specialists. But progress in laser engineering is advancing so rapidly that, we think, in a few years this temporary problem will be solved and we anticipate practical applications of the ideas and methods of laser spectroscopy without Doppler broadening to the γ region of the spectrum.

10.4 Selective Excitation of Atoms and Molecules with Overlapping Absorption Lines

Control of isotopically selective chemical reactions by selective excitation of atoms and molecules of a chosen type in a mixture is one promising application of laser radiation. To selectively control chemical reactions under laser radiation, it is necessary that the chosen atoms or molecules should be selectively excited. This complies with the requirement for selectivity of absorption, the primary photochemical process. Selective excitation needs at least two nonoverlapping absorption lines for two different atoms or molecules, one of which is to be selected. In some cases of interest, for example, for alkali atoms and certain polyatomic molecules that differ only in isotopic composition and those that differ only in isomeric nuclei, this condition cannot be fulfilled because of overlapping of Doppler-broadened spectral lines. A problem then arises, whether selective chemical reactions can be induced in such cases by laser radiation. Resonant excitation of atoms and molecules in a standing light wave, which allows the production of narrow nonlinear resonances in the density of excited molecules may be useful in these cases.

It is shown in Section 3.3 that the density of excited particles in the standing-wave field has a resonant minimum at exact tuning to the center of the Doppler profile. At strong saturation, the depth of the minimum may be as great as 30%. In a mixture of A and B particles with overlapping lines, the density of excited particles of each type as a function of the standing-wave frequency ν has the form given in Fig.10.17a. It is evident that, when the laser frequency is tuned to the center of the transition line of the particles that should not be excited, their density can be decreased 30% or more. The excitation-selectivity factor for particular particles, for instance A particles, expressed by

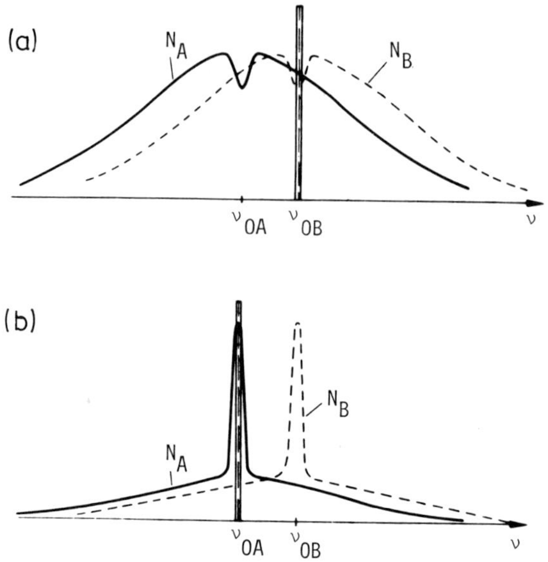

Fig.10.17a,b Selective excitations of atoms of molecules with overlapping Doppler-broadened absorption lines (a) during saturation of the one-quantum absorption and during two-quantum absorption (b) in laser standing wave

$$S = \frac{N^A_{exc}}{N^B_{exc}} - 1 \ , \tag{10.60}$$

has its maximum $S_{max} = \sqrt{2} - 1$ in this case according to (3.65) and (3.66). Therefore, this method displays rather limited selectivity with respect to excitation of definite particles.

The method of two-photon particle excitation in a standing light wave, considered in Chapter 4, is more efficient. In this case, the density of excited particles at the center of the Doppler line is increased by a considerable value, about (ku/Γ). When the frequency of the standing wave is tuned to the center of the absorption line of particular particles, their density in the excited state is greatly increased, compared to the particles of the other sort (Fig.10.17b). The selectivity factor of two-quantum absorption in the standing-wave field, with respect to excitation of particles of a particular sort, for example, A particles, is

$$S = \frac{N^A_{exc}}{N^B_{exc}} - 1 \simeq \frac{4ku}{\sqrt{\pi}\Gamma} \ , \tag{10.61}$$

that is, it may have values much greater than unity.

Detailed theoretical treatments of the two-photon isotopically selective excitation of atoms and molecules were presented in [10.66-68]. In particular, selective two-photon excitation of molecular vibrational-rotational lines was considered by CHEBOTAYEV et al. [10.66] and KELLEY et al. [10.67]. The greatest difficulty arises from low dipole moment, which requires a quasi-resonant intermediate level for high rate of excitation and the power shift of two-photon resonance (see Subsec.4.1.3). Two-photon and single-photon excitation for atomic isotope separation was compared by SHIMODA [10.68]. Two-photon transition is generally very weak, because it is a higher-order process than single-photon transition. In the case of irradiation with two laser beams travelling in opposite directions, however, the different velocities of molecules cumulatively contribute to Doppler-free two-photon resonance. The total probability of the two-photon transition is larger than that of the single-photon transitions when the incident laser power appreciably saturates the single-photon transition, whereas the two-photon transition probability of atoms or molecules in a narrow velocity range is still smaller than that of the single-photon transition.

Probably two-photon isotopically selective excitation will be important for atoms that have very small isotopic shifts (K, Rb, Ca, etc.) and some polyatomic molecules (see, for details, the review on laser isotope separation by LETOKHOV and MOORE [10.69]). This method makes it possible, in principle, to remove the limitations of selectivity that arise from overlapping of absorption lines.

References

10.1 C.H. Townes: *Advances in Quantum Electronics*, ed. by J. Singer (Columbia University Press, New York 1961)

10.2 L.O. Hocker, A. Javan, D. Ramachandra Rao, L. Frenkel, T. Sullivan: Appl.Phys.Lett. 10, 147 (1967)

10.3 L.O. Hocker, D. Ramachandra Rao, A. Javan: Phys.Lett. 24A, 690 (1967)

10.4 L.O. Hocker, J.G. Small, A. Javan: Phys.Lett. 29A 321 (1969)

10.5 N.G. Basov, V.S. Letokhov: Uspekhi Fiz.Nauk 96, 585 (1968)

10.6 V. Daneu, D. Sokoloff, A. Sanchez, A. Javan: Appl.Phys.Lett. 15, 398 (1969)

10.7 D.R. Sokoloff, A. Sanchez, R.M. Osgood, A. Javan: Appl.Phys.Lett. 17, 257 (1970)

10.8 L.M. Matarrese, K.M. Evenson: Appl.Phys.Lett. 17, 8 (1970)

10.9 K.M. Evenson, J.S. Wells, L.M. Matarrese, L.B. Elwell: Appl.Phys.Lett. 16, 159 (1970)

10.10 K.M. Evenson, J.S. Wells, L.M. Matarrese: Appl.Phys.Lett. 16, 251 (1970)
10.11 K.M. Evenson, G.W. Day, J.S. Wells, L.O. Mullen: Appl.Phys.Lett. 20, 133 (1972)
10.12 K.M. Evenson, F.R. Petersen: In *Laser Spectroscopy of Atoms and Molecules*, ed. by H. Walther, Topics in Applied Physics, Vol.2 (Springer-Verlag, Berlin, Heidelberg, New York 1976), p.349
10.13 K.M. Evenson, J.S. Wells, F.R. Petersen, B.L. Danielson, G.W. Day, R.L. Barger, J.L. Hall: Phys.Rev.Lett. 29, 1346 (1972)
10.14 R.L. Barger, J.L. Hall: Appl.Phys.Lett. 22, 196 (1973)
10.15 Comite Consultatif pour la Definition du Metre, 5th Session, Rapport (Bureau International des Poids et Mesures, Sevrés, France 1973)
10.16 P. Giacomo: 5th Session of the Comite Consultatif pour la Definition du Metre, BIPM, Sevres, France 1973
10.17 K.M. Baird, D.S. Smith, W.E. Berger: Opt.Commun. 7, 107 (1973)
10.18 H.P. Layer, R.D. Deslattes, W.G. Schweitzer, Jr.: Appl.Opt. 15, 734 (1976)
10.19 B.N. Taylor, W.H. Parker, D.N. Langenberg: *The Fundamental Constants and Quantum Electrodynamics* (Academic Press, New York, London 1969)
10.20 T.W. Hänsch, M.H. Nayfeh, S.A. Lee, S.M. Curry, I.S. Shahin: Phys. Rev.Lett. 32, 1336 (1974)
10.21 W.E. Lamb, Jr., R.C. Retherford: Phys.Rev. 79, 549 (1950)
10.22 T.W. Hänsch, I.S. Shahin, A.L. Schawlow: Nature Phys.Sci. 235, 63 (1972)
10.23 E.V. Baklanov, V.P. Chebotayev: Opt.Commun. 12, 323 (1974)
10.24 T.W. Hänsch, S.A. Lee, R. Wallenstein, C. Wieman: Phys.Rev.Lett. 34, 307 (1975)
10.25 S.A. Lee, R. Wallenstein, T.W. Hänsch: Phys.Rev.Lett. 35, 1262 (1975)
10.26 G. Herzberg: Proc.Roy.Soc.Ser.A 234, 516 (1956)
10.27 A.P. Mills, S. Berko, K.F. Canter: Phys.Rev.Lett. 34, 1541 (1975)
10.28 E.D. Theriot, Jr., R.H. Beers, V.W. Hughes, K.O.H. Ziock: Phys.Rev. A2, 707 (1970)
10.29 V.S. Letokhov, V.G. Minogin: Zh.Eksp.i Teor.Fiz. 71, 135 (1976)
10.30 A.I. Alekseev: Zh.Eksp.i Teor.Fiz. 34, 1195 (1958); 36, 1839 (1959)
10.31 V.I. Gol'danskii: *Physical Chemistry of Positron and Positronium* (Russian) (Izd "Nauka", Moscow 1968)
10.32 K.F. Canter, A.P. Mills, Jr., S. Berko: Phys.Rev.Lett. 33, 7 (1974)
10.33 K.F. Canter, A.P. Mills, Jr., S. Berko: Phys.Rev.Lett. 34, 177 (1975)
10.34 P.A.M. Dirac: Proc.Roy.Soc.London A165, 199 (1938)
10.35 R.H. Dicke: Rev.Mod.Phys. 34, 110 (1962)
10.36 R.H. Dicke: Science 129, 621 (1959)
10.37 K. Shimoda: Japan.J.Appl.Phys. 12, 1393 (1973)
10.38 S.N. Bagayev, V.P. Chebotayev: Pis'ma Zh.Eksp.i Teor.Fiz. 16, 614 (1972)
10.39 J.J. Snyder, J.L. Hall: In *Laser Spectroscopy*, Proc.2nd Intern.Conf., June 23-27, 1975, Megeve, France; Lecture Notes in Physics, Vol.43 (Springer-Verlag, Berlin, Heidelberg, New York 1975), p.6
10.40 A.J. Greenberg, D.S. Ayres, A.M. Cormack, R.W. Kenney, D.O. Cadwell, V.B. Elings, W.P. Hesse, R.J. Morrison: Phys.Rev.Lett. 23, 1267 (1969)
10.41 A.P. Kol'chenko, S.G. Rautian, R.I. Sokolovskii: Zh.Eksp.i Teor.Fiz. 55, 1864 (1968)
10.42 E.V. Baklanov: Opt.Commun. 13, 54 (1975)
10.43 S. Stenholm: J.Phys.B 7, 1235 (1974)
10.44 C.G. Aminoff, S. Stenholm: Phys.Lett. 48A, 483 (1974)
10.45 C.G. Aminoff, S. Stenholm: J.Phys.B 9, 1039 (1976)
10.46 J.L. Hall, C.J. Borde, K. Uehara: Phys.Rev.Lett. 37, 1339 (1976); V.R. Chebotayev: Report on Second Symposium of Frequency Standards, Boulder, July 1976
10.47 V.S. Letokhov: Phys.Lett. 41A, 333 (1972)
10.48 K. Siegbahn, ed.: *Alpha-, Beta- and Gamma-Ray Spectroscopy*, Vol.1 (North-Holland Publishing Co., Amsterdam 1965)

10.49 O.N. Kompanetz, A.R. Kookoodjanov, V.S. Letokhov, V.G. Minogin, V.L. Mikhailov: Zh.Eksp.i Teor.Fiz. 69, 32 (1975)
10.50 V.S. Letokhov: Phys.Lett. A53, 275 (1975)
10.51 O.N. Kompanetz, V.S. Letokhov, A.R. Kookoodjanov, L.L. Gervitz: Opt. Commun. (in press)
10.52 V.S. Letokhov: Pis'ma Zh.Eksper.i Teor.Fiz. 16, 428 (1972)
10.53 V.S. Letokhov: Phys.Rev.Lett. 30, 729 (1973)
10.54 V.S. Letokhov: Phys.Lett. 43A, 179 (1973)
10.55 V.S. Letokhov: In *Proceedings of Conference "Methodes de Spectroscopie sans largeur Doppler de niveaux excited de systems moleculaires simples"*, May 1973 (Publ.No. 217 CNRS, Paris 1974), p.127
10.56 V.S. Letokhov: In *Laser Spectroscopy*, Proc.2nd Intern.Conf., June 23-27, 1975, Megeve, France; Lecture Notes in Physics, Vol.43 (Springer-Verlag, Berlin, Heidelberg, New York 1975), p.18
10.57 V.S. Letokhov: Phys.Lett. 46A, 481 (1974)
10.58 L.N. Ivanov, V.S. Letokhov: Zh.Eksp.i Teor.Fiz. 68, 1748 (1975)
10.59 V.S. Letokhov: Phys.Lett. 46A, 257 (1974)
10.60 V.S. Letokhov: Phys.Rev. A12, 1954 (1975)
10.61 V.S. Letokhov, V.G. Minogin: Zh.Eksp.i Teor.Fiz. 70, 794 (1976)
10.62 V.S. Letokhov, V.G. Minogin: Zh.Eksp.i Teor.Fiz. 69, 1569 (1975)
10.63 V.S. Letokhov: Phys.Lett. 49A, 275 (1974)
10.64 E. Zavattini: In *Laser Spectroscopy*, Proc.2nd Intern.Conf., June 23-27, 1975, Megeve, France; Lecture Notes in Physics, Vol.43 (Springer-Verlag, Berlin, Heidelberg, New York 1975), p.370
10.65 V.S. Letokhov: *"Crossings" of Laser and Nuclear Spectroscopies* (North-Holland Publishing Co., Amsterdam 1976)
10.66 V.P. Chebotayev, A.L. Golger, V.S. Letokhov: Chem.Phys. 7, 316 (1975)
10.67 P.L. Kelley, H. Kildal, H.R. Schlossberg: Chem.Phys.Lett. 27, 62 (1974)
10.68 K. Shimoda: Appl.Phys. 9, 239 (1976)
10.69 V.S. Letokhov, C.B. Moore: Kvantovaya Elektronika 3, 248 (1976); 3, 485 (1976)
10.70 J.L. Hall, C.J. Borde, K. Uehara: Phys.Rev.Lett. 37, 1339 (1976)

11. Conclusion

Nonlinear laser Doppler-free spectroscopy is an example of basic transformations of the methods of atomic and molecular spectroscopy which became practicable with the advent of lasers. This approach to laser spectroscopy is one of the most efficient and promising for fundamental and applied research.

Finishing this monograph, we would like to stress again that at least four quite different approaches can be used as the basis for nonlinear laser Doppler-free spectroscopy.

1) Variations of the velocity-distribution function at levels n and m when the transition n-m is acted upon by a coherent light wave.

2) Two-quantum transition in the field of two waves of the same frequency propagating in opposite directions.

3) Variations of the velocity-distribution function of particles, irrespective of their quantum state, that is, variations of the velocity distribution function for all quantum levels at the same time.

4) Resonances in separated optical fields.

Approach 1 forms the basis for most methods of laser spectroscopy inside the Doppler profile considered in our monograph. Among them are three basic methods for production of narrow resonances: 1) resonances of saturated absorption (inverted Lamb dip) at a two-level transition; 2) absorption and emission resonances at transitions coupled with the levels of the transition under saturation; 3) density resonances of the total numbers of particles at the levels n or m that interact with the field. This approach has been experimentally developed for ten years, and at present it is the best known.

Approach 2 employs, not saturation of transition-level population but, compensation for Doppler shift when two photons of the same frequency are absorbed from counter-running waves. It has been practically realized quite recently, but it is now under rapid progress and is given much consideration in our monograph.

Approach 3 is based on variations of the distribution of real velocities of absorbing (radiating) particles under coherent laser radiation. The monograph deals with cooling of particles and trapping of particles that have small absolute velocities, in a three-dimensional field (Subsec.6.2.4). This approach has not been realized in experiments yet, but it is of principal importance because it promises to get rid of Doppler broadening at a great number of quantum transitions at once, not only at the resonance transition and its associated levels. It is very significant that the approach based on radiative cooling and trapping of particles will make it possible to overcome the limitation imposed by the finite transit time of atoms or molecules through a light beam and the second-order Doppler effect on resolution power. Because of this it will enable us to achieve resolution conditioned by the natural width of vibrational molecular transitions and forbidden electron transitions of atoms and molecules.

Approach 4 is associated with use of resonances in separated optical fields. Resonances arise owing to spatial transfer of polarization by excited particles. Though the method of separated fields is well known in the radio-frequency range, the possibility of its use in the optical band was rejected over a long period of time. This is due to the influence of a Doppler effect that is not taken into account in the radio-frequency range. Only two years ago it was shown that elimination of the influence of the Doppler effect owing to nonlinear interaction of fields in three-beam (Subsec.3.2.4) systems or at two-photon absorption gives rise to resonance with a width that is in inverse proportion to the time of flight of a particle between beams. The method opens up possibilities to obtain resonances with widths of several tens of Hertz in the optical band. Even in the nearest future one can obtain important results using this method.

Author Index

Subject Index

Titles of Related Interest

Laser Spectroscopy
of Atoms and Molecules

Topics in Applied Physics, Vol.2
ed. by *H. Walther*
137 figures. XVI,383 pages. 1975
ISBN 3-540-07324-8

H. Walther: Atomic and Molecular Spectroscopy with Lasers

E.D. Hinkley, K.W. Nill, F.A. Blum: Infrared Spectroscopy with Tunable Lasers

K. Shimoda: Double-Resonance Spectroscopy of Molecules by Means of Lasers

J.M. Cherlow, S.P.S. Porto: Laser Raman Spectroscopy of Gases

B. Decomps, M. Dumont: Linear and Nonlinear Phenomena in Laser Optical Pumping

K.M. Evenson, F.R. Petersen: Laser Frequency Measurements, the Speed of Light, and the Meter

High-Resolution Laser Spectroscopy

Topics in Applied Physics, Vol.13
ed. by *K. Shimoda*
91 figures. XII,378 pages. 1976
ISBN 3-540-07719-7

K. Shimoda: Introduction

K. Shimoda: Line Broadening and Narrowing Effects

P. Jacquinot: Atomic Beam Spectroscopy

V.S. Letokhov: Saturation Spectroscopy

J.L. Hall, J.A. Magyar: High Resolution Saturated Absorption Studies of Methane and Some Methyl-Halides

V.D. Chebotayev: Three-Level Laser Spectroscopy

S. Haroche: Quantum Beats and Time-Resolved Fluorescence Spectroscopy

N. Bloembergen, M.D. Levenson: Doppler-Free Two-Photon Absorption Spectroscopy

Laser Monitoring of the Atmosphere

Topics in Applied Physics, Vol.14
ed. by *E.D. Hinkley*
84 figures. XIII,380 pages. 1976
ISBN 3-540-07743-X

E.D. Hinkley: Introduction

S.H. Melfi: Remote Sensing for Air Quality Management

V.E. Zuev: Laser-Light Transmission through the Atmosphere

R.T.H. Collis, P.B. Russell: Lidar Measurement of Particles and Gases by Elastic Backscattering and Differential Absorption

H. Inaba: Detection of Atoms and Molecules by Raman Scattering and Resonance Fluorescence

E.D. Hinkley, R.T. Ku, P.L. Kelley: Techniques for Detection of Molecular Pollutants by Absorption of Laser Radiation

R.T. Menzies: Laser Heterodyne Detection Techniques

Laser Spectroscopy

Proceedings of the Second International Conference, Megève, June 23-27, 1975
Lecture Notes in Physics, Vol.43
ed. by *S. Haroche, J.C. Pebay-Peyroula, T.W. Hänsch, S.E. Harris*
230 figures. X,468 pages. 1975
ISBN 3-540-07411-2

Springer-Verlag
Berlin Heidelberg New York